Principles and Applications of ESR Spectroscopy

Anders Lund · Masaru Shiotani ·
Shigetaka Shimada

Principles and Applications of ESR Spectroscopy

 Springer

Prof. Anders Lund
Department of Physics, Chemistry
  and Biology, IFM
Linkoping University
The Emeritus Academy, Linkoping
  University
SE-581 83, Linkoping
Sweden
alund@ifm.liu.se, Anders.Lund@liu.se

Prof. Shigetaka Shimada
Graduate School of Engineering
Nagoya Institute of Technology
44-66 Midorigaoka
Midori-Machi
Owari-Asahi 488-0822
Japan
shimada.shigetaka@nitech.ac.jp

Prof. Masaru Shiotani
Graduate School of Engineering
Hiroshima University
4-26-8 Takamigaoka, Takaya
Higashi-Hiroshima 739-2115
Japan
mshiota@hiroshima-u.ac.jp

ISBN 978-94-007-8986-9   ISBN 978-1-4020-5344-3 (eBook)
DOI 10.1007/978-1-4020-5344-3
Springer Dordrecht Heidelberg London New York

Springer is part of Springer Science+Business Media (www.springer.com)

Preface

ESR or Electron Spin Resonance is a spectroscopic method for studies of paramagnetic species. Species of this kind are of interest both from fundamental viewpoints and for a broad range of applications in materials and polymer sciences, physical chemistry and chemical physics, bio-chemistry and medicine, catalysis and environmental sciences, radiation dosimetry and geological dating, as well as radiation physics and chemistry. The magnetic properties, mainly attributed to the electron spin, can be employed to determine both the structure (at a molecular level) and the amount of paramagnetic species in a sample with the ESR method. This method has been presented from different viewpoints in the past. Previous monographs on ESR spectroscopy have focused on the technique as such, while the general textbooks in physics, chemistry or spectroscopy only briefly explain the applications, which are mainly presented in specialist reviews and in the original literature. The present book is based upon the authors' long experience of teaching the subject to a mixed audience, with backgrounds ranging from physics to biology. It aims both at providing the principles of continuous wave and pulsed ESR techniques and to illustrate the potential of the method by examples of applications. The principles of ESR, multi-resonance and pulsed ESR methods, the analysis of spectra, and multi-frequency and high field ESR techniques treated in the first four chapters are thus followed by five chapters exemplifying recent applications in molecular science, in catalysis and environmental science, in polymer science, in spin labeling and molecular dynamics, and in quantitative ESR.

Theoretical derivations are in general left out, as they are presented repeatedly in previous works. The necessary theory is instead illustrated by practical examples from the literature. Commonly used computer codes to evaluate experimental ESR data are described with examples. Internet addresses to download the software are given, whenever possible. Formulae employed in those programs are reproduced in appendices, when the original literature references are not easily available. The theory and the application parts are to a large extent independent of each other to allow study of a special subject. For reasons of easy access of data and diagrams several examples from the authors' own work were employed to illustrate certain applications. Exercises included in the theoretical part are mainly concerned with spectra interpretations, as this is the key issue in the analysis of experimental data.

Our intention has been to prepare a textbook with the following issues in mind:

1. The book "Principles and applications of ESR spectroscopy" is for students and scientists planning to use the method without necessarily becoming experts. The book provides sufficient knowledge to properly apply the technique and to avoid mistakes in the planning and performance of the measurements.
2. The theoretical part is adapted for a non-specialist audience. Derivations of well-known formulae are left out. The non-specialist does not want the derivations, while the physicist does not need them. The necessary theory is instead illustrated by practical examples from the literature. The potential of the method is demonstrated with applications selected from the authors' wide experience.
3. Commonly used software to evaluate experimental ESR data is described with examples. Addresses to download the software are given.

Linköping, Sweden   Anders Lund
Higashi-Hiroshima, Japan   Masaru Shiotani
Owari-Asahi, Japan   Shigetaka Shimada

Contents

Part I
Principles

Chapter 1
Principles of ESR

Abstract The ESR (electron spin resonance) method is employed for studies of paramagnetic substances most commonly in liquids and solids. A spectrum is obtained in continuous wave (CW) ESR by sweeping the magnetic field. The substances are characterized by measurements of the g-factor at the centre of the spectrum and of line splittings due to hyperfine structure from nuclei with spin $I \neq 0$. Zero-field splitting (or fine structure) characteristic of transition metal ion complexes and other substances with two or more unpaired electrons ($S \geq 1$) can be observed in solid samples. Concentration measurements with the CW-ESR method are common in other applications. High field and multi-resonance (e.g. ENDOR) methods employed in modern applications improve the resolution of the g-factor and of the hyperfine couplings, respectively. Pulse microwave techniques are used for measurements of dynamic properties like magnetic relaxation but also for structural studies.

1.1 Introduction

ESR or Electron Spin Resonance is a spectroscopic method for studies of paramagnetic species. This method and the related NMR or nuclear magnetic resonance technique were both established around 1945 as a result of research in the Soviet Union and in the United States. Both methods make use of magnetic properties, in the first case of electrons in the second of nuclei.

The microwave technique in the frequency range 10^9 Hz and higher that is normally employed in ESR instruments had been developed for Radar detection units during the Second World War. Systematic studies of transition metal salts began in England at the end of the 1940s. These studies have formed the basis for continued research on complex biochemical systems that is now one of the most active areas of applications. Other fields have been opened by the ongoing development of resolution and sensitivity. Free radicals – molecules with an unpaired electron – have been tracked and identified and their reactions followed both in the liquid and the solid state in many instances. Such studies are still ongoing, nowadays more as one tool in conjunction with other methods to obtain chemical information. Thus, the electronic

A. Lund et al., *Principles and Applications of ESR Spectroscopy*,
DOI 10.1007/978-1-4020-5344-3_1, © Springer Science+Business Media B.V. 2011

and geometrical structures of free radicals have been clarified in many instances by correlation with quantum chemistry calculations. Intermediate species in the triplet state, i.e. excited molecules with two unpaired electrons appear in many photochemical reactions. ESR spectroscopy has become a valuable tool in the study of such species, recently also in systems of biological importance like in photosynthesis.

Free radicals are in most cases very reactive. By lowering the temperature or by inclusion in solid matrices the lifetime can be increased allowing observation: the recording time of an ordinary ESR spectrum is of the order of minutes. Special methods to obtain spectra of radicals with short lifetimes have, however, been developed.

1.2 Paramagnetism

A paramagnetic substance contains atoms, molecules or ions with permanent magnetic dipoles. Consider the case of the hydrogen atom that contains a single electron. A magnetic dipole μ arises by the motion of the electron about the nucleus.

Fig. 1.1 Classical model illustrating relationship between angular momentum $L= m_e \cdot v \cdot r$ of electron, e, moving around a nucleus N and magnetic moment μ.

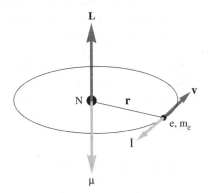

For a current loop (Fig. 1.1) one has $\mu = I \cdot A$; A is the enclosed area of the current (I) loop. The current is determined by the electronic charge, $-e$, and speed v. For a circular orbit the enclosed area is $A = \pi r^2$, the current $I = -e \cdot v/(2\pi r)$. The magnetic moment becomes $\mu = -e \cdot v \cdot r/2 = -e \cdot L/(2m_e)$, where L is the orbital angular momentum of the electron. The result is generally valid, classically, and can be written in vector notation as:

$$\mu = -\frac{e}{2m_e}\mathbf{L} \tag{1.1}$$

The angular momentum of atoms is quantised and can only take the discrete values

$$L = \frac{h}{2\pi}\sqrt{l(l+1)}, \tag{1.2}$$

where h is the Planck constant and l is the orbital quantum number that can take the values 0, 1, 2,..... Also the components of the vector **L** are quantised. For the z component one has:

$$L_z = \frac{h}{2\pi} m_l, \quad m_l = \pm l, \pm (l-1), \pm (l-2), \ldots, \pm 1, 0. \tag{1.3}$$

The energy E_m of a magnetic dipole in a magnetic field B directed along z is given by the product:

$$E_m = -\mathbf{\mu} \cdot \mathbf{B} = \mu_B \cdot B \cdot m_l, \quad \mu_B = eh/4\pi m_e = \text{Bohr magneton} \tag{1.4}$$

The energy thus becomes split in $(2l+1)$ levels.

1.2.1 Which Substances Are Paramagnetic?

Atoms with an odd number of electrons are paramagnetic. Chemical substances are on the contrary diamagnetic in most instances. This can be illustrated with the electron shell of rock salt, NaCl, compared to the shells of the free atoms:

$$Na^+ : 1s^2 2s^2 2p^6 3s^0 \quad Cl^- : 1s^2 2s^2 2p^6 3s^2 3p^6$$
$$Na : 1s^2 2s^2 2p^6 3s^1 \quad Cl : 1s^2 2s^2 2p^6 3s^2 3p^5$$

An s-orbital ($l = 0$) can contain two electrons with different spins, $m_S = \pm \frac{1}{2}$. There are three p-orbitals ($l = 1$) with different $m_l = \pm 1, 0$. Each can contain two electrons with different spins. The ions obviously have completely filled electron shells. Closed shells cannot possess spin angular momentum, since the electrons with different spins are paired. There is no orbital angular momentum as well. In fact, most non-metallic substances are diamagnetic. Important exceptions are the compounds of certain transition metal ions. The iron group in the periodic table is one such example, with partly filled 3d-shells ($l = 2$). This is repeated for the 4d shell in the fifth period, while in the next period partly filled 4f shells appear. The actinide group constitutes an analogous case. Certain ions of these metals are paramagnetic – in fact the earliest ESR investigations were entirely devoted to studies of compounds with such ions.

1.2.1.1 Free Radicals

Free radicals are molecules that have been modified to contain an unpaired electron. They are paramagnetic. Some are stable for instance the one that was first prepared (in 1900) according to

$$(C_6H_5)_3C - C(C_6H_5)_3 \leftrightarrow 2(C_6H_5)_3C\cdot$$

hexaphenylethane   triphenylmethyl

Most free radicals are, however, unstable, and appear as intermediate products in various chemical reactions such as during photolysis and by the influence of ionizing radiation and in certain enzymes. Under suitable conditions free radicals can have sufficient lifetime to be observed in the liquid. The appearance of free radicals in liquid hydrocarbons during radiolysis was first demonstrated in 1960. In the solid state free radicals can become stabilized almost indefinitely in many materials.

1.2.1.2 Triplet State Molecules

Some molecules contain two unpaired electrons with parallel spins. This can be caused, for example, by illumination with visible and ultraviolet light, so that the energy of the system becomes excited. The total spin quantum number is then $S = $ ½ + ½ = 1. Studies of triplet state molecules by ESR are usually made in single crystal or amorphous matrices, "glasses".

1.2.1.3 Transition Metal Ions

The ESR properties of transition metal ions are affected by the environment to a larger extent than is usually the case with free radicals and triplet state molecules. This is attributed to the interaction with surrounding ions or electric dipoles which causes a splitting of the energy levels, e.g. of the five originally degenerate 3d-orbitals of the iron group. A closer analysis shows that the properties depend on the strength and symmetry of the interaction, often described as a "crystal field". We refer to an example in Chapter 3 with a single 3d electron in tetragonal symmetry where the magnetism due to orbital motion is largely "quenched" by the crystal field.

1.3 Resonance

The hydrogen atom in its ground state has $l = 0$, i.e. no orbital angular momentum. Yet, it is paramagnetic. The explanation is that the spin angular momentum, S, of the electron gives rise to a magnetic moment similar to that from the orbit. The spin angular momentum is quantized as well and has the magnitude $S = \frac{h}{2\pi}\sqrt{s(s+1)}$ where $s = $ ½ for an electron. The energy in a magnetic field becomes:

$$E_m = g\mu_B B m_s, \quad m_s = \pm 1/2 \tag{1.5}$$

where an additional g-factor has been introduced. For the hydrogen atom $g = 2.00228$, i.e. close to 2 as expected from a deeper theoretical analysis. Thus, the energy becomes split in $(2s+1) = 2$ "Zeeman levels", Fig. 1.2.

A transition between the two Zeeman levels can be induced if the system takes up energy in the form of electromagnetic radiation of a suitable frequency ν. The resonance condition then applies:

$$h\nu = E(m_s = 1/2) - E(m_s = -1/2) = g\mu_B B \tag{1.6}$$

Fig. 1.2 Zeeman splitting of
energy levels of an electron
placed in a magnetic field (**B**)

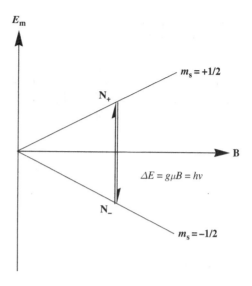

The frequency is usually kept constant, and the magnetic field strength is varied so
that resonance occurs. The system will then absorb energy from the electromagnetic
field. This field contains both an electric and a magnetic component. The magnetic
dipole transitions are induced by the magnetic field vector B_1. Figure 1.3 illustrates
how this field is oriented perpendicular to the static field B, while the B_1 field vector
rotates about the direction of the static field. Such a field is said to be circularly
polarized. According to this classical picture (Fig. 1.3), the magnetic moment pre-
cesses about the static field with an angular frequency $\omega_L = \frac{e}{2m_e} \cdot B$, the Larmor
precession, or more specifically for the precession of the electron spin magnetic
moment, $\omega_L = g\frac{e}{2m_e}B = \gamma \cdot B$, where the gyromagnetic ratio γ has been introduced.
Resonance occurs – in this view – when the rotating magnetic field follows the

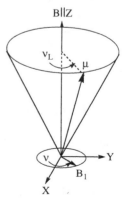

Fig. 1.3 Larmor precession
of magnetic moment μ about
stationary field B

magnetic moment, providing a torque that makes the moment flip its orientation, that is when $v = v_L$ (or $\omega = \omega_L$; $\omega = 2\pi v$). This condition is identical with that obtained by quantum mechanics.

A simple electron spin resonance spectrometer (Fig. 1.4) consists in principle of a microwave generator G that transmits electromagnetic energy via a waveguide W through a cavity C containing the sample to a detector D. The cavity is located between the poles N and S of a magnet.

Fig. 1.4 A simple ESR spectrometer

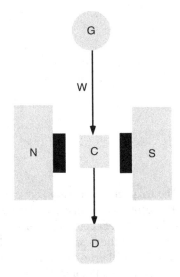

When the magnetic field has a value corresponding to resonance, energy is absorbed and the power reaching the detector decreases. Absorption can occur as long as the number of particles – atoms, radicals etc. – in the lower state, N_- exceeds the number N_+ of the upper. At equilibrium the ratio predicted by the Boltzmann distribution is $N_+/N_- = e^{-\Delta E/kT}$. To get a strong signal, ΔE must be large, which in turn implies that the sensitivity increases with increasing energy difference ΔE, that is with the microwave frequency. Many spectrometers operate at "X-band" with a frequency of about 10^{10} Hz which has been found appropriate for the study of several classes of materials. The recent trend to employ even higher frequencies (and fields) is mainly driven by the efforts to increase the spectral resolution rather than to achieve high sensitivity. Also the temperature influences the signal strength, a lowering of the temperature from 300 to 77 K would for instance increase the absorption by a factor four.

1.3.1 Relaxation – Saturation

At resonance the spin state changes from the lower to the higher state by the absorption of energy from the microwaves (Fig. 1.2). If the energy of the higher state could not be released, the number of particles in the two states would soon become equal, $N_+ = N_-$. The absorption of energy would stop after some time. This does not occur, however, because the surrounding "lattice" takes up the energy of the excited spin states by relaxation. The energy flow is therefore:

Electromagnetic field → Spin system → Lattice vibrations

The rate of the relaxation process is often expressed as a relaxation time, τ, that gives a measure how fast the higher spin level returns to the lower. The lattice vibrations of a crystalline solid are quantized phonons with energy $\hbar\omega$ and mechanisms of the following types may occur:

$$\uparrow \rightarrow \downarrow + \hbar\omega$$
$$\uparrow + \hbar\omega \rightarrow \downarrow + \hbar\omega' \, (\omega' < \omega)$$

Saturation occurs if the relaxation rate is too slow (τ is long) to dissipate the energy provided by the electromagnetic field to the spin system. This can be observed experimentally when the absorption does not increase and/or a line broadening occurs at increased microwave power. This unwanted situation can be avoided by decreasing the power, P, normally until the absorption becomes proportional to $\sqrt{P}$.

1.3.2 Hyperfine Structure

Many nuclei possess a spin angular momentum, and therefore have magnetic properties like the electron. Magnetic properties of the stable isotopes are summarised in Table G7 of General Appendix G. The component of the magnetic moment along the direction (z) of the magnetic field is $\mu_z = g_N \mu_N m_I$ in analogy with the electronic case. The real magnetic field at the electron will be composed of two contributions, one from the external magnetic field B, the other from the field caused by the magnetic moment of the nucleus. The resulting magnetic field depends on the orientation of the nuclear dipole, i.e. on the quantum number m_I. The energy can be written:

$$E(m_s, m_I) = g\mu_B B m_S + a m_I m_S - g_N \mu_N B m_I \tag{1.7}$$

Here a is the hyperfine coupling constant and $\mu_N = eh/4\pi m_p$ is the nuclear magneton (Table G2). Each m_S level for the hydrogen atom will be split in two, corresponding to $m_I = +\frac{1}{2}$, and $m_I = -\frac{1}{2}$ for $I = \frac{1}{2}$ of the proton, Fig. 1.5.

As the magnetic field is changed two transitions occur determined by the selection rules:

$$\Delta m_S = 1, \; \Delta m_I = 0$$

The measured separation between the two lines, directly gives the hyperfine splitting, 50.5 mT, in magnetic field units. The coupling when expressed in frequency units as is usual in ENDOR studies (Chapter 2) is converted to field units by:

$$a(\text{mT}) = 0.0714477 \frac{a(\text{MHz})}{g} \tag{1.8}$$

Fig. 1.5 Energy levels for an H atom with $I = \frac{1}{2}$. The *arrows* show the allowed $\Delta m_I = 0$ transitions induced by the microwave field radiation

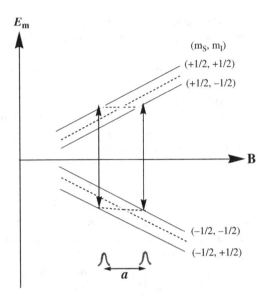

In a polyatomic paramagnetic species the electronic magnetic moment may inter-act with the moments of several magnetic nuclei. This interaction can give rise to a large number of lines and the ESR spectra can therefore become quite complex. A special case occurs when several nuclei are chemically equivalent as is the case for the anion radical of benzene, $C_6H_6^-$. It contains six hydrogen nuclei that interact equally with the unpaired electron. These give rise to the level splitting shown in Fig. 1.6: one proton gives rise to a doublet, two protons a triplet since two levels

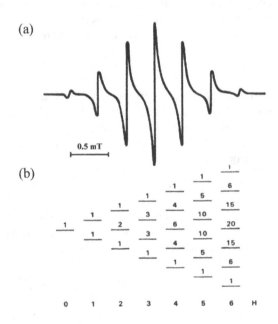

Fig. 1.6 ESR 1st derivative spectrum of the benzene anion. The relative intensities of the seven lines due to six equivalent H nuclei are obtained from the Pascal triangle

coincide, three protons give a quartet etc. The levels that coincide are said to be degenerate, and the degree of degeneracy can be obtained from the Pascal triangle, Fig. 1.6. The relative intensities of the lines depend on the degeneracy of the levels; since $\Delta m_I = 0$ transitions can only occur between the hyperfine levels of the $m_S = +\frac{1}{2}$ and $m_S = -\frac{1}{2}$ states with the same m_I value. The intensity is the binomial 1:6:15:20:15:6:1 as shown in the spectrum recorded as the first derivative of the absorption in Fig. 1.6.

1.3.3 Fine Structure or Zero-Field Splitting

For a total spin of $S = 1$ or higher, an additional interaction appears that gives rise to a "fine structure" or "zero-field splitting" in the ESR spectrum, Fig. 1.7.

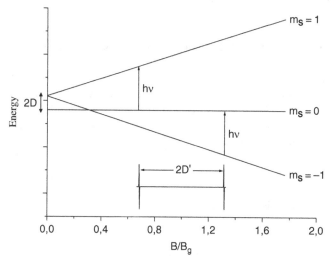

Fig. 1.7 Energy levels for a triplet state molecule as a function of the magnetic field in units of $B_g = h\nu/g\mu_B$. The magnetic field is in the direction (z) corresponding to the maximum zero-field splitting, $2 \cdot D$, observed as the line separation $2 \cdot D'$ in the ESR spectrum

In the case of two radicals separated by a fixed distance in the range 0.5–1 nm the spectrum is a doublet – the separation depends on the magnetic dipolar coupling between the two unpaired electrons. In contrast to the hyperfine coupling it has no isotropic part. As a result it can only be observed for species in the solid state; the coupling is averaged to zero by the rapid tumbling usually occurring in liquids. The splitting depends on the angle θ between the magnetic field and the line joining the two radicals:

$$F = D(3\cos^2\theta - 1) \tag{1.9}$$

The value of the fine structure constant D depends on the distance between the radicals of the pair. This property is employed for distance measurements by pulsed ESR in biochemical applications.

Fine structure or *zero-field splitting* also occurs for triplet state molecules. Also the previously mentioned ions of transition metal ions can have $S \geq 1$ and therefore fine structure appears also there. It arises, however, by a different mechanism, spin orbit interaction.

1.3.4 Quenching of Orbital Angular Momentum

The paramagnetism of transition metal ions is mainly attributed to the magnetic moment due to the electron spin as for free radicals and triplet state molecules. The contribution due to orbital angular momentum is to a large extent "quenched". In other words the orbital angular momentum is "locked" by the crystal field and cannot reorient in a magnetic field.

Quenching of Orbital Angular Momentum by the Crystal Field

The reason that orbital angular momentum does not contribute to the paramagnetism of transition metal ions can be explained by considering the electrostatic field (crystal field) effect on the free ion. Assume that the Schrödinger wavefunction is a complex function $\Psi = \varphi_1 + i\varphi_2$ and that the corresponding energy E is non-degenerate. Thus,

$$H\Psi = H\varphi_1 + iH\varphi_2 = E\varphi_1 + iE\varphi_2$$

The Hamiltonian H is real, so that

$$H\varphi_1 = E\varphi_1$$
$$H\varphi_2 = E\varphi_2$$

Two wave-functions with the same energy are obtained, contradicting the assumption of a non-degenerate state. Thus, the wave-function can be taken real ($\varphi_2 = 0$). The orbital angular momentum is also a real quantity. On the other hand, the corresponding quantum mechanical operator of e.g. the x-component, $L_x = i\hbar(y\frac{\partial}{\partial z} - z\frac{\partial}{\partial y})$ is imaginary. Thus, the expectation value is zero, $\langle \Psi | L_x | \Psi \rangle = 0$.

1.4 Instrumentation

Modern ESR spectrometers are much more complex than the simple instrument shown in Fig. 1.4, which would be too insensitive to be useful. Several designs are available commercially, while the development of special equipment is still made in several laboratories.

The *continuous wave* (CW) spectrometers are still mostly employed. The cavity is placed in a microwave bridge, thus avoiding the microwaves to reach the detector except at resonance (Fig. 1.8). The magnetic field is modulated at high frequency, 100 kHz is commonly employed.

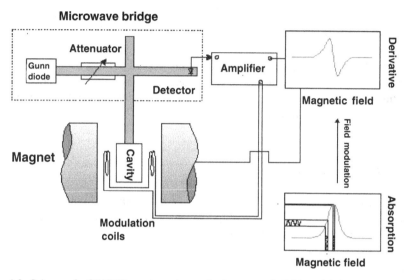

Fig. 1.8 Scheme of a CW ESR spectrometer employing magnetic field modulation. The method is employed to improve the signal/noise ratio by phase-sensitive detection. The principle for obtaining the derivative of the absorption is shown. The figure is reproduced from http://www.helmholtz-berlin.de/forschung/enma/si-pv/analytische-methoden/defektanalytik/esr/index_en.html   with permission from Dr. Klaus Lips

As a result of the field modulation and the phase-sensitive detection, the spectrum is recorded as the first derivative of the absorption, with the x-axis synchronized to the magnetic field that is swept across the resonance. For accurate measurements of the resonance parameters it is customary to calibrate the field, either with a field meter based on proton NMR, or to use a standard sample with known g-factors and hyperfine splittings. The microwave frequency is measured with a frequency meter attached to the microwave bridge. Modern instruments are computer controlled; spectra with instrument parameters are saved and stored digitally.

Measurements of the concentrations are most accurately performed by comparison with a standard sample. A double cavity can be used, with the sample in one half, the reference in the other. Other designs have a built in reference that is recorded simultaneously with the sample. Absolute calibration is complicated and large deviations often occur in comparisons between different laboratories. The highest reported sensitivity of commercial instruments is ca. $10^9 \cdot \Delta B$ spins, where ΔB is the width of the signal in mT. This applies to instruments operating at the normal "X-band" wavelength, $\lambda = 3$ cm. As mentioned the sensitivity increases at higher frequencies, but the sample volume decreases, and the instrument also becomes more difficult to use. The commercial instruments for relative measurements of concentrations that have been developed, e.g. as a means to measure radiation doses usually employ X-band, while higher frequencies like in W-band spectrometers (10^{11} Hz) find applications in structural studies.

1.5 Spin Distribution

Free radicals often give rise to spectra with many hyperfine lines. This reflects that the unpaired electron is delocalised and can interact with several atomic nuclei. The hyperfine coupling is isotropic, i.e. independent of the field orientation in liquids, while it is generally anisotropic in solids.

1.5.1 Isotropic Coupling

This coupling is referred to as the *contact coupling* since it appears when the unpaired electron has a finite probability to appear at the nucleus. This is expressed by the following equation, due to Fermi:

$$a = \frac{2}{3}\mu_0 g \mu_B g_N \mu_N \rho(X_N) \tag{1.10}$$

Here $\rho(X_N)$ is the unpaired electron density at a nucleus located at X_N. The expression shows that the experimentally measurable hyperfine coupling constant a is directly proportional to the spin density $\rho(X_N)$, related to the electronic structure of the radical. Some values for the free atoms are quoted in Table 1.1.

Table 1.1 Hyperfine couplings of some nuclei with the unpaired electron completely localized to an s-orbital (a_0) or to a p-orbital (B_0)

Nucleus	Spin (I)	Natural abundance (%)	a_0 (MHz)	B_0 (MHz)
^{1}H	1/2	99.985	1,420	
^{2}H	1	0.015	218	
^{13}C	1/2	1.11	3,110	91
^{14}N	1	99.63	1,800	48
^{19}F	1/2	100.0	52,870	see [1]
^{31}P	1/2	100.0	13,300	see [1]

Isotropic couplings occur for s-orbitals, while p-, d- and f-orbitals are characterised by a zero density at the nucleus (Fig. 1.9) and thus give no contribution to a_0. The values of a_0 and B_0 in the table are obtained when the unpaired electron is entirely localised to an s-orbital and a p-orbital, respectively.

Fig. 1.9 Angular dependence
of (a) 1s and (b) 2p orbitals.
The probability to find the
electron within the contour
is 0.9

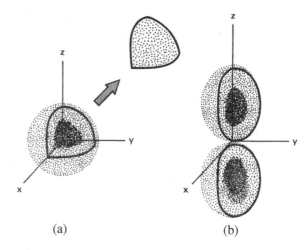

(a)   (b)

1.5.2 Anisotropic Coupling

If the unpaired electron is localised at a p-, d- or f-orbital of an atom, the hyperfine coupling is anisotropic, that is measurements on a species in a single crystal
give different values depending on the orientation of the magnetic field. When the
magnetic field is parallel to the z-axis of the 2p orbital in Fig. 1.9, one obtains the
value $2B_0$, while the coupling is $-B_0$ in the xy-plane. This kind of coupling is called
magnetic dipole coupling.

Some values for B_0 of light elements are given in Table 1.1. Values of a_0 as well
as of B_0 for the heavier atoms have been provided by J.A. Weil and J.R. Bolton
[1]. It should be pointed out that the values are not measured quantities in many
instances, but have been computed as quantum mechanical expectation values.

1.5.3 An Example

According to single crystal ESR measurements the unpaired electron of the
hydrazine cation, $N_2H_4^+$ (Fig. 1.10) is equally distributed over the two nitrogen
atoms, $B_{N1} = B_{N2} = 23.3$ MHz.

Fig. 1.10 Hydrazine cation,
$N_2H_4^+$ with the unpaired
electron equally distributed
over the two nitrogen atoms
in $2p_z$ orbitals

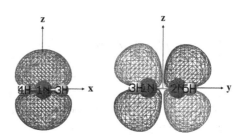

From Table 1.1 we can then calculate the probability for the unpaired electron to be located at each of the nitrogen atoms:

$$\rho_N 1(2p_z) = \rho N2(2p_z) = 23.3/48 = 0.49.$$

This agrees quite well with simple molecular orbital theory. We assume that the molecule is planar, and that the atoms are joined by σ-bonds in sp^2 hybridization. Ten electrons are used up for these bonds, and additionally four are at inner 1s shells of the two nitrogen atoms. There are totally seventeen electrons. The remaining three are placed in π-orbitals. These are constructed as a linear combination of atomic orbitals according to the LCAO method:

$$\pi = \frac{1}{\sqrt{2}}(2p_z(1) + 2p_z(2))$$

$$\pi^* = \frac{1}{\sqrt{2}}(2p_z(1) - 2p_z(2))$$

The π-orbital is binding and contains two electrons with paired spins. The anti-bonding π^*-orbital contains the unpaired electron. The probability to find the unpaired electron in each of the atomic $2p_x$-orbitals is obtained as the square of the corresponding amplitude:

$$\rho_{N1}(2p_z) = \rho_{N2}(2p_z) = (1/\sqrt{2})^2 = 0.5$$

The value is very close to that estimated from the anisotropic hyperfine coupling of the nitrogen nuclei. There is, however, also a certain probability for the unpaired electron to be located in s-orbitals at the nitrogen and hydrogen atoms, giving rise to the isotropic hyperfine couplings $a_N = 32.2$ MHz, $a_H = 30.8$ MHz. With use of Table 1.1 one obtains:

$$\rho_{N1}(2s) = \rho_{N2}(2s) = 32.2/1,800 = 0.018$$

$$\rho_{H1}(1s) = \rho_{H2}(1s) = 30.8/1,420 = 0.022$$

The term "spin density" is traditionally used both for this probability and for the unpaired electron density in Eq. (1.10) using the same symbol. The intended meaning is usually clear from the context.

1.5.4 Complex Spectra

The method to correlate theory with experimental results has turned out to be extremely useful for the analysis of complex ESR spectra.

ESR spectra of aromatic radical-ions in liquid solution were studied in early work. The ions give well resolved spectra with a large number of hyperfine lines.

The only magnetic nuclei that occur in these species are the protons. The most abundant carbon isotope, ^{12}C, has no magnetic moment. The unpaired electron is in these radicals located in the π-electron system, and is therefore delocalised over the entire molecule. The hyperfine coupling structure appears in an indirect process. Figure 1.11 illustrates that the hyperfine interaction of an H-atom bonded to a carbon atom with an unpaired electron occurs by spin polarisation. Structure (a) is favoured over (b) according to Hund's rule, so that the hyperfine coupling becomes negative – negative spin density is induced in the 1 s H orbital.

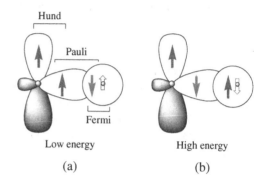

Fig. 1.11 Electron spin polarization mechanism for hyperfine interaction

The π-electron spin density ρ(C) at the carbon atom can be estimated from the measured coupling constant of the adjacent H atom by the McConnell relation.

$$a(H) = Q \cdot \rho(C) \tag{1.11}$$

A value of $Q \approx (-)\,29.0$ G gives the best correlation between experimental values and those calculated with Hückel molecular orbital theory for several aromatic ions. For the naphthalene radical anion in Fig. 1.12 with coupling constants $a(H_a) = (-)4.95$ G and $a(H_b) = (-)1.83$ G one obtains $\rho(C_a) = 0.171$, $\rho(C_b) = 0.063$, in good agreement with the theoretical values given in the Figure.

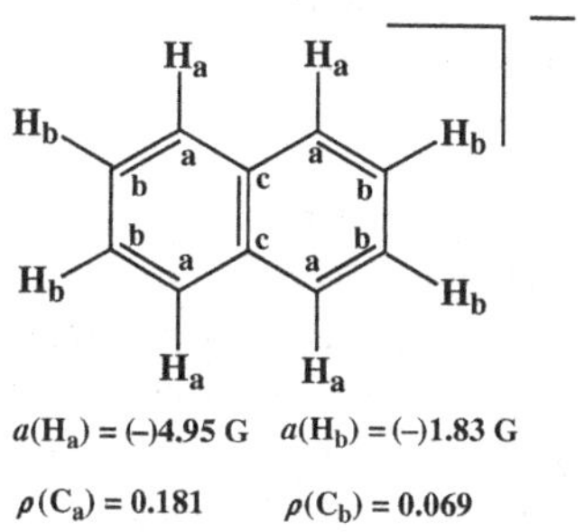

Fig. 1.12 The naphthalene radical anion. The magnitudes of the hyperfine couplings have been determined experimentally. The spin densities $\rho(C_a)$ and $\rho(C_b)$ were calculated by Hückel molecular orbital theory

$a(H_a) = (-)4.95$ G $a(H_b) = (-)1.83$ G

$\rho(C_a) = 0.181$ $\rho(C_b) = 0.069$

The coupling constants calculated with Eq. (1.11) is often used to assign the coupling constants to specific protons in aromatic radicals.

Spin delocalization can also occur in the σ-electron system. One example is the ions of aliphatic fluorocarbon molecules discussed in Chapter 5 that have been investigated by ESR in frozen matrices. Theoretical assignments of coupling constants to specific ^{19}F nuclei are in these and many other recent studies made by calculations based on methods summarized in Chapter 5.

1.5.5 Units, Constants and Symbols in ESR

SI units are employed in recent literature. In older publications the Gauss unit was used for the magnetic field, $1\ G = 1 \cdot 10^{-4}$ T. The units in the original works are as a rule retained in the illustrations of this and the following chapters. Thus, hyperfine couplings obtained from older literature are therefore often given in Gauss rather than in mT. Zero-field splittings are traditionally given in cm^{-1}, while data obtained by ENDOR are naturally specified in frequency units. A list of common units, constants and symbols in the SI system is given in Tables G2 and G5 of Appendix G, while conversion factors are in Table G3.

1.6 Historical and Modern Developments

One reason for the sensitivity increase at higher frequencies comes from the corresponding increased population difference between the upper and lower states of an ESR transition. In a sample with $S = \frac{1}{2}$ containing N_0 unpaired electron spins the excess population of the lower state is approximately $\Delta N \approx \frac{1}{4}N_0 \cdot h\nu/kT$, when the temperature is not too low, i.e. when $h\nu/kT << 1$. The first ESR experiments were accordingly reported soon after microwave equipment became available around 1945.

In the first experiments by Zavoisky [2] a wavelength of 25 m was used, corresponding to a resonance field of only 0.4 mT. As the line-width was about 5 mT, the observation was difficult. In the second experiment on a Cu^{2+} salt the frequency was 133 MHz. In 1946 microwaves were employed giving a resonance at about 100 mT.

Soon after the first experiments, studies on Mn^{2+} salts were performed by employing 10 cm microwaves, fed into the cavity containing the sample. The resonance was detected by readings on a d.c. instrument connected to the detector. In some following papers around 1948, single crystal studies were first reported. Systematic ESR studies of transition metal ions were initiated in a paper titled "Paramagnetic Resonance in Salts of the Iron Group" [3]. Many of these ions had very short relaxation times, which caused line broadening – a short relaxation time τ results in an uncertainty in the energy, and consequently in the line width Δ according to the uncertainty relation $\Delta \cdot \tau \geq \hbar$. The spin relaxation time τ becomes longer at lower temperature; by working with the sample and cavity immersed in liquid hydrogen ($T = 27$ K) or oxygen ($T = 88$ K) it became possible to obtain narrower lines, and thus better resolution.

In 1949 the first observation of hyperfine structure was reported [4]. The structure observed was due to Cu^{2+} ions ($I = 3/2$) from a dilute (5%) crystal of $Cu(NH_4)_2$ $(SO_4)_2 \cdot 6H_2O$ in the isomorphic $Mg(NH_4)_2(SO_4)_2 \cdot 6H_2O$. The dilution was necessary in order to reduce the broadening caused by the interaction between the electron spins of neighbour Cu^{2+} ions. This meant too that the absorption at resonance was diminished by a factor of 20 due to the dilution. The method of field modulation, introduced by British researchers Bleaney, Penrose, and Ingram provided an increased sensitivity, compensating for the dilution loss. By applying 50 Hz modulation on the cavity the signal could be traced on an oscilloscope. Modulation at 100 kHz (Fig. 1.8) is employed in most of the modern instruments. In combination with phase-sensitive detection a derivative signal rather than an absorption is traced on a recorder or computer screen.

1.6.1 High Frequency

ESR spectrometers are available commercially in several frequency ranges, while instruments at very high frequencies first reported in 1977 [5] have been built in specialised laboratories. The spectrometers are usually designated by a "band", as shown in Table 1.2.

Table 1.2 Band designation of commonly employed ESR-spectrometers

Band	Frequency (GHz)	Wavelength (cm)	Field (T), ($g = 2$)
S	3.0	10.0	0.107
X	9.5	3.15	0.339
K	23	1.30	0.82
Q	35	0.86	1.25
W	95	0.315	3.39

The sample size depends on the wavelength, largest at S- smallest at W-band. S-band spectrometers are therefore used for bulky samples, one application is ESR-imaging treated in Chapter 9. X-band is the most commonly used. Regular electromagnets can be used up to Q-band frequencies and corresponding magnetic field, while at W-band and higher superconducting magnets are employed. The advantages of high frequency ESR may be summarised:

- Increase in spectral resolution due to the higher magnetic field
- Increase in orientation selectivity in the investigation of disordered systems

- Simplification of spectra due to suppression of second order effects
- Accessibility of spin systems with large zero-field splitting
- Increase in detection sensitivity for samples of limited quantity

With refined measurement technology, several paramagnetic species have been detected in samples that previously were thought to contain just one component. The resonances of two species with different g-factors, g_1 and g_2 are separated by

$$\Delta B = B_2 - B_1 = \frac{h}{\mu_B} \cdot \frac{g_1 - g_2}{g_1 \cdot g_2} \cdot \nu \qquad (1.12)$$

The field separation and thus the resolution increases with the microwave frequency. Anisotropy of the g-factor occurring in solid materials is also better resolved; g_1 and g_2 could correspond to the axial ($g_{||}$) and perpendicular ($g_\perp$) components in a cylindrically symmetric case, or two of the three g-factors of a general g-tensor (Chapter 4). Splittings due to hyperfine- or zero-field interactions are (to first order) independent of the microwave frequency. Measurements at different frequencies can therefore clarify if an observed spectrum is split by Zeeman interactions (different g-factors) or other reasons.

Analysis of ESR-spectra is facilitated when they are "first order", i.e. when $h\nu \gg D$ and/or $h\nu \gg a$ in systems featuring zero-field ($S > \frac{1}{2}$) and/or hyperfine splittings. In several cases this approximation does not strictly apply at X-band due to large hyperfine couplings that occasionally are observed in free radicals (see Chapter 5 for examples) and more generally occur in transition metal ion complexes. The approximation can break down completely at X-band for systems with large zero-field splittings. The complications caused by deviations from first order can be avoided by measurements at high frequency. The analysis of ESR spectra is treated in Chapter 3.

1.6.2 ENDOR

ENDOR, Electron Nuclear Double Resonance, is a combination of electron- and nuclear magnetic resonance developed in 1956. The method was apparently originally intended for applications in physics to achieve nuclear polarisation [6], e.g. to obtain more nuclei with $m_I = -\frac{1}{2}$ than with $m_I = +\frac{1}{2}$. But its main application became to resolve hyperfine structure that is unresolved in regular ESR in liquid and solid samples. Two lines appear in the ENDOR spectrum of a radical ($S = \frac{1}{2}$), containing a single nucleus, one corresponding to the $m_S = +\frac{1}{2}$ the other to the $m_S = -\frac{1}{2}$ electronic level.

For the system in Fig. 1.13 the magnetic field strength is first adjusted to resonance. The microwave power is then increased to saturate the transition AA'; the ESR signal decreases as the population difference of the states decreases. Next the radio frequency ν_{RF} is swept, to induce a transition AB. The population difference between A and A' is restored. This is observed as an increase of the ESR signal when $\nu_{RF} = \nu$. A similar increase occurs when $\nu_{RF} = \nu'$. The electronic spin lattice

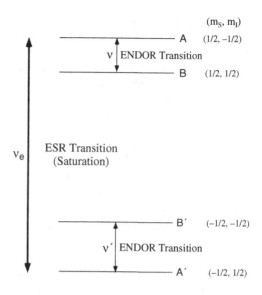

Fig. 1.13 Magnetic energy levels of an unpaired electron interacting with a nucleus, $I = \frac{1}{2}$, through an isotropic hyperfine interaction with a coupling constant $a > 0$, $g_N > 0$, and $g_N \mu_N B > 1/2a$

(m_S, m_I)

— A $(1/2, -1/2)$

ν | ENDOR Transition

— B $(1/2, 1/2)$

ν_e ESR Transition (Saturation)

— B′ $(-1/2, -1/2)$

$\nu′$ | ENDOR Transition

— A′ $(-1/2, 1/2)$

relaxation time must be sufficiently long to saturate the transition AA′, and $\nu \neq \nu′$, otherwise the populations tend to be equal. Feher investigated F-centres in KCl; F-centres are "holes" after a missing negative ion; an electron is trapped in the hole that is surrounded by neighbour Cl^- and K^+ ions. The hyperfine couplings from chlorine and potassium that are hidden in the line-width of the ESR spectrum could be observed at $T = 1.2$ K. The hyperfine structure of F and Li ions surrounding an F-centre could be resolved up to the seventh shell in a LiF crystal measured at 1.3 K. Silicon single crystals doped with P, As and Sb were also investigated at an early stage. The extra electron from the donor atom showed hyperfine structure from ^{29}Si atoms at different distances from the donor.

ENDOR has several advantages over regular ESR:

- Increase in spectral resolution due to the reduction of the number of lines
- Ability to assign transitions to a specific kind of nucleus due to the dependence on the nuclear frequency ν_N
- Signs of coupling constants by applying two radiofrequencies (TRIPLE)
- Differentiation between two species present in a single sample by applying field swept ENDOR

The main advantage of the method is the increased resolution at the expense of lower sensitivity compared to regular ESR. Modern applications of the method in liquid and solid samples are described in Chapter 2.

1.6.3 Short-Lived Radicals

Several methods have been developed to study the structure and reactions of short-lived free radicals in solution.

Recording of the ESR spectrum under simultaneous irradiation of the sample by an electron beam was used in the first experiments [7]. In later studies pulsed ESR was used in pulse radiolysis experiments [8]. This detection method has a high time resolution of ca. 10 ns for studies of reaction kinetics. Radicals formed by steady state photolysis were also studied at an early stage. Pulsed ESR is commonly used with laser flash photolysis to obtain high time resolution in modern applications [9]. For studies of radicals formed in chemical reactions a common method is by mixing of the reagents in a flow system before entering the ESR cavity. In these in situ experiments the ESR lines can occur in emission rather than absorption because the radicals are not in thermal equilibrium immediately after formation. A chemically induced electron polarization, CIDEP, occurs.

Reactive radicals can be stabilised by several methods. Techniques to increase the life-time by rapid freezing of samples have been employed for a long time [10]. The matrix isolation method discussed in Chapter 5 is a useful technique to obtain the ESR spectra of species that are too reactive to be studied in the liquid state. Trapping of radiolytically produced radicals in the solid state is a common method in radiation biophysics (Chapter 2) and polymer science (Chapter 7).

Short-lived free radicals can also be trapped chemically by so called spin trap molecules to produce a more stable radical detectable by ESR. The hyperfine structure of this radical is used to identify the radical initially trapped. The first experiments were reported in 1968–1969 [11–13]. The method is much used in biological applications.

1.6.4 Pulsed ESR

In pulsed ESR samples are exposed to a series of short intense microwave pulses, e.g. the 2-pulse sequence shown in Fig. 1.14 in place of the continuous radiation with microwaves at low power in the traditional continuous wave spectrometer.

Fig. 1.14 2-pulse sequence

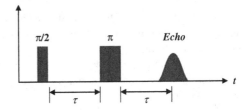

The method has been in use with home-built equipment for several decades, Measurements are now possible to perform also with commercial equipment that has become available from at least two manufacturers. The method can be applied to measure dynamic parameters, like relaxation times, but also to obtain structure information, e.g. hyperfine coupling constants, the latter usually by applying Fourier transformation (FT) to the time-domain signals. A full survey of the method and its applications for liquid and solid samples has recently been published, while we illustrate some applications to solids in Chapter 2. The basic principles are in fact the same as in pulsed NMR that has completely superseded CW NMR for several reasons:

- Sensitivity; a FT spectrometer measures signal all the time while a CW spectrometer measures baseline most of the time.
- 2-dimensional spectroscopy; interactions may be isolated and correlations detected that are not observable by CW methods.
- Time resolution; dynamic phenomena may be directly measured.

The development has been slower in ESR as faster relaxation rates (three orders of magnitude) and wider spectral ranges (three-four orders of magnitude) have provided technical obstacles. The free induction decay (FID) can be employed to obtain spectra of paramagnetic species in liquids, analogous to the procedure in NMR. For solid samples the FID is usually too fast to obtain a spectrum. The electron spin echo modulation (ESEEM) method is the common technique to obtain spectra of paramagnetic species in solids.

1.6.4.1 ESEEM

ESEEM is a special pulsed variant of ESR spectroscopy. A series of microwave pulses is applied to the sample and the amplitude of the transient response from the sample is monitored. In a basic experiment, Fig. 1.14, two pulses separated by a time τ are applied. The second pulse is twice as long as the first. A time τ after the last pulse a transient response appears from the sample, the so called spin echo. By monitoring the echo amplitude as a function of the time τ, a spin echo envelope can be recorded. The decay time of the envelope is called the phase memory time T_M. If weak anisotropic hyperfine structure (hfs) is present modulations can be superimposed on the envelope. The hfs is obtained either by trial-and-error simulations of decay curves or by a Fourier transform in modern instruments. In the latter case a spectrum in the frequency domain is obtained. Lines occur at the same frequencies as in ENDOR. The pioneering experimental and theoretical developments were done by Mims [14]. The spectral resolution of ESEEM is even greater than for ENDOR but in practice there is at present an upper limit on the magnitude of the couplings that can be measured of about 30–40 MHz. This method has been particularly applied to studies of solvation structures around paramagnetic species in disordered systems, e.g. trapped electrons in frozen glasses and of transition metal ions in metalloproteins for which no single crystals are available for X-ray diffraction.

1.6.4.2 Relaxation Times

Relaxation times are more directly measured by pulse experiments measure than by the CW saturation technique (Chapter 2). Physical applications often involve measurements made at different temperatures to obtain data regarding the dynamics in the solid state.

1.6.4.3 Distance Measurements

Distances between paramagnetic centers in the range 2–8 nm can be measured by specialized pulse techniques that record modulations due to the magnetic coupling between them. The coupling is of the same type as for the fine-structure in radical pairs. By measuring the fine-structure constant D the distance is obtained. This method has been used to obtain distances within "spin-labelled" proteins, where the spin labels are free radicals that have been attached to amino acids at two known positions.

1.7 Summary

The ESR (electron spin resonance) method established in 1945 is employed for studies of free radicals, transition metal ion complexes, and other paramagnetic substances most commonly in liquids and solids. An applied magnetic field causes a splitting of the energy levels. In free radicals the splitting is mainly caused by the magnetic moment of the spin ($S = \frac{1}{2}$) of the unpaired electron. A transition occurs from the lower ($m_S = -\frac{1}{2}$) to the higher ($m_S = +\frac{1}{2}$) energy state by absorption of energy from an applied electromagnetic field in the microwave region. A spectrum is obtained in continuous wave (CW) ESR by sweeping the magnetic field B to obtain resonance when $h\nu = g\mu_B B$ at a fixed microwave frequency ν. Isotopes with nuclear spin $I \neq 0$ give rise to hyperfine structure in the ESR spectrum determined by the selection rule $\Delta m_S = 1$, $\Delta m_I = 0$. The values of the g-factor and particularly of the hyperfine coupling are used to identify the species in a sample. Light exposure or other excitation can produce species containing two unpaired electrons with parallel spin, resulting in a zero-field splitting (or fine structure) in the ESR spectra in crystalline or amorphous matrices. The zero-field splitting is also characteristic for transition metal ion complexes with electron spin $S \geq 1$. The splitting is anisotropic, i.e. depends on the direction of the magnetic field. The g-factor and the *hfc* may also be anisotropic in solids.

The sensitivity and resolution of modern ESR spectrometers have been increased by employing high microwave frequencies, multi-resonance designs, and pulse microwave methods. High-frequency ESR is employed to improve the resolution of g-factors. The main advantage of the electron-nuclear resonance method, ENDOR, is the increased resolution of hyperfine couplings (*hfc*). Pulsed techniques are used for measurements of dynamic properties like magnetic relaxation but also for structural studies. The method provides a direct method for the measurement of relaxation rates. A variant known as ESEEM, the electron spin echo modulation technique, can resolve weaker *hfc* than is possible even by ENDOR. The ENDOR method still has an advantage of being applicable also to resolve isotropic *hfc*, e.g. in the liquid state where pulse spin-echo methods fail. A recent application of pulsed ESR involves the measurements of distances between paramagnetic centres, particularly in biological samples.

The ESR method that previously has been a specialty for physicists with good knowledge in electronics, is now increasingly more used as one tool in combination with other methods to solve different problems in chemistry, biochemistry and medicine. A main aim particularly in the past was to determine the structure of the paramagnetic species present in the sample under investigation. The electronic and geometric structure of paramagnetic substances can thereby be obtained by analysis of the measured parameters. Examples of the kind of information that one can obtain was presented in the chapter. Short-lived species may be investigated by generation in situ, by stabilization at low temperature and by reaction with other molecules known as spin traps. The new species formed in the latter case are stable free radicals. The chemical structure reminds of the nitroxide radicals used as labels on other molecules to give information about molecular dynamics and to provide means to determine distances, the former typically with CW-ESR, the latter with pulse methods. ESR measurements of the amount of paramagnetic species in a sample have applications in radiation dosimetry, in geological dating, in characterisation of metal proteins and in ESR imaging where a concentration profile can be mapped. Some of the applications below will be further treated in the following chapters while other are treated in the specialized works listed below.

Structures of transition metal ion complexes: Measurements performed on single crystals, powders and frozen solutions give information about the geometric and electronic structure by analysis of the strengths and symmetries of the g-, zero-field- $(S > \frac{1}{2})$ and hyperfine interactions. The field is summarised in a modern treatise [15].

Liquid systems: Structural studies of free radicals in the liquid state using regular ESR are classical fields, and some monograph treatments are accordingly relatively old [16, 17]. Studies of radicals in natural products have been summarised by Pedersen [18]. Applications of ENDOR to organic and biological chemistry are given in the book by Kurreck et al. [19].

Radiation effects: A vast material has been collected, about the formation of free radicals by ionising radiation and by photolysis in chemical and biochemical systems. The treatise [20] dealing with studies of primary radiation effects and damage mechanisms in molecules of biological interest is also a valuable source of information of ENDOR spectroscopy in solids. Specialised treatises of radiation effects in solids involve studies of inorganic [21] and organic systems [22], radiation biophysics [20], radical ionic systems [23], radicals on surfaces [24], and radicals in solids [25].

Polymeric systems: The book by Rånby and Rabek [26] about ESR spectroscopy in polymer research contains more than 2,500 references. Advanced ESR methods in polymer research are summarised in a recent treatise [27].

Biology and Medicine: This topic was recently treated in the book by Halliwell and Gutteridge [28].

Pulse ESR: The standard modern treatment is by A. Schweiger and G. Jeschke [29]. ESEEM is treated by Dikanov and Tsvetkov [30].

Reviews and data bases: Literature reviews are contained in the series of books issued annually by the Royal Society of Chemistry [31]. The tabulated data in the Landolt Börnstein series [32] provide useful structural information.

Textbooks: The works [1, 33, 34] by Weil and Bolton, by Rieger and by Brustolon and Giamello are recent general textbooks. Two textbooks in Japanese have been published [35, 36].

References

1. J.A. Weil, J.R. Bolton: *'Electron Paramagnetic Resonance: Elementary Theory and Practical Applications'*, 2nd Edition, J. Wiley, New York (2007).
2. E. Zavoisky: J. Phys. USSR **9**, 211 (1945).
3. M.S. Bagguley, B. Bleaney, J.H.E. Griffiths, R.P. Penrose, B.I. Plumpton: Proc. Phys. Soc. **61**, 542 (1948).
4. R.P. Penrose: Nature **163**, 992 (1949).
5. A.A. Galkin, O.Ya. Grinberg, A.A Dubinskii, N.N. Kabdin, V.N. Krimov, V.I. Kurochkin, Ya.S. Lebedev, L.G. Oranskii, V.F. Shovalov: Instrum. Exp. Tech. (Eng. Transl.) **20**, 284 (1977).
6. G. Feher: Phys. Rev. **103**, 500 (1956).
7. R.W. Fessenden, R.H. Schuler: J. Chem. Phys. **33**, 935 (1960).
8. A.D. Trifunac, J.R. Norris, R.G. Lawler: J. Chem. Phys. **71**, 4380 (1979).
9. K.A. McLauchlan: In *'Foundations of modern EPR'* ed. by G. Eaton, S.S. Eaton, K.M. Salikhov, World Scientific, Singapore (1998).
10. R.C. Bray: J. Biochem. **81**, 189 (1961).
11. C. Lagercrantz, S. Forshult: Nature **218**, 1247 (1968).
12. G.R. Chalfront, M.J. Perkins, A. Horsfield: J. Am. Chem. Soc. **90**, 7141 (1968).
13. E.G. Janzen, B.J. Blackburn: J. Am. Chem. Soc. **91**, 4481 (1969).
14. W.B. Mims Phys: Rev. B **5**, 2409 (1972).
15. J.R. Pilbrow: *'Transition Ion Paramagnetic Resonance'*, Clarendon Press, Oxford (1990).
16. F. Gerson: *Hochauflösende ESR-spektroskopie dargestellt anhand aromatischer Radikal-Ionen'*, Verlag Chemie, Weinheim (1967).
17. K. Scheffler, H.B. Stegmann: *'Elektronenspinresonanz, Grundlagen und Anwendung in der organischen Chemie'*, Springer-Verlag, Berlin (1970).
18. J.A. Pedersen: *'Handbook of EPR Spectra from Natural and Synthetic Quinones and Quinols'*, CRC Press, Boca Raton, FL (1985).
19. H. Kurreck, B. Kirste, W. Lubitz: *'Electron Nuclear Double Resonance Spectroscopy of Radicals in Solution - Application to Organic and Biological Chemistry'*, VCH publishers, Weinheim (1988).
20. H.B. Box: *'Radiation effects, ESR and ENDOR analysis'*, Academic Press, New York (1977).
21. P.W. Atkins, M.C.R. Symons: *'The structure of inorganic radicals. An application of ESR to the study of molecular structure'*, Elsevier publishing company, Amsterdam (1967).
22. S.Ya. Pshezhetskii, A.G. Kotov, V.K. Milinchuk, V.A. Roginski, V.I. Tupilov: *'EPR of free radicals in radiation chemistry'*, John Wiley & Sons, New York (1974).
23. A. Lund, M. Shiotani (eds.): *'Radical Ionic systems'*, Kluwer academic publishers, Dordrecht (1991).
24. A. Lund, C. Rhodes (eds.): *'Radicals on surfaces'*, Kluwer academic publishers, Dordrecht (1995).
25. A. Lund, M. Shiotani (eds.): *'EPR of Free Radicals in Solids'*, Kluwer academic publishers, Dordrecht (2003).
26. H.B. Rånby, J.F. Rabek: *'ESR spectroscopy in polymer research'*, Springer-Verlag, Berlin (1977).

27. S. Schlick (ed.): '*Advanced ESR Methods in Polymer Research*', Wiley, Hoboken, NJ (2006).
28. B. Halliwell, J.C. Gutteridge: '*Free Radicals in Biology and Medicine*', Third Ed., Oxford Univ. Press (1999).
29. A. Schweiger, G. Jeschke: '*Principles of pulse electron paramagnetic resonance*', Oxford University Press, New York (2001).
30. S.A. Dikanov, Yu.D. Tsvetkov: '*Electron Spin Echo Envelope Modulation (ESEEM) Spectroscopy)* ' CRC Press, Inc. - Boca Raton, FL (1992).
31. '*Electron paramagnetic resonance*', Royal Society of Chemistry, Vol. **1** (1973) to **21** (2008) Thomas Graham House, Cambridge UK.
32. H. Fischer (ed.): '*Landolt-Börnstein, Numerical data and functional relationships in science and technology: Magnetic properties of free radicals*', Springer-Verlag, Berlin (1965-89).
33. P.H. Rieger: '*Electron Spin Resonance: Analysis and Interpretation*', RSC Publishing, Cambridge, UK (2007).
34. M. Brustolon, E. Giamello (eds.): '*Electron Paramagnetic Resonance: A Practitioner's Toolkit*', Wiley, Hoboken, NJ (2009).
35. J. Yamauchi: '*Magnetic Resonance - ESR: Spectroscopy of Electron Spin*', Saiensu Sya (2006). (In Japanese)
36. Japanese Society of SEST: '*Introduction to Electron Spin Science and Technology*', Nihon Gakkai Jimu Center (2003). (In Japanese)

Exercises

E1.1 Naturally occurring chemical substances are in general diamagnetic. (a) Which of the substances CO, NO, CO_2, and NO_2 might be paramagnetic? (NO is a special case, see Chapter 6.) (b) Can any of the hydrocarbon species shown below be paramagnetic? Can ions of the species be paramagnetic?

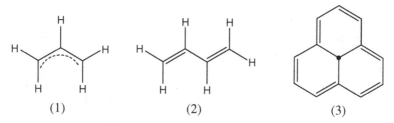

      (1)   (2)   (3)

E1.2 Ions of organic compounds have been extensively studied by ESR as discussed in Chapter 5.

(a) The spin density is equally distributed over the carbon atoms of the cyclo-octatetraene anion radical. What is the expected hyperfine pattern (number and intensities of hyperfine lines due to hydrogen)?

(b) Hyperfine couplings due to three groups of protons were observed from the radical cation produced by the treatment of anthracene with $SbCl_5$ in liquid solution. The coupling constants were 6.44, 3.08 and 1.38 G, respectively [L.C. Lewis, L.S. Singer: J. Chem. Phys. **43**, 2712 (1965)]. Assign the couplings to specific positions using the Hückel spin densities given below.

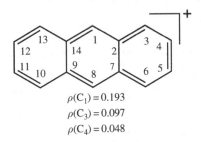

$$\rho(C_1) = 0.193$$
$$\rho(C_3) = 0.097$$
$$\rho(C_4) = 0.048$$

E1.3 The anisotropic hyperfine coupling of the NO_3 radical trapped in a nitrate salt was found to be axially symmetric, $A_{||} = a + 2b$, $A_\perp = a-b$, with $a = 0.7$ G, $b = 0.7$ G [21]. The radical is assumed to be planar.

(a) Estimate the spin density in the nitrogen $2p_z$ orbital with its axis (z) perpendicular to the plane, the spin density at the nitrogen nucleus, and the total spin density on nitrogen by a procedure analogous to that in Section 1.5.3.

(b) Estimate the spin density on the oxygen atoms by assuming that they are equivalent and that $\rho(N) + 3\cdot \rho(C) = 1$.

E1.4 Consider the influence of the microwave frequency on the spectrum in Fig. 1.7 showing fine structure or zero-field splitting (zfs).

(a) Would it be possible to directly measure the zero-field splitting with a microwave frequency that is 1/3 of that in the figure, for instance by measuring at S- rather than at X-band?

(b) Do"first order"conditions strictly apply for the spectrum in Fig. 1.7? (The spectrum is first order when $2\cdot D \ll h\nu$ in energy units or $2\cdot D \ll B_g$ in field units. The zfs can then be directly measured as shown in the figure.) Suggest an experimental method to satisfy this condition. (A theoretical procedure was applied to obtain the zfs of excited naphthalene in the $S = 1$ triplet state at an early stage, [H. van der Waals, M.S. de Groot: Mol. Phys. **2**, 333 (1959)].

Chapter 2
Multi-Resonance and Pulsed ESR

Abstract Multi-resonance involves ENDOR, TRIPLE and ELDOR in continuous-wave (CW) and pulsed modes. ENDOR is mainly used to increase the spectral resolution of weak hyperfine couplings (*hfc*). TRIPLE provides a method to determine the *signs* of the *hfc*. The ELDOR method uses *two* microwave (MW) frequencies to obtain distances between specific spin-labeled sites in pulsed experiments, PELDOR or DEER. The electron-spin-echo (ESE) technique involves radiation with two or more MW pulses. The electron-spin-echo-envelope-modulation (ESEEM) method is particularly used to resolve weak anisotropic *hfc* in disordered solids. HYSCORE (Hyperfine Sublevel Correlation Spectroscopy) is the most common two-dimensional ESEEM method to measure weak *hfc* after Fourier transformation of the echo decay signal. The ESEEM and HYSCORE methods are not applicable to liquid samples, in which case the FID (free induction decay) method finds some use. Pulsed ESR is also used to measure magnetic relaxation in a more direct way than with CW ESR.

2.1 Introduction

Multi-resonance and pulsed ESR techniques can provide better spectral resolution than conventional ESR. Multi-resonance involves ENDOR, TRIPLE and ELDOR. In an ENDOR experiment a radiofrequency (RF) field is applied in addition to the microwave (MW) employed in standard continuous wave (CW) ESR. ENDOR is mainly used to increase the spectral resolution, so that overlapping or unresolved hyperfine structure in the ESR spectra can be detected. In the classical work by Feher [1] the radiofrequency was continuously swept. CW X-band spectrometers with an ENDOR attachment have been commercially available for a long time. Accessories for other frequency bands and for pulsed ESR have been developed more recently. In a TRIPLE experiment two RF fields are applied [2, 3]. A theoretical application has been to determine the relative signs of two hyperfine couplings. In an ELDOR experiment two MW frequencies are applied. Early applications using

A. Lund et al., *Principles and Applications of ESR Spectroscopy*,
DOI 10.1007/978-1-4020-5344-3_2, © Springer Science+Business Media B.V. 2011

CW spectrometers are summarised in a book by Kevan and Kispert [4]. The technique has been revived in pulsed ELDOR experiments to obtain distances between specific spin-labelled sites in polymers and biopolymers [5, 6].

Pulsed ESR can also yield better spectral resolution than conventional ESR. The electron-spin-echo (ESE) technique involves radiation with two or more MW pulses followed by measurement of the echo signal decay. The electron-spin-echo-envelope-modulation (ESEEM) method provides a means to resolve weak (anisotropic) hyperfine (*hf*) structure from the modulation of the decay curve [7–9]. Following the seminal work by Mims, ESEEM has been particularly successful for studies of solvation structures of paramagnetic species in disordered solids [10–13]. It is less used than ENDOR for single crystal measurements as the measurements are time-consuming. The HYSCORE (Hyperfine Sublevel Correlation Spectroscopy) method [14] is also time-consuming, but permits a safe assignment of the nucleus giving the pattern observed. The analysis is usually made after Fourier transformation (FT) of the echo decay signal. In favorable cases the (anisotropic) hyperfine coupling can be read directly from the FT pattern, in other cases simulation is needed. Pulsed ESR is also used to measure magnetic relaxation in a more direct way than with continuous-wave ESR.

2.2 ENDOR

Enhanced resolution as compared to standard ESR can be obtained by the electron nuclear double resonance (ENDOR) technique [1]. We discuss CW-ENDOR applications for studies of free radicals and other species with electron spin S = ½ in this section, while pulsed techniques are treated in connection with pulsed ESR.

CW-ENDOR is performed by setting the magnetic field of the ESR spectrometer to induce an ESR transition while simultaneously applying a radiofrequency (RF) field that is swept over the range of nuclear resonance frequencies (typically 1–100 MHz) of the system. The intensity of the ESR response is measured as a function of the RF frequency to register the changes in ESR intensity that accompanies the passage through nuclear resonance. Sufficient microwave power must be applied to partially saturate the ESR transition. Nuclear resonance between the nuclear states induced by the RF-field will desaturate the ESR signal causing detection of the ENDOR transitions. ENDOR can be applied to liquid samples, to single crystals and to disordered solids.

2.2.1 ENDOR in Liquids

ESR spectra of organic radicals in solution [15] are often composed of many partially overlapping lines particularly due to *hf* structure with groups of equivalent ^{1}H nuclei. A group of *n* equivalent hydrogen atoms gives $2 \cdot n + 1$ ESR *hf* lines, but only two ENDOR lines. The number of ENDOR lines is therefore usually less than in an ESR spectrum. The biphenyl radical cation discussed in Section 2.3 containing three groups of equivalent H atoms with $n_1 = 2$, $n_2 = 4$ and $n_3 = 4$ would for example give $3 \cdot 5 \cdot 5 = 75$ ESR lines but only $2+2+2 = 6$ ENDOR lines.

The appearance of an ENDOR spectrum depends on the relative size of the hyperfine coupling (*hfc*) and the NMR frequency ν_N. For a ^{1}H nucleus with nuclear g-factor $g_H = 5.586$ a frequency of $\nu_H = \frac{g_H \mu_N B}{h} \approx 15$ MHz is characteristic for free radicals at X-band ($B \approx 339$ mT at $g = 2.0023$). The ^{1}H hyperfine couplings of aromatic radicals are usually relatively small, $|a| < 2\nu_H$. The ENDOR spectrum can in this case be accounted for by considering the hyperfine coupling as an additional field that can add or subtract to the applied magnetic field acting on the nucleus depending on the orientation of the electron spin. The ENDOR transitions with $\Delta m_I = 1$ occur at $\nu_\pm = \nu_H \mp \frac{1}{2}a$ for the $m_S = \pm\frac{1}{2}$ states. In this case the two ENDOR lines are thus separated by the hyperfine coupling a, and the spectrum is centered at the nuclear frequency as in Fig. 2.1(a). The same result is obtained by quantum mechanics using the equation given in Fig. 2.2, where a' denotes the *hfc* in energy units. In the opposite case with $|a| > 2\nu_H$ a spectrum like that in Fig. 2.1(b) appears.

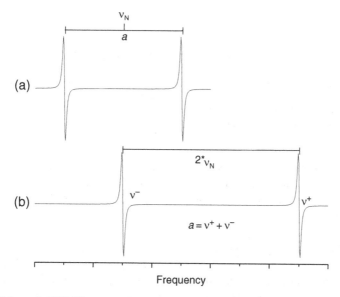

Fig. 2.1 Schematic ENDOR spectra for an $S = \frac{1}{2}$, $I = \frac{1}{2}$ system. (**a**) With a small hyperfine coupling $|a| < 2$ $g_N \mu_N B/h = 2 \cdot \nu_N$. (**b**) With $|a| > 2 \cdot \nu_N$

2.2.1.1 Hyperfine Couplings with Protons

Equivalent nuclei: A group of n equivalent ^{1}H or other $I = \frac{1}{2}$ nuclei gives rise to $2 \cdot n + 1$ equally spaced nuclear energy levels for each m_S value according to the Pascal triangle method (Chapter 1). The ENDOR spectrum for a free radical containing a group of equivalent $I = \frac{1}{2}$ nuclei accordingly consists of only two lines with positions given by the energy separations between adjacent nuclear levels for $m_S = \pm\frac{1}{2}$.

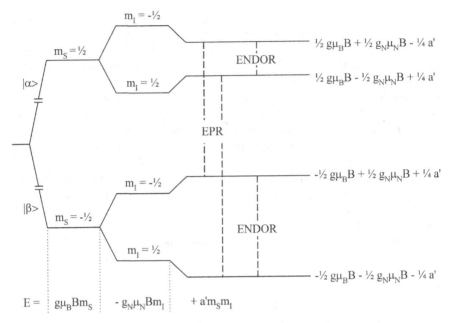

Fig. 2.2 ESR and ENDOR transitions for an $S = \frac{1}{2}$, $I = \frac{1}{2}$ system with a small ($a' < 2\, g_N \mu_N B$) positive hyperfine coupling. The ESR ($\Delta m_S = 1$, $\Delta m_I = 0$) and the ENDOR ($\Delta m_I = \pm 1$, $\Delta m_S = 0$) transitions are marked with dashed lines between the corresponding states. The ENDOR lines occur at $\nu_{\pm} = \nu_N \mp \frac{1}{2} a'/h$ for $m_S = \pm\frac{1}{2}$, the ESR lines at $B_{\pm} = B_0 \mp \frac{1}{2} a'/g\mu_B$ for $m_I = \pm\frac{1}{2}$ with $B_0 = h\nu_e/g\mu_B$. The diagram is reproduced from [T.A. Vestad, Dissertation, University of Oslo (2005)] with permission from Dr. Vestad

A radical containing several sets of equivalent nuclei gives two lines for each set as in Fig. 2.3. Information about the number of equivalent nuclei belonging to each set is, however, usually not obtainable from the intensities of the lines, for the following reasons.

- Hyperfine-enhancement effects: The hyperfine interaction causes a perturbation so that the wave-functions are not pure (m_S, m_I) states. The intensities of the ENDOR transitions are accordingly affected by the magnitude of the hyperfine coupling. The high frequency line of a set is often stronger than that at low frequency. This hyperfine-enhancement effect is more pronounced for the larger hyperfine coupling with $a_H = 22.18$ MHz in Fig. 2.3.
- Relaxation effects: The intensity of the different sets of ENDOR lines is largely determined by the relaxation properties of the particular nuclei. We refer to the literature for procedures that take relaxation into account on the intensity of ENDOR lines [15].

One may conclude that the increased resolution of classical ENDOR in the liquid state also implies loss of information regarding the numbers of equivalent nuclei.

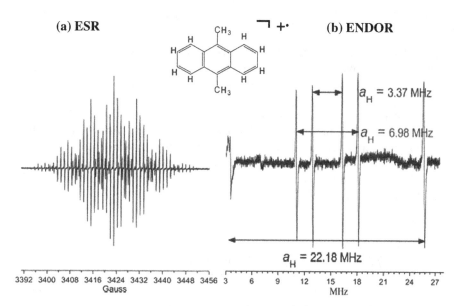

Fig. 2.3 (**a**) ESR and (**b**) ENDOR spectra of the 9,10-dimethylanthracene cation radical in fluid solution. Diagram is reproduced from [16] with permission from The Royal Society of Chemistry

Experimental remedies to obtain these numbers are the TRIPLE CW and pulsed ENDOR methods described in this chapter. In some cases the number of equivalent nuclei belonging to a particular hyperfine coupling constant can be determined visually from a sufficiently resolved ESR spectrum. Thus, the main structure with $a_H = 22.18$ MHz in the ESR spectrum of the 9,10-dimethylanthracene cation radical in Fig. 2.3 is a septet due to the six equivalent ^{1}H of the methyl groups. Assignment of the couplings of the ring protons is often made, experimentally by selective deuteration, and theoretically by quantum chemistry. Simulations are particularly employed for the interpretation of ENDOR spectra of disordered solids discussed later in this chapter and in Chapter 3.

^{1}*H hyperfine couplings in S > ½ systems*: ENDOR measurements of ^{1}H hyperfine structure studies have proved possible in fluid solution of paramagnetic species with $S > ½$ [15]. Measurements of the hyperfine structure due to the ligands in transition metal complexes in frozen solution are discussed in Section 2.2.3.3.

2.2.1.2 Spin Density

ENDOR measurements of isotropic hyperfine couplings are extensively used to obtain the spin density with the aim to clarify the electronic and geometric structures of paramagnetic species. The isotropic or contact hyperfine (*hfc*) coupling, a_{iso}, is proportional to the unpaired electron density $\rho(N)$ at a nucleus (N) with $I \neq 0$ [17a], see [17b] for a nearly classical treatment.

$$a_{iso} = \frac{2}{3}\mu_0 g\mu_B g_N \mu_N \rho(N)$$

The word "contact" is used to emphasize that the coupling is due to a finite unpaired electron density at the nucleus. Thus, the wave-function must have some s-character ($l = 0$), since orbitals with angular momentum quantum numbers $l > 0$ are identically zero at the origin.

The unpaired electron density (or spin density) in this equation is measured in units of electrons/volume. A related dimensionless quantity giving the probability of finding the unpaired electron at a specific atom is also commonly used employing the same symbol ρ and referred to as "spin density". The McConnell equation [18] applying to the hyperfine coupling constant of an H atom in a π-*electron* radical, $\rangle\dot{C}_\alpha$ –H provides an example:

$$a_\alpha = Q \cdot \rho_C^\pi \text{ with } Q \approx -70 \text{ MHz}.$$

The quantity ρ_C^π is the 2p orbital spin density at the $\dot{C}_\alpha$ atom. The intended meaning of spin density is usually clear from the context.

TRIPLE measurements discussed in the next section are also commonly employed for structure determination of radicals in liquid solution.

2.2.1.3 TRIPLE

Two RF-frequencies are employed in two different variants of TRIPLE.

- Special TRIPLE [2] can give the number of equivalent $I = \frac{1}{2}$ nuclei. The two RF-frequencies are swept to simultaneously correspond to the two ENDOR transitions for the same nucleus, Fig. 2.4(c). The method is employed particularly to radicals in liquids showing hyperfine structure due to ^{1}H nuclei. The method is more sensitive than the usual ENDOR experiment. The theoretical background is described in original [2] and review literature [15, 16].
- General TRIPLE [3, 15, 16] can give the sign of the hyperfine coupling constant of one nucleus relative to another of different magnitude from another nucleus. One RF-frequency is adjusted to an ENDOR transition for the first nucleus, while the other is swept through the ENDOR transitions of other nuclei. An increase in the signal intensity is observed, Fig. 2.4(d). The relative signs are obtained by analysis of the increase of intensity compared to the regular ENDOR signal.

As an example, the TRIPLE resonance experiments for the phenalenyl radical [15, 16] are shown in Fig. 2.4. Two different hyperfine coupling constants are expected due to three and six equivalent ^{1}H nuclei in accordance with the radical structure shown in the figure. The magnitudes of the two couplings are obtained from the regular ENDOR spectrum, but an assignment to specific positions is not possible. The special ENDOR gives rise to a *single* line for each set of equivalent nuclei, occurring at a frequency deviating by a/2 from the frequency of a free ^{1}H

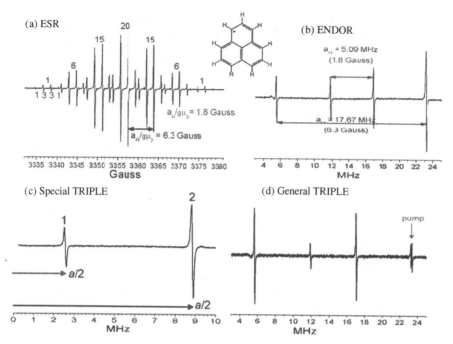

Fig. 2.4 (**a**) ESR, (**b**) ENDOR, (**c**) special TRIPLE and (**d**) general TRIPLE resonance spectra of the phenalenyl radical recorded at room temperature in mineral oil. The special TRIPLE frequency axis shows the deviation from the frequency of a free ^{1}H nucleus. The intensities of signals 1 and 2 are proportional to the number of equivalent ^{1}H nuclei. The relative signs of the couplings are obtained by analysis of the intensity change of the general TRIPLE compared to the regular ENDOR signal. The figure is adapted from [16] with permission from The Royal Society of Chemistry

nucleus. The intensity of signals 1 and 2 are proportional to the number of equivalent ^{1}H nuclei in the ratio 3:6. An analysis of the line intensity patterns of the general TRIPLE compared to regular ENDOR makes it possible to conclude that the two hyperfine couplings have opposite signs. Thus one has $a_H(6H) = -17.67$ MHz for the set with six equivalent ^{1}H, and $a_H(3H) = +5.09$ MHz for the other set. In this example the assignment of the hyperfine couplings to specific ^{1}H nuclei could have been made directly from the ESR spectrum as suggested in exercise E2.6. Well-resolved ESR spectra are not always available, however. The absolute signs are assigned on theoretical grounds as indicated in Chapter 1, and in more detail in the section on quantum chemistry calculations in Chapter 5.

2.2.1.4 Hyperfine Interactions with Other Nuclei

We refer to a textbook dealing with ENDOR resonance spectroscopy of radicals in solution [15] for a review of measurements on radicals showing hyperfine structure due e.g. to ^{2}D, ^{13}C and ^{15}N in isotopic- enriched samples, ^{14}N, ^{23}Na, and ^{29}Si in natural abundance. ENDOR studies in the solid state frequently involve measurements of samples containing nuclei with $I > \frac{1}{2}$ and are discussed in the Sections 2.2.2.2 and 2.2.3.2.

2.2.2 ENDOR in Single Crystals

Several of the advantages with ENDOR measurements in liquid solution also apply for solid samples. Single crystal measurements are particularly informative.

- Hyperfine couplings are measured with higher accuracy than with ESR, in part due to the reduction of number of lines as in liquids, in part to the narrower line-widths in ENDOR (ESR line-widths in solids are typically one order of magnitude wider than the ENDOR lines).
- Hyperfine structure due to nuclei of the crystal matrix can be resolved to provide information about the trapping site as demonstrated already in early ENDOR studies of solids [1]. The distances from the paramagnetic centre to nuclei of the matrix can be measured.
- Overlapping spectra due to different species can be differentiated by ENDOR-induced ESR (EIE) .
- ^{1}H anisotropic hyperfine structure of magnitude comparable to the nuclear Zeeman energy (about 15 MHz at X-band) gives rise to "forbidden" transitions in ESR spectra that complicates the ESR analysis. The ENDOR spectra are not affected by such effects.
- Quadrupole interactions for $I > \frac{1}{2}$ nuclei that usually cannot be obtained from ESR are measured by ENDOR in single crystals.

Commonly employed measurement procedures are summarized below. Detailed accounts have appeared in general textbooks [19–21], and in specialized treatises of radiation effects and of free radicals in solids [22, 23].

2.2.2.1 The Measurement of ^{1}H Hyperfine Coupling (*hfc*) Tensors

The improved resolution in ENDOR spectra makes it possible not only to measure the hyperfine couplings with high accuracy, but also to observe splittings that are not resolved by ESR. An example is shown in Fig. 2.5 for the malonic acid radical, $H\dot{C}_\alpha(COOH)_2$ [24]. The ESR spectrum is a main doublet due to hyperfine coupling with the ^{1}H at the $\dot{C}_\alpha$ position. The resolution is limited by the line-width and the occurrence of forbidden and so-called spin flip lines discussed in Chapter 4. The ENDOR lines denoted ν_α^+ and ν_α^- are narrower than the ESR lines by more than an order of magnitude. As in the liquid state the intensities between the pair differ due to hyperfine enhancement and relaxation factors. The additional lines in the ENDOR spectrum were examined using the ENDOR Induced ESR (EIE) method described in Section 2.2.2.4. Additional lines A and C correspond to EIE spectra (A and C) that are not apparent in the ESR spectrum and are accordingly attributed to another radical species. The features #2–#5 were assigned to ^{1}H hfc at neighbour molecules weakly interacting with the malonic acid radical, based on the observed similar EIE and ESR spectra in Fig. 2.5.

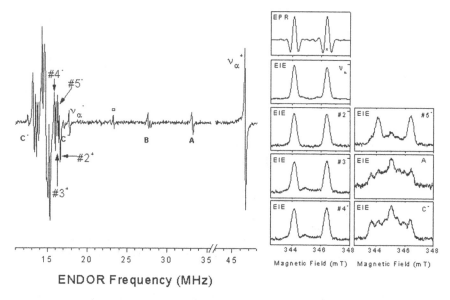

ENDOR Frequency (MHz)

Fig. 2.5 ENDOR spectrum of the malonic acid radical, $H\dot{C}(COOH)_2$ in a single crystal of malonic acid. The $\nu_{\alpha}^{\pm}$ lines corresponding to $m_S = \pm\frac{1}{2}$ for the H_α nucleus differ in intensity due to hyperfine enhancement and relaxation factors. Additional lines A, B and C are attributed to other species that are not apparent in the ESR spectrum, while the overlapping features #2–#5 represent weak couplings due to ^{1}H at neighbour molecules based on the EIE spectra to the right. The line marked by a square is an instrumental artifact. The ENDOR spectrum was recorded at a magnetic field value marked by * in the ESR spectrum. The figure is reproduced from [24] with permission from the American Chemical Society.

The method originally suggested by Schonland to evaluate g-tensor anisotropy from ESR measurements is also the most common to obtain hyperfine coupling tensors in ENDOR [23]. The method, described in Chapter 3, involves measurements in three crystal planes by rotating the crystal perpendicular to the magnetic field, commonly about the crystallographic axes. Two ENDOR lines corresponding to the two values of $m_s = \pm\frac{1}{2}$ for an $S = \frac{1}{2}$ species are expected to occur. In practice the lines can differ much in intensity as for the H_α nucleus of the malonic acid radical, $H\dot{C}_\alpha(COOH)_2$ so that only one transition, usually that at high frequency is measured as a function of the angle of rotation with respect to the magnetic field, Fig. 2.6. The low-frequency lines in the diagram are attributed to H-atoms at neighbour molecules.

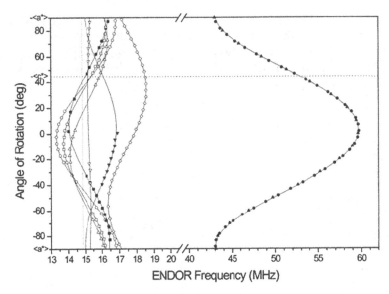

Fig. 2.6 Angular variation plot (rotation axis **<b>** perpendicular to the magnetic field) for the ENDOR lines of the malonic acid radical, $H\dot{C}_\alpha(COOH)_2$ in a single crystal of malonic acid. The fully drawn curves are calculated using the hyperfine coupling tensors obtained from measurements in three planes, according to the Schonland procedure. The curve at high frequency corresponds to H_α in Table 2.1, while the low-frequency lines are attributed to H-atoms at neighbour molecules. The figure is reproduced from [24] with permission from the American Chemical Society

The hyperfine coupling tensors obtained from measurements in three planes according to the Schonland procedure are usually summarized as in Table 2.1, reporting the principal values of the hyperfine coupling tensors and the direction cosines of the principal directions with respect to the crystallographic axes. The signs of the principal values A_i, $i = 1$–3, can usually not be obtained experimentally but are often assigned on theoretical grounds as in the table. It is thus known that $A_i < 0$ for α-^{1}H couplings [18] while $A_i > 0$ for β-^{1}H in carbon-centred π-electron radicals. Thus, for the $H\dot{C}_\alpha(COOH)_2$ species the high-frequency line ν_α^+

Table 2.1 Principal values and direction cosines of α-^{1}H hyperfine coupings in the malonic acid radical $H\dot{C}(COOH)_2$ in a single crystal of malonic acid [24]

Principal values (MHz)	Isotropic value (MHz)	Dipolar coupling (MHz)	Direction cosines with respect to		
			<a*>	<b'>	<c*>
−90.25 (7)		−32.19	0.0406(24)	0.4074(1)	0.9124(2)
−56.52 (1)	−58.06 (4)	1.54	0.9991(1)	−0.0051(6)	−0.0423(6)
−27.42 (3)		30.64	−0.0128(8)	0.9133(2)	−0.4072(1)

Numbers in parentheses are uncertainties in the last significant digit(s) of the corresponding number [24].

is assigned to $m_s = +\frac{1}{2}$ on theoretical grounds. The hyperfine couplings due to weakly interacting protons or other nuclei at a longer distance than ca. 2 Å from the radical centre are usually mainly caused by dipolar interactions, and consequently the tensors are nearly axial symmetric with different relative signs of the parallel and perpendicular components. The absolute signs are not experimentally obtained, since the electron quantum number m_s for an observed ENDOR frequency can as a rule not be established.

The experimental procedure of measurements in three crystal planes is also applicable for systems with appreciable g-tensor anisotropy. The technical details to obtain the hyperfine coupling tensors from ENDOR single crystal measurements are discussed in Chapter 3.

The malonic acid radical is a carbon-centred π-electron radical. The information about the electronic and geometric structure obtained from single crystal studies is of relevance in polymer science (Chapter 7) and other applications where π-electron radicals are frequently observed. Radicals of the π-electron type are planar about the atom where the unpaired electron is localized in a carbon 2p orbital. A negative sign for the isotropic *hfc* was predicted theoretically in a classical paper [18]. The experimental value for the malonic acid radical, –58.06 MHz is obtained as the average of the principal values $A_1 = -90.25$, $A_2 = -56.52$, $A_3 = -27.42$ MHz in Table 2.1. These values are the sum of the isotropic and dipolar couplings, *i.e.* $A_i = a_{iso} + B_i$. Calculations at an early stage [25] have verified that the principal values of the dipolar *hfc* are directed as shown in Fig. 2.7 with the largest coupling (B_1) along the C_α-H bond, near zero along the 2p-orbital axis (B_3) perpendicular to the radical plane, and with a negative value (B_2) along the in-plane direction perpendicular to the bond.

The values are typical also for other π-electron radicals with the unpaired spin localised to a carbon atom. The orientation of a π-electron radical in a crystal lattice can thus in general be deduced by comparing the directions determined experimentally for the B_1, B_2, and B_3 axes (the principal directions) with crystallographic data.

The *hfc* due to ^{1}H in the β-position of π-electron radicals is significant in species like the succinic acid radical, HOOC-$C_\beta H_2$-$\dot{C}_\alpha$H–COOH [26]. The isotropic

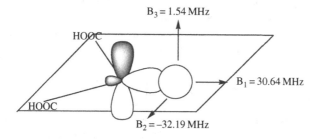

Fig. 2.7 Malonic acid π-electron radical showing principal values and directions for α-^{1}H dipolar hyperfine coupling

coupling is positive as expected for a hyperconjugation mechanism [20]. A relation
of the type [27] (Fig. 2.8):

$$a_\beta = \rho_C(A + B\cos^2\theta)$$

with $A \approx 0$, $B \approx 140$ MHz is frequently employed to estimate the dihedral angle θ
from the measured isotropic hfc, a_β.

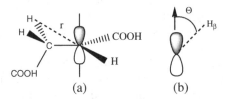

(a)   (b)

Fig. 2.8 (a) Succinic acid radical with major spin density at carbon 2p-orbital, $\rho_C \approx 1$, yielding
dipolar hfc with β-H at a distance $r \approx 2.15$ Å from the radical centre. (b) The isotropic hfc is
estimated with the dihedral angle θ viewed in the plane perpendicular to C_α—C_β bond as $a_\beta =$
$\rho_C(A + B\cos^2\theta)$

The maximum value of the β-H dipolar hfc occurs along the C_α-H_β bond according to the point-dipole approximation. This crude assumption is often sufficient for
the assignment of the radical structure in single crystal studies by comparison of the
C_α-H_β direction determined from the hyperfine coupling tensor with that obtained
from crystal structure data. The dipolar couplings can be estimated more accurately
by taking account of the extension of the electron wave function for the unpaired
electron [26]. Computation of the electronic structure with evaluation of the coupling tensors from first principles in quantum chemistry programs is an alternative
that is frequently used in modern applications discussed in Chapter 5.

2.2.2.2 The Measurement of Nuclear Quadrupole Coupling (*nqc*) Tensors

The interaction between the electric *quadrupole moment* of a nucleus with $I > \frac{1}{2}$
and the electric field gradient from the surrounding medium is usually difficult to
extract from ESR spectra. By contrast, the influence of quadrupole couplings is
clearly shown by the appearance of more than one ENDOR line for each electron
spin quantum number m_S of a species with electron spin S and nuclear spin $I > \frac{1}{2}$.
In the common case of an $S = \frac{1}{2}$ species with an *nqc* that is small compared to the
hfc the frequencies are given by an equation of the type (see Chapter 3):

$$\nu_\pm = G_\pm + 3P_\pm\left(m_I - \frac{1}{2}\right)$$

The subscripts correspond to the two electron spin quantum numbers, $m_S = \pm\frac{1}{2}$.
For each m_S value $2\cdot I$ lines with different frequencies occur depending on the value
of the nuclear quantum number m_I for the transition $(m_S, m_I) \rightarrow (m_S, m_I - 1)$.

The analysis in Chapter 3 shows that fittings to the quantities $G_\pm^2$ and $G_\pm^2 P_\pm$ as a function of crystal orientation with respect to the magnetic field according to the Schonland method yield the elements of the hyperfine and nuclear quadrupole coupling tensors. A simple method therefore involves separate measurements of $G_\pm$ and $P_\pm$ from the observed ENDOR frequencies.

The procedure to obtain the hyperfine and quadrupole couplings from the orientation dependent variation of ENDOR frequencies is schematically illustrated in Fig. 2.9 for the case of an $S = \frac{1}{2}$ system containing a single $I = 1$ nucleus. The g-factor is isotropic. Four ENDOR lines occur, two for each m_S. The quantities $G_\pm$ used to determine the hyperfine coupling tensor are the averages of v_1 and v_2 ($m_S = -\frac{1}{2}$) and v_3 and v_4 ($m_S = +\frac{1}{2}$), respectively, in Fig. 2.9. The values of $P_\pm$ are obtained from the splittings between v_1 and v_2 and between v_3 and v_4. Measurements of $G_\pm$ and $P_\pm$ as a function of orientation in three crystal planes are required to obtain the tensors, see Chapter 3 for the mathematical details used to calculate the nqc tensor. The tensor is usually presented similarly as the hyperfine coupling tensor, giving the principal values and the direction cosines of the principal directions with respect to the crystallographic axes as in Table 2.2. Theoretically the nuclear quadrupole coupling is completely anisotropic with no isotropic part. The sum of the principal values is therefore zero.

An alternative method to determine A and Q coupling tensors simultaneously by fitting directly to the experimental ENDOR frequencies [28] provides a means to estimate the uncertainties of the principal values and directions, similarly as for the hyperfine coupling tensor in Table 2.1.

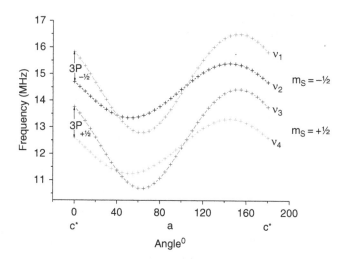

Fig. 2.9 Schematic variation of ENDOR frequencies due to ^{14}N hfc and nqc in a single crystal. Experimental data in Table 2.2 for the radical $H_3N^+CH_2\dot{C}HOSO_3^-$, were used to obtain the angular variation in the ac* crystal plane of 2-aminoethyl hydrogen sulfate X-irradiated at 295 K

Table 2.2 Principal values (in MHz) and axes of ^{14}N hyperfine and quadrupole coupling tensors observed in the radical $H_3N^+CH_2\dot{C}HOSO_3^-$, in single crystals of 2-aminoethyl hydrogen sulfate X-irradiated at 295 K. The $\pm$ and $\mp$ signs refer to two symmetry related sites in the crystal. The direction cosines refer to the orthogonalized **a**, **b**, and **c*** axes of the monoclinic crystal system, where **c*** is the normal to the (**a**, **b**) crystallographic plane

Tensor	Principal value (MHz)	Direction cosines with respect to		
		$\langle a\rangle$	$\langle b\rangle$	$\langle c^*\rangle$
	30.624	0.468	$\pm$0.381	−0.798
$\mathbf{A}(^{14}N)$	24.313	0.094	$\mp$0.919	−0.383
	24.070	−0.879	$\pm$0.104	−0.466
	0.482	0.268	$\pm$0.388	−0.882
$\mathbf{Q}(^{14}N)$	−0.264	0.535	$\mp$0.821	−0.199
	−0.218	0.801	$\pm$0.418	0.428

In practical applications more than one paramagnetic species may be present in the sample. This is the case in the ENDOR spectrum in Fig. 2.10 due to two forms of the radical $H_3N^+CH_2\dot{C}HOSO_3^-$ present after X-Irradiation at 295 K of 2-aminoethyl hydrogen sulphate. Overlap of spectra occurs in the region where lines due to ^{14}N

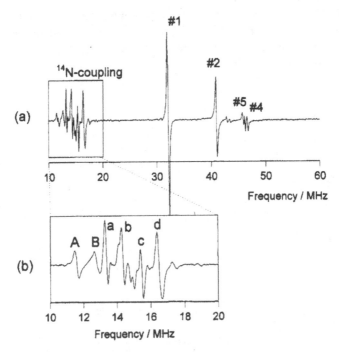

Fig. 2.10 ENDOR spectra from single crystals of 2-aminoethyl hydrogen sulfate X-irradiated at 295 K. The ENDOR resonances labelled 1, 2, 4, and 5 result from proton couplings. The four resonances labelled *a–d* result from *hfc* and *nqc* due to the ^{14}N nucleus in the radical $H_3N^+CH_2\dot{C}HOSO_3^-$. The resonances labelled *A* and *B* are attributed to another radical. The figure is reproduced from [28] with permission from the American Chemical Society

appear. The procedure to separate spectra of different species and to assign ENDOR signals of two nuclei to a specific species with the EIE (ENDOR-induced ESR) technique is discussed in Section 2.2.2.4. The signals "a", "b", "c", and "d" could be assigned to a single species. Similar EIE spectra obtained with the radiofrequency locked at the ENDOR lines due to ^{1}H and ^{14}N in Fig. 2.10 show that they are both assigned to the radical $H_3N^+C_\beta H_2\dot{C}_\alpha HOSO_3^-$. The ^{1}H ENDOR signals are due to the hydrogen atoms in α and β positions.

A common complication not only in ENDOR but also in other single crystal studies is the occurrence of different spectra due to identical paramagnetic species that are differently oriented due to crystal symmetry. The species are said to be located in different sites, and the overlap of spectra due to different orientations is referred to as site splitting. In the triclinic crystal system, e.g. for the malonic acid discussed in Section 2.2.2.1, there is only one molecule in the unit cell and no site-splitting occurs. For higher crystal symmetries several sites are available. For the 2-aminoethyl hydrogen sulfate of the previous section two sites occur. The $\pm$ and $\mp$ signs of the direction cosines in Table 2.2 refer to the different orientations of the symmetry related sites in the crystal. No site splitting occurs in the ac* plane, where the direction cosine components with respect to $\langle b \rangle$ do not influence the ENDOR frequencies, while two sites appear in the ab and bc* planes. The subject is further treated in Chapter 3, Section 3.3.3.

2.2.2.3 Resolution of Hyperfine Structure

Hyperfine structure that is hidden in the ESR line width often originates from ions or molecules surrounding the paramagnetic species. Due to its higher resolution ENDOR measurements on single crystals can provide a picture of the trapping site of the species as shown already in initial studies [1] of hyperfine structure due to alkali ions surrounding paramagnetic centers in halide crystals. Studies of this type give information about the nature of surrounding ions, and their distance to the paramagnetic center, i.e. the local geometry.

In the example shown in Fig. 2.11 [29] the paramagnetic species is the carbon dioxide radical anion, $\dot{C}O_2^-$, in an X-irradiated lithium formate crystal. ENDOR lines from several ^{7}Li nuclei ($I = 3/2$) at different lattice positions occur. The principal values and directions of the hyperfine coupling tensors provide a means to obtain the geometric arrangement of the Li^+ ions about $\dot{C}O_2^-$. The analysis is based on the point dipolar approximation, in which the maximum value, b_{max}, in MHz of the dipolar coupling is given by

$$b_{max} = 2 \cdot \frac{\mu_0}{4\pi h} \frac{g\mu_B g_N \mu_N}{r^3}$$

The distance r between the atom, which the electron is centered on, and the nucleus is obtained by identifying b_{max} with the numerically largest principal value of the dipolar coupling tensor determined experimentally. The formula applies to the case of a localized unpaired electron. The method can be elaborated to cases

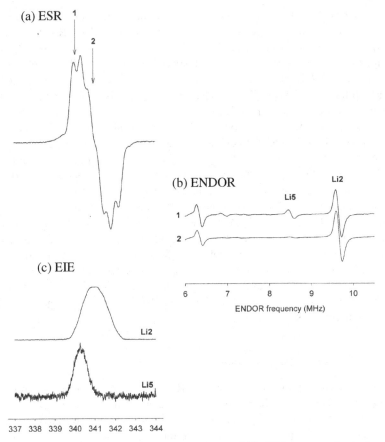

Fig. 2.11 (**a**) ESR, (**b**) ENDOR, and (**c**) ENDOR-induced ESR (EIE) spectra of a single crystal of $HCO_2Li \cdot H_2O$, X-irradiated to a dose of about 25 kGy and measured with the external magnetic field oriented along crystal axis <a>. The ENDOR spectra are recorded with the external magnetic field locked at the positions indicated by the *arrows* and the labels in the EPR spectrum. The EIE spectra are recorded with the ENDOR frequency locked to the ENDOR transitions Li2 and Li5, as indicated by the labels. The figure is reproduced from [29(c)] with permission from Dr. T.A. Vestad

where the unpaired electron is delocalized over several atoms (Appendix A2.1) and to systems with anisotropic g-factors, see Chapter 6.

The overall dominating $\dot{C}O_2^-$ radical species is trapped in the crystal matrix at an orientation not very different from that of the parent CO_2 fragment in the unir-radiated matrix. Four lithium couplings and four proton couplings were assigned to specific matrix nuclei of the trapping site of the $\dot{C}O_2^-$ radical as illustrated in Fig. 2.12.

The difference between the EIE spectra in Fig. 2.11(c) indicates the presence of an additional species associated with Li5. The use of this technique is further discussed in the next section.

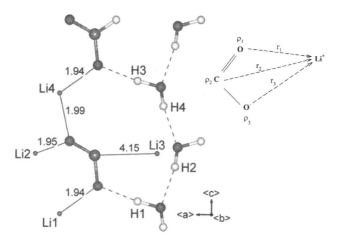

Fig. 2.12 The hydrogen bonding and lithium ion coordination of the CO_2^- radical in the crystal structure of $HCO_2Li \cdot H_2O$, in a projection viewed down <b>. The lithium ions and the protons giving rise to the hyperfine couplings reported by Vestad [29] are explicitly marked. All distances are in Å (10^{-10} m). The dipolar *hfc* is composed of contributions from spin densities at three atoms as indicated by the the sketch to the right. The figure is reproduced from [29(c)] with permission from Dr. T.A. Vestad

2.2.2.4 Resolution of Paramagnetic Species by ENDOR-Induced ESR (EIE)

ENDOR-induced ESR (EIE) is a technique originally developed for free radicals in liquids [30] to separate the ESR signals due to different paramagnetic species. The RF frequency is in this method locked to a particular line in the ENDOR spectrum while the magnetic field is swept. The obtained EIE spectrum is then attributed to a particular species. Comparison of EIE spectra obtained at different ENDOR lines permits assignment to the same or different species depending on the shape of the EIE spectra. For couplings due to $I = \frac{1}{2}$ nuclei the ESR and EIE spectra of a single species are similar except that the EIE lines for technical reasons usually are in absorption rather than the first derivative mode. The method has been applied to identify several paramagnetic species in single crystal samples, e.g. of three different types of radicals in X-irradiated *l*-alanine, the established material for ESR-dosimetry [31]. A closer analysis shows that an EIE spectrum recorded with the RF frequency locked at ENDOR lines due to $I > \frac{1}{2}$ nuclei can differ from the corresponding ESR spectrum in the number of lines. Examples are given in the literature [28].

In the example shown in Fig. 2.13, different EIE spectra were obtained by locking the RF frequency at different ENDOR lines from an X-irradiated single crystal of the Rochelle salt, [$^-$OOC-CHOH-CHOH-COO$^-$, Na$^+$, K$^+$] 4H$_2$O.

The EIE measurements in this study accordingly led to the conclusion that three species with different ESR spectra shown in Fig. 2.13 are present in the irradiated crystal.

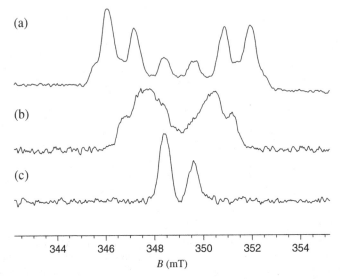

Fig. 2.13 EIE spectra from partially deuterated crystals of Rochelle salt, [⁻OOC-CHOD-CHOD-COO⁻, Na⁺, K⁺]4D₂O, X-irradiated and measured at 10 K. The spectra were recorded with the RF set at three different ENDOR lines and with the magnetic field in (**a**) and (**c**) at B ∥ <b> crystal axis, and in (**b**) at B ∥ <a>. E. Hole and E. Sagstuen are acknowledged for permission to reproduce spectra prior to publication

2.2.3 ENDOR in Disordered Solids

Single crystal ENDOR measurement is an informative but not always applicable method to identify paramagnetic species in solids. It is for instance difficult to obtain single crystals of biochemical materials. In other cases the paramagnetic species are intentionally trapped in a disordered matrix or in a frozen solution. The ENDOR lines are then usually broadened by the anisotropy of the magnetic couplings. Some data that a single crystal analysis can provide are difficult to extract from a powder sample. However, a considerable amount of information can often be obtained from ENDOR spectra of disordered systems.

- Principal values of hyperfine coupling tensors for ^{1}H (and other $I = \frac{1}{2}$ nuclei) can often be obtained by analysis of the ENDOR "powder" spectra. Visual analysis is sometimes sufficient, in other cases computer simulation is required to refine the analysis.
- Nuclear quadrupole couplings due to nuclei with $I > \frac{1}{2}$ are obtained by fitting simulated ENDOR powder spectra to experimental. The procedure can be quite tedious.
- Paramagnetic species with appreciable g-anisotropy give different ENDOR spectra at different settings of the magnetic field due to the angular selection effect. At magnetic field settings corresponding to $g_{\parallel}$ and $g_{\perp}$ of an axially symmetric species, single crystal like ENDOR spectra are obtained. Simulation procedures that take angular selection in account are described in Chapter 3.

2.2.3.1 Hyperfine Coupling Tensors

Experimentally, the ESR lines of the paramagnetic species in a solid matrix are quite broad compared to the ENDOR lines from the same sample. This applies quite generally, and the ENDOR spectra of solids are therefore usually much better resolved than the corresponding ESR spectra. The ENDOR spectrum of the biphenyl cation in Fig. 2.14 provides an example. Hyperfine lines from ^{1}H in the *para* and *ortho* positions are apparent. The small coupling from the H atoms at *meta*-positions is not resolved in the CFCl$_3$ frozen matrix employed. Some ^{19}F matrix lines are indicated by arrows. The experimental spectrum can be regarded as a superposition of the two hypothetical spectra with ^{1}H hyperfine structure in *para* and *ortho* positions, respectively. The contribution of a particular nucleus to the hyperfine structure is therefore practically independent of the other nuclei in an ENDOR spectrum of a disordered solid. Information about the different directions for the extreme values of the anisotropic couplings – the *principal directions* – of these atoms is in other words lost. It is therefore difficult to obtain the geometric structure of a paramagnetic species in a disordered solid, except in instances when angular selection occurs as discussed below. The *principal values* of the individual couplings can, however,

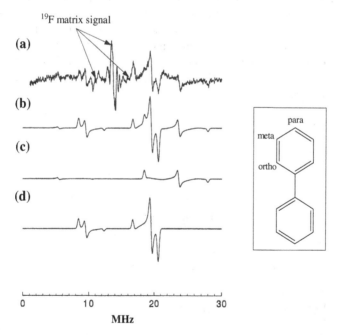

Fig. 2.14 Experimental (**a**) and simulated (**b**)–(**d**) powder X–band ENDOR spectra of the biphenyl radical cation in CFCl$_3$ matrix at 120 K. Lines marked by *arrows* are due to *hfc* from ^{19}F in the matrix. Simulation (**b**) shows the total spectrum while (**c**) and (**d**) show the individual contributions from the para and ortho proton groups respectively. The influence of matrix fluorine and meta protons were not taken into account in the simulation. The figure is adapted from [32] with permission from Elsevier

be read quite accurately from the positions of more or less pronounced lines corresponding to these values as observed in Fig. 2.14. ENDOR hyperfine enhancement effects are often observed experimentally as in Fig. 2.14(a). Procedures to include this effect in simulations are described in Chapter 3.

Applications of ENDOR spectroscopy to the study of radical cations in halocarbon matrices have been reviewed [33]. Radicals on surfaces [34] and of paramagnetic species in frozen solutions of biomolecules [35] have also been examined.

2.2.3.2 Nuclear Quadrupole Couplings (*nqc*) (*I* > 1/2)

Paramagnetic species in biological systems often feature ENDOR signals due to ^{14}N, with nuclear spin $I = 1$. In certain cases, like that of ^{14}NO-ligated *ferrocytochrome c heme*, signals due to ^{14}N(Histidine) and ^{14}N(NO) appear in separate spectral regions and can be assigned in terms of their different hyperfine and nuclear quadrupole couplings [32]. The powder ENDOR spectrum of the $H_2\dot{C}NHCOC_6H_5$ radical in X-irradiated hippuric acid is likewise affected by nuclear quadrupole coupling (*nqc*) in the ^{14}N spectral region as shown in Fig. 2.15. As seen in the theoretical spectrum (c) an analysis that ignores the *nqc* cannot account for the experimental powder spectrum (a). The assumptions made in Section 2.2.2.2 for the measurement of *nqc* are also not satisfactory for this species. The reason is that the *nqc* is approximately of the same magnitude as the *hfc*. These parameters were originally obtained from a single crystal analysis, made with an exact method. An alternative but still accurate method developed by Erickson [32] to analyze powder ENDOR spectra in the presence of large *nqc* was used for the simulation in Fig. 2.15(b). The method is described in Chapter 3.

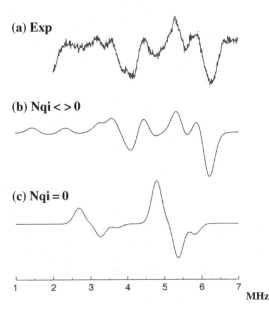

(a) **Exp**

(b) **Nqi < > 0**

(c) **Nqi = 0**

1   2   3   4   5   6   7
MHz

Fig. 2.15 Experimental (**a**) and simulated (**b**), (**c**) powder ENDOR spectra at 110 K of $H_2\dot{C}NHCOC_6H_5$ radical in X-irradiated hippuric acid. Only the region of ^{14}N-signals is shown. The simulations were made including (**b**) and excluding (**c**) the *nqc*. The figure is adapted from [32] with permission from Elsevier

2.2.3.3 Angular Selection

The ESR spectrum of a disordered solid is a superposition of intensity from all orientations of the paramagnetic species with respect to the magnetic field. A strong g-anisotropy can, however, be employed to obtain single crystal like ENDOR spectra even from a disordered solid [36]. In initial studies hyperfine structure from ^{14}N corresponding to molecules in a single orientation with respect to the magnetic field could be obtained from a Cu^{2+} complex with practically axial symmetric g- and hfc-tensors. The new information in this case was thus that the ^{14}N hyperfine and quadrupole couplings along the symmetry axis of the complex (the parallel direction), and in the plane perpendicular to it (the perpendicular plane) could be obtained by ENDOR measurements on a frozen solution. The ability to perform ENDOR at several different magnetic fields is useful in studying systems of this type, including also frozen solutions of proteins and enzymes and heterogeneous systems that contain transition metal ions.

An application to clarify the ligand structure of an aqueous Cu^{2+} complex in ZSM-5 zeolite is shown in Fig. 2.16. The ENDOR spectrum obtained with the

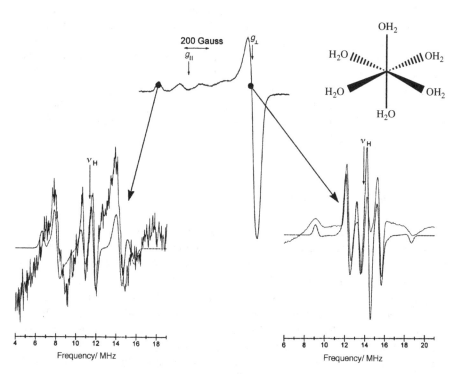

Fig. 2.16 Powder ESR and single-crystal like ENDOR spectra of $Cu^{2+}(H_2O)_6$ in ZSM-5, recorded at 4 K: (*full line*) experimental spectrum; (*dashed line*) simulation. The figure is adapted from [37] with permission from the Royal Society of Chemistry

magnetic field set at the $g_{||}$ region in the figure resembles a single crystal spectrum with the field along the symmetry axis of the complex, while that obtained from $g_{\perp}$ is expected to be an average over directions in the equatorial plane. To observe a strong ENDOR effect very low temperature is often needed. ENDOR spectra of the Cu^{2+} complex with water in Fig. 2.16 were measured at 4 K. Methods to take angular selection into account in the simulated spectra of the type shown in Fig. 2.16 are discussed in Chapter 3.

2.3 Pulsed ESR

In pulsed ESR samples are exposed to a series of short intense microwave pulses in place of the continuous radiation with microwaves at low power in the traditional CW (continuous wave) spectrometer. The basic principles are the same as in pulsed NMR that has completely superseded CW NMR for reasons of sensitivity, flexibility and time resolution. The development has been slower in ESR, however, as faster relaxation rates (three orders of magnitude) and wider spectral ranges (three-four orders of magnitude) have provided technical obstacles. The state of the technique up to 2001 including many examples of applications was described in the work by Schweiger and Jeschke [6], while methods to measure hyperfine couplings in solid materials were summarised in [9, 38] following initial work by Mims [7]. Commercial equipment is now available, allowing experiments to be done by the non-specialist. The method can be applied to measure dynamic parameters, like relaxation times, but also to obtain structure information, e.g. hyperfine coupling constants, the latter usually by applying Fourier transformation (FT) to the time-domain signals. Up to now, spin-echo methods have been most commonly applied. We illustrate some of the applications below.

2.3.1 Pulsed ESR in Solution

Free induction decay, originally developed for NMR, involves a single microwave 90° or $\pi/2$ pulse. The magnetic moment is tipped into the xy-plane by the torque exerted by the microwave magnetic field with a strength B_1 and duration t_p adjusted so that the tipping angle $\theta = \frac{g\mu_B B_1 t_p}{h} \cdot 2\pi = \frac{\pi}{2}$ radians, Fig. 2.18(b). The signal response obtained after the pulse varies in time as shown schematically in Fig. 2.17 and eventually decays when the magnetization returns to equilibrium. It is referred to as a free induction decay or FID. The precessing motion in the xy-plane gives rise to modulations and Fourier transformation (FT) gives the spectrum. The method is useful for studies of photochemically induced short-lived paramagnetic species with synchronised signal detection and laser flash excitation [11–13].

The FID decays rapidly in solid samples, for which spin echo measurements are better suited.

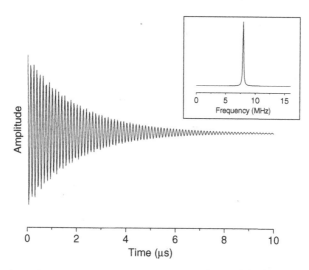

Fig. 2.17 Free induction decay (FID) after a microwave pulse and the spectrum (*inset*) obtained by Fourier transformation

2.3.2 Pulsed ESR in Solids

The *2-pulse or Hahn-echo* experiment, originally developed in NMR is extensively used in pulsed ESR in solids for measurements of magnetic relaxation and of anisotropic *hfc*.

The experiment can be qualitatively understood by a vector model for the reorientation of the electron spins by the microwave pulses. The first pulse rotates the spins by 90° to the *y*-axis (Fig. 2.18(b)). During time τ between the first and second pulse, the different spins precess in the xy-plane at different rates. The second pulse inverts the orientations of the spins causing them to realign after a time 2τ and an echo is observed (Fig. 2.18(e)).

2.3.2.1 ESEEM

ESEEM, Electron Spin Echo Envelope Modulation, is a special application of the spin-echo technique for the measurement of anisotropic hyperfine couplings (and of quadrupole couplings for nuclei with $I > \frac{1}{2}$). The transient response from the sample, obtained after a series of microwave pulses is applied, contains modulations superimposed on the envelope. The spectral resolution of ESEEM illustrated in Fig. 2.19 is even greater than that for ENDOR. This increased resolution has been employed to deduce the structure of the trapping sites of paramagnetic species by the observation of hyperfine couplings from surrounding molecules of the rigid matrix [10, 11].

The pioneering work by Mims [7] has been extended and summarized in recent monographs and reviews [6, 9, 38].

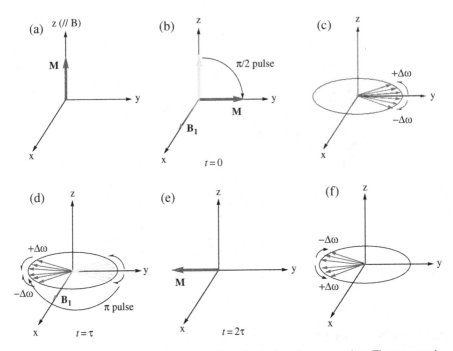

Fig. 2.18 Vector model for the reorientation of the spins by the microwave pulses. The sequence is employed to measure the phase-memory relaxation time from the decay of Hahn or primary echo signal and of anisotropic hyperfine and nuclear quadrupole couplings observable as modulations on the echo decay curve. The latter is referred to as the ESEEM method

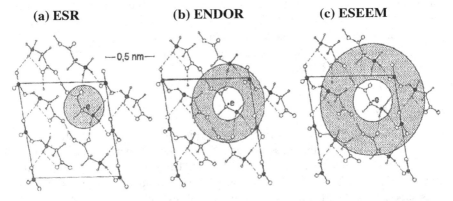

Fig. 2.19 Hyperfine couplings from the *shadowed* regions give information at different resolution with (**a**) ESR, (**b**) ENDOR and (**c**) ESEEM about the structure of the trapping sites of paramagnetic species in rigid matrices. The figure is reproduced from [Elektronowy Rezonans Paramagnetyczny, S.K Hoffmann and W. Hilczer, Editors, Wydawnictwo, NAKOM, Poznan (1997)] with permission from Prof. S. Hoffmann

The modulations depend on the combined influence of an anisotropic hyperfine coupling and the nuclear Zeeman interaction and for $I > \frac{1}{2}$ also the nuclear quadrupole interaction. This gives rise to both the allowed ($\Delta m_I = 0$) and the forbidden ($\Delta m_I = 1$) transitions discussed in Chapter 4 that are prerequisites for the appearance of modulations on the echo signal [6, 7, 9, 10]. The hyperfine and/or quadrupole coupling constants can be determined through simulation of the ESEEM decay curves. Equations used to analyse the ESEEM signals for the most commonly employed methods described below are given in Chapter 3. The reader is referred to standard texts for derivations of equations and of qualitative explanations of the modulation phenomenon [9–11]. Fourier transformation (FT) of the time-domain modulation pattern is an alternative approach employed using software normally available with commercial spectrometers. The FT spectra have lines at frequencies corresponding to the nuclear transitions observed in ENDOR. Some limitations of the method should be noted.

- The method is not suited for measurements of isotropic hyperfine couplings that may occur even in solid samples. Such couplings do not give rise to modulations.
- The hyperfine couplings must be within the frequency content (bandwidth) of the applied pulses. Application of short pulses to obtain the necessary bandwidth requires proportionally higher microwave power to reorient the spins according to the discussion in Section 2.3.1. The magnitude of the largest couplings that at present can be measured with commercial equipment is of the order 30–40 MHz.
- Echo decay times must be sufficiently long to obtain a satisfactory spectrum. Low temperatures are often required to increase the decay time. Rapid echo decay is still often a problem with the two-pulse method.

The methods most widely used by non-specialists are briefly described below. Surveys of advanced methods and applications have been described in several specialist reports [6, 9–11].

Two-Pulse ESEEM

In the basic two-pulse or primary echo experiment, two pulses separated by a time τ are applied. The second pulse is twice as long as the first. At time τ after the last pulse a transient response appears from the sample, the so called spin echo. By monitoring the echo amplitude as a function of the time τ, a spin echo envelope can be recorded. The hyperfine couplings are obtained either by trial-and-error simulations to reproduce the modulations superimposed on the decaying echo amplitude (the original procedure) or by a Fourier transform to obtain nuclear frequencies in modern instruments as in Fig. 2.20. The frequencies are the same as obtained in ENDOR. Contrary to ENDOR, combination peaks at the sum and difference frequencies may also occur.

The method has been particularly applied to studies of solvation structures around paramagnetic species in disordered systems, e.g. trapped electrons in frozen glasses and of transition metal ions in metalloproteins for which no single crystals

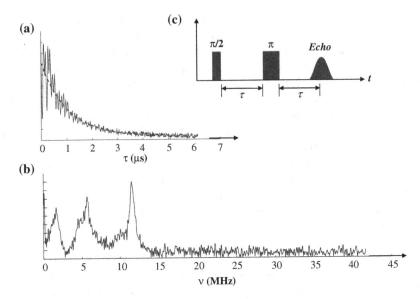

Fig. 2.20 (a) 2-pulse ESEEM from $\dot{C}O_2^-$ radicals in an X-irradiated powder of lithium formate. The echo envelope shows modulations due to Li ions adjacent to the $\dot{C}O_2^-$ radicals. The smooth dashed curve shows an exponential decay with the phase memory time T_M. (b) FT-transform of (a). (c) 2-pulse sequence employed in the experiment (Data provided by Dr. H. Gustafsson)

are available for X-ray diffraction [10, 11, 35]. The 2-pulse ESEEM from $\dot{C}O_2^-$ radicals in an X-irradiated powder of lithium formate in Fig. 2.20(a) shows modulations attributed to ^7Li nuclei. Peaks in (b) observed after Fourier transformation (FT) are in the expected frequency range for Li ions adjacent to the $\dot{C}O_2^-$ radicals based on previous assignment by ENDOR [29]. The FT transform (b) was obtained after correcting for the exponential echo decay with the phase memory time T_M, shown by the smooth dashed curve in (a).

The echo decays due to spin-spin and spin-lattice relaxation. The corresponding *phase memory time T_M* is in most cases of the order of a few microseconds. This is sometimes too short to obtain a good frequency spectrum after Fourier transformation. The decay time is longer in the three-pulse sequence.

Three-Pulse ESEEM

The two-pulse echo decay is sometimes too fast to obtain a satisfactory frequency spectrum after Fourier transformation. In this case the three-pulse sequence shown in Fig. 2.21(a) is an alternative. It gives rise to a *stimulated* echo at time τ after the third $\pi/2$ pulse. The decay rate is limited by the electron spin-lattice relaxation time T_1, which is usually longer than the phase memory relaxation time T_M for the two-pulse decay.

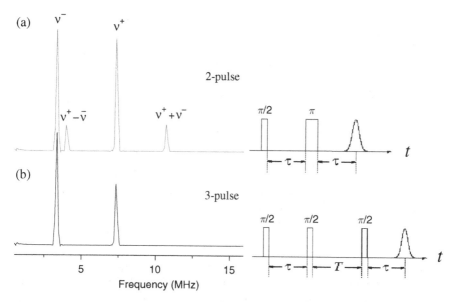

Fig. 2.21 Schematic (**a**) 2-pulse and (**b**) 3-pulse ESEEM spectra for an $S = \frac{1}{2}$ species with main lines v^+ and v^- due to anisotropic ^{1}H *hfc*. In addition weak combination peaks occur at $v^+ \pm v^-$ in the 2-pulse spectrum

In a three-pulse ESEEM experiment the time T between the second and the third pulse is increased while the time τ between the first and second pulse is kept constant. In contrast to the two-pulse ESEEM experiment, the three-pulse ESEEM spectra do not contain sum and difference frequencies as illustrated schematically in Fig. 2.21 for an $S = \frac{1}{2}$ species with anisotropic hyperfine coupling due to a proton. Both spectra contain lines with nuclear frequencies v^+ and v^- expected for $m_S = \pm \frac{1}{2}$. The combination lines at $v^+ \pm v^-$ seen as satellites in the two-pulse spectrum do not appear in the corresponding 3-pulse spectrum. On the other hand lines can escape detection in the 3-pulse spectrum for certain values of the time τ between the first and second pulse at so called blind spots. It is therefore customary to record several 3-pulse spectra with different values of τ.

Nuclear Quadrupole Couplings

ESEEM spectra obtained after Fourier transformation (FT) are similar to those obtained by ENDOR, and therefore allow determination of hyperfine and nuclear quadrupole coupling tensors from species containing nuclei with $I > \frac{1}{2}$. Although single crystal ESEEM measurements have been made to obtain these data the method has been particularly employed for measurements on disordered systems [9–11] even though the resolution is lower than in single crystals. Effects of the quadrupole coupling upon the ESEEM spectra are typically observed for nuclei like ^{27}Al ($I = 5/2$) and ^{23}Na ($I = 3/2$) with relatively large electric quadrupole

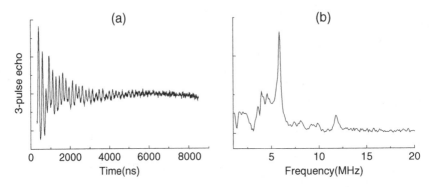

Fig. 2.22 (**a**) 3-pulse echo signals corrected for echo decay and (**b**) the ESEEM spectrum after FT for nitric oxide dimer with $S = 1$ trapped in Na-A zeolite. The figure is adapted from [D. Biglino et al. Chem. Phys. Lett. **349**, 511 (2001)] with permission from Elsevier

moments. Elements of this type occurring for instance in oxides and zeolites can interact with paramagnetic species trapped in the material. Information about the trapping site geometry is in this case obtainable from both the hyperfine and the nuclear quadrupole coupling tensors.

The spectra can become quite complex when the nuclear quadrupole coupling is appreciable, as for nitric oxide (NO) introduced into Na-A zeolite discussed in Chapter 6. It is also known that NO tends to dimerize forming an $S = 1$ species under these conditions [39]. A triplet state complex is formed interacting with one or more ^{23}Na nuclei in the zeolite [40]. The spectrum obtained after FT of the three-pulse ESEEM signal in Fig. 2.22 is difficult to analyze by visual inspection. Methods to obtain the hyperfine and quadrupole couplings by simulations are described in Chapter 3.

HYSCORE

Overlap of lines can make analysis difficult when several nuclei contribute in the one-dimensional (1D) two- and three-pulse ESEEM spectra. Following the development in NMR, methods to simplify the analysis involving two-dimensional (2D) techniques have therefore been designed. The Hyperfine Sublevel Correlation Spectroscopy, or HYSCORE method proposed in 1986 [14] is at present the most commonly used 2D ESEEM technique. The HYSCORE experiment has been applied successfully to study single crystals, but is more often applied to orientationally disordered systems. It is a four-pulse experiment (Fig. 2.23(a)) with a π pulse inserted between the second and the third $\pi/2$ pulse of the three-pulse stimulated echo sequence. This causes a mixing of the signals due to the two nuclear transitions with $m_S = \pm \frac{1}{2}$ of an $S = \frac{1}{2}$ species. For a particular nucleus two lines appear at (ν^-, ν^+) and (ν^+, ν^-) in the 2D spectrum as shown most clearly in the contour map (d) of Fig. 2.23. The lines of a nucleus with a nuclear Zeeman frequency

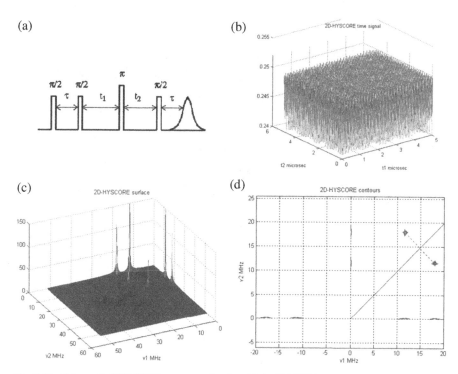

Fig. 2.23 Schematic HYSCORE spectrum showing (**a**) the HYSCORE sequence, (**b**) the 2D time-domain modulation signal, (**c**) the 2D HYSCORE spectrum and (**d**) the contour plot of a single crystal sample for an $S = \frac{1}{2}$ species containing a ^{1}H nucleus with an axially symmetric hyperfine coupling. The magnetic field is at an angle $\theta = 10°$ with the $A_{\parallel}$ axis. The nuclear Zeeman frequency $\nu_H \approx 15$ MHz is larger than the hyperfine coupling, i.e. $|A_{\perp}| < |A_{\parallel}| < 2\,\nu_H$

that is larger than the hyperfine coupling occur in the right quadrant. Lines can also appear in the left quadrant for nuclei with large hyperfine couplings. The pattern can become more complex for nuclei with large *nqc* ($I \geq 1$) see e.g. Fig. 2.24 in the next section.

To acquire the 2D time-domain modulation signal, the stimulated echo amplitude is observed as a function of pulse separations t_1 and t_2 (with fixed τ). The recorded modulation signal is then 2D Fourier transformed to obtain the spectrum as a function of the two frequencies ν_1 and ν_2. The procedure is schematically illustrated in Fig. 2.23. The nuclear transitions corresponding to a particular nucleus are identified most conveniently by a contour map of the type shown in Fig. 2.23(d). The map is obtained as a projection of the spectrum on the frequency plane. Projected peaks that are symmetrically disposed about the diagonals in the diagram for an $S = \frac{1}{2}$ species are attributed to nuclear transitions for the $m_S = +\frac{1}{2}$ and $-\frac{1}{2}$ electronic states, respectively, (often denoted α and β) of a particular nucleus.

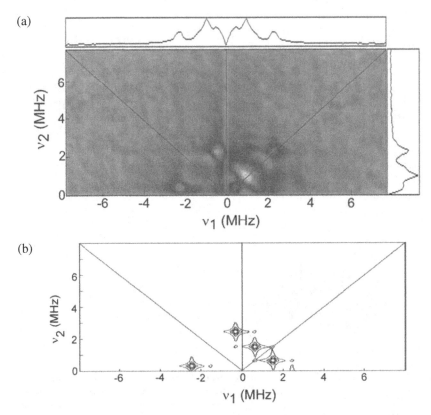

Fig. 2.24 Contour plot of single crystal HYSCORE spectrum (**a**) and simulation (**b**) showing interaction between ^{14}N of released ammonia and deaminated radical H$_3$CĊHCOO$^-$ in irradiated *l*-alanine. The figure is reproduced from [41] with permission from Elsevier

Single Crystals

The HYSCORE spectrum and the contour plot schematically shown in Fig. 2.23 is for a crystalline sample showing a signal from an $S = \frac{1}{2}$ species with an axially symmetric *hfc* due to ^{1}H. Note that the magnetic field orientation must differ from the parallel and perpendicular directions of the *hfc* to obtain a signal, like for 1D ESEEM as discussed in Exercise E2.9. The nuclear Zeeman frequency is larger than the hyperfine coupling $|A_\perp| < |A_\|| < 2 \cdot v_H$. The HYSCORE spectrum is therefore confined to the (+, +) quadrant. The frequencies of the α and β transitions are obtained directly from the diagram, while the nuclear frequency ($v_N \approx 15$ MHz) read at the intersection of the diagonal and the line joining the peaks serves to identify the nucleus (^{1}H). A nuclear transition with a hyperfine coupling that is large compared to the nuclear Zeeman frequency appears with a negative frequency in the HYSCORE spectrum. It is therefore customary to display both the (+, +) and the (−, +) quadrants of the contour plot as shown in Figs. 2.23 and 2.24.

HYSCORE has a higher sensitivity in the region of low nuclear frequencies than that of ENDOR spectroscopy. Single crystal measurements have therefore been applied to studies of nitrogen-containing paramagnetic species (nuclear Zeeman frequency $v(^{14}N) \approx 1$ MHz at X-band) of interest in biochemical and fundamental applications. The local structure around the species may be obtained as discussed below for the radical $H_3C\dot{C}HCOO^-$ in irradiated l-alanine.

HYSCORE single crystal spectra of irradiated l-alanine in Fig. 2.24(a) show an interaction with ^{14}N nuclei ($I = 1$) that is too weak to be resolved in CW-ESR and ENDOR spectra. As for $I = \frac{1}{2}$ the nuclear transitions are identified most conveniently by a contour map of the type shown in Fig. 2.24. Software for the data processing to obtain the map is usually provided with commercial instruments. The interpretation is, however, complicated by the quadrupole interaction of the ^{14}N nucleus. The procedure to analyse the data employed in [41] was analogous to that applied for measurements of *hfc* and *nqc* tensors by ENDOR (Section 2.2.2.2). The analysis was supported by simulations (Fig. 2.24(b)) which also accounted for the lines observed in the left quadrant due to the low nuclear frequency of ^{14}N at X-band. The results were interpreted as due to a deaminated radical, $H_3C\dot{C}HCOO^-$, weakly interacting with the released ammonia group.

Disordered Systems

HYSCORE spectra take longer to record than 2- and 3-pulse ESEEM. As mentioned above, the technique is therefore more commonly applied to orientationally disordered systems, e.g. frozen solutions in chemical and biochemical applications, heterogeneous samples in applications to catalysis and environmental sciences.

The bonding of transition metal ions in zeolites can be elucidated as exemplified below for vanadium (VO^{2+}) exchanged ZSM-5. ^1H-*hfc* due to several water ligands contribute to the HYSCORE spectrum of the hydrated sample, Fig. 2.25. The nuclear Zeeman frequency (X-band), $v_H \approx 15$ MHz, is larger than the hyperfine coupling, $|A_\perp| < |A_{||}| < 2 \cdot v_H$. The HYSCORE spectrum is therefore confined to the (+, +) quadrant. The spectrum feature that is off the frequency axes in the 2D-HYSCORE spectrum appears as a ridge symmetrically displaced about the diagonal of the (+, +) quadrant in the contour plot. The spots observed in single crystals are thus replaced by contours reflecting the anisotropy of the hyperfine couplings. Methods to extract the principal values of the hyperfine coupling tensors from these contours have been worked out for cases of practical interest. The procedure is more complex in the presence of several, inequivalent ^1H nuclei. A method to obtain the parameters graphically has been described, see [42] for an application and references for further details. The ridge observed in the HYSCORE spectrum of hydrated VO_2^+-ZSM-5 was attributed to four different protons by this method. The spectrum in Fig. 2.25 was thus interpreted as composed of four overlapping ridges as indicated by the solid, dashed, and dotted lines, each corresponding to a different proton A, B, and C and D [42].

A review of 1D and 2D ESEEM applications to the metal ions in biological systems is given in a work by Deligiannakis et al. [43]. The analysis can become

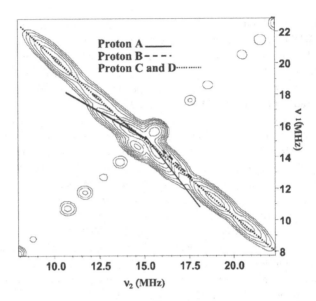

Fig. 2.25 HYSCORE contours of hydrated VO^{2+}-ZSM-5 showing the proton ridge. This spectrum is interpreted as composed of four overlapping ridges as indicated by the *solid*, *dashed*, and *dotted lines*, each corresponding to a different proton A, B, and C and D. The figure is reproduced from [42] with permission from the American Chemical society

complex for $I > \frac{1}{2}$, due to an appreciable nuclear quadrupole interaction occurring frequently e.g. for ^{14}N. Isotopic labelling with ^{15}N ($I = \frac{1}{2}$) can in this case simplify the analysis [35]. The reader is referred to the original papers for details [41, 43–45]. Simulation methods that have been developed are described in Chapter 3.

2.3.3 Pulsed ENDOR

An intensity change of the electron spin echo signal occurs when a RF pulse with a frequency corresponding to a nuclear spin transition is applied. An ENDOR spectrum is obtained by sweeping the RF frequency. Some well-established methods are named after their inventors. Mims-ENDOR [46] is a stimulated echo sequence with a RF pulse inserted between the second and third mw-pulses. The method is particularly used for measurements of small hyperfine couplings. Another pulse technique for performing ENDOR devised by Davies [47] is also commonly employed. The pulse sequence for this method is shown in Fig. 2.26.

Advances in the methodology have been reviewed [48]. Commercial equipment is available up to W-band, while pulsed ENDOR spectrometers at higher microwave frequencies have been constructed in specialized laboratories. Microwave frequencies higher than X-band are of advantage in applications where the amount of sample is limited. The high sensitivity at W-band (microwave frequency: 95 GHz)

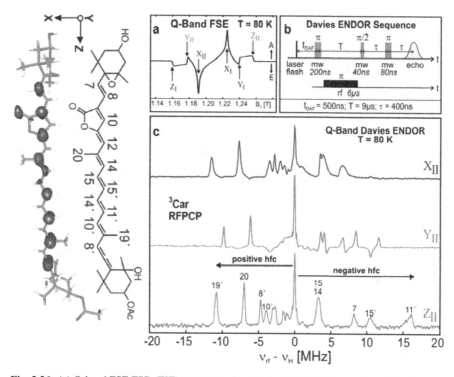

Fig. 2.26 (**a**) Q-band FSE EPR (EIE) spectrum of peridinin triplet (A) absorption, (E) emission; (**b**) Davies ENDOR pulse sequence; (**c**) Q-band ^{1}H ENDOR spectra recorded at the three canonical orientations X$_{II}$, Y$_{II}$, Z$_{II}$, which are marked with *arrows* in the ESR spectrum of **panel** (**a**) using the conditions in **panel** (**b**). At the proton Larmor frequency ν$_H$ a narrow and intense line is visible resulting from nuclear transitions in the $m_S = 0$ manifold. The frequency axis gives the deviation from ν$_H$ in the respective spectra. The excitation wavelength was 630 nm. *Left*: numbering and spin density plot of peridinin in its excited triplet state. The orientation of the ZFS tensor axes X, Y, and Z is also given. The figure is adapted from [51] with permission from the American Chemical Society

has made it possible to study small-size single crystals of e.g. proteins [49]. Measurements at high microwave frequency are also of advantage to observe ENDOR from nuclei that have low nuclear frequencies at X-band. Hyperfine structure due to ^{14}N and to ^{15}N-enriched samples can be resolved by ENDOR at high microwave frequency in single crystals and frozen solutions of biochemical substances [50]. Pulsed ENDOR has also been employed for the study of surface adsorption complexes, see Chapter 6 for a study of nitric oxide in a zeolite matrix.

Pulsed ENDOR complements ESEEM and conventional ENDOR in several respects:

- Hyperfine couplings in liquids and other isotropic systems can be measured with pulsed ENDOR but not with ESEEM.

- The signal intensity in pulsed ENDOR is less affected by nuclear relaxation than in conventional ENDOR. Nuclear transitions that are not visible in conventional ENDOR due to unsuitable relaxation conditions can be detected.
- Pulsed ENDOR can be employed to obtain relaxation times.

In addition to structural studies on stable paramagnetic complexes like those mentioned in Chapter 6, pulsed ENDOR can also be applied in investigations of short-lived species, for which CW methods are unsuitable. This is exemplified by work on photosynthetic systems where ENDOR spectra due to triplet state species ($S = 1$) appear on laser flash photolysis. Studies of photosynthetic systems like the carotenoid peridin serving as a light-harvesting pigment [51] provide an example. The Q-band field-swept ENDOR spectrum with zero-field splitting parameters $|D| = 48.2$ mT and $|E| = 4.7$ mT appeared partly in emission, partly in absorption after a laser flash as shown in Fig. 2.26(a). At least 13 ^{1}H hyperfine couplings were resolved in the pulsed Davies ENDOR spectrum at Q-band. Different Davies ENDOR spectra were recorded with the magnetic field set at $X_{||}$, $Y_{||}$ and $Z_{||}$. These fields correspond to the "canonical" or principal axes directions of the zero-field tensor in Fig. 2.26(c). A spin density plot of peridinin in its excited triplet state could be obtained from the data.

2.3.4 Relaxation Times

Relaxation times are more directly measured by pulse experiments than by the CW saturation technique. Physical applications often involve measurements made at different temperatures to obtain data regarding the dynamics in the solid state. The relevance in ESR applications to the biological field has been summarised [52]. Several of the points discussed there are also of relevance in fundamental studies and for other applications e.g.

- The spin-lattice relaxation rate depends upon the electronic structure of the paramagnetic center. Analysis of data provides information e.g. of presence of low-lying excited states of the species under study.
- Motions shorten spin-lattice relaxation times. Analysis can provide insight into molecular motions. Dynamics of a radicals can be studied by analysis of contributions due to rotational modulation of hyperfine and Zeeman anisotropies.
- A knowledge of electron spin relaxation permits predictions concerning utility of a paramagnetic species for instance as an MRI (Magnetic Resonance Imaging) contrast.
- The local concentration of spins affects relaxation. Local concentrations can be distinguished from bulk concentrations. Spin echo decays can also be used to determine local concentrations of nuclear spins.

Eaton and Eaton [52] have discussed the terminology, which in some cases is not uniform in the literature. Table 2.3 is a summary of the terms with their suggested definitions.

Table 2.3 Relaxation terms with definitions proposed in [52]

Symbol	Name	Description
T_1	Spin-lattice	Time constant for equilibration of populations between two electron spin energy levels. Released energy is absorbed by lattice
T_2	Spin-spin	Time constant for mutual electron spin interchange, causing dephasing of precession in xy-plane
	Spectral diffusion	Time constant for all processes that move magnetization between positions in the ESR spectrum of a single species
	Spin diffusion	Time constant for all processes that move magnetization between positions in the ESR spectra of two different species
	Cross relaxation	Transfer of energy between spins with different Zeeman frequencies
	Nuclear spin diffusion (Spatial diffusion)	Time constant for mutual electron spin interchange, contributing to spectral diffusion and spin-echo dephasing
T_M	Phase-memory (Spin-echo dephasing)	Time constant for all processes that cause loss of spin coherence

2.3.4.1 Spin-Lattice Time, Phase Memory Time, and Pulsed ESR Spectrum

The spin-latice relaxation time (T_1) can be determined by the inversion recovery technique. The echo signal grows exponentially with time τ between the pulses as

$$I(\tau) = I_0 \left(1 - 2e^{-\frac{\tau}{T_1}}\right).$$

The factor two appears because the echo recovers from an inverted intensity $-I_0$ at $\tau = 0$. In the corresponding saturation recovery experiment the echo signal is zero at time $\tau = 0$. The echo grows as:

$$I(\tau) = I_0 \left(1 - e^{-\frac{\tau}{T_1}}\right).$$

The choice between these or other methods depends on the system, and the reader is referred to other works [6, 52] for a discussion of proper methods for instance in cases where spin diffusion occurs. An example of a T_1 measurement using the inversion recovery method is shown in Fig. 2.27(a).

The decay of the echo intensity after a 2-pulse sequence can be used to determine the phase-memory time (T_M) as shown in Fig. 2.27(b). The decay is usually approximated by an equation of the form:

$$I(\tau) = I_0 e^{-\frac{2\tau}{T_m}}$$

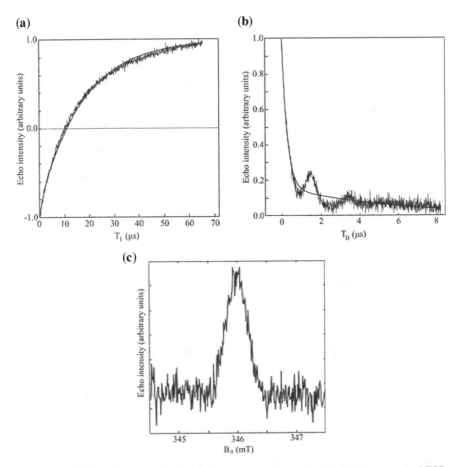

Fig. 2.27 (**a**) Spin-lattice time T_1, (**b**) phase memory time T_M, and (**c**) field sweep pulsed ESR measured at room temperature on an X-irradiated $K_2S_2O_6$ polycrystalline sample. (**a**) T_1 measurement was performed using an inversion recovery pulse sequence. (**b**) T_M measurement was performed using a Hahn echo sequence. (**c**) The pulsed ESR spectrum was obtained by measuring the Hahn echo intensity as a function of the magnetic field. The data were provided by Dr. H. Gustafsson

or as a sum of such terms with different values of T_M and the intensity I_0 at time $\tau = 0$. A two-pulse echo sequence is also of use to measure the echo intensity as a function of the magnetic field yielding a pulsed ESR spectrum, Fig. 2.27(c).

2.3.4.2 Relaxation Times by CW Microwave Saturation

Procedures to obtain relaxation times T_1 and T_2 from microwave (CW) saturation curves of the type shown in Fig. 2.28 were developed many years ago [53–56]. The method has recently been applied to characterize saturation properties of samples

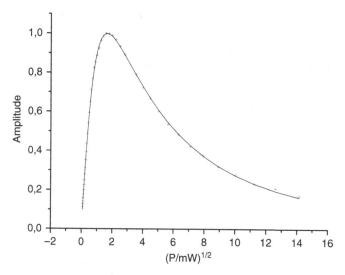

Fig. 2.28 Experimental data (+) and fitted saturation curve (*solid line*) to obtain relaxation times T_1 and T_2 of a crystalline material used for ESR-dosimetry. The experimental data were provided by Prof. E. Sagstuen

employed in applications involving determination of concentrations of paramagnetic species as in ESR dosimetry and other applications of quantitative ESR, see further Chapter 9.

2.4 Distance Measurements

Distance measurements by ESR have recently attracted attention particularly in biological systems. Detailed accounts of pulsed- and CW-ESR methods to determine electron-electron distances are given in [52]. Measurements of electron-nucleus distances have been an essential part in the characterization of trapping sites of paramagnetic species first by ENDOR, later also by ESEEM. Table 2.4 provides a limited overview of methods and applications.

Methods based on the electron-electron and electron-nucleus magnetic dipolar coupling are often employed. The point-dipolar approximation discussed in the next section provides a procedure to obtain distances from the observed coupling in pulse or CW-ESR spectra.

2.4.1 The Point-Dipole Approximation

The interaction between the spin magnetic moments of two unpaired electrons, or of an electron and a nucleus gives rise to an anisotropic coupling. The point-dipole approximation applies when the distance R between the *electron spins* or the *electron and nuclear* spins is large compared to the extension of the orbital(s) of the

Table 2.4 Measurement methods for electron-electron and electron-nucleus distances by ESR, see [52] for details

Type	Measurement method	Range (Å)	Measured quantity	Theory
Electron-nucleus	ESR, ENDOR, ESEEM	1–2	Electron-nucleus dipolar coupling	Quantum average
Electron-nucleus	ENDOR, ESEEM	2–4	Electron-nucleus dipolar coupling	Point dipole
Electron-electron	ESR	5–15	Electron-electron dipolar coupling	Point dipole
Electron-electron	Pulsed ELDOR	15–100	Electron-electron dipolar coupling	Point dipole
Electron-electron	Pulsed ESR	12–40	T_1, T_m	See [52]

unpaired electron(s). The approximation is often applied when the electron-electron and electron-nuclear distances are longer than ca. 5–6 and 2.5 Å, respectively. The coupling is then proportional to $1/R^3$. Distances can therefore be deduced from the measured couplings. The ranges for different methods are given in Table 2.4. At shorter distances the interaction energy is given by a quantum mechanical average.

Electron-electron coupling: The dipolar coupling F between the magnetic moments of two unpaired electrons is in the point-dipole approximation given by the expression

$$F = D(1 - 3\cos^2 \theta)$$

where θ is the angle between the magnetic field and the line joining the spins. The numerical values of the quantity D, denoted D_u and D_c depend on if the electrons are treated as uncorrelated or correlated particles. The first case applies when the electrons are at relatively long distance from each other, for instance in a doubly spin-labelled biomolecule. The second case has been assumed for radical pairs at a distance $R \approx 5$–6 Å [57].

$$D_u = \frac{\mu_0}{4\pi h} \frac{g^2 \mu_B^2}{R^3} = 12.98 \frac{g^2}{R^3} \ (R \text{ in nm}; D \text{ in MHz})$$

$$D_c = \frac{3}{2} \frac{\mu_0}{4\pi h} \frac{g^2 \mu_B^2}{R^3} = 19.47 \frac{g^2}{R^3}$$

Electron-nucleus coupling: The interaction between the magnetic moments of an unpaired electron and a magnetic nucleus gives rise to an anisotropic hyperfine coupling (A_{dip}). The coupling A_{dip} calculated with the point-dipolar approximation is (Appendix A2.1):

$$A_{dip} = b(1 - 3\cos^2 \theta)$$

$$b = -\frac{\mu_0}{4\pi h} \frac{g\mu_B g_N \mu_N}{R^3} = -7.0692 \cdot 10^{-3} \frac{g g_N}{R^3} \ (R \text{ in nm}; b \text{ in MHz})$$

2.4.2 Electron-Electron Distances

The distances of interest in pulsed ESR are usually sufficiently long to use the point dipole approximation for uncorrelated spins described in the section above. The magnitude of the coupling is then proportional to $1/R^3$ where R is the electron-electron distance. By measuring the constant D_u the distance R is accordingly obtained.

Distances between paramagnetic centers in the range 3–13 nm can be measured by pulse techniques that record modulations due to the magnetic coupling between them. The modulations are caused by the interaction between the magnetic moments of the unpaired electrons. This method has been used to obtain distances within "spin-labelled" proteins, where the spin labels are free radicals that have been attached to amino acids at two known positions, and to dimeric biomolecules, each monomer containing a free radical.

A commonly employed method is by pulsed ELDOR, see [5] for a review. The modulation frequency measured from the ELDOR spectrum equals the value of D_u of the previous section in frequency units i.e. $\nu_{ELDOR} = \frac{\mu_0}{4\pi h} \frac{g^2 \mu_B^2}{R^3}$. Typical applications of the method are e.g. measurement of radical-radical distances within an enzyme as exemplified in Fig. 2.29 [58], and to probe conformational changes in spin-labelled DNA [59]. The distance between paramagnetic centers is often not fixed at a specific value, but shows a distribution. Software to obtain the distance and its distribution is described in ref. [60].

Distances shorter than *ca.* 1–2 nm can be measured as a splitting or as a line broadening in the CW-ESR spectra. The splitting can be measured even from the powder spectrum as indicated schematically in Fig. 2.30 using the pronounced "perpendicular" features separated by D. The features correspond to $\theta = 90°$, i.e. with the magnetic field in the xy-plane.

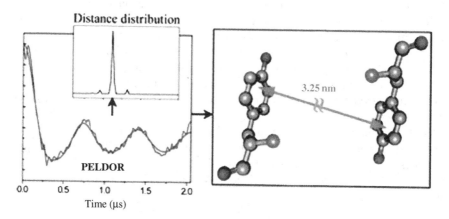

Fig. 2.29 Four-pulse X-band PELDOR of mouse R2 ribonucleotide reductase protein: experimental (*blue line*) and simulated (*red line*) spectrum. The modulations are due to the interaction between two tyrosyl radicals at a distance of 3.25 ± 0.05 nm. The figure is reproduced from Ref. [58] with permission from the Royal Society of Chemistry

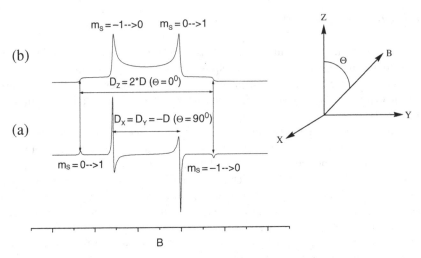

Fig. 2.30 (**a**) Schematic 1st order powder spectrum showing fine structure (*zfs*) with axial symmetry ($E = 0$) for an $S = 1$ species. (**b**) The absorption spectrum increases stepwise at the magnetic fields corresponding to the $m_S = 1 \leftrightarrow 0$ and $0 \leftrightarrow -1$ transitions with the magnetic field along Z, which is the orientation for maximum dipolar coupling between the two electron spins. Transitions cannot occur at lower or higher fields, and the absorption is zero. Two features separated by $2\,D$, referred to as the parallel features occur in the XY plane ($\theta = 0°$)

2.4.3 *Electron-Nucleus Distances*

The trapping site structure with surrounding ions and atoms of neighbour molecules in crystals and the solvation structure in disordered systems can in many cases be elucidated by ENDOR and ESEEM measurements. The point-dipole approximation is frequently used to assign the observed *hfc* to specific nuclei. More elaborate models are needed when the unpaired electron is delocalized over several atoms, exemplified below, and in presence of g-factor anisotropy, see Chapter 6.

Point-dipole approximation with delocalized electron spin: This method is an extension of the point-dipole approximation, applicable to paramagnetic species with spin density distributed over several atoms. Hyperfine (*hf*) interactions in hydrogen-bonded systems and trapping site structure with surrounding ions can be elucidated. An example briefly discussed in Section 2.2.2 is the model deduced from ENDOR measurements on X-irradiated Li-formate for the trapping of $\dot{C}O_2^-$ ion radicals in a crystal matrix. The dipolar *hfc* is composed of contributions from spin densities at three atoms as indicated by the sketch to the right in Fig. 2.12. A procedure to add the contributions described in Appendix A2.1 involves the following steps.

- For a given lithium nucleus the dipolar *hfc* due to each of the atoms with spin densities (ρ_1, ρ_2, ρ_3) in the $\dot{C}O_2^-$ radical is estimated using the point dipole approximation and distances (r_1, r_2, r_3) obtained from crystallographic data.

- The couplings due to each atom are transformed into a common reference system, thus allowing for their individual contributions to be added. The transformation is made using the directions for each r_k, $k = 1$–3, obtained from crystallographic data.
- The tensors are added to a resulting *hfc* tensor. The tensor is diagonalized, thus providing principal values and corresponding eigenvectors.
- The calculated eigenvector for the maximal principal value of a specific lithium ion is compared with the maximal values and vectors for all ^{7}Li *hfc* tensors determined experimentally. The ^{7}Li *hfc* tensors are assigned to specific Li$^+$ ions in the crystal structure by this comparison [29].

The point dipole method is less accurate or more difficult to apply in several cases of practical importance discussed below.

Dipolar hyperfine coupling of β- ^1H: The point dipole method is a common approximation for the ^{1}H dipolar *hfc* in β-position of π-electron radicals (see Fig. 2.8). The dipolar couplings can be estimated more accurately by taking account of the extension of the electron wave-function for the unpaired electron [26]. Computation of the electronic structure with evaluation of the coupling tensors from first principles in quantum chemistry programs is an alternative that is frequently used in modern applications.

α -^{1}H in π-electron radicals: The point-dipole approximation does not apply. A quantum mechanical average is necessary because of the short C_α-H_α bond length. The distance, ca. 0.1 nm in π-electron radicals of the type H_α—$\dot{C}\langle$ is comparable to the extension of the orbital of the unpaired electron, see [25] for details.

Nuclei other than hydrogen: The point dipolar method to estimate distances is difficult to apply for atoms carrying appreciable spin density. This applies to the centre atom of a radical or a transition metal ion complex, but often also to atoms or ligands other than hydrogen. This is for instance the case for ^{19}F in the fluorocarbon anions discussed in Chapter 5. Theoretically computed values are in these cases often sufficiently accurate to support or reject an assignment. In favourable cases modern DFT or ab initio methods can even be of predictive value to reproduce experimental spectra as exemplified in Chapter 5.

2.5 Summary

Multi-resonance and pulsed ESR techniques provide better spectral resolution than conventional ESR, although usually with a lower sensitivity. Multi-resonance involves ENDOR, TRIPLE and ELDOR in CW and pulsed modes. ENDOR is mainly used to increase the spectral resolution, so that overlapping or unresolved hyperfine structure in the ESR spectra can be resolved. The techniques are applicable both to liquid and solid samples. CW X-band spectrometers with an ENDOR attachment that also allows TRIPLE experiments have been commercially available for a long time. In a TRIPLE experiment *two* RF fields are applied. A theoretical application is to determine the relative *signs* of two hyperfine couplings. Accessories for other frequency bands and for pulsed ENDOR have been developed more recently. In an ELDOR experiment *two* MW frequencies are applied. The technique has been revived in pulsed ELDOR experiments to obtain distances between specific spin-labeled sites in polymers and biopolymers.

The electron-spin-echo (ESE) technique is the common pulsed ESR method for solid samples. It involves radiation with two or more MW pulses followed by measurement of the *echo* signal decay. The electron-spin-echo-envelope-modulation (ESEEM) method provides a means to resolve weak anisotropic hyperfine structure from the modulation of the decay curve. Following seminal work by Mims, ESEEM has been particularly successful for studies of solvation structures of paramagnetic species in disordered solids. It is less used than ENDOR for single crystal measurements as the measurements are time-consuming. The HYSCORE (Hyperfine Sublevel Correlation Spectroscopy) method is also time-consuming, but permits a safe assignment of the nucleus giving the spectral pattern observed. The analysis is usually made after Fourier transformation (FT) of the echo decay signal. In favorable cases the (anisotropic) hyperfine coupling can be directly read from the Fourier pattern, in other cases simulation is needed. The ESEEM and HYSCORE methods are not applicable to liquid samples with overall isotropy, in which case the FID (free induction decay) method finds some use. Pulsed ESR is also used to measure magnetic relaxation in a more direct way than with CW-ESR. Pulsed ELDOR measurements are employed to obtain distances in bio-molecules.

Appendix

A2.1 Dipolar Hyperfine Couplings (hfc) with Delocalized Electron Spin

In a simplified model the energy of an electron and a nuclear spin interacting with each other is obtained by the point dipole approximation as shown in Fig. 2.31 below. The result is anticipated from the classical expression for the energy of the spin magnetic moments μ_S and μ_I when the electron and nuclear spins are both oriented along the applied field that makes an angle θ with the line joining the spins S and I.

$$E_d = \frac{\mu_0}{4\pi} \cdot \left[\frac{\mu_S \cdot \mu_I}{r^3} - 3\frac{(\mu_S \cdot \mathbf{r}) \cdot (\mu_I \cdot \mathbf{r})}{r^5} \right] \propto \frac{\mu_0}{4\pi} \frac{\mu_S \cdot \mu_I}{r^3}(1 - 3\cos^2\theta) \quad (2.1)$$

The interaction between the magnetic moments of an unpaired electron and a magnetic nucleus gives rise to an anisotropic *hfc*. The electron magnetic moment is negative. The coupling b in frequency units calculated with the point-dipolar approximation is thus obtained as:

$$b = -\frac{\mu_0}{4\pi h} \frac{g\mu_B g_N \mu_N}{r^3}(1 - 3\cos^2\theta) \quad (2.2)$$

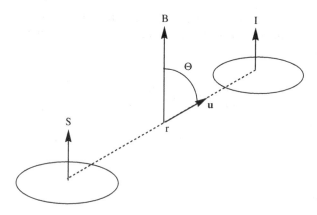

Fig. 2.31 Simplified model for dipolar coupling energy $E_d = -\frac{\mu_0}{4\pi} \cdot \frac{g\mu_B g_N \mu_N}{r^3}(1-3\cos^2\theta)$ between an electron (S) and a nuclear (I) spin in a magnetic field (B) making an angle θ with line r joining the spins

The approximation holds when the distance r between the electron and nuclear spins is large compared to the extension of the orbital of the unpaired electron. At short distance the interaction energy has to be calculated by a quantum mechanical average. A procedure for the common situation of a hydrogen atom in a carbon-centred π-electron radical $\rangle\dot{C}_\alpha$-H was given in a classical paper [25].

The point-dipole approximation with delocalized electron spin is an extension of the usual point-dipole approximation described above and is applicable to paramagnetic species with spin density distributed over several atoms. Hyperfine interactions in hydrogen-bonded systems and trapping site structure with surrounding ions have been elucidated with this method. An example is the model deduced from ENDOR measurements on X-irradiated Li-formate for the trapping of $\dot{C}O_2^-$ ion radicals in a crystal matrix. The dipolar *hfc* is composed of contributions from spin densities at three atoms as indicated by the sketch. The *hfc* perpendicular and parallel to the r_n, ($n = 1$–3) directions are given by the expressions (2.3).

$$b_\perp^n = -\rho_n \frac{\mu_0}{4\pi h} \frac{g\mu_B g_N \mu_N}{r_n^3} \qquad b_\parallel^n = -2 \cdot b_\perp^n \qquad (2.3)$$

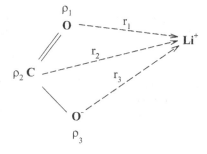

The contributions from spin density ρ_n at each atom are added after transformation to the crystallographic axes system. Due to the assumed axial symmetry the simple formula (2.4) applies.

$$b_{ij} = \frac{\mu_0}{4\pi} \cdot \mu_B \mu_N g g_N \sum_{n=1}^{3} \rho_n \frac{3 u_i^n u_j^n - \delta_{ij}}{r_n^3} \qquad (2.4)$$

Here u_j^n are the direction cosines (j = x, y, z) of axis r_n in the crystallographic system [29].

References

1. G. Feher: Phys. Rev. **103**, 834 (1956).
2. K.P. Dinse, R. Biehl, K. Möbius: J. Chem. Phys. **61**, 4335 (1974).
3. R. Biehl, M. Plato, K. Möbius: J. Chem. Phys. **63**, 3515 (1975).
4. L. Kevan, L.D. Kispert: '*Electron Spin Double Resonance Spectroscopy*', Wiley, New York, NY (1976).
5. A.D. Milov, A.G. Maryasov, Yu.D. Tsvetkov: Appl. Magn. Reson. **15**, 107 (1998).
6. A. Schweiger, G. Jeschke: '*Principles of Pulse Electron Paramagnetic Resonance*', Oxford University Press, New York, NY (2001).
7. W.B. Mims: Phys. Rev. B **5**, 2409 (1972); Phys. Rev. B **6**, 3543 (1972).
8. E.J. Reijerse, S.A. Dikanov: J. Chem. Phys. **95**, 836 (1991).
9. S.A. Dikanov, Y.D. Tsvetkov: '*Electron Spin Echo Envelope Modulation (ESEEM) Spectroscopy*', CRC Press, Boca Raton, FL (1992).
10. L. Kevan, R.N. Schwartz (eds.): '*Time Domain Electron Spin Resonance*', Wiley, New York, NY (1979).
11. L. Kevan, M.K. Bowman (eds.):'*Modern Pulsed and Continuous-Wave Electron Spin Resonance*', Wiley, New York, NY (1990).
12. C.P. Keijzers, E.J. Reijerse, J. Schmidt (eds.): '*Pulsed EPR: A New Field of Applications*', North Holland, Amsterdam (1989).
13. A.J. Hoff (ed.): '*Advanced EPR Applications in Biology and Biochemistry*', Elsevier, Amsterdam (1989).
14. P. Höfer, A.Grupp, H. Nebenführ, M. Mehring: Chem. Phys. Lett. **132**, 279 (1986).
15. H. Kurreck, B. Kirste, W. Lubitz: '*Electron Nuclear Double Resonance Spectroscopy of Radicals in Solution*', VCH Publishers, New York, NY (1988).
16. D.M. Murphy, R.D. Farley: Chem. Soc. Rev. **35**, 249 (2006).
17. (a) E. Fermi: Z. Phys. **60**, 320 (1930). (b) M. Bucher: Eur. J. Phys. **21**, 19 (2000).
18. H.M. McConnell, D.B. Chesnut: J. Chem. Phys. **28**, 107 (1958).
19. J.A. Weil, J.R. Bolton: '*Electron Paramagnetic Resonance: Elementary Theory and Practical Applications*', 2nd Edition, Wiley, Hoboken, NJ (2007).
20. N.M. Atherton: '*Principles of ESR*', Ellis Horwood and Prentice Hall, London (1993).
21. M. Brustolon, E. Giamello (eds.): '*Electron Paramagnetic Resonance: A Practitioner's Toolkit*', Wiley, Hoboken, NJ (2009).
22. H.C. Box: '*Radiation Effects, ESR and ENDOR Analysis*', Academic Press, New York, NY (1977).
23. A. Lund, M. Shiotani (eds.): '*EPR of Free Radicals in Solids, Trends in Methods and Applications*', Kluwer Academic Publishers, Dordrecht (2003).
24. E. Sagstuen, A. Lund, Y. Itagaki, J. Maruani: J. Phys. Chem. A **104**, 6362 (2000).

25. H.M. McConnell, J. Strathdee: Mol. Phys. **2**, 129 (1959).
26. H.C. Box, H.G. Freund, E.E Budzinski: J. Chem. Phys. **57**, 4290 (1972).
27. C. Heller, H.M. McConnell: J. Chem. Phys. **32**, 1535 (1960).
28. A.R. Sørnes, E. Sagstuen, A. Lund: J. Phys. Chem. **99**, 16867 (1995).
29. (a) T.A. Vestad, E. Malinen, A. Lund, E.O. Hole, E. Sagstuen: Appl. Radiat. Isot. **59**, 181 (2003). (b) T.A.Vestad, H. Gustafsson, A. Lund, E.O. Hole, E. Sagstuen: Phys. Chem. Chem. Phys. **6**, 3017 (2004). (c) T.A. Vestad: '*On the Development of a Solid-State, Low Dose EPR Dosimeter for Radiotherapy*', Dissertation, University of Oslo (2005).
30. J.S. Hyde: J. Chem. Phys. **43**, 1806 (1965).
31. E. Sagstuen, E.O. Hole, S.R. Haugedal, W.H. Nelson: J. Phys. Chem. A **101**, 9763 (1997).
32. R. Erickson: Chem. Phys. **202**, 263 (1996).
33. F. Gerson: Acc. Chem. Res. **27**, 63 (1994).
34. R.B. Clarkson, R.L. Belford, K. Rothenberger, H. Crookham: J. Catalysis **106**, 500 (1987).
35. B.M. Hoffman: Acc. Chem. Res. **24**, 164 (1991); Acc. Chem. Res. **36**, 522 (2003).
36. G.H. Rist, J.S. Hyde: J. Chem. Phys. **52**, 4633 (1970).
37. D. Biglino, H. Li, R. Erickson, A. Lund, H. Yahiro, M. Shiotani: Phys. Chem. Chem. Phys. **1**, 2887 (1999).
38. M. Brustolon, A. Barbon: In '*EPR of Free Radicals in Solids: Trends in Methods and Applications*' ed. by A. Lund, M. Shiotani, Kluwer Academic Publishers, Dordrecht (2003), Chapter 2.
39. (a) P.H. Kasai, R.J. Bishop Jr.: In '*Zeolite Chemistry and Catalysis*' ed. by J.A. Rabo, ACS Monograph 171, American Chemical Society, Washington, DC, (1976), p. 350. (b) P.H. Kasai, R.M. Gaura: J. Phys. Chem. **86**, 4257 (1982).
40. H. Li, D. Biglino, R. Erickson, A. Lund: Chem. Phys. Lett. **266**, 417 (1997).
41. B. Rakvin, N. Maltar-Strmečki: Chem. Phys. Lett. **415**, 161 (2005).
42. J. Woodworth, M.K. Bowman, S.C. Larsen: J. Phys. Chem. B **108**, 16128 (2004).
43. Y. Deligiannakis, M. Louloudi, N. Hadjiliadis: Coord. Chem. Rev. **204**, 1 (2000).
44. S.A. Dikanov, M.K. Bowman: J. Magn. Reson. **116**, 125 (1995).
45. A.G. Maryasov, M. Bowman: J. Magn. Reson. **179**, 120 (2006).
46. W.B. Mims: Proc. R. Soc. Lond. Ser. A **283**, 452 (1965).
47. E.R. Davies: Phys. Lett. A **47**, 1 (1974).
48. C. Gemperle, A. Schweiger: Chem. Rev. **91**, 1481 (1991)
49. J.W.A. Coremans, O.G. Poluektov, E.J.J. Groenen, G.W. Canters, H. Nar, A. Messerschmidt: J. Am. Chem. Soc. **118**, 12141 (1996).
50. D.L. Tierney, H. Huang, P. Martasek, B. Sue, S. Masters, R.B. Silverman, B.M. Hoffman: Biochemistry **38**, 3704 (1999).
51. (a) J. Niklas, T. Schulte, S. Prakash, M. van Gastel, E. Hofmann, W. Lubitz: J. Am. Chem. Soc. **129**, 15442 (2007). (b) W. Lubitz, F. Lendzian, R. Bittl: Acc. Chem. Res. **35**, 313 (2002).
52. L.J. Berliner, S.S. Eaton, G.R. Eaton: '*Biological Magnetic Resonance*', Volume 19: Distance Measurements in Biological Systems by EPR, Kluwer Academic Publishers, Dordrecht (2002).
53. A.M. Portis: Phys. Rev. **91**, 1071 (1953).
54. T.G. Castner, Jr.: Phys. Rev. **115**, 1506 (1959).
55. J. Maruani: J. Magn. Reson. **7**, 207 (1972).
56. S. Schlick, L. Kevan: J. Magn. Reson. **22**, 171 (1976).
57. Y. Kurita: J. Chem. Phys. **41**, 3926 (1964).
58. D. Biglino, P.P. Schmidt, E.J. Reijerse, W. Lubitz: Phys. Chem. Chem. Phys. **8**, 58 (2006).
59. G. Sicoli, G. Mathis, O. Delalande, Y. Boulard, D. Gasparutto, S. Gambarelli: Angew. Chem. Int. Ed. **47**, 735 (2008).
60. G. Jeschke: 'DEER Analysis 2009': http://www.epr.ethz.ch/software/index

Exercises

E2.1 The X-band ESR spectrum of the duroquinone anion in liquid solution contains 13 equidistant lines separated by 0.93 mT. The structure of the duroquinone anion is:

(a) How many lines are expected in the ENDOR spectrum with the magnetic field at the centre (339 mT) of the X-band ESR spectrum?

(b) What is the separation in MHz between the lines?

(c) What is the expected frequency at the centre of the ENDOR spectrum?

E2.2 The Fermi equation for the hyperfine coupling $a = \frac{2}{3}\mu_0 g\mu_B g_N\mu_N\rho(0)$ contains the factor $\rho(0) = |\psi(0)|^2$, the spin density at the nucleus. Its value for the hydrogen atom is obtained with the wave-function $\psi(r) = \frac{1}{\sqrt{\pi a_0^3}}e^{-\frac{r}{a_0}}$, and $a_0 = 0.52918$ Å.

(a) Which of the two interpretations of "spin density" applies in this calculation?

(b) The experimental *hfc* for the H atom was determined to be $a = 1{,}420.4$ MHz. What value is obtained theoretically?

E2.3 Explain using the example in Fig. 2.2 why the nuclear Zeeman term in the energy expression for a liquid $E(m_s, m_I) = g\mu_B Bm_s + am_s m_I - gN\mu_N Bm_I$ must be taken into account in ENDOR, but can be neglected in ESR.

E2.4 The angular variation of the ENDOR frequencies at X-band ($B = 339$ mT) of a free radical in a single crystal with anisotropic *hfc* due to ^{1}H is schematically shown in the figure. The data were accidentally obtained in the principal axes system (X, Y, Z).

(a) Is the *hfc* smaller or larger than $2\cdot\nu_H$, i.e. does case (a) or (b) in Fig. 2.1 apply?

(b) The ENDOR lines at $\theta = 45°$ are centred at $\nu^+ = 51.8$ and $\nu^- = 23.8$ MHz. Is the line separation exactly equal to $2\cdot\nu_H$ like in Fig. 2.1? If not suggest a reason! Hint: The ENDOR frequencies in the XY plane of the principal axes system depend on the angle θ of the magnetic field with the X-axis as:

$$\left(v^{\pm}\right)^2 = \left(\pm 1/2 A_X - v_H\right)^2 \cos^2 \theta + \left(\pm 1/2 A_Y - v_H\right)^2 \sin^2 \theta$$

(c) Confirm that the line separations are $2 \cdot v_H$ when $\theta = 0$ (B $||$ X) or $\theta = 90°$ (B $||$ Y), i.e. when the magnetic field is parallel to the principal directions X and Y of the *hfc* tensor!

(d) Is it reasonable to obtain the principal *hf*-couplings in powder ENDOR from spectrum features like those in Fig. 2.14?

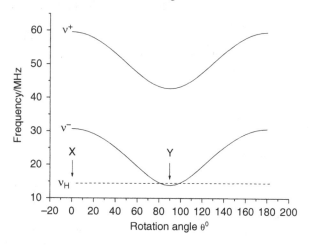

E2.5 A complete set of ENDOR measurements was made of the proton hyperfine couplings of the *π-electron radical* HOOC-$C_\beta H_2$-$\dot{C}_\alpha$H-COOH produced by X-rays in single crystals of succinic acid [26]. The isotropic α-^{1}H coupling was −60.7 MHz, and the anisotropic components were (−33.0, +1.5, +31.5) MHz.

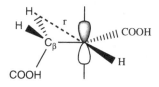

(a) Estimate the spin density on $\dot{C}_\alpha$ using McConnell's relation [18] $a_\alpha = Q \cdot \rho_C^\pi$, with $Q \approx -70$ MHz. Which of the two interpretations of spin density applies?

(b) The isotropic couplings of the β-^{1}H were 106.4 and 78.2 MHz. Suggest probable values for the angles θ in McConnell's relation [27] for the two β-H, $a_\beta = \rho_C^\pi(A + B\cos^2 \theta)$ with $A \approx 0$, $B = 140$ MHz. (The result shows that the β-^{1}H conformation of the methylene group is fixed as is often observed for π-electron radicals trapped in solids.)

(c) The assumption commonly made for the assignment of β-couplings to specific H atoms is that the maximum dipolar coupling occurs with

B || $\dot{C}_\alpha$–H_β direction (r in figure) of the magnetic field. What is the explanation?

(d) The maximum coupling calculated with the point dipolar approximation is: $2 \cdot b(\text{MHz}) = 2 \cdot \frac{\mu_0}{4\pi h} \frac{g\mu_B g_N \mu_N}{r^3} = 14.14 \cdot 10^{-3} \frac{g g_N}{r^3}$. What value of $2\,b$ is obtained theoretically for $r = 0.215$ nm, see [26] for the experimental value. Better agreement is obtained by taking account of the extension of the electron wave-function for the unpaired electron. The procedure is more commonly employed to obtain the distance from the measured value of $2 \cdot b$.

(e) Describe in words (no formulae) how delocalization of spin density to C_β might be taken into account for the dipolar β-coupling. The issue was addressed by Derbyshire many years ago [W. Derbyshire: Mol. Phys. **5**, 225 (1962)].

E2.6 The ENDOR spectrum of the phenalenyl radical in fluid solution is due to two sets of equivalent ^{1}H nuclei.

(a) Use information from the ESR spectrum in Fig. 2.4 to obtain the number of equivalent nuclei in each group.

(b) A Hückel MO calculation predicts the spin densities at the carbon atoms as shown in the structure to the left, supporting that the large coupling is due to six equivalent ^{1}H nuclei. A calculation with the modified Hückel method due to McLachlan [A.D. McLachlan, Mol. Phys. 3, 233 (1960)] predicts numerically small negative spin densities at the three positions shown to the right.

Hückel   McLachlan, λ = 1.0

Estimate the *hfc* theoretically using McConnell's relation [18] $a_\alpha = Q \cdot \rho_C^\pi$, with $Q \approx -70$ MHz.

(c) What are the predicted signs of the couplings? Suggest experimental methods to determine the relative signs of the two couplings.

E2.7 The frequencies of the ENDOR lines in a crystal depend on the direction of the magnetic field specified by the direction cosines (ℓ_x, ℓ_y, ℓ_z) according to the equation $\nu^2 = (\mathbf{T}^2)_{xx}\ell_x^2 + (\mathbf{T}^2)_{yy}\ell_y^2 + (\mathbf{T}^2)_{zz}\ell_z^2 + 2(\mathbf{T}^2)_{xy}\ell_x\ell_y + 2(\mathbf{T}^2)_{xz}\ell_x\ell_z + 2(\mathbf{T}^2)_{yz}\ell_y\ell_z$, see Chapter 3. The "ENDOR tensor" $T_{ij} =$

$m_S A_{ij} - \delta_{ij} \nu_N$, with ν_N equal to the nuclear Zeeman frequency is usually determined by the Schonland procedure. Assume that two sites 1 and 2 are symmetry related by $x(2) = x(1)$, $y(2) = -y(1)$ and $z(2) = z(1)$. Does site splitting occur in the xy, xz and yz planes?

Hint: A magnetic field direction (ℓ_x, ℓ_y, ℓ_z) for site 1 corresponds to the direction $(\ell_x, -\ell_y, \ell_z)$ for site 2. The variation of ν^2 in the xy, xz and yz planes of the two sites can then be compared.

E2.8 ^{14}N ENDOR single crystal measurements were made on a species assigned to the $N_2 D_4^+$ cation radical present after X-irradiation of $Li(N_2 D_5) SO_4$ with the magnetic field in the *bc*, *ca* and *ab* crystal planes of the orthorhombic crystal.

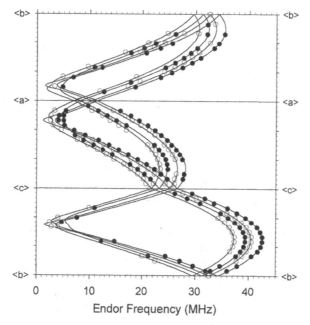

Endor Frequency (MHz)

Experimental high-frequency ($m_s = -\frac{1}{2}$) ^{14}N ENDOR data from the $N_2 D_4^+$ cation. The diagram is reproduced from [Y. Itagaki et al.: J. Phys. Chem. A, **106**, 2617 (2002)] by permission of the American Chemical Society

(a) More than one site can appear in the orthorhombic $Li(N_2 D_5) SO_4$ crystal How many sites are observed in the *bc* plane?

(b) Curves indicated by "•" were assigned the to ^{14}N ENDOR transitions for $m_S = -\frac{1}{2}$. Suggest a reason for the appearance of two such curves for each site. How many would occur for $^{15}N_2 D_4^+$?

(c) Is the *hfc* smaller or larger than $2 \cdot \nu_N$, the ^{14}N nuclear Zeeman frequency, i.e. does case (a) or (b) in Fig. 2.1 apply?

(d) Methods to obtain the ^{14}N hyperfine- (**A**) and nuclear quadrupole- tensor (**Q**) from the data are discussed in Chapter 3. The simplest procedure is when the *hfc* >> *nqc* (the nuclear quadrupole coupling. Examine the figure to determine if this case applies.

E.2.9 An ESEEM signal occurs in the presence of an anisotropic hyperfine interaction that is of a magnitude comparable to the nuclear Zeeman energy. The schematic ESR spectrum in the left part of the figure below contains allowed and "forbidden" lines due to a mixing of nuclear states. For an $I = \frac{1}{2}$ nucleus the ESEEM amplitude is proportional to $k = \sin^2 \alpha/2 \cdot \cos^2 \alpha/2 = I_0 \cdot I_i$, where α is the angle between the effective fields acting on the nucleus for $m_S = \pm\frac{1}{2}$ and I_0 and I_i are the amplitudes of the outer and inner ESR lines.

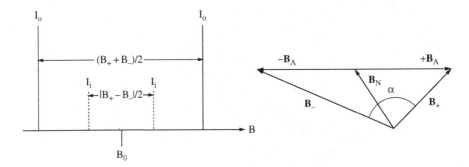

(a) The amplitude is zero when the field is along one of the principal axes of the hyperfine coupling tensor. Suggest an explanation using the right part of the figure or the equations given in Chapter 3.

(b) Why is ESEEM absent for an isotropic hyperfine coupling due to an $I = \frac{1}{2}$ nucleus? Can ESEEM be obtained under special conditions from ^{14}N with an isotropic hyperfine coupling?

(c) Suggest an explanation why S-band ESEEM spectroscopy sometimes give better signals than at X-band, especially for "weakly coupled" nuclei [R.B. Clarkson, D.R. Brown, J.B. Cornelius, H.C. Crookham: Pure & Appl. Chem. **64**, 893 (1992)]. Hint: The modulation depth factor for $I = \frac{1}{2}$ nuclei, $k \propto \sin^2 \frac{\alpha}{2} \cdot \cos^2 \frac{\alpha}{2} \propto \sin \alpha$, has a maximum for $\alpha = 90°$, in which case $I_i = I_0 = \frac{1}{2}$, i.e. the "allowed" and "forbidden" lines have the same intensity. Would this situation depend on the microwave frequency?

Chapter 3
Analysis of Spectra

Abstract The analysis of ESR, ENDOR, and ESEEM data to extract the resonance parameters is treated. Free radicals in solution are mainly identified by their hyperfine couplings (*hfc*). The analysis of ESR and ENDOR spectra by visual inspection and by computer simulation is discussed. The Schonland method to obtain the principal values and directions of the anisotropic *g*- and *hfc*- tensors from single crystal ESR and ENDOR data is presented. The modifications needed when $S > \frac{1}{2}$ or $I > \frac{1}{2}$ to obtain zero-field splitting (*zfs*) or nuclear quadrupole coupling (*nqc*) tensors are considered. Examples of simulations to extract *g*-, *hfc*-, *zfs*-, and *nqc*-tensors from ESR and ENDOR spectra of disordered systems are presented. Simulation methods in pulsed ESR (1- and 2-dimensional ESEEM) studies are exemplified. Internet addresses for down-loading software for the simulation of ESR, ENDOR, and ESEEM spectra are provided. Software for the analysis of single crystal data by the Schonland method is also available.

3.1 Introduction

The assignment of spectra to specific paramagnetic species in a sample is a central aim of ESR spectroscopy. In liquids the species under study are most often free radicals. For liquid samples the main information for the assignment thus comes from measurement of the *g*-factor, and above all from the hyperfine coupling constants. The main experimental methods used for studies of liquid systems are CW-ESR, CW- and pulsed ENDOR. In solids these parameters are usually anisotropic, *i.e.* their values depend on the orientation of a crystal sample in the magnetic field. For systems with $S > \frac{1}{2}$ one may also need to take into account the zero-field splitting as well as other interactions typical for the solid state, exemplified in Chapters 4–8. Multiple frequency CW-ESR (Chapter 4), and pulsed methods are employed in addition to ENDOR for measurements of single crystal and polycrystalline or disordered systems. In many instances the experimental studies are complemented by quantum chemistry calculations to support the assignments. Examples are given in Chapter 5.

A. Lund et al., *Principles and Applications of ESR Spectroscopy*,
DOI 10.1007/978-1-4020-5344-3_3, © Springer Science+Business Media B.V. 2011

The analysis of spectra from liquids, single crystals and disordered samples differ considerably and are treated separately.

Coupling constants are reported in different units, depending on the application. For free radicals it is convenient to employ magnetic field units, (mT) in CW-ESR studies. The g-factors are then usually near $g_e = 2.00232$, the free electron value. The Gauss unit with $10\ G = 1\ mT$ is standard in older works, and is also used in several illustrations in this chapter. Magnetic field units are convenient to use, but ambiguities in the conversion to energy units can occur if the g-factors deviate from the free electron value and are anisotropic as is often the case for transition metal ions. Energy units are therefore often employed in this case, traditionally in cm^{-1} units for the zero-field splitting ($S > \frac{1}{2}$) but also for hyperfine couplings. In the latter case MHz units are also employed In ENDOR and pulsed ESR studies coupling constants are directly obtained in energy (MHz) units. In theoretical studies MHz, cm^{-1}, and atomic units (Hartrees) are commonly employed. Couplings in Gauss units are sometimes calculated with the implicit assumption that $g = g_e$. Conversion factors are given in Table G3 of Appendix G.

3.2 Liquids

The spectral analysis of paramagnetic species featuring hyperfine structure is emphasized in textbooks [1–4] since it helps identify the chemical and electronic structure of the species under study. A majority of the studies has involved free radicals. ESR analysis of hyperfine structure is discussed in Sections 3.2.2–3.2.6. ENDOR spectra of liquid samples are usually analysed visually, see Chapter 2 for examples, and [2] for a detailed account, including a description of a simulation program referenced in Section 3.2.7.

3.2.1 Hyperfine Couplings

While g-factors are measured by ESR, there are several methods like ESR, CW and pulsed ENDOR, special and general TRIPLE to determine the strength of the magnetic coupling (*i.e.* hyperfine coupling) between the electronic and nuclear magnetic moments in liquid solution. The method chosen therefore depends both on the system and the available equipment. The various possibilities are summarised in Table 3.1.

A nucleus with spin I gives rise to $2 \cdot I + 1$ hyperfine lines in ESR while only two lines occur in the ENDOR spectrum corresponding to the electronic quantum numbers $m_S = \pm \frac{1}{2}$ for an $S = \frac{1}{2}$ species. The relative signs of several isotropic hyperfine coupling constants in a radical can be determined by the general TRIPLE method as described in Chapter 2. The number of equivalent nuclei in the radical systems discussed below can be obtained from the line intensities in ESR, Special TRIPLE, and pulsed ENDOR, but usually not from CW-ENDOR for reasons discussed in the previous chapter.

Table 3.1 Methods
employed to determine
hyperfine couplings (a_{iso})
in solution

Method	Sign of a_{iso}	Number of equivalent nuclei	Sensitivity	Resolution
ESR	No	Yes	High	Low
CW-ENDOR	No	No	Low	High
Pulsed ENDOR	No	Yes	Low	High
Special TRIPLE	No	Yes	Low	High
General TRIPLE	Yes	No	Low	High

3.2.2 Equivalent Nuclei

ESR spectra of free radicals in solution often show hyperfine structure due to several nuclei, particularly ^{1}H, with the same coupling constant. The ESR spectrum of the methyl radical in Fig. 3.1 shows a line pattern with three equivalent protons ($I = \frac{1}{2}$). The analysis by the diagram under the spectrum shows the construction of a stick pattern with an intensity ratio 1:3:3:1.

The so called Pascal triangle shown below is practical to apply when the number of equivalent nuclei with $I = \frac{1}{2}$ is not too high, up to six in Table 3.2.

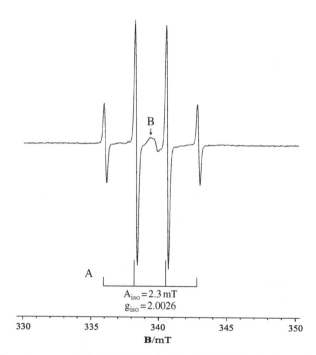

Fig. 3.1 ESR spectrum recorded at 210 K of γ-irradiated H-rho/CH$_4$ zeolite. *A*: signal of methyl radical, *B*: signal from radiation induced damage in the zeolite framework. The figure is reproduced from [M. Danilczuk et al.: J. Phys. Chem. B **110**, 24492 (2006)] with permission from the American Chemical Society

Table 3.2 Pascal triangle.
The numbers and relative
intensities (R) of the
hyperfine lines for up to
$N = 6$ equivalent $I = \frac{1}{2}$
nuclei

N					R				
0					1				
1				1		1			
2				1	2	1			
3			1	3		3	1		
4		1	4		6		4	1	
5	1	5		10		10		5	1
6	1	6	15		20		15	6	1

For a certain N the R values are calculated by adding the numbers immediately to the left and the right in the row for N–1. Thus, not only the strength of the coupling but also the number of nuclei is obtained from the analysis. When the number of equivalent nuclei is very large it may be difficult to observe all lines as shown in the next example, the duroquinone anion radical.

The duroquinone anion has totally 12 equivalent protons due to the four methyl groups. By extending the triangle to $N = 12$ or using the procedures in Section 2.7, 13 lines are expected with an intensity ratio given in Fig. 3.2. Only the nine

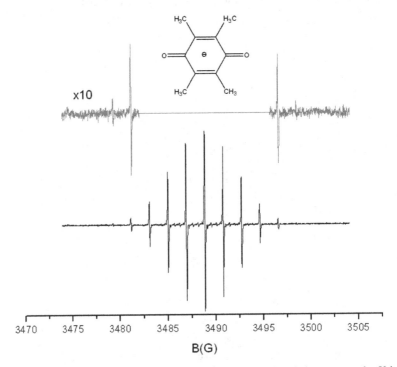

Fig. 3.2 ESR spectrum of the duroquinone anion spectrum in liquid solution, measured at X-band. The observed lines are due to hyperfine structure with 12 equivalent protons with an expected intensity ratio 1 : 12 : 66 : 220 : 495 : 792 : 924 : 792 : 495 : 220 : 66 : 12 : 1, but the weakest lines on each wing are not detected. The spectrum was provided by Bruker Biospin

strongest lines are clearly seen in the lower trace of the spectrum, and additionally two at 10 times higher amplification. The theoretical amplitudes for the N-value that best match the relative intensities of the observed lines can help to identify the species present in the sample.

3.2.3 Inequivalent Nuclei

The procedure to analyse spectra with inequivalent nuclei by means of stickplots is exemplified in Fig. 3.3, upper trace for the galvinoxyl radical in liquid solution. The spectrum is a doublet of quintets. The quintet structure with approximate intensity ratio 1:4:6:4:1 is caused by the four equivalent ring protons, while the doublet splitting is due to the proton of the CH-group that joins the rings. The two quintets slightly overlap at the centre of the spectrum. The tertiary butyl protons are not resolved. The spectrum in the lower trace is obtained from the radical with two of the tertiary butyl groups replaced by methyl. The spectrum can be analysed by assuming that the protons of the methyl groups are equivalent, see Exercise E3.1.

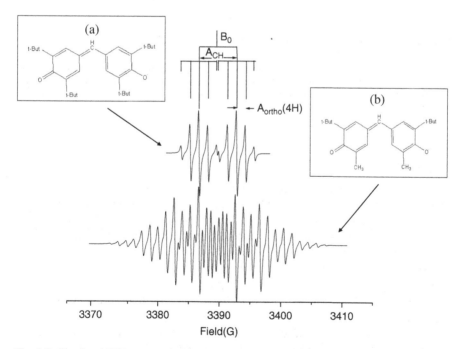

Fig. 3.3 Simulated ESR spectra of (**a**) galvinoxyl radical with $a_{CH} = 0.61$ mT, $a_{ortho}(4H) = 0.144$ mT, and of (**b**) CH$_3$-substituted galvinoxyl radical with $a_{CH} = 0.595$ mT, $a_{ortho}(4H) = 0.137$ mT, $a_{CH3}(6H) = 0.395$ mT, see paper by C. Besev et al. [Acta Chem. Scand. **17**, 2281 (1963)] for experiments and analysis

The overlap of lines can be pronounced as shown in Fig. 3.4 below.

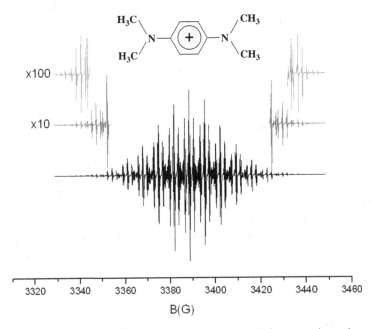

Fig. 3.4 ESR spectrum of the Wurster'blue cation, synthesized from experimental parameters $a_H(12H) = 6.774$ G, $a_N(2\ ^{14}N) = 7.052$ G, $a_H(4H) = 1.989$ G, line width = 0.1 G, see [J.R. Bolton, A. Carrington, J. dos Santos-Veiga: Mol. Phys., **5**, 615 (1962)] for an analysis of the ESR spectrum of the unsubstituted Wurster blue cation, and [G. Grampp et al.: Monatshefte für Chemie **136**, 519 (2005)] for ESR, ENDOR, and theoretical investigations on various Wurster' radical cations in solution

3.2.3.1 Nuclei with I > ½

Hyperfine structure due to nuclei with $I > ½$, particularly to ^{14}N is frequently observed in the ESR spectra of free radicals. ESR measurements of nitrogen-containing radicals with $I(^{14}N) = 1$ are made for instance in applications using nitrogen-containing compounds for spin trapping of radicals in solution and in spin-labelling studies of bio-molecules. The analysis is often complicated by near equal hyperfine couplings of ^{14}N and ^1H in the studied radicals. The previously discussed hydrazine cation, $N_2H_4^+$, for example, is reported to have hyperfine structure with two equivalent ^{14}N nuclei with $a_N = 1.10$ mT and with four protons with $a_H = 1.15$ mT in liquid solution. As in the previous example the protons will give a 1:4:6:4:1 hyperfine pattern. The Pascal triangle method does not apply to nuclei with spins $I > ½$, see Exercise E3.2 for the construction of a stick diagram for two ^{14}N nuclei with $I = 1$, to yield a quintet hyperfine pattern in the 1:2:3:2:1 ratio. The two quintets, due to the four equivalent protons, and the two ^{14}N nuclei ($I = 1$) are not well separated, see spectrum in Exercise E3.2.

Even more complex ESR spectra can occur for nitrogen-containing aromatic radicals of the type in Fig. 3.4. A stickplot analysis is not easily performed. At this level of complexity a combination of ESR and ENDOR measurements, simulation of the observed spectra, and theoretical calculations of hyperfine coupling constants is often applied to obtain a reliable assignment. The procedure to obtain the coupling constants from the ENDOR spectrum of the Wurster blue cation is indicated in Exercise E3.3. ENDOR lines due to ^{1}H can be observed more easily than signals due to ^{14}N, a phenomenon that is quite typical in CW-ENDOR studies. The hyperfine couplings due to ^{14}N nuclei may therefore have to be deduced from simulation of the ESR spectrum.

3.2.4 Large Hyperfine Couplings

The analysis of hyperfine couplings in ESR spectra of liquids differs from the first order treatment in Sections 3.2.2, 3.2.3 and 3.2.4 when the high field approximation $a << g\mu_B B$ does not apply. One of the first applications was of radicals formed during radiolysis of liquid fluorocarbons [5] where second and third order perturbation theory had to be used in the analysis of the hyperfine structure of X-band ESR spectra of fluorinated methyl radicals. A difference compared to a "first order" spectrum is that the hyperfine lines are shifted towards lower field. The displacement is proportional to a^2/B and therefore needs to be taken into account when the hyperfine coupling is relatively large compared to the electron Zeeman energy. Due to this displacement the g-factor is not obtained at the centre of the spectrum, but at a higher field as discussed in Exercise E3.4. The separation of the hyperfine lines for nuclei with $I > \frac{1}{2}$ is not constant but increases towards high field.

Another difference compared to the first order spectrum occurs with large hyperfine couplings of a group of equivalent nuclei [5]. It is then also necessary to use the effective nuclear spin values, I_{eff}, for the group – an individual treatment of the hyperfine coupling for each nucleus does not give the correct result. As a result the Pascal triangle method does not apply even for equivalent $I = \frac{1}{2}$ nuclei when the high-field approximation does not strictly hold. A method that makes use of the addition of angular momentum – in the present case nuclear spins – to a resultant value is convenient to apply [5]. The procedure is exemplified in Exercise E3.6. The resulting values of I_{eff} and the number of times a specific value appears (N_I) are then used to obtain the line positions and intensities by 2nd and 3rd order formulae [5] reproduced in Chapter 5 (Section 5.2.1.1).

The line positions for each I_{eff} are not centred at the field B_0 corresponding to the g-factor, but are displaced as shown in the stickplot of Fig. 3.5 for a radical with a large hyperfine coupling of six equivalent $I = \frac{1}{2}$ nuclei (data are for $C_3F_6^-$ in Chapter 5) with effective nuclear spin values, $I_{\text{eff}} = 0, 1, 2$, and 3 which occur 5, 9, 5 and 1 times. The $2 I_{eff} +1$ hyperfine lines for a specific value of $I_{eff} > \frac{1}{2}$ are not equally spaced but the line splittings increase towards higher fields. The

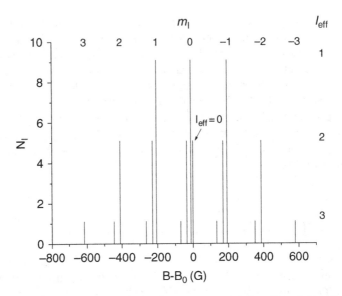

Fig. 3.5 Stickplot for a radical with hyperfine coupling = 199 G at X-band (data for $C_3F_6^-$ in Chapter 5). Effective nuclear spin values, $I_{eff} = 0, 1, 2,$ and 3 occur 5, 9, 5 and 1 times (= N_I values). B_0 is the field corresponding to the g-factor

most accurate value of the hyperfine coupling constant is obtained as the splitting between the lines labelled with the nuclear quantum numbers m_I and $-m_I$, $a = 1/2|B(m_I) - B(-m_I)|/m_I$. The second order effects then cancel out.

The stickplot can be obtained by manual calculation, but is also conveniently computed with a numeric procedure to obtain the effective spin values I_{eff} and their multiplicities N_I in a computer algorithm described in Section 3.2.7. The procedure to obtain the g-factor when 2nd order effects occur is described in Section 3.2.7, see also Exercise E3.4. Examples of spectra showing second order effects are given in Chapter 5 for perfluoro-cycloalkane radical anions, with several equivalent ^{19}F nuclei with large hyperfine couplings.

The reason for the failure of the first order analysis for these species is that the hyperfine coupling constant is a relatively large fraction, ca 5% of the energy separation between the electronic $m_S = \pm\frac{1}{2}$ levels. One reason to use high microwave frequencies and corresponding high fields is that the analysis becomes simpler – the first order analysis can be applied to measure larger hyperfine couplings than is possible at X-band. This is shown schematically in Fig. 3.6 for X-band ($\nu \approx 9.5$ GHz) and W-band ($\nu \approx 95$ GHz) spectra of the c-$C_4F_8^-$ anion radical discussed in Chapter 5.

Not all ESR laboratories can afford the price and running costs of a high field ESR spectrometer equipped with a superconducting magnet. Applications like those considered in Chapter 5 are also most conveniently performed at X-band. Thus the need for analysis beyond first order remains.

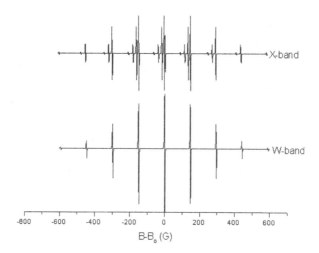

Fig. 3.6 Schematic X- and W-band ESR spectra for eight equivalent ^{19}F ($I = \frac{1}{2}$) nuclei with a_{iso} = 148 G for the c-$C_4F_8^-$ anion radical discussed in Chapter 5

A troublesome complication is caused by the overlap of spectra due to different paramagnetic species considered in the next section. Spectrum simulations described in Section 3.2.7 are often applied in this case to refine the hyperfine coupling constants.

3.2.5 Multi-Component Spectra

It is usually quite difficult to control the experimental conditions such that only one paramagnetic species is present. ESR-methods that are capable of discriminating between different species were discussed in Chapter 2. At present these methods have lower sensitivity than CW-ESR. An analysis of ESR-spectra containing several radicals illustrates the procedure when only ESR data are available.

When coronene, $C_{24}H_{12}$, (C) is treated with thallium(III) trifluoroacetate as oxidant under dried nitrogen atmosphere in 1,1,1,3,3,3-hexafluoropropane-2-ol solvent coronene radical species are formed. Depending on the sample preparation the ESR spectra are dominated by the monomeric ion $(C)^+$, or are superpositions of spectra due to $(C)^+$ and the dimeric ion $(C_2)^+$ or even the trimeric ion $(C_3)^+$. A stick-plot analysis of the spectrum of the nearly pure monomer is part of Exercise E3.5. It is also possible to accurately measure the hyperfine coupling constant (0.0766 mT for 24 H) and the corresponding peak-peak line-width ($\Delta H_{pp} = 0.009$ mT) for the $(C_2)^+$ species in Fig. 3.7(b), since the lines are quite sharp, compared to those from the monomer. The line shape of $(C_2)^+$ can therefore be obtained by simulation, that of $(C)^+$ from experiment The component spectra are then superimposed in a ratio appropriate for the experimental spectrum. This ratio is usually obtained by manual adjustment for best visual agreement with the experimental line shape; in the present

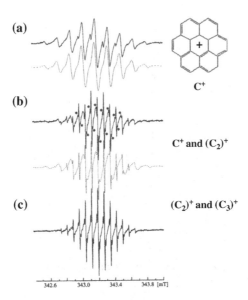

Fig. 3.7 ESR spectra observed at room temperature for 1,1,1,3,3,3-hexafluoropropen-2-ol (HFP) solution containing ca 5 mmol dm^{-3} coronene (C) and thallium trifluoroacetate (TlIII). (**a**) Spectrum attributed to the monomer radical cation, $(C)^+$ from a solution containing ~10 mmol dm^{-3} TlIII; Simulated spectrum calculated with an isotropic ^{1}H-hfs of 0.156 mT (12 H) and a peak-peak line width (ΔB_{pp}) of 0.035 mT; (**b**) Spectrum attributed to a mixture of $(C)^+$ and the dimeric radical, $(C_2)^+$, from a solution containing ~2 mmol dm^{-3} TlIII; Simulated spectrum calculated with the concentration ratio $[(C)^+] : [(C)_2^+] = 20 : 1$. The line shape of $(C_2)^+$ was calculated with an isotropic ^{1}H-hfs of 0.0766 mT (12 H) and ΔB_{pp} of 0.009 mT. The *satellite lines* denoted by an *asterisk*, flanking the sharp $(C_2)^+$ features, are attributed to the trimeric cation, $(C_3)^+$; (**c**) Spectrum attributed to a mixture of $(C_2)^+$ and $(C_3)^+$ formed in a solution containing $\leq$ 1 mmol dm^{-3} TlIII. The figure is reproduced from [K. Komaguchi et al.: Spectrochim. Acta A **63**, 76 (2006)] with permission from Elsevier

example the concentration ratio $[C^+] : [C_2^+] = 20 : 1$ is appropriate. Spread-sheet programs like Excel and dedicated software for graphic manipulations are often employed for the adjustment, but the superposition of spectra can also be carried out automatically using built-in commands for least-squares fit in commercial programs like Matlab.

3.2.6 Software

Simulation is often the final stage of analysis after an assignment by stickplots has been made. Stickplot analyses are usually sufficient for ESR spectra with well resolved hyperfine structure due to a few magnetic nuclei as e.g. in the ESR spectrum due to the galvinoxyl radical in Fig. 3.3. But when one of the tert-butyl groups in each ring is replaced by a methyl group, the spectrum becomes more complex. Although a stick plot analysis can give preliminary values of the coupling constants, it is advisable to refine the values by simulation in this and more complex cases. Software specially designed for ESR simulation in the liquid state is rarely published since the programming is straight-forward. Published codes primarily intended for wider applications can, however, be employed. Publically available electron spin resonance software tools from National Institute of Environmental Health Sciences, NIEHS are described at http://www.niehs.nih.gov/research/resources/software/tools/index.cfm. Simulation programs can also be downloaded from the International EPR (ESR) Society at http://www.epr-newsletter.ethz.ch/links.html. A limited number of computer programs from other sources are presented in Table 3.3, and are described below.

Table 3.3 Software for simulation of isotropic ESR and ENDOR spectra ($S = \frac{1}{2}$)

Program	Order	Available from	Code
EasySpin	Exact	http://www.easyspin.org/	MatLab
HRESOL	1–2	http://www.esr-spectsim-softw.fr/programs.htm	APL
Simfonia	1–3	http://www.bruker-biospin.de/EPR/software/emx.html	Exe
Iso	1–2	This work	Fortran77, Exe
Spin	1–2	This work	QB45 (Basic), Exe
ENDOR		[2]	

EasySpin: The tools for isotropic CW-ESR in EasySpin apply to $S = \frac{1}{2}$ species with arbitrary number of nuclei. Resonance fields are calculated exactly (no perturbation formulae). The magnetic field range is automatically determined. A least-squares fitting to an experimental spectrum can be made. The program can be downloaded at http://www.easyspin.org/. MatLab must be installed on the computer and is not provided with EasySpin. *EasySpin* is written and maintained by Dr. S. Stoll at the University of California, Davis.

HRESOL: High resolution spectra are calculated of radicals in fluid isotropic media with hyperfine coupling of several nuclei of any spin, taking into account the second order shift of resonance lines and the dependence of the line widths on the magnetic quantum number m_I of the nucleus showing the largest hyperfine coupling. The program is written by Dr. C. Chachaty and is available on request from the author who ensures the scientific and technical support as well as the adaptation of programs to new applications. The program is written in the APL language. The APL software itself cannot be provided but must be obtained commercially. Details are given at http://www.esr-spectsim-softw.fr/programs.

SimFonia: Spectra of radicals in fluid solution are calculated using up to third order perturbation theory for isotropic hyperfine couplings. Parameters for the simulation are given in an easy-to-use graphical interface. Instrumental effects such as modulation amplitude and receiver time constant can be included to account for line shape distortions or line broadening. Spectra from SimFonia can be transferred to WinEPR software for editing and output. The software is a commercial product but one version is available free of charge at http://www.bruker-biospin.de/EPR/software/emx.html.

Iso: For the isotropic simulations in this work a home-made code was employed. The hyperfine structure is calculated to second order employing the method of effective nuclear spins for groups of equivalent nuclei. Hyperfine structure due to arbitrary values of nuclear spin can be treated. The magnetic field range is automatically determined. Lorentz, Gauss, and Voigt line-shapes can be employed. The program was developed in two versions in Fortran77 and QB45 code. No additional software is required to run the executable codes.

Spin: The program calculates the relative intensities of the hyperfine lines due to equivalent nuclei with arbitrary spin. Effective nuclear spins are employed using the vector model for adding spins, $\mathbf{I}_{eff} = \mathbf{I}_1 + \mathbf{I}_2$ with $|I_1 - I_2| \leq I_{eff} \leq I_1 + I_2$. The procedure is equivalent to the Pascal triangle method when $I = \frac{1}{2}$. The intensity ratio can in this case also be calculated by the general formula: $R_i = \frac{N!}{(i-1)!(N-i+1)!}$. N is the number of equivalent protons or other $I = \frac{1}{2}$ nuclei, giving rise to totally N + 1 lines, with intensity R_i of line i. A procedure based on 2nd order perturbation theory is applied to obtain the line positions for each I_{eff} when the high-field approximation, $a << g\mu_B B$, does not strictly apply. The input is produced interactively. An example of an output is shown graphically in Fig. 3.5. No additional software is required to run the executable code.

The fitting by visual inspection is still not an objective method, and attempts have therefore been made even at an early stage to make the final refinement by automatically optimising the parameters, *i.e.* the coupling constants and the line width by a least squares fitting of the experimental spectrum to a computed one.

ESRCON: The program ESRCON optimizes approximated or guessed ESR parameters such as coupling constants, scaling factors, and line widths, using the method of least squares. From the computed "best" parameters, an absorption curve is constructed and then compared directly with the experimental spectrum. In addition to various other constants, a digitalized spectrum, which may be obtained either "by hand" or electronically, is required as part of the input data. The program written by Heinzer was the first to employ least-squares fitting methods to refine the values of the hyperfine coupling constants of an experimental spectrum. The original program and documentation [6] may no longer be available, but a slightly upgraded version has been used in the preparation of this book.

EPRFT, HFFIT, HFFITS: Three computer programs for the simulation and iterative least-squares fitting of high-resolution ESR spectra have been described. Program EPRFT assumes constant linewidths, programs HFFIT and HFFITS allow for linewidth variations. Program EPRFT offers three minimization techniques: evolutionary Monte Carlo, simplex and Marquardt method. Schemes for an automated analysis of EPR(ESR) spectra are investigated, using a combination of a search technique and iterative least-squares fitting. Details can be found in the literature [7].

Optimised values of hyperfine coupling constants can also be obtained with the publically available EasySpin and HRESOL suite of programs referred to in Table 3.3.

3.2.7 The g-Factor

A majority of studies in liquid solution concerns spectra of radicals with g-factors that deviate only slightly from the free-electron value of 2.00232. Measurement of the g-factor to five decimal places or more is often achieved in high resolution ESR spectra. The measurement method depends on the system and the available experimental equipment.

Direct measurement of magnetic field and microwave frequency: An absolute measurement is possible by simultaneously measuring the microwave frequency and the magnetic field at the centre of the ESR signal using the equation $g = 0.0714477 \cdot \nu_e(\text{MHz})/B_0(\text{mT})$, where B_0 is the field at the centre of the signal and ν is the microwave frequency. An accurate microwave frequency meter is often standard equipment of commercial instruments, while the calibration of the magnetic field has to be checked by a field meter usually based on NMR.

Example: Estimate the uncertainty in g of a single-line spectrum in solution assuming that the centre can be located to within 0.1 of a line-width of 0.01 mT at X-band. The calculated uncertainty $\left|\frac{\Delta g}{g}\right| = \left|\frac{\Delta B}{B_0}\right| = \frac{0.1 \times 0.01}{350} = 3 \cdot 10^{-6}$ at $B_0 \approx$ 350 mT (X-band) suggests that the g-factor can be measured to five decimal places ($g \approx 2.00$). The error in the microwave frequency is neglected.

Comparison method using a reference sample: In this method the g-factor is obtained by simultaneously measuring the centre of a reference sample with a known g-factor, g_{ref} and that of the sample separated in field by ΔB from B_{ref} as shown in Fig. 3.8.

Fig. 3.8 g-factor by *comparison method*. The g-factor is obtained as $g = g_{\text{ref}}(1 + \Delta B/B_{\text{ref}})$ by simultaneously measuring the centre of a reference sample with a known g-factor, g_{ref} and the sample. The reference may also be used for magnetic field calibration, using known hyperfine splittings e.g. between the three ^{14}N hyperfine lines of Fremy's salt in aqueous solution [8] shown in the figure

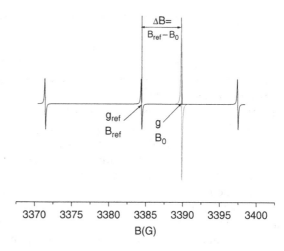

To obtain high accuracy the g and g_{ref} factors should be close to each other, see Exercise E3.7. A table of reference samples with accurately known g-factors is given in [4]. The di-anion radical of Fremy's salt in aqueous solution [8] is also suitable for magnetic field calibration, by the ^{14}N hyperfine coupling constant

$A_0(t) = 12.978 + 0.00311 \cdot t$, when the temperature, t, is given in °C. Samples like Mn:MgO powder are also used as a combined g-factor and field calibration sample, especially for high field applications.

Multiple-line spectra: To first order the g-factor of a spectrum with several lines can be calculated by assuming that B_0 is at the center of the spectrum. This first order analysis is valid when the hyperfine coupling a is small compared to B_0; in the X-band spectrum in Fig. 3.1 of the methyl radical one has $2.3 = a \ll B_0 = 339$ mT. For the X-band spectrum of the hydrogen atom in Exercise E3.4 the g-factor is not exactly at the center of the spectrum The displacement of the "1st order" calculated spectrum compared to the experimental one is due to a shift in the line positions that needs to be taken into account when the hyperfine splitting is relatively large compared to the electron Zeeman energy, *i.e.* when the high field approximation $a \ll g\mu_B B$ does not apply, see Exercise E3.4 for an application.

3.3 Analysis of Single Crystal Spectra

Paramagnetic species trapped in solid materials usually possess anisotropic g- and hyperfine couplings. Zero-field splittings occur when $S > \frac{1}{2}$. The spin Hamiltonian formalism described in Appendix A3.1 is a convenient means to summarise the different interactions. The following spin-Hamiltonian is adequate to illustrate most aspects of the analysis.

$$H = \mu_B \mathbf{B} \cdot \mathbf{g} \cdot \mathbf{S} + \mathbf{S} \cdot \mathbf{D} \cdot \mathbf{S} + \mathbf{S} \cdot \mathbf{A} \cdot \mathbf{I} - \mu_N g_N \mathbf{B} \cdot \mathbf{I} + \mathbf{I} \cdot \mathbf{Q} \cdot \mathbf{I}$$

The first three (electron Zeeman, zero-field and hyperfine coupling) terms are usually the most relevant for the ESR analysis. The two last (nuclear Zeeman and quadrupole) terms do not affect the ESR spectra, unless they are of comparable magnitude to the hyperfine coupling. They always affect the ENDOR and ESEEM spectra, however. $\mathbf{D}$, $\mathbf{A}$ and $\mathbf{g}$ are assumed to be symmetric tensors. Each tensor is specified by three principal values giving the strength and three principal directions giving the orientation of the coupling. They can be thought of as the lengths and the directions of the axes of an ellipsoid, see Appendix A3.2. The goal of the single crystal analysis is to obtain these values and directions. The directions are specified in a laboratory system, most commonly the crystal axes system. The principal axes do not in general coincide with the crystal axes. The tensors are represented as $3 \cdot 3$ symmetric matrices in this system. The procedures to obtain those data by single crystal measurements and to obtain the principal values and directions are similar for all tensors. ESR is the appropriate method for determining the principal values and directions of the g and (for $S > \frac{1}{2}$) $\mathbf{D}$ tensors, and also for the $\mathbf{A}$ tensor when the resolution of the hyperfine structure is good, as e.g. for transition metal ions. ENDOR and ESEEM methods are suitable for smaller hyperfine coupling tensors and also for the determination of the nuclear quadrupole tensor of nuclei with $I > \frac{1}{2}$.

3.3.1 The Schonland Method

The Schonland method [9] originally developed for the determination of the principal *g*-values in ESR is applicable also for the determination of **A** and **D** tensors under the condition that the corresponding couplings in field units are much less than the applied magnetic field or, equivalently, that the couplings in MHz are much smaller than the microwave frequency. The Schonland method can therefore be applied to larger zero-field and hyperfine couplings at the high frequency bands (Q,W...) than in X-band. The spectra are then referred to as "first order" , implying that the analysis is based on first order perturbation theory. The Schonland method can also be modified to obtain hyperfine and nuclear quadrupole tensors from ENDOR and ESEEM measurements and is consequently the basis for nearly all analyses of single crystal data by ESR techniques.

3.3.2 The g-Tensor

In the original Schonland procedure [9] the *g*-factor was measured by ESR as a function of the angle of rotation of the crystal with respect to the magnetic field in three different mutually orthogonal planes xy, yz and zx. Data are frequently collected over a full period of 180° but are often summarised in diagrams of the type in Fig. 3.9 to more clearly display the match of the *g*-factor along the <X>, <Y>, and <Z> axes.

It is customary to use the same symbol for the measured *g*-factor and the tensor **g**. For the rotation in the xy-plane one has:

$$
\begin{aligned}
g^2 &= (\mathbf{g}^2)_{xx}\cos^2\theta + (\mathbf{g}^2)_{yy}\sin^2\theta + 2(\mathbf{g}^2)_{xy}\sin\theta\cos\theta \\
&\equiv \frac{(\mathbf{g}^2)_{xx} + (\mathbf{g}^2)_{yy}}{2} + \frac{(\mathbf{g}^2)_{xx} - (\mathbf{g}^2)_{yy}}{2}\cos 2\theta + (\mathbf{g}^2)_{xy}\sin 2\theta
\end{aligned}
\tag{3.1}
$$

where θ is angle of the magnetic field with the x-axis. The expressions for the orientation dependence of g^2 in the yz and zx planes are obtained by cyclic permutations of the subscripts. The second form was employed in Schonland's original treatment to obtain a fitting of the equations to the data for each plane. The elements of the g^2 tensor obtained from the fit were then employed to calculate the principal values and directions by diagonalization. The principal *g*-values were finally obtained as the square root of the corresponding g^2 values. The principal directions for $\mathbf{g}^2$ and $\mathbf{g}$ coincide.

The procedure can give rise to an ambiguity in the determination of the *g*-tensor when the rotation is done in opposite senses between the different planes. An ambiguity of this type occurs for instance when different crystals are used for the measurements in the different planes. It is then difficult to experimentally determine the sense of rotation, a crystal may for instance have been mounted "upside

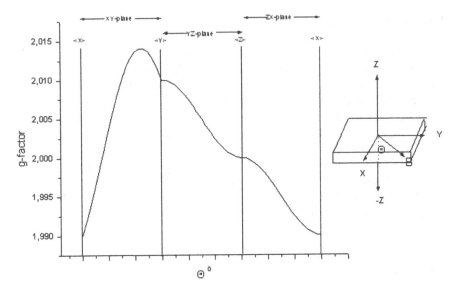

Fig. 3.9 Schematic variation of the *g*-factor with the magnetic field in the xy-, yz-, and zx-planes in the Schonland method to determine the *g*-tensor. The variation is shown over a 90° interval for the angle Θ specifying the orientation of the magnetic field in each plane. For a perfect fit the measured *g*-factor should agree along the <X>, <Y>, and <Z> axes in the different planes. A sign error for $(g^2)_{xy}$ occurs if the crystal is unintentionally mounted for rotation about −z in place of +z, see text for methods to resolve this "Schonland ambiguity"

down", implying wrong sense for angle θ specifying the direction of the magnetic field, see Exercise E3.8.

The ambiguity gives rise to just two different tensors and can be resolved by measurements in a fourth "skew" plane [9]. As a rule only one of the two tensors reproduces the experimentally determined *g*-factor variation in the skew plane. A simpler procedure is applicable when the principal values can be read from the experimental powder spectrum. Only one of the two sets of principal values obtained from the analysis usually matches the experimental data.

3.3.3 Sites

A common complication in ESR and ENDOR single crystal studies is the occurrence of different spectra due to identical paramagnetic species that are differently oriented due to crystal symmetry. The species are said to be located in different sites, and the overlap of spectra due to different orientations is referred to as site splitting. In the triclinic crystal system, e.g. for the malonic acid discussed in Chapter 2, there is only one molecule in the unit cell and no site-splitting occurs. Two symmetry-related sites with coordinates (x, y, z) and (−x, y, −z) give identical ESR spectra when the crystal is rotated about the y axis. The result is plausible since the magnetic field is confined to the xz-plane, but can also be formally calculated, see Exercise E2.7. Site-splittings are expected in the xy- and yz-planes. This situation is characteristic for crystals of monoclinic symmetry. The different possibilities for all types of crystal symmetry have been analyzed by Weil et al. [4, 10]. Only eleven different cases can occur, depending on the point group symmetry of the crystal. Some of the computer programs described below for the calculation of g- and other tensors by the Schonland method accept input data for several sites (Fig. 3.10). A method that uses the coordinates of the sites is employed e.g. in [11] and is exemplified in Exercise E3.12.

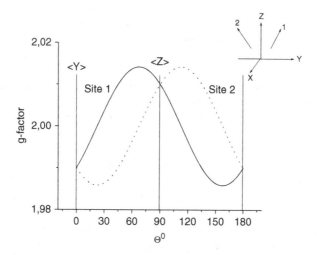

Fig. 3.10 g-factor variation in a plane showing site-splitting due to a single species trapped in two equivalent but differently oriented positions, "sites" in the crystal

The Schonland procedure is also applicable for the determination of hyperfine coupling tensors by ESR and ENDOR, for the zero-field splitting tensor $(S > \frac{1}{2})$ by ESR, and the nuclear quadrupole couplings for $I > \frac{1}{2}$ by ENDOR and ESEEM, discussed in the following sections.

3.3.4 The Hyperfine and Nuclear Quadrupole Coupling Tensors

The Schonland procedure to obtain the hyperfine coupling tensor differs in some details depending on the magnitude of the coupling and on the applied methods, usually ESR or ENDOR.

(1) *ESR measurements of hyperfine coupling much larger than the nuclear Zeeman energy*: The conditions usually apply for the hyperfine structure due to transition metal ions. The g-factor of such systems is usually also anisotropic, leading to a modification of the Schonland procedure. For the xy-plane one has:

$$K^2 g^2 = (\mathbf{g}\mathbf{A}^2\mathbf{g})_{xx} \cos^2\theta + (\mathbf{g}\mathbf{A}^2\mathbf{g})_{yy} \sin^2\theta + 2(\mathbf{g}\mathbf{A}^2\mathbf{g})_{xy} \sin\theta \cos\theta \qquad (3.2)$$

Formulae for the yz-, and zx-planes are obtained as in Section 3.3.2 by cyclic permutations of the x, y, and z indices. Thus, to obtain the tensor $\mathbf{g}\mathbf{A}^2\mathbf{g}$ the separation K between adjacent hyperfine lines and the g-factor are both measured at each angle θ in the three planes. The $\mathbf{A}^2$ tensor is then obtained by an algebraic procedure [12] using the g-tensor determined from the same set of measurements. The analysis is simpler for isotropic g, see Exercise E3.9.

This procedure (neglect of nuclear Zeeman term) is sometimes also adopted to obtain hyperfine coupling tensors from ESR measurements of free radicals. The method is, however, not suitable for the analysis of hyperfine structure due to α-H in π-electron radicals of the type $\rangle\dot{\mathbf{C}}_\alpha$-H at X-band, and other cases where the anisotropic hyperfine coupling and the nuclear Zeeman energy are of comparable magnitudes, as discussed for case 3 below.

(2) *ESR measurements of hyperfine coupling much smaller than the nuclear Zeeman energy*: This case with nuclear Zeeman energy >> hyperfine coupling is rare at X-band and lower frequencies, but is of interest for measurements at the higher frequency bands (Q,W..) which have become commercially available. The variation of the hyperfine coupling can be analyzed by a procedure discussed in Exercise E3.10.

(3) *ESR measurements of anisotropic hyperfine couplings comparable to the nuclear Zeeman energy:* When none of the two extreme cases 1 and 2 is applicable an ESR analysis is usually complicated. This case occurs frequently in X-band measurements on carbon-centred radicals with hyperfine structure from a hydrogen atom bonded to the carbon atom carrying the spin density, *i.e.* at the α-position. The reason for the difficulty is the breakdown in the $\Delta m_I = 0$ selection rule, resulting in four instead of two hyperfine lines appearing as an "inner" and an "outer" doublet for each ^{1}H nucleus. Procedures to analyse this case when the splittings of both the inner and the outer doublets can be measured at different directions of the magnetic field have been devised by Poole and Farach [13]. The ENDOR method discussed in the next section is a simpler and more generally applicable method for this case.

(4) *ENDOR measurements of hyperfine coupling tensors of $I = \frac{1}{2}$ nuclei:* In this case it is not necessary to make any assumption regarding the relative magnitudes of the nuclear Zeeman and hyperfine energies. For an $S = \frac{1}{2}$ species two lines with frequencies $\nu_\pm$ appear corresponding to the two values of the electron quantum number $m_S = \pm\frac{1}{2}$. For an isotropic g the variation of the ENDOR frequencies with the orientation of the magnetic field in the xy-plane is analogous to Eq. (3.1):

$$\nu_\pm^2 = (\mathbf{T}_\pm^2)_{xx} \cos^2\theta + (\mathbf{T}_\pm^2)_{yy} \sin^2\theta + 2(\mathbf{T}_\pm^2)_{xy} \sin\theta\cos\theta \qquad (3.3)$$

with the difference that the "ENDOR tensor" in frequency units with components $(\mathbf{T}_\pm)_{ij} = \left(\pm 1/2 A_{ij} - \delta_{ij} \cdot g_N\mu_N B\right)/h$ is used in place of the g-tensor.

The static magnetic field is normally constant or nearly so during the measurements. The tensor $\left(\mathbf{T}^\pm\right)^2$ can then be obtained by a Schonland type fit. After diagonalization the magnitudes of the principal values of the ENDOR tensor are obtained as $|t_\pm|_k = |\pm 1/2 A_k - \nu_N|$ where the nuclear frequency $\nu_N = g_N\mu_N B/h$ and the principal values in frequency units A_k, $k = 1$–3, have been introduced. For

many systems of practical interest the signs are known from theoretical considerations, for instance of α-H ($A_k < 0$) and β-H ($A_k > 0$) couplings in carbon-centred
π-electron radicals. In less obvious cases the ambiguity in evaluating the principal
values A_k can be reduced if both of the two ENDOR transitions corresponding to
$m_S = \pm\frac{1}{2}$ can be observed. In this case one has $|A_k| = \left|(t_+)_k^2 - (t_-)_k^2\right| / (2\nu_N)$.
The absolute signs are not obtained experimentally, since one cannot assign specific
values of the electrion spin quantum number m_S to the observed transitions.

(5) *ENDOR measurements of nuclear quadrupole coupling tensors of $I > \frac{1}{2}$
nuclei*: A slightly lengthier analysis is required in the presence of nuclear quadrupole
interactions, (*nqi*), from $I > \frac{1}{2}$ nuclei. When the hyperfine and nuclear Zeeman interactions dominate over that caused by the *nqi* a simple method involves separate
measurements of the parameters, G_+ and P_+ and/or G_- and P_- for $m_S = \pm\frac{1}{2}$ of a
free radical, as illustrated with the energy diagram in Fig. 3.11 for $I = 1$.

The magnitudes of $G_\pm$ and $P_\pm$ are obtained from the ENDOR frequencies ν_1 and
ν_2 or ν_3 and ν_4 (Fig. 3.11) as a function of orientation in three crystal planes. The
quantities can therefore be determined separately. When g is isotropic an analysis

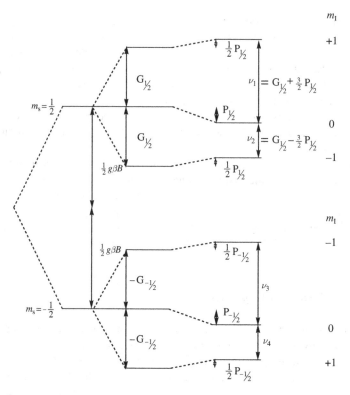

Fig. 3.11   Energy level diagram for a radical with $S = \frac{1}{2}$ and $I = 1$. Theoretical expressions for the
ENDOR frequencies ν_1 and ν_2 are given. The diagram is reproduced from [12] with permission
from Springer

according to the Schonland method can be made. Thus, the variation of $G_\pm$ makes it possible to obtain the hyperfine coupling tensor as for the $I = \frac{1}{2}$ case, employing $G_\pm$ in place of $v_\pm$ in Eq. (3.3). The expression (3.4) for the xy-, and analogous for the yz-, and zx-planes finally makes it possible to obtain the quadrupole tensor $\mathbf{Q}$:

$$P_\pm G_\pm^2 = (\mathbf{T}_\pm \mathbf{Q} \mathbf{T}_\pm)_{xx} \cos^2 \theta + (\mathbf{T}_\pm \mathbf{Q} \mathbf{T}_\pm)_{yy} \sin^2 \theta + 2(\mathbf{T}_\pm \mathbf{Q} \mathbf{T}_\pm)_{xy} \sin \theta \cos \theta \quad (3.4)$$

The ENDOR tensor $\mathbf{T}_\pm$ is in other words first obtained from the Schonland fit to the experimental $G_\pm^2$ data. Next the tensor $\mathbf{T}_Q = \mathbf{T}_\pm \mathbf{Q} \mathbf{T}_\pm$ is calculated from a similar Schonland fit to the experimental $G_\pm^2 \cdot P_\pm$ data. Finally $\mathbf{Q} = (\mathbf{T}_\pm)^{-1} \mathbf{T}_Q (\mathbf{T}_\pm)^{-1}$ is calculated and diagonalised to obtain the principal nuclear quadrupole coupling values and the directions of the principal axes.

When the g-factor is anisotropic this simple approach fails. A Schonland fitting procedure can be applied when the ENDOR lines for both $m_S = \frac{1}{2}$ and $-\frac{1}{2}$ are observable, while non-linear least squares methods are more generally applicable. The latter method can be adopted e.g. to simultaneously determine hyperfine and nuclear quadrupole tensors. Detailed procedures may be obtained from the literature [12, 14].

Similar treatments are applicable to the analysis of the lines in frequency-domain ESEEM spectra. This method is particularly suited to resolve weak couplings that frequently occur in paramagnetic complexes of biological interest. It may then happen that the quadrupole energy is of the same magnitude as the combined hyperfine and nuclear Zeeman energy, in which case the perturbation treatment leading to Eq. (3.4) cannot be applied. This case occurs frequently in paramagnetic complexes with weak ^{14}N hyperfine couplings. Complexes of this type can, however, usually be studied only in disordered systems, see Section 3.4 for treatments employed to analyse powder spectra.

(6) *Higher order and exact analysis*: Higher order perturbation theory can be applied to the case when the first order analysis is inadequate, e.g. when the hyperfine coupling is large. The second order corrections given in Appendix A3.2.2 are in a suitable form to be applied in single crystal analysis e.g. in ENDOR measurements of ^{14}N hyperfine and quadrupole couplings. A summary of this and other procedures, suggested in the literature, is given in our previous work [12].

3.3.5 The Zero-Field Splittng

Zero-field splittings occur in ESR spectra of paramagnetic species with electron spin $S > \frac{1}{2}$. The energy for the transition $(m_S \leftrightarrow m_S - 1)$ is $\Delta E = g\mu_B B + \frac{3}{2} F \cdot (m_S - 1/2)$ provided that the electron Zeeman term is the dominant one, $g\mu_B B >> F$. The ESR spectrum then contains $2 \cdot S$ lines, separated by $3 \cdot F$. With the field in the xy-plane one has:

$$Fg^2 = (\mathbf{g} \mathbf{D} \mathbf{g})_{xx} \cos^2 \theta + (\mathbf{g} \mathbf{D} \mathbf{g})_{yy} \sin^2 \theta + 2(\mathbf{g} \mathbf{D} \mathbf{g})_{xy} \sin \theta \cos \theta \qquad (3.5)$$

The tensor **gDg** can accordingly be obtained by measuring the separation $3 \cdot F$ between adjacent zero-field lines and the g-factor at each angle θ in this and two additional planes, e.g. yz and zx. The principal values and the principal axes of the D-tensor are calculated by the same algebraic procedure as for the Q-tensor in the analysis of ENDOR single crystal data of systems with $I \geq 1$ described above and in [15] employing a slightly different notation.

The analysis is simpler when the g-anisotropy is small, as for organic triplet state molecules. The zero-field splitting obtained from the separation between the two zero-field lines corresponding to $m_S = 1 \leftrightarrow 0$ and $m_S = 0 \leftrightarrow -1$ can then be analysed by the Schonland method, using the equations:

$$F = D_{yy} \cos^2 \theta + D_{zz} \sin^2 \theta + 2D_{yz} \sin \theta \cos \theta$$

$$F = D_{zz} \cos^2 \theta + D_{xx} \sin^2 \theta + 2D_{xz} \sin \theta \cos \theta$$

$$F = D_{xx} \cos^2 \theta + D_{yy} \sin^2 \theta + 2D_{xy} \sin \theta \cos \theta$$

in the yz-, zx- and xy-planes, respectively.

The data in Fig. 3.12, slightly modified from those in the literature for triplet state trimethylene methane, show a variation from positive to negative values of the zero-field splitting, in contrast to the previously discussed cases, where normally the magnitude, but not the sign of the splitting is determined. Due to this behaviour the relative (but not the absolute) signs of the principal values can accordingly be obtained by single crystal measurements of the zero-field splitting.

Theoretically the sum of the principal values should be zero, $D_X + D_Y + D_Z = 0$, i.e. the zero-field splitting has no isotropic component. The usual zero-field splitting

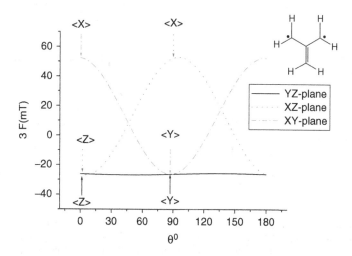

Fig. 3.12 Schematic angular variations of the line separation $3 \cdot F$ due to zero-field splitting in the yz, xz, and xy planes of a single crystal. The curves were reconstructed from literature data for triplet state trimethylene methane in methylene cyclopropane with zero-field splitting parameters, $D = 0.0248$ cm^{-1} and $|E| \leq 0.0003$ cm^{-1} [O. Claesson et al.: J. Chem. Phys. **72**, 1463 (1980)]

parameters are therefore given by: $D = (3/2)\,D_Z$, $E = (D_X - D_Y)\,/2$, see Exercise E3.11. The variation of the zero-field splitting from a positive to negative values in Fig. 3.12 might give the impression that the relative signs of D and E can be determined. But since the enumeration of the principal values is arbitrary the sign of D/E depends on how the principal axes X, Y, Z are specified. By convention D_Z is assigned to the numerically largest principal value.

3.3.6 Combined Zero-Field and Hyperfine Couplings

As first pointed out by Minakata and Iwasaki [16] an asymmetry in the hyperfine pattern between the $m_S = 1 \leftrightarrow 0$ and $m_S = 0 \leftrightarrow -1$ electronic transitions of a triplet state can be attributed to the influence of the nuclear Zeeman term. This makes it possible in principle to determine the relative signs of the hyperfine coupling and the zero-field splitting. An application to trimethylene methane is shown in Fig. 3.13.

The asymmetry of the ESR spectrum in Fig. 3.13 arises because of the different values for the nuclear frequencies v_1 and v_{-1}, while v_H is approximately equal to the ^{1}H nuclear frequency, or ca. 14 MHz. The hyperfine splitting in frequency units

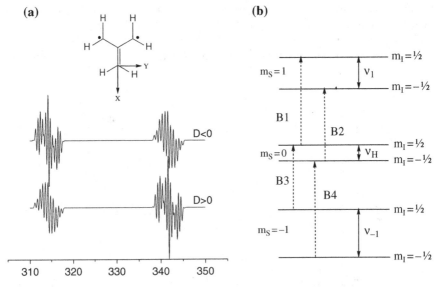

Fig. 3.13 (a) Schematic single crystal X-band spectra for triplet state trimethylenemethane in a rigid crystal matrix. The magnetic field is in the XY-plane at an angle of 13° with the X-axis. The experimental spectrum [O. Claesson et al.: J. Chem. Phys. **72**, 1463 (1980)] corresponds to the $D > 0$ case for the zero-field splitting. (b) Simplified triplet state energy diagram with hyperfine lines (*dashed*) due to a single ^{1}H ($I = \frac{1}{2}$) nucleus with non-negligible nuclear Zeeman energy. The hyperfine couplings for the transitions between the $m_S = 1 \rightarrow 0$ and $m_S = 0 \rightarrow -1$ levels differ because of the different values for the separations v_1 and v_{-1}. v_H is equal to the splitting of the free nucleus, ca. 14 MHz for ^{1}H at 330 mT

between the ESR lines B1 and B2 is $|\nu_1 - \nu_H|/2$ that between B3 and B4 $|\nu_{-1} - \nu_H|/2$, where for simplicity only the $\Delta m_I = 0$ ESR transitions are indicated in the diagram (b). The ^{1}H nuclear frequency is, however, of the same magnitude as the principal values of the hyperfine couplings, determined as $A_x = -14.0$, $A_y = -37.8$ and $A_z = -26.3$ MHz, implying that the spectra in (a) are also affected by $\Delta m_I = 1$ transitions. It is quite difficult to visually interpret spectra of the type shown in Fig. 3.13, and analysis by simulation is usually required. The shape of such spectra depends strongly on the microwave frequency due to the difference in the effective fields acting on the nuclei.

3.3.7 Software for Single Crystal Analysis

Software for this purpose is often developed in laboratories specialized in single crystal measurements, but the programs are not always available. The computer programs listed in Table 3.4 that the authors are aware of contain several steps, (1) to provide the experimental data (g-factors, hyperfine couplings, zero-field splittings, ENDOR or ESEEM frequencies) as function of crystal orientation, (2) to perform a least squares fitting of a theoretical model to the data, and (3) to make an error analysis of the parameter values, *i.e.* the principal values and direction cosines of the principal axes of the coupling tensors.

Table 3.4 Software for single crystal analysis

Program	Order	Available from	Code
MAGRES	1	(a)	Fortran77
ENDPAQ	1	(b)	Fortran77
TENSOR		(c)	
X32	1–2	This work	QB45
AQ_Fit	1	This work	Fortran77

(a) W.H. Nelson: J. Magn. Reson. **38**, 71 (1980).
(b) A. Sornes: Ph.D Thesis Oslo University (1994).
(c) A. Ponti, C. Oliva: Computers & Chemistry, **16**, 233 (1992).

The last step is model independent and similar code can be applied in all cases. The first step usually requires only minor modification depending on the type of data. We refer to the original literature (Table 3.4) for the technical procedures applied in the second step, but consider below some practices that can help to improve the accuracy of the analysis.

(1) The angle Θ specifying the magnetic field direction in the Schonland method is usually measured relative to the crystal axes (denoted by <X>, <Y>, and <Z> in Fig. 3.9). These positions must therefore be accurately determined. A satisfactory positioning is indicated in the Schonland method if the spectra obtained with the magnetic field along one of these axes are identical between two planes.

An exact match may be difficult to achieve, experimentally, however. Some programs therefore have a feature for automatic adjustment of the positions of the axes. In crystals of symmetry higher than triclinic, spectra due to different sites that become identical along the crystal axes makes it easier to accurately determine the obtain axes as shown in Fig. 3.10. Another difficulty, related to the Schonland ambiguity then appears, however, as discussed next.

(2) Some programs have a feature to reverse the sense of rotation. This causes a change in the sign of the angle Θ which is equivalent to a sign change e.g. of $(\mathbf{g}^2)_{xy} \rightarrow -(\mathbf{g}^2)_{xy}$ in Eq. (3.1). The feature is introduced to compensate for a wrong sense of rotation made either accidentally in case of low symmetry (no sites), or by combining curves due to different rather than the same site in cases of high symmetry. The latter ambiguity is difficult to resolve experimentally from measurements in three planes, and alternative tensors obtained by changing rotation senses are therefore usually tested against measurements in a fourth plane [9], or by measurement and simulation of the powder spectrum. This "Schonland ambiguity" first observed for the g-tensor, applies equally for other anisotropic couplings, determined by ESR. Surprisingly, the ambiguity can be resolved for hyperfine couplings of free radicals determined by ENDOR if the two lines corresponding to $m_S = \pm\frac{1}{2}$ can be observed [17].

(3) Computer programs available today make a simultaneous fit to all data [11]. It is then not necessary to employ orthogonal planes for the measurements. Moreover, error estimates of the principal values and directions are provided.

MAGRES, Schonland type analysis of the g and hyperfine tensors obtained by EPR (ESR) and ENDOR. The experimental data, e.g. the g-value or the hyperfine coupling from ESR, or the ENDOR frequencies are given as function of the rotation angle θ in one or several crystal planes. The planes need not be mutually orthogonal. A wrong sense of rotation that may occur experimentally (the Schonland ambiguity) can be analysed by reversing the sign of θ. It is possible to mix experimental data obtained at different microwave frequencies, and in ENDOR to include data for both the $m_S = \frac{1}{2}$ and $-\frac{1}{2}$ branches. The latter feature can be employed to resolve the Schonland ambiguity [17] in ENDOR studies. Data for equivalent sites, *i.e.* with the same principal values and symmetry related directions of the principal axes can be included in the fit. In this case the symmetry transformation appropriate for the crystal system must be specified.

The MAGRES program provides error estimates of the principal values and direction cosines and is the most extensively used and documented of the programs in Table 3.4. One advantage with this and similar programs based on the Schonland method is the simple input data making them particularly suited for analysis of g and hyperfine data. The program is written and maintained by Prof. W.H. Nelson, Dept. Physics and Astronomy, Georgia State University, Atlanta [11]. The nuclear quadrupole tensor of $I > \frac{1}{2}$ nuclei can, in principle be obtained by this software using the stepwise procedure described in Section 3.3.4, while the programs ENDPAQ and/or AQFit described below are directly applicable in this case.

ENDPAQ, Simultaneous least squares fitting of hyperfine and quadrupole tensors to experimental ENDOR data. ENDOR data of radicals containing $I > \frac{1}{2}$ nuclei are analysed to simultaneously obtain the hyperfine (**A**) and nuclear quadrupole (**Q**) coupling. **Q** may be confined traceless as a constraint. The elements of the the **A** and **Q** tensors are refined iteratively starting from an initial guess of parameters. Experimental data consist of crystal orientation, ENDOR frequencies, and the magnetic field strength in at least three planes. In addition the quantum numbers for the observed ENDOR lines must be assigned, *i.e.* the quantum numbers m_S ($\pm\frac{1}{2}$) and m_I for the electron and nuclear spins for a transition $|m_S, m_I > \leftrightarrow |m_S, m_I - 1>$ are given. Data input files may be constructed interactively while running the program, following directions given by the program. Data input in free format is accepted.

All observed transitions for different m_S and m_I quantum numbers can be included simultaneously in the fit. This feature helps to avoid the ambiguity that may occur in the analysis with the Schonland procedure discussed above and in Section 3.2. Error estimates of the parameters are computed. The program also simulates transition frequencies for the rotation planes. The simulation results are automatically stored in files of two columns, field orientation (deg.) and transition frequency (MHz). Both rotation senses ("sites" for orthorombic a,b,c and monoclinic crystals) are simulated. The program was developed by A. Sørnes, and can be obtained at [http://www.fys.uio.no/biofysikk/eee/esr.htm].

AQFIT, fit to ENDOR and ESEEM single crystal data for $S = \frac{1}{2}$ species with hyperfine and quadrupole tensors of comparable magnitudes. The hyperfine-, the nuclear Zeeman-, and the nuclear quadrupole- interactions are treated as a joint perturbation on the electronic Zeeman term. This allows analysis of experimental data for $S = \frac{1}{2}$ species featuring weak hyperfine coupling and an appreciable nuclear quadrupole coupling with $I > \frac{1}{2}$ nuclei, described in Section 3.4. Input and output files are modified from the ENDPAQ program. The observed lines are assigned to transitions between adjacent nuclear levels for a fixed electron spin quantum number m_S. The levels are mixed states of nuclear quantum numbers m_I, and are accordingly specified by integer numbers, (in ascending energy order) rather than by the m_I values employed in ENDPAQ. Identical tensors are computed with the AQFIT and ENDPAQ programs when the theoretical model for the latter is valid. This latter model fails when the hyperfine and nuclear Zeeman energies are comparable to the magnitude of the nuclear quadrupole interaction, in which case ordinary perturbation theory is invalid. This case often applies in the analysis of ^{14}N hyperfine couplings which may vary much in magnitude, while the principal values of the ^{14}N quadrupole tensor are typically of the order of 1 MHz, comparable to the nuclear Zeeman energy at X-band. Programs that overcome this limitation have been developed previously [18], but the software may not be available.

TENSOR. A program to determine hyperfine tensors from ESR or ENDOR single crystal data is described. The variation of hyperfine interaction upon rotating the crystal with respect to the static field is studied by a least-squares algorithm which affords various weighting methods. These take into account the fact that the model equations are not linear in the variables, so that equal measurement errors affect the calculation to different extents. An iterative procedure is available which accounts for experimental uncertainty in both dependent and independent variables. Fitting parameters are then cast into a matrix form of tensors. These are diagonalized by a Householder transform followed by application of Sturm bisection algorithm. This procedure is very accurate and matrix inversion is not needed. Extensive error analysis is performed and error propagation is followed throughout the calculation. The standard error of tensor eigenvalues and eigenvectors is estimated. TENSOR features a friendly menu-guided user interface, a data editor, a file manager and routines for printing and plotting. The program is described in [A. Ponti, C. Oliva: Computers & Chemistry **16**, 233 (1992)].

X32, *Schonland type analysis for S* > ½. The software is based on the 2nd order perturbation theory by Iwasaki [19] to obtain the principal values and directions of tensors for $S > ½$ species including zero-field splittings. The program is in obsolete code (Quickbasic), and is not maintained. It has been applied by the authors for the analysis of zero-field splitting tensors which requires a modification of the Schonland method in accordance with Eq. (3.5) that apparently is not implemented in the other programs in Table 3.4.

3.4 Analysis of Powder Spectra

Single crystal measurements, although informative, are not always applicable to determine the values of magnetic couplings in solids. It is for instance difficult to obtain single crystals of biochemical materials. In other cases the paramagnetic species are intentionally trapped in a disordered matrix or in a frozen solution. The ESR, ENDOR, or ESEEM lines are then usually broadened by the anisotropy of the magnetic couplings. Some data, e.g. the position of the paramagnetic species in the lattice that a single crystal analysis can provide, are difficult to extract from a powder. However, a considerable amount of information can often be obtained from an analysis of powder spectra of, for instance, free radicals or other $S = ½$ species with anisotropic g and/or hyperfine coupling, of $S > ½$ species with zero-field splittings, and of combined effects of several anisotropic couplings.

As exemplified below visual analysis can be sufficiently accurate to obtain the g-tensor (Section 3.4.1.1), the hyperfine coupling tensor (Section 3.4.1.3), and the zero-field splitting tensor (Section 3.4.1.6) for well-resolved ESR spectra. Computer methods for the analysis of less resolved spectra were developed and applied at an early stage, for instance in the simulation of Cu^{2+} ($I = 3/2$, $S = ½$) complexes in enzymes (Section 3.4.1.7). Simulations are also commonly necessary to analyse spectra of systems containing several nuclei. Definite procedures for the simulation of ESR spectra of radicals of this type in amorphous solid samples were published already in 1965 (Sections 3.4.1.2, 3.4.1.5, and 3.4.1.7). Advanced simulation methods are required for $S = ½$ species containing nuclei with $I > ½$ featuring hyperfine and nuclear quadrupole couplings of comparable magnitudes as exemplified for the ClSS· radicals discussed in Section 3.4.1.4. Recently developed free-ware and commercial software for simulations of systems with arbitrary electron and nuclear spin values are presented in Section 3.4.1.7. Several of the general programs can be applied equally for the analysis of ESR, ENDOR, and ESEEM powder spectra.

3.4.1 ESR Powder Spectra

ESR is applied for the analysis of anisotropies of g- hyperfine and (for $S > \frac{1}{2}$) the zero-field splitting tensors of powder samples under the same conditions as for single crystals.

3.4.1.1 g-Anisotropy

The g-anisotropy of organic free radicals is small, often more pronounced for inorganic radicals, and significant in many transition metal ion complexes. Whether or not g-anisotropy is resolved in a powder spectrum depends on the difference between the g-components, the linewidth, and the frequency band of the spectrometer.

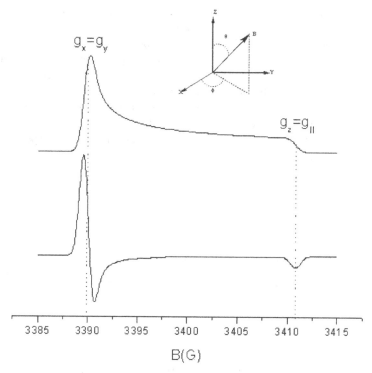

Fig. 3.14 Powder ESR spectra at X-band of an $S = \frac{1}{2}$ species showing axial g-anisotropy, $g_x = g_y = g_\perp = 2.00232$, $g_z = g_\parallel = 1.99$ in absorption and 1st derivative modes. The positions of $g_\perp$ and $g_\parallel$ are indicated

Axial symmetry: The simplest case to analyse is that with an axially symmetric g-tensor, which gives rise to a spectrum of the type shown in Fig. 3.14. It has several characteristic features:

- The intensity of the absorption spectrum is stronger around $g_\perp$ than at $g_\parallel$ because more radicals are oriented in the xy or perpendicular plane of Fig. 3.14 than along the z- or parallel direction. The reason for the difference in shape at $g_\perp$ and $g_\parallel$

was clarified many years ago, see e.g. [20], and is further discussed in advanced treatises [4].

- The spectrum around $g_\perp$ has the normal derivative shape, while the shape around $g_\parallel$ reminds of an absorption. This feature can be understood from the shape of the absorption spectrum.
- The field corresponding to $g_\perp$ is approximately at the center of the derivative shaped feature and that corresponding to $g_\parallel$ is at the absorption-like peak.

Approximate axial g-tensors are frequently observed for transition metal ion complexes. Inorganic radicals can also have appreciable axial g-anisotropy. This property is of value for the assignment of ESR powder spectra in applied studies. Carbon dioxide radical anions, $\dot{C}O_2^-$, and related species contribute for instance to the ESR signal used for geological dating [21] see Chapter 9. The ESR spectrum of this anion has also been employed as an indicator that a certain foodstuff has been irradiated, for dosimetric purposes in certain carboxylic acid salts, and as a component in tooth enamel samples used in retrospective dosimetry.

The X-band spectrometers employed in those applications result in powder spectra that are usually not completely resolved, resulting in less accurate visual determination of the anisotropic g-factors. Spectra of the type in Fig. 3.15 are usully analysed by line shape simulations (see Section 3.4.1.7) when more accurate anisotropic g-values are of interest. The values obtained by this method in Fig. 3.15 are in the range typical for $\dot{C}O_2^-$ radical anions, although also other species may contribute to the signal.

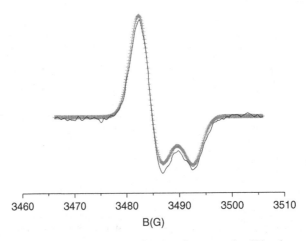

Fig. 3.15 X-band spectrum $(-)$ of an X-irradiated tooth measured at X-band, $\nu = 9{,}762.1$ MHz. The simulated spectrum $(+)$ was obtained with $g_\perp = 2.0023$ and $g_\parallel = 1.9970$

Rhombic symmetry: In the general case paramagnetic species in a solid are characterised by three g-factors. The $\dot{C}O_2^-$ radical anion discussed in the previous section is in fact characterised by a rhombic g-tensor according to single crystal measurements. This may be difficult to verify by visual inspection of the powder spectrum at X-band in Fig. 3.15 because of a small deviation from the axial symmetry, see

Exercise E3.14 for a discussion of this point and the use of high field ESR in Chapter 4 to increase the resolution.

In the schematic spectrum in Fig. 3.16 for an $S = \frac{1}{2}$ species the g-factors can be directly read on the positive and negative absorption-like peaks on the wings and at the center of the derivative peak in the middle.

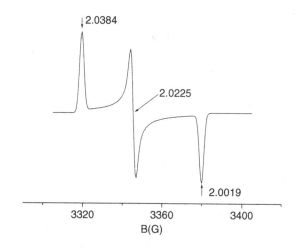

Fig. 3.16 Schematic ESR spectrum of an $S = \frac{1}{2}$ species with rhombic g-anisotropy. The g-factors are the same as those for the ClSS·radical discussed in Section 3.4.1.4 [27]. The ^{35}Cl and ^{37}Cl hyperfine and nuclear quadrupole couplings have been ignored in the schematic spectrum

3.4.1.2 Hyperfine Coupling Anisotropy

The line-broadening in powder spectra of free radicals with small g-anisotropy is usually caused by the anisotropy of the hyperfine coupling. This is exemplified in Fig. 3.17(a) for a so-called α-H hyperfine coupling tensor typical for radicals

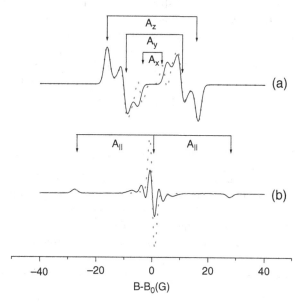

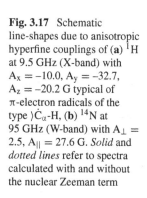

Fig. 3.17 Schematic line-shapes due to anisotropic hyperfine couplings of (**a**) ^1H at 9.5 GHz (X-band) with $A_x = -10.0$, $A_y = -32.7$, $A_z = -20.2$ G typical of π-electron radicals of the type $\rangle\dot{C}_\alpha$-H, (**b**) ^{14}N at 95 GHz (W-band) with $A_\perp = 2.5$, $A_\parallel = 27.6$ G. *Solid* and *dotted lines* refer to spectra calculated with and without the nuclear Zeeman term

of the type $)\dot{C}_\alpha$-H frequently observed in organic, polymeric and biological samples. The three principal values of an α-H coupling are all different [22], resulting in a smeared spectrum. The smearing can become more pronounced and/or additional lines can appear due to the direct-field effect [23], see Section 4.4.1. The hydrogen nucleus is exposed both to an internal (hyperfine) field from the electron spin and an external field from the applied magnet. The phenomenon is commonly but somewhat misleadingly described as a breakdown of the $\Delta m_I = 0$ selection rule giving rise to $\Delta m_I = 1$ or (nuclear) spin flip lines. The effect is most pronounced when the anisotropic coupling and the nuclear Zeeman energy are of comparable magnitudes, very often occurring at X-band for species with anisotropic hyperfine couplings due to ^{1}H and occasionally to ^{19}F, the two nuclei with the largest nuclear g-factors among the stable isotopes. The smearing and the appearance of additional lines could be accounted for as early as 1965 with the first general computer program for simulations of free radicals in disordered systems [24]. The effect is illustrated in Fig. 3.17(a) for simulated powder spectra at X-band taking into account (solid line) and neglecting (dotted line) the nuclear Zeeman energy. Typical anisotropic hyperfine couplings due to an α-H coupling were used. Features due to the principal values marked in the figure are less developed or even absent in the former more realistic case. Complications due to the direct field do not appear for isotropic couplings, and are usually not of major concern for so-called β-H in radicals of the type $)\dot{C}_\alpha$-$C_\beta H_2$- and other couplings with small anisotropy, and at lower microwave frequencies (e.g. X-band) also for ^{14}N and other nuclei with small nuclear g-factors.

^{14}N and ^{13}C hyperfine couplings in free radicals are often approximately axially symmetric, giving line-shapes that may be analysed by inspection to obtain $A_{||}$ as shown in Fig. 3.17(b). The magnitude of the coupling $A_x = A_y = A_\perp$ is usually small and difficult to estimate accurately. A typical powder line-shape at X-band is like the dotted spectrum in the figure. The influence of the nuclear Zeeman interaction is neglible. At W-band the Zeeman energy cannot in general be ignored as illustrated by the multi-line pattern (solid line) occurring in the central region. The pattern might mistakingly be attributed to hyperfine structure with additional nuclei, but is accounted for by the analysis made in an early paper for a single nucleus with $I = 1$ [24]. In theory $(2 \cdot I + 1)^2 = 9$ hyperfine lines are expected for a ^{14}N nucleus, although not all transitions may be visible due to low intensity. The combined effects of g- and hyperfine coupling anisotropy are considered in the following section.

3.4.1.3   Combined g- and Hyperfine Coupling Anisotropy

Paramagnetic species with g-factor and hyperfine coupling anisotropy were analysed by simulations at an early stage in frozen solutions of, for instance, copper enzymes [25]. In this section the powder spectra of two well-known species, the NO_2 molecule and the isotope labelled $^{13}\dot{C}O_2^-$ anion radical are used to illustrate visual and simulation procedures for the analysis. The reader is referred to the classical textbook by Carrington and McLachlan [1] for an account of the electronic structures of these isoelectronic molecules (23 electron system) based on ESR data.

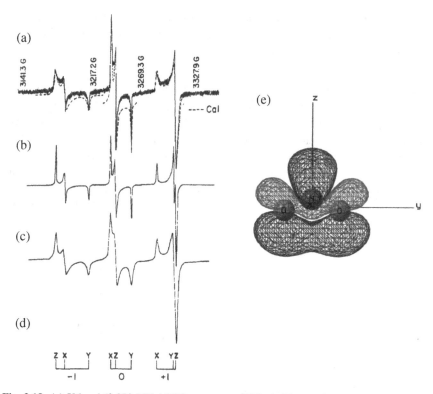

Fig. 3.18 (**a**) X-band (9,070 MHz) ESR spectrum of NO_2 in Vycor glass at 4.8 K. (**b**) and (**c**) Simulated spectra using $g_x = 2.0051$, $g_y = 1.9913$, $g_z = 2.0017$, $A_x = 50.0$ G, $A_y = 46.0$ G, $A_z = 65.5$ G, and Lorentian line widths of 0.6 G and 1.6 G, respectively. The *dotted lines* in (**a**) correspond to the simulated line shape of (**b**) at $m_I = 0$ and (**c**) at $m_I = \pm 1$ bands. (**d**) The stickplot shows the resonance line positions from a visual analysis. (**e**) Singly occupied molecular orbital (SOMO) computed for NO_2 with ground electronic state together with the xyz axes specifying the directions of the g- and hyperfine coupling components. The MO computations were carried out using the B3LYP/6-311+G(2df,p)//UHF/6-31+G(d,p) method. The ESR spectra were reproduced from [M. Shiotani, J.H. Freed: J. Phys. Chem. **85**, 3873 (1981)] with permission from the American Chemical Society

Nitrogen dioxide: NO_2 isolated in noble gas or zeolite matrices gives rise to an ESR spectrum, with $S = ½$ due to the unpaired electron of this odd-electron molecule. The spectrum in Fig. 3.18 shows features of both g-anisotropy and an anisotropic hyperfine coupling due to ^{14}N with $I = 1$. A visual analysis of the spectrum could proceed as follows:

(1) Assume that the axes for the g- and hyperfine coupling tensors coincide. This is reasonable in view of the symmetry of the molecule.
(2) Identify the three ^{14}N hyperfine lines with $m_I = 1, 0, -1$, by assigning the outermost $m_I = \pm 1$ lines as "z-features", the others as indicated in Fig. 3.18.

(3) Are the two spacings between $m_I = -1$ to 0 and 0 to 1 ^{14}N hyperfine lines (approximately) the same for each of the x, y, and z components? If not, go back to step 2 to reassign the x, y and z features.

$\dot{C}O_2^-$: In addition to the central features, the carbon dioxide anion spectrum also contains weak lines on the wings. These features are attributed to $^{13}\dot{C}O_2^-$, while the central parts are due to $^{12}\dot{C}O_2^-$. The ^{12}C carbon isotope with $I = 0$ is in 99% abundance, and consequently the hyperfine lines from the ^{13}C isotope ($I = \frac{1}{2}$, 1.1% natural abundance) can only be seen at high amplifier gain, as shown in Fig. 3.19.

The visual analysis of the spectrum in Fig. 3.19 is less accurate than that in Fig. 3.18 due to the lower resolution, but proceeds similarly. The procedure is shown in the figure. The large ^{13}C hyperfine coupling makes it necessary to include second

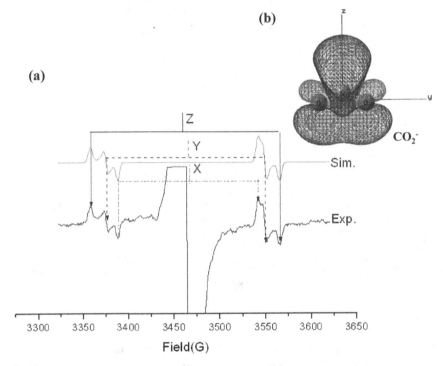

Fig. 3.19 (a) Experimental and simulated spectra of $^{13}\dot{C}O_2^-$ radical anion in γ-irradiated Li-formate powder. The experimental spectrum is amplified to show hyperfine lines due to ^{13}C in natural abundance (1.1%). The anisotropic hyperfine couplings were estimated as indicated, while the g-factors were obtained from the $^{12}\dot{C}O_2^-$ spectrum of the type shown in Fig. 3.15. The simulation was obtained with $g_x = 2.0037$, $g_y = 1.9975$, $g_z = 2.0017$, $A_x = 465.5$, $A_y = 447.5$, $A_z = 581.3$ MHz. (b) Singly occupied molecular orbital (SOMO) computed for $\dot{C}O_2^-$ in its ground electronic state together with the xyz axes specifying the directions of the g- and hyperfine coupling components. The MO computations were carried out using the B3LYP/6-311+G(2df,p)//UHF/6-31+G(d,p) method. The figure was adapted from [K. Komaguchi et al.: Spectrochim. Acta A **66**, 754 (2007)] with permission from Elsevier

order corrections in the analysis of the X-band spectrum. It is then not correct to read the g-factors at the centre of the doublet features corresponding to A_x, A_y and A_z. The anisotropic g-factors could in this particular example be obtained from the $^{12}CO_2^-$ spectrum which is similar to that shown in Fig. 3.15, but in general cases corrections have to be applied. Corrections for the axial symmetric case are relatively easy to apply manually using the equations in Appendix A3.2.1, while the formulas for rhombic symmetry, also reproduced in Appendix A3.2.2 have a more complex form. The corrections are included in the simulation in Fig. 3.19.

3.4.1.4 Combined Hyperfine and Nuclear Quadrupole Couplings

To a first approximation nuclear quadrupole couplings which are smaller than the hyperfine coupling do not influence the shape of the ESR spectra. Exceptions occur for species containing $I > \frac{1}{2}$ nuclei with large nuclear quadrupole moments. We refer to [26] for an analysis of a metal-ion complex by a combined experimental and computational study. Other examples involve species containing halogens such as Cl, Br or I. In the example shown in Fig. 3.20 the principal values of the hyperfine

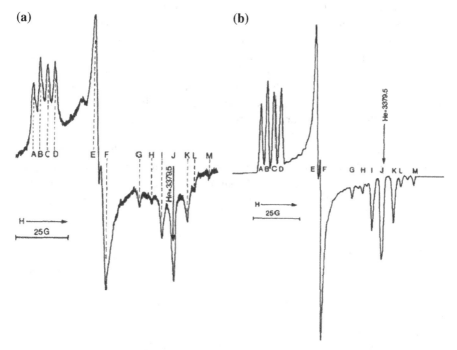

Fig. 3.20 Observed (**a**) and simulated (**b**) powder spectrum of the ClSS· radical. The simulation was obtained with $g_1 = 2.0019$, $g_2 = 2.0225$, $g_3 = 2.0384$, $A_1(^{35}Cl) = 5.9 \cdot 10^{-4}$ cm^{-1}, $A_2(^{35}Cl) = -1.3 \cdot 10^{-4}$ cm^{-1}, $A_3(^{35}Cl) = -3.5 \cdot 10^{-4}$ cm^{-1}, $D_Q = 3/2 \cdot Q_3 = -5.65 \cdot 10^{-4}$ cm^{-1}, $E_Q = |Q_1 - Q_2| = 0.10 \cdot 10^{-4}$ cm^{-1}. The letters A to M designate observed and calculated lines, differing in positions of at most 0.3 G. The figure is reproduced from [27] with permission from the American Institute of Physics

coupling and nuclear quadrupole tensors for ^{35,37}Cl ($I = 3/2$) in the ClSS· radical
are of comparable magnitude. Even though the couplings are much smaller than the
electronic Zeeman energy, the formulas given in Appendix A3.2.2 do not apply, and
programs based on these equations are unsuitable to simulate spectra of this type.
The analysis of this spectrum performed by spectrum simulations at an early stage
is therefore a remarkable achievement [27].

Software to simulate spectra of this type by exact methods and by treating the
nuclear quadrupole-, the nuclear Zeeman- and the hyperfine terms as a joint small
perturbation, are further discussed in Section 3.4.1.7.

3.4.1.5 Hyperfine Structure with Several Nuclei

Powder ESR spectra of species with hyperfine structure due to several nuclei are
often difficult to analyse by visual inspection, and complementary data and/or simulations are required, at least when the principal axes of the corresponding coupling
tensors do not coincide. Hyperfine coupling tensors obtained from single crystal
ESR or ENDOR measurements if available can be used to simulate the ESR powder
spectrum. This procedure is applicable e.g. as a test of the single crystal assignments
in complex systems [28]. Hyperfine coupling data can also be obtained from theory.
DFT (Density Functional Theory) calculations have proved to be particularly useful
to predict both isotropic and anisotropic hyperfine couplings. The theoretical data
are in certain applications quite accurate as exemplified by the agreement between
simulated and experimental spectra for the fluorocarbon anion radicals discussed in
Chapter 5. It is therefore remarkable that the X-band spectrum of the trifluoromethyl
radical, ĊF$_3$, in rigid noble gas matrix could be interpreted by simulations at an early
stage without theoretical guidance except by assuming different although symmetry
related principal directions for the three ^{19}F nuclei.

Later quantum chemistry calculations (see Chapter 5) have shown that the principal axes for the hyperfine coupling tensors due to α-F are in fact not along and perpendicular to the C$_\alpha$—F bond, [29, 30] contrary to the case for α-H in hydrocarbon
radicals [22]. The complex shape of the ESR spectrum in Fig. 3.21 was attributed

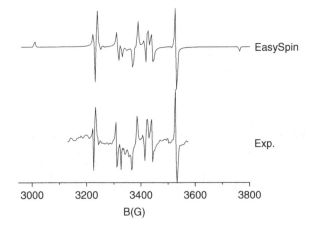

Fig. 3.21 Exp.: X-band ESR
spectrum of ĊF$_3$ radicals
trapped in an inert matrix at
4 K. EasySpin: Simulated
ESR spectrum using
EasySpin with parameters
given in [31]. The outer
features present in the
simulation are not visible in
the experimental spectrum

to the different orientations of the principal directions of the three ^{19}F hyperfine couplings tensors, with large anisotropy of the principal values [31]. Spectra of this type are most accurately simulated with an exact method, as further exemplified in Chapter 5. The freely available EasySpin program described in Section 3.4.1.7 was used to simulate the X-band ESR spectrum of $\dot{C}F_3$ radicals trapped in a noble gas matrix at 4 K in Fig. 3.21. We refer to the literature for an account of the complications occurring in an analysis based on perturbation theory [31].

In conclusion the analysis of complex hyperfine patterns in powder ESR spectra can now be performed more accurately due to the recent development of general simulation programs based on exact theory than was possible in early works.

3.4.1.6 Zero-Field Splitting

The zero-field splitting is a purely anisotropic coupling that can only occur in paramagnetic species with $S > \frac{1}{2}$. It is caused by two separate mechanisms, the electron-electron magnetic dipole coupling occurring e.g. in organic triplet states and the spin-orbit coupling which is the dominating mechanism for transition metal ion complexes.

Axial symmetry: The g-factor anisotropy can in many cases be neglected in the analysis of spectra of organic triplet states.

The zero-field splitting D can be measured from the powder spectrum as schematically shown in Fig. 3.22 for a triplet state ($S = 1$). The maximum separation between the outermost peaks of $2D$ are parallel features with an absorption-like line shape in the 1st derivative spectrum while the perpendicular lines separated

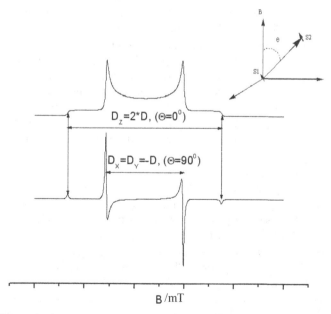

Fig. 3.22 Schematic ESR spectrum of an $S = 1$ species with zero-field splitting D in absorption and 1st derivative modes

by D are more similar to the usual derivative shape. The perpendicular peaks are stronger than the parallel as there are more triplet state molecules in the perpendicular plane, than in the parallel direction like for the spectrum showing g-anisotropy with axial symmetry (Fig. 3.14). The analysis is indicated in the figure. The D-value is assumed to be sufficiently small for a first order analysis to apply ($g\mu_B B >> D$).

Rhombic symmetry: The analysis of this case is discussed in Exercise E3.17 for triplet state naphthalene as a typical example with an asymmetry parameter of $E = -0.0137$ cm^{-1}, while $D = 0.1003$ cm^{-1}.

Simulations of the X-band spectra by approximate (2nd order perturbation theory) and exact (matrix diagonalization) methods differ slightly in this case, see Fig. 3.24 in Section 3.4.1.7. Some spectral features caused by the fact that the high-field approximation does not strictly apply are worth noting:

- The center of the outermost peaks separated by $2D$ can differ from that of the peaks separated by $D + 3E$ or $D - 3E$, even for isotropic g.
- The appearance of the $\Delta m_S = 2$ transition as a single line at approximately half field of the centre of the $\Delta m_S = 1$ powder spectrum can often be taken as evidence that the observed species is really a triplet state molecule. The intensity increases with increasing D as $(D/g\mu_B B)^2$. As seen in Fig. 3.24 its amplitude is quite strong for the naphthalene triplet at X-band because of the small anisotropy of this feature, see Exercise E3.18.

Combined zero-field and g-tensor anisotropy: The g-tensor anisotropy can be appreciable for transition metal ion complexes, but also for some triplet state molecules. The Q-band spectrum of an $(NO)_2$ surface complex in zeolite LTA has been analyzed to have rhombic symmetry for the g-tensor and the zero-field splitting, see Fig. 6.5 in Chapter 6. Complications due to overlap with another spectrum (in this example an NO surface complex) are common in practical applications. Variations of the experimental conditions, e.g. the sample composition (amount of nitric oxide in this example) and measurements at different microwave frequencies can then give support for the assignment. Refinement of the visual assignment by simulations as discussed in Section 3.4.1.7 is also frequently employed.

High spin species (S > 1): Several transition metal ion systems [32], atoms trapped in crystals, and artificially synthesised high spin molecules [12] are characterized by an electron spin $S > 1$.

The three $\Delta m_S = \pm 1$ transitions, ($m_S = 3/2 \leftrightarrow \frac{1}{2}$, $m_S = \frac{1}{2} \leftrightarrow -\frac{1}{2}$, $m_S = -\frac{1}{2} \leftrightarrow -3/2$) give rise to "perpendicular" and "parallel" features for the $S = 3/2$ species with axial symmetry in Fig. 3.23. The parallel lines are separated by $2D$, the perpendicular by D, as in the case of triplet state spectra, under the high-field approximation with the zero field splitting $D << h\nu$ (the microwave energy).

Large zero-field splittings: Visual analysis is less accurate when the high field approximation does not strictly apply. Shifts in the line positions can be taken into account by 2nd order perturbation theory with formulae given in Appendix A3.2. The difference in the positions of the $\Delta m_S = 2$ lines between the simulated X-band spectra of triplet state naphthalene with this method and an exact method in Fig. 3.24

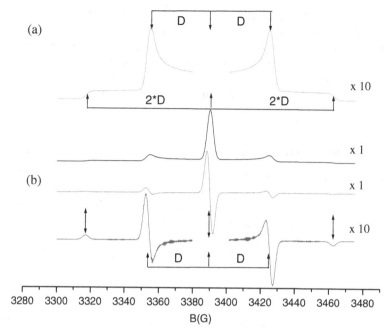

Fig. 3.23 Schematic X-band absorption (**a**) and 1st derivative (**b**) spectra of an $S = 3/2$ system. The weak and strong lines separated by $2D$ and D are parallel and perpendicular features for the electronic transitions ($m_S = 3/2 \leftrightarrow 1/2$, $m_S = 1/2 \leftrightarrow -1/2$, $m_S = -1/2 \leftrightarrow -3/2$)

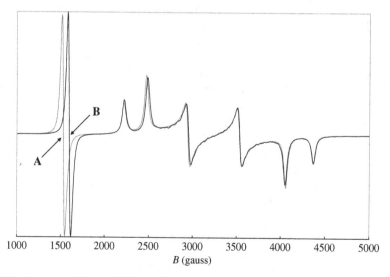

Fig. 3.24 Spectrum of the naphthalene triplet state computed with $D = 1,074$ G, $E = -164$ G, [C.A. Hutchison, B.W. Mangum, J. Chem. Phys. **34**, 908 (1961)]. *A*: diagonalization of the spin Hamiltonien matrix. *B*: second order perturbation treatment of the spin Hamiltonian. Spectra are adapted from http://www.esr-spectsim-softw.fr/programs with permission from Dr. C. Chachaty

is probably due to the relatively large ratio of $D(\text{MHz})/\nu(\text{MHz}) = 3,000/9,$ $500 \approx 0.3$, which limits the accuracy of the perturbation method. For even larger ratios analysis with perturbation theory is not applicable. Simulation methods involving numerical diagonalization of the Hamiltonian matrix exemplified in Table 3.5 are generally applicable. Experimental procedures to measure large zero-field splittings using high-field and multiple frequency methods are discussed in Chapter 4.

3.4.1.7 ESR Simulation Software

A large number of simulation methods have been developed during the last 40 years, the earlier ones, e.g. in [24, 25, 33] being based on perturbation theory. Methods using exact diagonalization of the spin Hamiltonian have been developed more recently and are now beginning to replace the perturbation methods [34, 35].

The simulated powder ESR spectrum has traditionally been assumed to have the form (3.6) [24], but see e.g. [32] for a more recent method of calculation in frequency space.

$$S(B,P) = C \int\limits_{0}^{\pi} \sin\Theta d\Theta \int\limits_{0}^{2\pi} d\varphi \sum_{p<q} I_{pq} f\left(\frac{B - B_{pq}}{w}\right) \tag{3.6}$$

The summation is over the transitions between the magnetic states p and q. I_{pq} is the transition probability and B_{pq} is the resonance field. The microwave frequency ν is constant and the spectrum $S(B,P)$ is calculated as a function of the applied magnetic field B. The shape of the spectrum is determined by a set of parameters P representing e.g. the components of g- and hyperfine coupling tensors, line width w, or total intensity C. The line shape $f(x)$ is approximated by Gauss, Lorentz or Voigt functions. The orientation of the magnetic field is specified by angles (Θ,φ), see Fig. 37 in Appendix A3.2. The spectrum is a superposition of spectra from paramagnets with different orientations.

Several commercial and free-ware program packages can be downloaded electronically. Examples of software for general use are given in Table 3.5, while we refer to a literature survey [36] for programs dedicated to the powder spectrum simulation of free radicals and triplet state molecules.

EasySpin is a Matlab toolbox for simulating ESR spectra. Tools are included for the following types of CW ESR spectra of solid samples (crystals or powders):

- Arbitrary number of electron and nuclear spins
- All interactions, including high-order operators and nuclear quadrupole
- Resonance fields are exact (no perturbation formulae)
- Broadening models: g, A and D strain, unresolved hyperfine splittings
- Non-equilibrium populations
- Perpendicular and parallel detection mode

Table 3.5 Programs for simulation of powder ESR spectra

Program	Order	Available from	Code
EasySpin (general)	Exact	http://www.easyspin.org/	MatLab 6.5.1
EPR-NMR (general)	Exact	http://www.chem.queensu.ca/eprnmr/	Fortran77, Exe
RIGMAT ($S = \frac{1}{2}$)	2	http://www.esr-spectsim-softw.fr/programs	APL
MULTIP ($S = 1$)	2	http://www.esr-spectsim-softw.fr/programs	APL
HMLT	Exact	http://www.esr-spectsim-softw.fr/programs	APL
Simfonia (general)	1–2	http://www.bruker-biospin.de/ EPR/software/emx.html	Exe (commercial)
XEMR	Exact	http://www.csun.edu/~jeloranta/xemr/	Source, binary
XSOPHE	Exact	http://www.bruker-biospin.com/xsohpe.html	Exe (commercial)
Weihe	Exact	Dr Høgni Weihe, weihe@kiku.dk	Exe

EasySpin is available free of charge. The current version of the program can be downloaded at http://www.easyspin.org/. MatLab version 6.5.1 or higher must be installed on the computer and is not provided with EasySpin. *EasySpin* is written and maintained by Dr. S. Stoll at the University of California, Davis. For bug reports, inquiries, additional examples and suggestions, contact the developer. Details are given in [35].

EPR-NMR: This program was written primarily to achieve generality and flexibility in handling magnetic resonance spectra of single crystals and powders, but spectra of low-viscosity liquids can easily be dealt with. The number of spins included and their values (non-negative half-integers and integers) is arbitrary, as is their assignment as either electronic or nuclear. The program sets up spin-hamiltonian matrices, and determines their eigenvalues (energies) using "exact" diagonalization. It is a versatile program, having many operating modes tailored to a variety of applications. These modes can be grouped into four categories, in increasing order of complexity as:

- Energy-level calculation
- Spectrum simulation
- Comparison with observed data
- Parameter optimization

The version (6.40) of EPR-NMR, including both the Fortran source code and utilities, and a much updated manual, is available. This version of the program contains many corrections and improvements, including:

- Capability for NMR as well as ESR calculations
- Capability to handle several unpaired electronic spins
- Availability of Boltzmann factors (but not relaxation effects)
- Revised and reorganized manual

The program runs on any computer capable of running 32-bit Fortran77. The program has been developed by Prof. J.W. Weil. It can be downloaded from http://www.chem.queensu.ca/eprnmr/.

EPR simulation and fitting programs for transition metal ion complexes: Software for simulation and fitting of ESR spectra has been developed during the last 20 years at the University of Copenhagen. Simulations are performed by diagonalization of the spin Hamiltonian used to model the complex under study. General Hamiltonians can be treated like exchange-coupled complexes of transition metal ions. Single crystal and powder spectra can be analysed. A particular feature is that the relative signs of the spin Hamiltonian parameters are determined by comparison of simulated and experimental spectra, see e.g. [P.S. Piligkos et al.: Mol. Phys., **105**, 2025 (2007)]. Excellent fits to experimental spectra are obtained by the least squares method. Further details can be provided by Dr. H. Weihe, Department of Chemistry, University of Copenhagen.

RIGMAT, MULTIP, HMLT: A suite of ESR simulation programs has been prepared by Dr. C. Chachaty and is available free of charge at http://www.esr-spectsim-softw.fr/programs.

RIGMAT is for the simulation of radicals or metal ions of electron spin $S = \frac{1}{2}$ in rigid glassy or polycrystalline matrices for the following cases:

- Anisotropy of **g** tensor, no hyperfine coupling.
- Anisotropy of **g** tensor and of hyperfine coupling tensor(s) **A** for 1 or 2 nuclei with second order shift of resonance lines. The principal axes of these tensors are assumed to be common.
- Any orientation in a molecular frame of the **g** tensor and of hyperfine coupling tensors **A** of several nuclei of any spin. For nuclei of spin $I = \frac{1}{2}$ the satellite lines resulting from the "forbidden" $\Delta m_I = 1$ transitions are taken into account. The spectral simulations resort to the second order treatment of the spin Hamiltonian by Iwasaki [19]. The optimized principal values of **g** and **A** tensors can be obtained from the least-squares fitting of spectra.

MULTIP simulates ESR spectra of triplet states, biradicals, radical or ion pairs in rigid matrices for $\Delta m_S = 1$, and $\Delta m_S = 2$, transitions. Several programs are available according to the system symmetry and to the existence or not of hyperfine coupling. When the g, electron dipole and hyperfine coupling tensors have different principal axes, the spectrum is computed utilizing the second order treatment of Iwasaki [19].

HMLT simulates spectra for cases when the hyperfine coupling of a nucleus or the dipole coupling between two electron spins exceeds about 10% of the magnetic field intensity. The computed positions of resonance lines by the second order perturbation treatment then lack in accuracy. More precise positions of the lines are given by the transitions between energy levels of the spin system computed as functions of the magnetic field intensity by diagonalization of the spin Hamiltonian matrix. Such a procedure is applied to free radicals in liquid phase and to triplet states, biradicals or ion pairs in rigid matrices. An example is given in Fig. 3.24.

XSOPHE: The XSophe-Sophe-XeprView computer simulation software suite enables scientists to easily determine spin Hamiltonian parameters from isotropic, randomly oriented and single crystal continuous wave electron paramagnetic(spin) resonance (CW-EPR(ESR)) spectra from radicals and isolated paramagnetic metal ion centers or clusters found in metalloproteins, chemical systems and materials science. XSophe provides an X-windows graphical user interface to the Sophe programme and allows: creation of multiple input files, local and remote execution of Sophe, the display of sophelog (output from Sophe) and input parameters/files. Sophe is a sophisticated computer simulation software programme employing a number of innovative technologies including; the Sydney OPera HousE (SOPHE) partition and interpolation schemes, a field segmentation algorithm, the mosaic misorientation linewidth model, parallelization and spectral optimisation. In conjunction with the SOPHE partition scheme and the field segmentation algorithm, the SOPHE interpolation scheme and the mosaic misorientation linewidth model greatly increase the speed of simulations for most spin systems. Results from Sophe are transferred to XeprView where the simulated CW-ESR spectra (1D and 2D) can be compared to the experimental spectra. Energy level diagrams, transition roadmaps and transition surfaces aid the interpretation of complicated randomly oriented CW-ESR spectra and can be viewed.

The simulations available include conventional experiments and special 2D experiments:

- CW-ESR spectra
- Orientation Dependent CW-ESR spectra
- Energy Level Diagrams
- Transition Surfaces Diagrams

Spectra are simulated based on full matrix diagonalization with the Sophe partitioning scheme. Further, the superhyperfine interactions may be treated with up to 3rd order perturbation theory. This methodology imposes no limitations on the spin systems to be simulated:

- Isolated Systems
- Magnetically Coupled Systems
- Unlimited Electron and Nuclear Spins including Multiple Nuclear Isotopes

The program is available commercially at http://www.bruker-biospin.com/xsohpe.html. Details are given in [34].

Xemr is an (ESR) EPR (Electron Paramagnetic Resonance), ENDOR (Electron Nuclear DOuble Resonance), and TRIPLE (electron-nuclear-nuclear TRIPLE resonance) spectrum manipulation and simulation package written for Linux systems. It should be noted that Xemr does not run under MicroSoft DOS/Windows environments and the porting would currently be a rather lengthy task. Xemr source and binary code is distributed under the GNU General Public License.

Xemr can calculate (ESR) EPR transitions using the first order simulation or the solution of fully numerical spin Hamiltonian. In the latter case the numerical transition moments can also be calculated. The first order simulation is restricted to $S = \frac{1}{2}$ and to electron Zeeman and hyperfine interaction whereas the numerical method can handle electron and nuclear Zeeman, hyperfine interaction, electron-electron interaction, and nuclear quadrupole interaction. The latter method can simulate both (ESR) EPR and ENDOR spectra. In addition a simple 1st order ENDOR simulation is also possible, so that the parameters can be extracted from the ENDOR spectra with better accuracy.

The currently available lineshape modes are: Stick, combination of Lorentz and Gauss, asymmetric Lorentz (m_I dependent anisotropy), and intra-molecular exchange. These lineshape models can be integrated over the angular variables to obtain powder spectra. Also absorption, first derivative and second derivative spectra can be generated. The powder spectrum integrator (based on Gaussian quadrature) can be added on top of these methods. Normally a regular mesh is used for integrations. Igloo partitioning can also be requested. The simulation routines accept G, mT and MHz units.

The parameters concerning the simulations are provided interactively. The program is written and maintained by Dr. J. Eloranta. The latest Xemr release is available in source form at http://www.csun.edu/~jeloranta/xemr/. For additional information see [37].

3.4.2 ENDOR Powder Spectra

ENDOR measurements of glassy, polycrystalline or amorphous samples usually aim at deducing structural properties from an analysis of hyperfine- and nuclear quadrupole interactions that are too small to be resolved by ESR. The subject has been summarized in recent textbooks [3, 4, 12], in reviews about radicals on surfaces [38], about radical ions in frozen matrices [39], and about paramagnetic species in biological systems [40–44].

Several factors unique for ENDOR affect the intensities, *i.e.* magnetic relaxation, hyperfine enhancement, and angular selection. The two first effects also affect spectra of liquid and crystalline samples, while the third is typical for powder spectra of species with anisotropic *g*-values. Methods that take the two latter effects into account have been developed and are usually incorporated in software developed for the simulation of ENDOR spectra in the solid state. Simulations that take magnetic relaxation effects into account have been employed only to analyse ENDOR spectra in the liquid state [2]. It is possible that the commonly observed poor agreement between experimental and simulated intensities in the solid state is at least in part due to relaxation effects that are not taken into account in any software we are aware of.

3.4.2.1 Hyperfine Coupling Anisotropy

Powder ENDOR lines are usually broadened by the anisotropy of the hyperfine couplings. The parameters of well resolved spectra can be extracted by a visual analysis analogous to that applied in ESR. The principle is indicated in Fig. 3.25 for an $S = \frac{1}{2}$ species with anisotropic ^1H hyperfine structure, where the hyperfine coupling tensor of axial symmetry is analysed under the assumption that $0 < A_{||} < A_\perp < 2 \cdot v_H$. The lines for electronic quantum numbers $m_S = \frac{1}{2}$ and $-\frac{1}{2}$, centered at the nuclear frequency $v_H \approx 14.4$ MHz at X-band, are separated by distances equal to the principal values of the hyperfine coupling tensor as indicated in the figure. Absorption-like peaks separated by $A_{||}$ in the 1st derivative spectrum occur due to the step-wise increase of the amplitude in the absorption spectrum, like in powder ESR spectra (Section 3.4.1). The difference in amplitude commonly observed between the $m_S = \pm\frac{1}{2}$ branches is caused by the "hyperfine enhancement" effect on the ENDOR intensities first explained by Whiffen [45a]. The effect of hyperfine enhancement is apparent in Figs. 3.25 and 3.26.

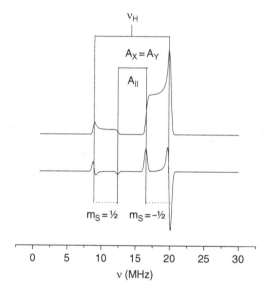

Fig. 3.25 Schematic X-band powder ENDOR spectrum of an $S = \frac{1}{2}$ species with anisotropic ^1H hyperfine structure. The hyperfine coupling tensor of axial symmetry is analysed under the assumption $0 < A_{||} < A_\perp < 2 \cdot v_H$ The lines for electronic quantum numbers $m_S = \frac{1}{2}$ and $-\frac{1}{2}$ are centred at the nuclear frequency, $v_H \approx 14.4$ MHz, and are separated by distances equal to the principal values of the hyperfine coupling as indicated in the figure. The difference in intensity of the $m_S = \frac{1}{2}$ and $-\frac{1}{2}$ branches is due to hyperfine enhancement. Absorption-like peaks separated by $A_{||}$ in the 1st derivative spectrum occur due to the step-wise increase of the amplitude in the absorption spectrum, like in powder ESR spectra (Section 3.4.1)

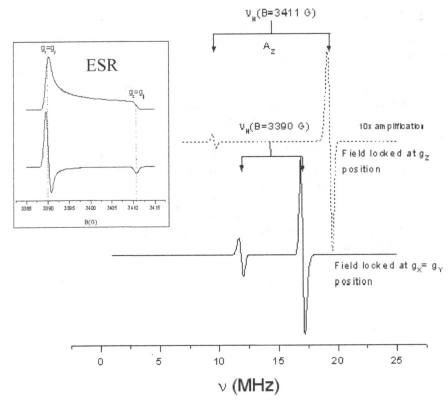

Fig. 3.26 Schematic powder ENDOR spectra of an $S = ½$ species with axially symmetric g and ^{1}H hyperfine structure. ENDOR spectra with the magnetic field locked at $g_{||}$ and $g_{\perp}$, respectively, are single-crystal like due to angular selection. The lines for electronic quantum numbers $m_S = ½$ and $–½$ are separated by distances equal to $A_{||}$ and $A_{\perp}$, the principal values of the hyperfine coupling tensor as indicated in the figure

Anisotropic hyperfine couplings of rhombic symmetry give rise to powder ENDOR spectra of the type shown in Exercise E3.20. Absorption-like peaks in the ENDOR spectrum occur also in this case.

3.4.2.2 Angular Selection

Angular selection can affect the powder spectrum shape as schematically shown in Fig. 3.26.

The recording of ENDOR spectra with the magnetic field locked at different positions of the ESR spectrum, e.g. at $g_{\perp}$ and $g_{||}$ in Fig. 3.26 is a common procedure to extract weak anisotropic hyperfine couplings from powder samples. Coupling constants corresponding to those obtained from a single crystal with the magnetic field oriented along the principal axes of the g-tensor can be obtained from an analysis of the ENDOR spectra. Experimental studies to elucidate the structure of metalloproteins have been reviewed, and software dedicated to the simulation of powder ENDOR spectra that take this angular selection effect into account has been developed [40].

3.4.2.3 Simulation of ENDOR Spectra

Technical procedures to simulate powder ENDOR spectra are given in Appendix A3.3. Simulation is employed in several applications, of which some are exemplified below.

- Confirmation of single crystal analysis
- Solvation structure of paramagnetic species by analysis of hyperfine couplings of surrounding nuclei
- Ligand structure of transition metal ions by angular selection analysis
- Surface complex structures by analysis of hyperfine *and* nuclear quadrupole couplings
- $S > \frac{1}{2}$ species

Confirmation of Single Crystal Analysis

This application may seem redundant, but is motivated e.g. by the so-called Schonland ambiguity in determining the hyperfine coupling tensor from single crystal measurements. It is only very recently that methods to eliminate the ambiguity [17] have been proposed, and the procedures are not always applicable. An investigation to clarify the radical structure of *l*-alanine is taken as an example in Fig. 3.27.

In Fig. 3.27(b) and (c) are shown powder ENDOR spectra obtained with applied magnetic field B at the centre (b) and the outermost high field (c) lines of the ESR spectrum in Fig. 3.27(a) of irradiated alanine together with their simulations. At the former position ENDOR signals of the two radicals $R1 = CH_3\dot{C}HCOOH$ and $R2 = H_3N^+\dot{C}(CH_3)COO^-$ contribute to the spectrum. One reason for the lack of agreement between the relative intensities of the experimental and simulated curves is that the influence of relaxation is not taken into account in the simulations. The line positions are, however, in good agreement, suggesting that ENDOR powder spectroscopy in conjunction with simulations is a usable alternative in cases where single crystal analysis is not feasible.

Solvation Structure of Paramagnetic Species

The resolution of hyperfine structure by ENDOR is a well-established method to obtain the structure of the trapping sites of paramagnetic species as shown already in seminal work on single crystals [46]. Information about the local geometry of ions or molecules surrounding the species can also be obtained from powder

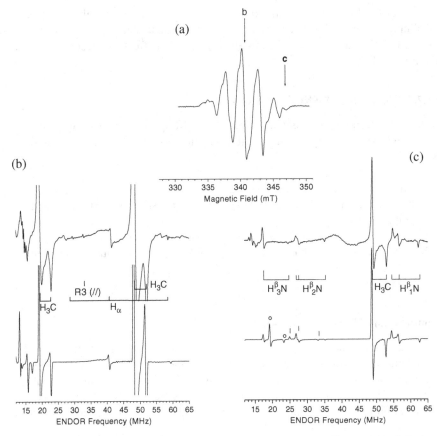

Fig. 3.27 (**a**) First-derivative X-band ESR spectrum from a polycrystalline sample of alanine. X-irradiated at 295 K and measured at 221 K. The *arrows indicate* field positions of ENDOR spectra in (**b**) and (**c**). (**b**) Experimental (*top*) and simulated powder ENDOR spectrum due to radical R1 at 221 K. The experimental spectrum was obtained by saturating the central ESR *line at arrow* "b" in (**a**). (**c**) Experimental (*top*) and simulated powder ENDOR spectrum due to radical R2 at 221 K. The experimental spectrum was obtained by saturating the ESR *line at arrow* "c" in (**a**). The figure is reproduced from [12] with permission from Springer

ENDOR measurements. The ESR spectrum of the radiation induced free radicals in polycrystalline lithium formate, proposed as an ESR-dosimeter (see Section 9.3, in Chapter 9), consists of a singlet with a line width of 0.92 mT. The g- and ^{13}C-hyperfine tensors obtained from ESR, $\mathbf{g} = (2.0037, 1.9975, 2.0017)$, and $\mathbf{A}(^{13}\text{C}) = (465.5, 447.5, 581.3)$ MHz identify the species as $\dot{\text{C}}\text{O}_2^-$ in agreement with an assignment made by single crystal ESR and ENDOR studies of samples with natural isotopic composition, discussed in Chapter 2. The observed stability of the CO_2^- radicals formally formed by H atom detachment from HCO_2^- is attributed to the solvation structure of surrounding molecules and Li^+-ions.

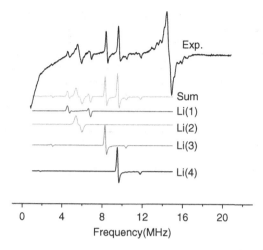

Fig. 3.28 Experimental (*top*) and simulated powder ENDOR spectrum due to $\dot{C}O_2^-$ ion radicals formed by X-irradiation of a polycrystalline lithium formate sample at room temperature. The experimental spectrum was obtained by saturating the central ESR resonance line. The simulated spectra represent the components due to the hyperfine couplings of ^7Li at four different positions in the vicinity of the $\dot{C}O_2^-$ ion. The sum spectrum is obtained by addition of the components. The low frequency branches of the ^7Li transitions are weak or not observable in experimental and theoretical spectra probably due to different hyperfine enhancement effects for $m_S = \frac{1}{2}$ and $-\frac{1}{2}$. The figure is reproduced from [K. Komaguchi et al.: Spectrochimica Acta Part A **66**, 754 (2007)] with permission from Elsevier

The anisotropic hyperfine couplings estimated according to the procedure outlined in Fig. 3.28 (axial symmetry assumed) agree with those from the single crystal measurements discussed in Chapter 2.

A limitation of the analysis of powder ENDOR is that the directions of the principal axes for the hyperfine couplings cannot in general be obtained unless the g-anisotropy is sufficient for angular selection to occur as discussed in Section 3.4.2.2. The shape of the experimental spectrum in Fig. 3.28 is for example reproduced by simulating the individual contributions from each ^7Li and is insensitive to the relative orientation of the symmetry axes of the nearly axially symmetric tensors. An alternative procedure to assign the couplings to specific nuclei is to combine experimental measurements with theoretical modelling. This is exemplified in Table 3.6 by a comparison of the experimental ^7Li- hyperfine coupling tensors with the theoretical values obtained by density functional theory (DFT) for $\dot{C}O_2^-$ occupying a relaxed structure in the intact lithium formate crystal matrix (Fig. 3.29). By this association a probable solvation structure for the surroundings of CO_2^- could be deduced. Theoretical modeling is thus a useful complement for the interpretation of powder ENDOR spectra. In this particular case the trapping site for CO_2^- had previously also been established by single crystal ENDOR measurements.

Table 3.6 ^{7}Li-hfs couplings (in MHz) of an irradiated LiOOCH·H$_2$O polycrystalline sample observed by ENDOR together with theoretical values (in parentheses) calculated for the surrounding ^{7}Li ions in the vicinity of the $\dot{C}O_2^-$ radical. The calculation was done with DFT, at the B3LYP/3-21G* level for a model structure consisting of 113 atoms, where the central $\dot{C}O_2^-$ component was optimized and other atoms were fixed in the same positions as in the undamaged LiOOCH·H$_2$O crystal. For comparison the *hfs* values obtained by single crystal ENDOR meaurements are shown in the second line of each entry

	a_{iso}	b_1	b_2	b_3
Li(1)	9.80 (8.09)	−1.588 (−1.739)	−1.588 (−1.366)	3.176 (3.105)
	9.70	−1.495[a]	−1.495[a]	2.99
Li(2)	7.07 (6.24)	−1.467 (−1.593)	−1.467 (−1.332)	2.934 (2.925)
	6.85	−1.39[a]	−1.39[a]	2.78
Li(3)	0.0 (0.174)	−0.533 (−0.0021)	−0.420 (−0.0021)	0.953 (0.0042)
	0.01	−0.445[a]	−0.445[a]	0.89
Li(4)	−0.972 (−1.006)	−1.523 (−1.524)	−1.159 (−1.207)	2.683 (2.731)
	−0.92	−1.265[a]	−1.265[a]	2.53

[a] The observed single crystal data are averaged to be axially symmetric.

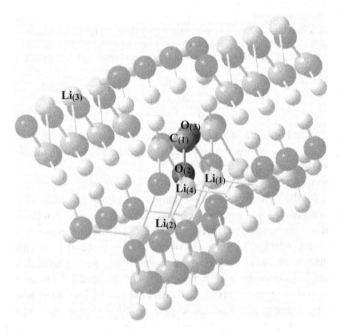

Fig. 3.29 Schematic representation of trapping site of $\dot{C}O_2^-$ radical anion generated in an orthorhombic LiOOCH·H$_2$O crystal. The whole structure composed of 113 atoms was used as a model for the calculation, where the $\dot{C}O_2^-$ fragment, O$_{(2)}$—C$_{(1)}$—O$_{(3)}$, was geometrically optimized and the surroundings were kept at the same positions as in the undamaged crystal. The figure is reproduced from [K. Komaguchi et al.: Spectrochimica Acta Part A **66**, 754 (2007)] with permission from Elsevier.

Ligand ENDOR of Transition Metal Complexes

The angular selection method established by Rist and Hyde [47] for the analysis of ligand ENDOR of metal complexes in powders has been further developed and applied for biological systems. Measurements at X- and Q-band are often adequate due to an appreciable g-anisotropy. We refer to recent reviews for further account of this application [48].

Nitrogen-containing ligands usually give poorly resolved ESR spectra as exemplified in Fig. 3.30 for ^{14}NO-ligated ferrocytochrome c heme a3. An analysis of the corresponding ENDOR spectrum must take into account the effect of nuclear quadrupole coupling of the ^{14}N nucleus. The two bands observed in the experimental spectra, assigned to ^{14}N histidine and ^{14}NO, could be reproduced by simulations with an exact method [49].

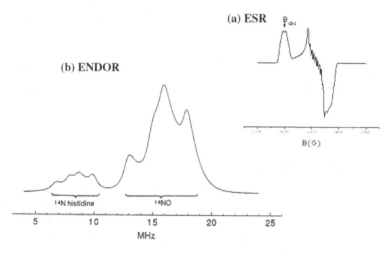

Fig. 3.30 Simulated powder ENDOR spectrum (in absorption) of ^{14}NO-ligated ferrocytochrome c heme a3, at the field setting (g = 2.079) marked in the X-band (ν = 9.32 GHz) ESR spectrum. The parameters **g** = (2.082, 1.979, 1.979) **A**(^{14}N-His) = (16.5, 16.1, 19.3) MHz, Q(^{14}N-His) = (+0.67, −1.12, + 0.45) MHz, **A**(^{14}N-NO) = (30.56, 30.56 59.90) MHz, Q(^{14}N-NO) = (+1.03, −0.51, −0.52) MHz were employed for the simulation, using a method taking angular selection into account. For experimental spectra see [R. LoBrutto et al., J. Biol. Chem. 258 (1983) 7437], for simulation with an exact method see [49]. The spectrum is adapted from [R. Erickson, Chem. Phys. **202**, 263 (1996)] with permission from Elsevier

Surface complex structures

When the g-anisotropy is small the analysis of the orientation-selective ENDOR experiments benefits from the improved resolution of the different g-components at high field, e.g. at W-band. This procedure is well illustrated by work to clarify the structures of the adsorption complexes of nitric oxide (NO) interacting with metal ions in zeolites [50]. These structures are of interest from an applied view to elucidate the catalytic decomposition of NO into N_2 and O_2 over transition metal ion-exchanged zeolites [51]. Complexes with alkali-metal ions such as Na^+–NO have been examined in early ESR studies and more rececently with high field ESR, pulsed ENDOR and ESEEM methods. The readers are referred to Section 6.2.1.3 (^{14}N and ^{23}Na hyperfine couplings and structure of Na^+–NO complex) in Chapter 6 for more details.

S > ½ Species

ENDOR studies of paramagnetic species with $S > ½$ reported in the literature cover a wide range: semiconductor materials, organic triplet state molecules, artificially made high-spin molecules, and biological samples. Molecules can be excited to the triplet state ($S = 1$) by illumination. Recent examples involve studies of photosynthetic systems like the carotenoid peridin serving as a light-harvesting pigment discussed in Chapter 2 [52]. The Q-band field-swept (FSE) ENDOR spectrum with zero-field splitting parameters $|D| = 48.2$ mT and $|E| = 4.7$ mT appeared partly in emission, partly in absorption after a laser flash as shown in Fig. 2.27(a).

A notable feature of this and other systems with $S > ½$ is that angular selection is achieved, not only by g-anisotropy but is mainly due to zero-field splitting. Not all simulation software described below can handle this case, which can more easily be taken into account by the general simulation programs exemplified in Table 3.5.

3.4.2.4 ENDOR Simulation Software

Due to the lower resolution of powder ENDOR spectra compared to spectra from liquids and single crystals, computer assisted analysis by simulation is often required, see Figs. 2.14–2.16 in Chapter 2 for examples. Brief descriptions of dedicated software for ENDOR simulation we are aware of are given below, while general programs for ESR and ENDOR are listed in Table 3.5. In general, the positions of the ENDOR lines can be accurately simulated. The experimental intensities are more difficult to reproduce even taking into account the different intensities of the ENDOR lines due to the hyperfine enhancement effect. It is possible that the commonly observed poor agreement between experimental and simulated intensities in the solid state is at least in part due to the neglect of relaxation effects. These effects have, as far as we know, been taken into account only in software to analyse ENDOR spectra in the liquid state [2].

In the ENDOR spectra of transition metal ions and other systems with pronounced g-factor anisotropy one can take advantage of orientational selectivity in the analysis [47]. In this case the ENDOR spectrum obtained at a specific magnetic field setting contains contributions from a limited range of orientations, e.g. along a line or in a plane, for an axially symmetric system with the field set at $g_{||}$ and $g_\perp$. This effect is automatically accounted for in the general simulation programs using exact diagonalization of the spin Hamiltonian, and in some of the programs based on perturbation methods. Programs we are aware of are summarised below, based on the descriptions by the developers.

GENDOR is a program for the simulation of ENDOR powder spectra featuring angular selection due to g-factor anisotropy. The program is available free of charge in executable code for PC and can be downloaded at http://chemgroups.northwestern.edu/hoffman/endor_files/simulationprograms.htm.

A new version of the previous ENDOR spectrum simulation program, GENDOR, has been made much friendlier by combining data input and computation into a single Visual Basic program with a gui interface. Simulations are output in ASCII X,Y format and can therefore be read into nearly any spread sheet/graphing program. Simulations can be performed either at uniform intervals in g-value or at an input list of specified fields. An updated instruction manual is being assembled. The recent version, GENDOR50 may be downloaded and unzipped to obtain the executable ENDOR simulation program, GENDORVB.exe. The program is developed and maintained in the group led by Professor Brian Hoffman, Department of Chemistry, Northwestern University, Evanston, IL. For further information, please contact the developers.

MSPEN/MSGR: Simultaneous EPR and ENDOR powder spectra synthesis by direct Hamiltonian diagonalization. Powder-type ESR and ENDOR spectra are calculated by direct diagonalization of an appropriate spin Hamiltonian. Orientational selectivity is taking into account in ENDOR with a weighting function for the number of spin packets around the field setting. The performance of the program is demonstrated for several published examples, among other for the ESR and ENDOR simulations for NO-ligated ferrocytochrome c heme a3, discussed in section 3.4.2.3, Fig. 3.30. The program developed by Kreiter and Hüttermann (Fachrichtung Biophysik der Universität des Saarlandes, D-66421 Homburg, Germany) is described in [49]. The procedure to obtain orientational selectivity developed for MSPEN/MSGRA has later been adopted in the ENDOR F2 program described below.

ENDOR F2: In many free radical systems the g-anisotropy is quite small and it is then impossible to obtain single crystal like ENDOR spectra, which instead are made up of a large number of orientations. In this case the ESR line shape function (in absorption) may be employed as a weighting function to select the transitions that contribute to the ENDOR signal [49], and thus give rise to (partial) angular selection at the magnetic field of the ENDOR spectrum. A program employing this idea has been used in the simulations for this chapter [53]. Several nuclei with any value of nuclear spin I can be handled by an unconventional 1st order perturbation method, where the electronic Zeeman term is the dominating one, while the hyperfine, quadrupole and nuclear Zeeman interactions are treated, simultaneously, as a joint perturbation. Thus, no assumptions were made regarding their relative magnitude or the relative orientation of the principal axes of the tensors. The ENDOR frequencies for the $m_S = \pm\frac{1}{2}$ electron states are obtained by diagonalisation of the perturbation matrices of each nucleus. The formulas for the ENDOR intensities by this perturbation method predict enhancement effects analogous to those given in classical treatments [45]. The reader is referred to the original paper [53] for details and Appendix A3.3.2 for a summary.

3.4.3 ESEEM Powder Spectra

The subject has been treated in recent monographs and reviews for applications like solvation structure of electrons, atoms and radicals [54], coordination of ligand molecules to transition metal complexes, metalloproteins and photosynthetic centers [55, 56].

The hyperfine couplings can be obtained by direct analysis of the modulations on the echo decay curve. The analysis must then generally be made by fitting a simulated curve to the experimental. This procedure was usually employed in early work. Analysis of frequency domain spectra obtained by Fourier transformation (FT) is more common in recent studies. The resulting FT or frequency domain spectrum has lines with the same frequencies as in ENDOR. Like in ENDOR visual analyses of FT spectra are often followed by simulation to obtain accurate values for the anisotropic hyperfine coupling.

ENDOR and FT ESEEM spectra differ mainly in the intensities of the lines, which in ESEEM are given by a factor related to the ESR transition probabilities. A necessary prerequisite for modulations in the time domain spectrum is that the "allowed" $\Delta m_I = 0$ and "forbidden" $\Delta m_I = 1$ hyperfine lines have appreciable intensities in ESR. The zero ESEEM amplitude thus predicted with the field along the principal axes of the hyperfine coupling tensor is of relevance for the analysis of powder spectra. Analytical expressions describing the modulations have been obtained for nuclear spins $I = \frac{1}{2}$ and $I = 1$ [54, 57] by quantum mechanical treatments that take into account the mixing of nuclear states under those conditions. Formulae are reproduced in Appendix A3.4.

Allowed and forbidden ESR transitions can also occur by the nuclear quadrupole interaction for $I > \frac{1}{2}$. Analytical formulas are applicable when the quadrupole energy is small compared to the combined effect of hyperfine and nuclear Zeeman interactions [57]. Numerical solutions have been applied when this approximation does not hold [58]. Systems with $S > \frac{1}{2}$ require special treatments [59, 60].

Methods frequently employed in the analysis of ESEEM and HYSCORE data are summarized in Sections 3.4.3.1–3.4.3.7. Advanced methods briefly mentioned in Section 3.4.3.8 are described in dedicated works [54, 61]. Simulation programs are referenced in Section 3.4.3.8, while mathematical treatments have been placed in Appendix A3.4.

3.4.3.1 The Spherical Model

The model was applied in early studies of solvation structures of paramagnetic species in frozen glassy matrices. Frozen aqueous and alcohol solutions can stabilize excess electrons by solvation due to hydroxyl groups. The excess electrons were usually formed by ionizing radiation. The structure of the solvation shell has been elucidated by ESEEM studies by measurements of the anisotropic hyperfine coupling of the hydrogens in the hydroxyl groups. The electron-hydrogen distance and the number of hydrogen atoms in the solvation shell were obtained by an analysis of the modulations on the echo decay curve. Two assumptions were made:

(1) the dipolar hyperfine coupling is given by the point-dipole formula. (2) the hydrogen atoms surrounding the electron are randomly oriented relative to each other. The first assumption makes it possible to use only two parameters, the electron-hydrogen distance and the isotropic hyperfine coupling, instead of the six parameters that are generally needed to specify the hyperfine coupling tensor. The second assumption decreases the computation times in numerical calculations. Tests of the method against the spatial configuration model discussed in the next section have shown that the assumptions are satisfactory for electron-hydrogen distances $r_{e-H} > 2.5$ Å [54]. An example of software for analysis of spectra by the spherical model is given in Section 3.4.3.9.

The ESEEM due to to ^{2}H hyperfine couplings of specifically deuterated samples were more easily analysed than those from ^{1}H in early applications because of the deeper modulations. Examples of experimental results, leading to a suggested interpretation by a geometric model, are shown in Fig. 3.31.

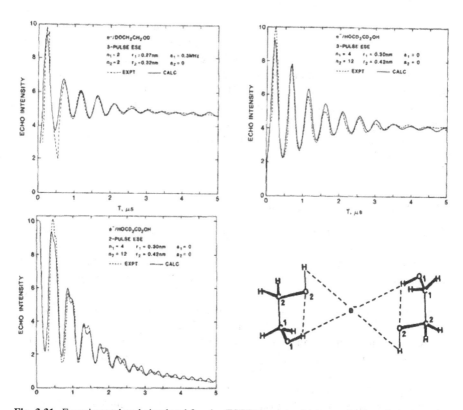

Fig. 3.31 Experimental and simulated 3-pulse ESEEM spectra for trapped solvated electrons in DOCH$_2$CH$_2$OD (*left, above*) and HOCD$_2$CD$_2$OH (*right, above*) frozen glass (*above*), and 2-pulse ESEEM for DOCH$_2$CH$_2$OD (*below*). The suggested solvation structure (*right, below*) has $r_{e-H(O1)}$ = 0.27 and $r_{e-H(O2)}$ = 0.32 nm. The figure is adapted from [M. Narayana et al.: J. Chem. Phys. **81**, 2297 (1984)] with permission from the American Institute of Physics

3.4.3.2 Spatial Configuration

The difference in orientations of e.g. H-O1 and H-O2 in Fig. 3.31 is not taken into account in analyses based on the spherical model. For a structure with several nuclei at fixed positions relative to the paramagnetic centre other averaging methods have been employed. An integration procedure that is applicable when the point dipole approximation applies (axial symmetry) is described in Chapter 6 of Ref. [54]. A procedure based on theory developed by Iwasaki and Toriyama [57] is applicable also with rhombic hyperfine anisotropy. The hyperfine coupling tensors of the different nuclei are then expressed in a common laboratory system. The averaged modulation signal is obtained by an integration over all orientations of the magnetic field, usually specified by the polar angles (θ, φ), similar to the case of powder ESR and ENDOR.

3.4.3.3 ESEEM due to Nuclei with I ≥ 1

An exact analysis of polycrystalline ESEEM including quadrupole interactions that are smaller than the energy due to hyperfine and nuclear Zeeman interactions has been developed [57]. The equations are relatively complex and are therefore best suited for computer simulations. This was first done in an analysis of modulations due to ^{2}H for methyl radicals trapped in $CH_3COOLi \cdot 2D_2O$, where inclusion of the deuterium quadrupole coupling improved the fit of the simulated modulation pattern to the experimental, Fig. 3.32. Neglect of this coupling, as was generally made in early studies, can lead to poor fits to experiment as seen e.g. in the 2-pulse ESEEM for $DOCH_2CH_2OD$ in Fig. 3.31.

3.4.3.4 Large Nuclear Quadrupole Effects

The assumption that the quadrupole coupling is smaller than the energy due to hyperfine and nuclear Zeeman interactions is not always satisfactory. For paramagnetic bio-molecules containing ^{14}N for example the hyperfine couplings can be small at certain directions of the magnetic field, while the nuclear Zeeman frequency at X-band ca. 1 MHz, is of the same magnitude as the nuclear quadrupole energy. The general analysis of ^{14}N $(I = 1)$ ESEEM powder spectra in [58] is applicable to systems with hyperfine and nuclear quadrupole coupling tensors with arbitrary relative orientations and strengths. A demonstration of the method applied to the ESEEM of the nitrogenase MoFe protein was made to determine the hyperfine and nuclear quadrupole tensors of ^{14}N nuclei interacting with the $S = 3/2$ FeMo-cofactor. The Hamiltonian matrix of the hyperfine, nuclear Zeeman, and nuclear quadrupole interactions was diagonalised, to obtain accurate values for the nuclear frequencies and the depth parameters entering the formulas for ESEEM modulation. Angular selection due to g-anisotropy was also taken into account. The treatment is similar to that applied in the analysis of powder ENDOR for systems featuring large nuclear quadrupole interactions described in Section 3.4.2.

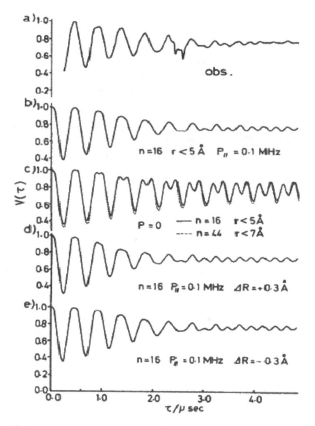

Fig. 3.32 Modulation pattern on the two-pulse ESE decay from D_2O in $CH_3COOLi \cdot 2D_2O$, (**a**) observed, (**b**) simulated modulation including and (**c**) excluding nuclear quadrupole interaction. Simulations with 16 D atoms shifted closer to (**d**) and away from (**e**) the positions assumed in (**b**) do not agree with experiment. The figure is reproduced from [57] with permission from the American Institute of Physics

3.4.3.5 Exact Methods

Following seminal work at an early stage [59] non-perturbative simulation methods have been further developed following a recent trend. The exact methods are based on the diagonalization of the Hamilton or Liouville matrices, see e.g. [60, 62]. This eliminates the risk of errors due to approximations used in the methods based on perturbation theory. The case with large nuclear quadrupole couplings is an example where approximate methods can fail, although the analytic treatment in the previous paragraph for nuclei with $I = 1$ can in principle be extended to higher nuclear spins with a numeric method. Such special methods are not always easy to apply by non-specialists, in which case simulation with an exact method provides a safer alternative.

3.4.3.6 HYSCORE

As mentioned in Chapter 2 overlap of lines can make analysis difficult when several nuclei contribute in the one-dimensional (1D) ESEEM spectra. The HYSCORE method is at present the most commonly used two-dimensional (2D) ESEEM technique to simplify the analysis. Contour maps obtained after 2D Fourier transformation of the echo decay signal followed by projection on the frequency plane are mainly employed for visual or computer analysis to obtain the anisotropic hyperfine couplings. Software for the data processing to obtain the contour is often provided with commercial instruments. Tools for 1D and 2D Fourier transforms are also available in commercial software like Matlab.

Contour Maps

The nuclear transitions corresponding to a particular orientation of a crystalline sample are identified most conveniently by a contour map of the type shown in Fig. 3.33 for an $S = \frac{1}{2}$ species. Projected peaks that are symmetrically disposed about the diagonals are attributed to nuclear transitions for the $m_S = \frac{1}{2}$ and $-\frac{1}{2}$ electronic states, respectively, (denoted α and β in Fig. 3.33) of a particular nucleus. The frequencies of the α and β transitions are obtained directly from the diagram, while the frequency of the free nucleus read at the intersection of the diagonal and the line

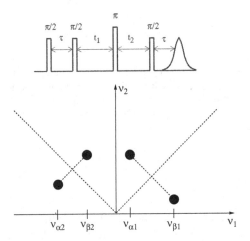

Fig. 3.33 Schematic HYSCORE contour plot for an $S = \frac{1}{2}$ single crystal species with an anisotropic hyperfine coupling due to two $I = \frac{1}{2}$ nuclei. The spots symmetrically displaced from the diagonals correspond to nuclear frequencies with electron quantum number $m_S = \pm\frac{1}{2}$. The correlation between the nuclear transitions for a particular nucleus is achieved by the π pulse inserted between the second and the third $\pi/2$ pulse of the three-pulse stimulated echo sequence. The spots in the right quadrant are due to a nucleus with frequencies ($\nu_{\alpha1}$, $\nu_{\beta1}$) for the two nuclear transitions corresponding to $m_S = \pm\frac{1}{2}$. The frequencies to the right are small compared to the nuclear Zeeman frequency, while the spots in the left quadrant are for the opposite case with large frequencies ($\nu_{\alpha2}$, $\nu_{\beta2}$)

joining the peaks serves to identify the nucleus. The schematic contour in Fig. 3.33 therefore demonstrates the presence of two magnetic nuclei. Spots due to the nuclear transitions for $m_S = \frac{1}{2}$ and $-\frac{1}{2}$ lines in the right quadrant occur for nuclei with a nuclear Zeeman frequency that is large compared to the hyperfine coupling while those in the left quadrant are for a large hyperfine coupling. Both quadrants are therefore generally shown.

HYSCORE spectra take longer to record than 2- and 3-pulse ESEEM. Although HYSCORE experiments have been applied to study single crystals, the technique is therefore more commonly applied to orientationally disordered systems [54–56]. The 2D time-domain modulation signal, the HYSCORE spectrum and the contour plot of a disordered sample schematically shown in Fig. 3.34, are for a single ^{1}H nucleus with an axially symmetric hyperfine coupling. The nuclear Zeeman frequency is larger than the hyperfine coupling, $|m_S \cdot A_\perp| < |m_S \cdot A_\parallel| < \nu_H$. The HYSCORE spectrum is therefore confined to the $(+, +)$ quadrant. The spots appearing in the contour maps of single crystal spectra are replaced by a ridge symmetrically displaced about the diagonal of the $(+, +)$ quadrant.

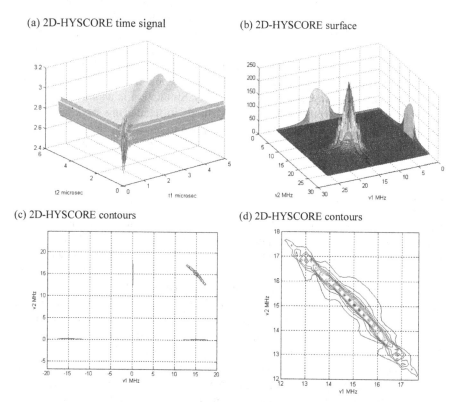

(a) 2D-HYSCORE time signal

(b) 2D-HYSCORE surface

(c) 2D-HYSCORE contours

(d) 2D-HYSCORE contours

Fig. 3.34 (a) Schematic 2D time-domain modulation signal, (b) HYSCORE spectrum, (c) contour plot and (d) expanded contour of a disordered sample for an $S = \frac{1}{2}$ species containing a ^{1}H nucleus with an axially symmetric hyperfine coupling of $A_\perp = -4.4$ MHz, $A_\parallel = 6.6$ MHz. The nuclear Zeeman frequency $\nu_H \approx 15$ MHz is larger than the hyperfine coupling, i.e. $|A_\perp| < |A_\parallel| < 2 \cdot \nu_H$

The spots observed in single crystals are thus replaced by contours reflecting the anisotropy of the hyperfine couplings. The nuclear frequencies ν are in the range $(\nu_H - \frac{1}{2} \cdot A_{||}) < \nu < (\nu_H + \frac{1}{2} \cdot A_{||})$, i.e. at most 11.7–18.3 MHz with the parameters $A_\perp$ = −4.4 MHz, $A_{||}$ = 6.6 MHz, ν_H =15 MHz. The range is shorter in the expanded contour plot (Fig. 2.25d), partly because the ESEEM amplitude is zero along the parallel direction of the *hfc* tensor. The amplitude is zero also in the perpendicular plane, *cf.* E2.9. Direct measurement of the *hfc* from the contour plot is therefore not an accurate method. Better methods to extract the principal values of the hyperfine coupling tensors from these contours have been worked out for cases of practical interest, e.g. for $I = \frac{1}{2}$ [63–65]. The analysis is more complex in the presence of several, inequivalent ^{1}H nuclei as in the case discussed in Chapter 2 for samples of hydrated VO_2^+ in ZSM-5 zeolite. The HYSCORE spectrum in Fig. 2.26 contains a ridge attributed to four different protons with axially symmetric hyperfine couplings of the form, $A_\perp = a_{iso} + T$, $A_{||} = a_{iso} - 2 \cdot T$. The contour lineshape of cross-peaks could be described by a simple equation in which T and a_{iso} were calculated from the slope and intercept of the straight line obtained when ν_α^2 was plotted as a function of ν_β^2, see [66] for details. The analysis is even more complex for nuclei with $I > \frac{1}{2}$ due to appreciable nuclear quadrupole interaction e.g. for ^{14}N ($I = 1$). HYSCORE has a higher sensitivity in the region of low nuclear frequencies than that of ENDOR spectroscopy. The method has therefore been applied to studies of nitrogen-containing paramagnetic species ($\nu(^{14}$N) ≈ 1 MHz at X-band) in biochemical applications [48, 67]. Isotopic labelling with ^{15}N ($I = \frac{1}{2}$) can in this special case simplify the analysis. Simulation is another alternative for the analysis of systems with $I > \frac{1}{2}$. Methods to handle this case have been developed. Analysis of 1D and 2D ESEEM applied to metal ions in biological systems is discussed in a recent review. The reader is referred to the original papers for details [58, 63–65].

3.4.3.7 S > ½ species

Analysis of ESEEM data for $S > \frac{1}{2}$ species, e.g., photoexcited triplet state molecules [54], radical pairs in photosynthetic reaction centres [68], high electron spin transition metal ion centres like Mn^{2+} ($S = 5/2$) interacting with ^{15}N and ^{31}P nuclei ($I = \frac{1}{2}$) [69], and paramagnetic complexes on surfaces usually requires simulation. General simulation programs have been described [59, 70, 71]. Exact diagonalization was employed to obtain the magnetic energies and wavefunctions of a Hamiltonian constructed from electron and/or nuclear Zeeman interactions, exchange and hyperfine couplings, zero-field splittings and nuclear quadrupole interactions. In this case no assumptions had to be made about the relative magnitude of the various interactions. Perturbation methods are applicable when the zero-field splitting is smaller than the electronic Zeeman energy, *i.e.* $D << g \cdot \mu_B \cdot B$. This "high field "approximation has been applied, for instance, for the analysis of X-band ESEEM spectra of organic triplet state molecules [54]. Quadrupole interactions for nuclei with $I > \frac{1}{2}$ may be accounted for analytically or numerically [71]. Approximate analysis procedures therefore tend to be specific depending of the properties of the systems. We illustrate with an example, the $(NO)_2\text{-}Na^+$ complex [72].

The analysis of the ESEEM signals required consideration of the effects of the triplet state ground state, of angular selection due to zero-field and g-tensor anisotropy, and of modulation from ^{23}Na ($I = 3/2$) with relatively large nuclear quadrupole interaction. The experimental data were analysed by simulations to obtain the hyperfine and nuclear quadrupole couplings by an extension of the method described in [58]. The perturbation caused by the combined action of hyperfine, nuclear Zeeman and nuclear quadrupole terms of the ^{23}Na nucleus was thus calculated by diagonalization of the corresponding matrices for each electronic state. These magnetic triplet state energies were calculated exactly. The corresponding 3- and 4-pulse time domain spectra were calculated by an extension of the procedure described in Appendix A3.4.5 for $S = \frac{1}{2}$ to $S = 1$ [72, 73]. An angular selection factor, proportional to the ESR intensity at the applied magnetic field was included in an analogous fashion as in the ENDOR simulation program by Erickson [53]. The simulated spectra obtained after Fourier transformation were fitted to the experimental by trial-and-error adjustment of the anisotropic hyperfine- and quadrupole- couplings, as shown in Fig. 3.35.

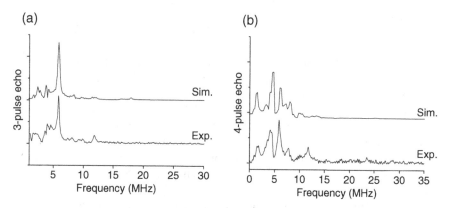

Fig. 3.35 Experimental and simulated 3-pulse (**a**) and 4-pulse (**b**) ESEEM spectra of $(NO)_2$ dimer in Na-LTA zeolite. The 4-pulse spectrum was obtained with the pulse sequence shown in Fig. 3.33 by varying t_1, while t_2 was kept equal to t_1. The simulated curves were obtained with $A_\perp = 4.6$ MHz, $A_{||} = 8.2$ MHz, $Q_\perp = -0.3$ MHz, $Q_{||} = 0.6$ MHz attributed to ^{23}Na in an $S = 1$ complex of the type $(NO)_2\text{-}Na^+$. The figure is adapted from [D. Biglino et al.: Chem. Phys. Lett. **349**, 511 (2001)] with permission from Elsevier

3.4.3.8 Advanced Methods

The analysis of ESEEM data is usually made under the assumption that the applied pulses are "ideal". All ESR transitions are induced, *i.e.* the whole spectrum is "excited". The microwave pulses must then have a large amplitude and a short duration [54]. This may be difficult to realise for broad ESR spectra. The effect of non-ideal pulses is considered in standard textbooks on pulsed ESR [54, 61], and in reviews [70]. Specially designed pulse sequences are also discussed in this context.

3.4.3.9 ESEEM Simulation Software

Programs for general use that have been described in the literature are mainly concerned with 1D ESEEM and HYSCORE simulations and are particularly employed in the analysis of powder data. The earliest program of this type, MAGRES [59], allows for orientation selection and has been extended for HYSCORE simulations. The approach by Benetis et al. [71] for ESEEM and HYSCORE simulations is applicable for $S > \frac{1}{2}$ systems containing nuclei with $I > \frac{1}{2}$. Systems with $I > \frac{1}{2}$ nuclei including the nuclear quadrupole coupling are also treated by an approximate method in works from the group of Pöppl [74]. The effect of partial excitation by non-ideal microwave pulses is taken into account in a program for frequency-domain simulation of HYSCORE spectra [75]. The recent program developed in the group of Arthur Schweiger (deceased) is a most general one, with implemented procedures for angular selection non-ideal pulses and accurate calculations of spectra [70]. Apparently, the program was intended to be integrated in a future version of the EasySpin software.

Software for special applications has been developed e.g. to analyse ESEEM data in studies of solvation structures in disordered systems and for the study of biological systems. The spherical model is of interest to apply when the solvation shell about the paramagnetic centre is disordered, as occurring in frozen glassy matrices. The approach is documented for $S = \frac{1}{2}$ species interacting with hydrogen or deuterium nuclei of neighbour molecules [76]. The analysis aims at obtaining the number of nuclei in the solvation shell and the average distance between a nucleus and the paramagnetic centre. The procedure is computationally efficient. Analysis of ^{14}N ($I = 1$) ESEEM applied to metallo-proteins may require a treatment that is applicable even when the nuclear quadrupole and hyperfine couplings are of comparable magnitudes. Angular selection due to g-anisotropy may also need to be taken into account. Trial-and-error adjustment of parameters by simulation tends to be very time-consuming. The very recent announcement of software (OPTESIM) for automatic optimization of ESEEM spectra is therefore of interest.

Summaries of selected software based on the descriptions given by the developers are given below.

MAGRES – a general program for ESR, ENDOR and ESEEM: The program package MAGRES (MAGnetic RESonance) is able to calculate ESR, ENDOR (without relaxation) and ESEEM spectra for, in principle, any spin system in single crystals as well as in powders. The spin Hamiltonian may be constructed from electron and/or nuclear Zeeman interactions, exchange and hyperfine couplings, zero-field splittings and nuclear quadrupole interactions. For the calculation of eigenvalues and eigenvectors the program uses exact diagonalization, hence no assumptions have to be made about the relative magnitude of the various interactions. Comparison of the calculated frequencies with the experimental ones permits the optimization of the interaction tensors. Simulation of ESEEM spectra is possible for the two- and three-pulse sequence, and the effect of the dead-time may be included. Hyscore spectra can be simulated by an updated program version. The program can also be used for the calculation of c.w. NMR spectra. The program is described in [59].

Numerical Simulation of One- and Two-Dimensional ESEEM Experiment: Automatic orientation selection, grouping of operator factors, and direct selection and elimination of coherences can be used to improve the efficiency of time-domain simulations of one- and two-dimensional electron spin echo envelope modulation (ESEEM) spectra. The program allows for the computation of magnetic interactions of any symmetry and can be used to simulate spin systems with an arbitrary number of nuclei with any spin quantum number. Experimental restrictions due to finite microwave pulse lengths are addressed and the enhancement of forbidden coherences by microwave pulse matching is illustrated. A comparison of simulated and experimental HYSCORE (hyperfine sublevel correlation) spectra of ordered and disordered systems with varying complexity shows good qualitative agreement. The program is described in [70]. It was written in the group of Prof. A. Schweiger, Physical Chemistry, ETH Zurich, CH-8093 Zurich, Switzerland.

Automatic Spin-Hamiltonian Diagonalization for Electronic Doublet Coupled to Anisotropic Nuclear Spins Applied in One- and Two-Dimensional Electron Spin-Echo Simulation: Earlier theoretical inventions in electron paramagnetic resonance were put together into an efficient method leading to accurate explicit expressions of echo modulation spectra in one and two dimensions. The resulting "automatic" diagonalization of the high-field spin-Hamiltonian gives a quick method for electron spin-echo modulation computations. It accounts accurately for the effects of nonaxial electron Zeeman and hyperfine coupling tensors to an arbitrary system of nuclei in any relative orientations, including also the secular part of the quadrupole tensors in the computation of the modulation frequencies. The program is particularly useful in economizing powder simulations of multipulse sequences like hyperfine sublevel correlation spectroscopy. The program is described in [60].

Characteristics of ESEEM and HYSCORE Spectra of $S > \frac{1}{2}$ Centers in Orientationally Disordered Systems: One- and two-dimensional electron-spin echo envelope modulation (ESEEM) spectra of Kramers multiplets in orientationally disordered systems are simulated using a simple mathematical model. A general g-tensor and arbitrary relative orientations between all tensors involving the electron-spin S, the nuclear spin I, and their interaction are assumed. The zero field splitting and the nuclear quadrupole interactions are, however, approximated by their respective secular part in a way that retains all orientation dependencies and it is assumed that the nuclear quadrupole interaction is smaller than the hyperfine interaction. These approximations yield an effective sublevel nuclear Hamiltonian for each ESR transition and are sufficient to account for the most important characteristics of the ESEEM spectra of high electronic multiplets in orientationally disordered systems. The pulses are considered as ideal and selective with respect to the different ESR transitions. The contributions of the latter to the echo intensity are weighed according to their different nutation angles and equilibrium Boltzmann populations. Experimental results obtained from $Mn(D_2O)_6{}^{2+}$ and $VO(D_2O)_5{}^{2+}$ in frozen solutions are presented, compared, and analyzed in light of the theoretical part. The program is described in [71].

General Analysis of ^{14}N (I = 1) Electron Spin Echo Envelope Modulation: General equations for the nuclear interactions in an electron spin system where the ESR signal arises from an isolated Kramers doublet are used to obtain the nuclear frequencies for $I = 1$ associated with such a system. These are incorporated into equations for single-crystal ESEEM amplitudes, which in turn are incorporated into general equations for the orientation-selective ESEEM that arises when the ESR envelope of a frozen-solution (powder) sample is determined by g anisotropy. This development leads to a general picture of the response of the $I = 1$ modulation amplitude to variations in the nuclear hyperfine and quadrupole coupling constants, relative to the nuclear Zeeman interaction. Strong modulation occurs not only in the well-known regime where the "exact/near cancellation" condition ($A/2 \approx \nu_N$) is satisfied, but also when the nuclear hyperfine interaction is much larger than the nuclear Zeeman interaction. The orientation-selective ^{14}N ESEEM frequency-domain patterns in the presence of anisotropic (rhombic) hyperfine and electron Zeeman interactions for both coaxial and noncoaxial cases are computed using analytical solutions when the g-, hyperfine, and nuclear quadrupole tensors are coaxial. The method was applied to the ESEEM of the nitrogenase MoFe protein (Av1) to determine the full hyperfine and nuclear quadrupole tensors of ^{14}N nuclei interacting with the FeMo-cofactor. The program is described in [58]. It was written in the group of Prof. B.M Hoffman, Department of Chemistry, Northwestern University, Evanston, Illinois, 60208, USA.

ESEEM Spherical Model: A program written in IGOR Pro language (WaveMetrics Inc.) that runs on Windows or Mac operating systems has been developed for ESEEM simulations using the spherical model. The nuclear quadrupole interaction for $I > \frac{1}{2}$ is ignored. Fitting of a simulated modulation pattern to an experimental signal is performed. Data are entered interactively from the menu "ESEEM". Submenus are provided for fitting of the decay curve ("Decay curve fitting") and for the shells of nuclei ("Shell parameter"). The type of nuclei is selected by atomic weight in the menu, e.g. 1 for ^1H, 2 for D, and 14 for ^{14}N. The number of nuclei in a shell, the nucleus – electron spin distance (nm), the isotropic hyperfine coupling constant (MHz), and the nuclear spin I are entered in the same menu. The g-factor and the magnetic field strength is entered in the menu "ESR parameter Setting". To run the program choose "Do Simulation" from the main menu. The program written in version 4 of Igor Pro is available from Prof. Jun Kumagai, Department of Applied Chemistry, Graduate School of Engineering, Nagoya University, Nagoya 464-8603, Japan. Some changes in commands are required for simulations with higher versions of Igor Pro.

OPTESIM: The OPTESIM toolbox enables automated numerical simulation of powder ESEEM for arbitrary number (*N*) and type (*I*, g_N) of coupled nuclei, and arbitrary mutual orientations of the hyperfine tensor principal axis systems. The toolbox is based on the Matlab environment, and includes the following features:

(1) A fast algorithm for the computation of spin Hamiltonian into ESEEM
(2) Variety of optimization methods that can be hybridized to achieve an efficient coarse-to-fine grained search of the parameter space and convergence to a global minimum
(3) Statistical analysis of the simulation parameters, which allows the identification of simultaneous confidence regions at specific confident levels

The toolbox includes a geometry preserving spherical averaging algorithm as default for $N > 1$, and global optimization over multiple experimental conditions, such as the dephasing time (τ) for three-pulse ESEEM, and external magnetic field values. In addition, a Java-RMI based distributed computational framework is included in OPTESIM. This framework allows users to build a distributed system of ESEEM simulation on their own PC computer hardware resources. The software is developed in the group of Dr. K. Warncke, Department of Physics, Emory University, Atlanta, GA 30322, USA, and is described in [77].

3.5 Summary

The identity of a paramagnetic species can be determined by measurements of parameters like the g-factor, the hyperfine coupling and other resonance parameters. Free radicals in solution are mostly identified by their hyperfine couplings. ESR spectra can in this case often be analysed visually by stick-plots. The splitting between the lines gives the coupling constant, the relative intensity the number of equivalent nuclei. The assignment of species containing several magnetic nuclei is simplified by simulation of the ESR spectra. ENDOR provides higher resolution, often allowing visual analysis, but has lower sensitivity. Procedures to analyse ESR spectra of samples with large hyperfine couplings or containing several species are discussed. Spectra of paramagnetic species in solids are usually anisotropic. Single crystal measurements are most informative. ESR, ENDOR, or ESEEM measurements in three different crystal planes are usually required to obtain the anisotropic parameters like the g-, the hyperfine-, the zero-field- ($S > \frac{1}{2}$) and nuclear quadrupole coupling tensors. In the majority of cases the Schonland method is applicable for the data analysis to obtain the principal values and directions of all these tensors. Measurements on powders or frozen solutions are common in applied studies. The ESR, ENDOR, or ESEEM lines are then broadened by the anisotropy of the coupling tensors. The principal values but not the directions can in favourable cases be deduced by visual inspection. In general, analysis by simulation is required. Examples are provided for the analysis of anisotropic g, hyperfine couplings, nuclear quadrupole couplings and zero-field splittings in ESR. The angular selection technique in ENDOR is applicable for species with appreciable g-anisotropy to obtain hyperfine couplings with a resolution comparable to that in single crystals. Examples of the analysis of powder ENDOR spectra to obtain the solvation structure of paramagnetic species, the ligand structure of transition metal ions, the structure of surface complexes and of biomolecules are presented. ESEEM is a pulsed ESR method to resolve weak anisotropic hyperfine couplings. Analysis of time domain spectra by simulation was the dominant technique in early studies to obtain solvation structures of trapped electrons in irradiated materials and for characterisation of micro- and mesoporous materials. A procedure based on the spherical approximation is often applied for such systems. In recent applications the analysis is usually made using the frequency domain spectra obtained by Fourier transformation. Simulation methods of ESEEM spectra due to weakly coupled nitrogen nuclei featuring nuclear quadrupole couplings, of relevance in biochemical studies, are exemplified. Exact methods based on the diagonalization of the Hamiltonian and Liouville matrices are briefly reviewed. Methods to simplify the analysis by two-dimensional (2D) techniques have been developed. The so-called HYSCORE method is commonly employed. Procedures to obtain anisotropic hyperfine couplings from 2D contour maps are illustrated by examples. A considerable effort has been made to find suitable software for the simulation of powder ESR, ENDOR and ESEEM spectra. Addresses for down-loading of these programs from the Internet are provided when known. Software for the analysis of single crystal data by the Schonland method is also available.

Appendices

A3.1 The Spin Hamiltonian

Consider a paramagnetic transition metal ion in a crystal when a magnetic field is applied. The total Hamiltonian may in this case be represented by:

$$H = H_{ion} + H_{S\text{-}O} + H_{HFS} + H_{C\text{-}F} + H_m$$

The first three terms account for the interactions within the free ion. With the approximations usually made, H_{ion} represents electrostatic interactions, $H_{S\text{-}O}$ is an internal magnetic interaction, "spin-orbit coupling", and H_{HFS} is the hyperfine structure term. The $H_{C\text{-}F}$ term represents the interaction with surrounding ions or electric dipoles according to the crystal-field approximation, while H_m is the interaction with an external magnetic field. Representative values of the energy splittings due to the various terms for the transition metal ions of the 3d "iron" group are: $E_{ion} \approx 10^5$, $E_{S\text{-}O} \approx 10^2$, $E_{HFS} \approx 10^{-2}$, $E_m \approx 1$, $E_{C\text{-}F} \approx 10^4$ cm^{-1}. The situation with $E_{C\text{-}F} > E_{S\text{-}O}$ is referred to as the medium field case. There are five different orbitals for an electron in the 3d shell. Each can accommodate two electrons with different spin quantum numbers. We consider the example case with a single 3d electron, treated in detail in the classical monograph by Pake [78]. The crystal field splits the energy levels as shown in Fig. 3.36 for cubic and tetragonal symmetries. The ground state is with the 3d electron in the orbital Ψ_1.

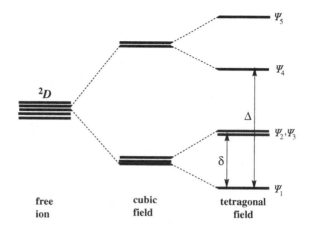

Fig. 3.36 Splitting of $3d$ electron energy levels in a transition metal ion with tetragonal distortion of the ligands

We temporarily neglect the effect of the $H_{S\text{-}O}$ term and consider next the splitting of the Ψ_1 state due to an applied magnetic field, $H_m = \mu_B \mathbf{B}(\mathbf{L} + g_e \mathbf{S})$. The orbital angular momentum $\mathbf{L}$ will then not affect the energy, since the orbitals are 'locked' by the crystal field. The magnetic splitting $\Delta E_m = g_e \mu_B B$ is due entirely to the electron spin in this approximation. Now, of course the spin-orbit interaction, $H_{S\text{-}O} = \lambda \mathbf{L} \cdot \mathbf{S}$, where λ is the spin-orbit coupling, should have been considered before the magnetic interaction. The magnetic splitting of the Ψ_1 state is obtained by first order perturbation theory as $\Delta E_m = (g_e - 8\frac{\lambda}{\Delta})\mu_B B$ when $B \parallel z$. From the resonance condition we define a g_z value, $h\nu = g_z \mu_B B$, with $g_z = (g_e - 8\frac{\lambda}{\Delta})$. Similar calculations with B along the x and y axes yield $g_x = g_y \neq g_z$. The result can be summarized with a spin Hamiltonian using a fictitious spin $S' = \frac{1}{2}$.

$$H = \mu_B(g_x B_x S'_x + g_y B_{y_x} S'_y + g_z B_z S'_z) \tag{3.7}$$

The x, y and z axes were in this example obtained from symmetry. This is possible for species with high symmetry as often assumed for transition metal complexes. A safer procedure is to determine the orientation of the axes experimentally, e.g. by the Schonland procedure described in the main text. For this procedure a more general form of the spin Hamiltonian is employed.

$$H = \mu_B \sum_{i,j=x,y,z} g_{ij} B_i S_j \equiv \mu_B \mathbf{B} \cdot \mathbf{g} \cdot \mathbf{S} \tag{3.8}$$

The g-anisotropy is expressed by a 3·3 matrix, referred to as the g-tensor. The tensor is assumed to be symmetric ($g_{xy} = g_{yx}$, etc.). It is usually specified by three principal values giving the strength and three principal directions giving the orientation of the axes. They can be thought of as the lengths and the directions of the axes of the ellipsoid in Fig. 3.37.

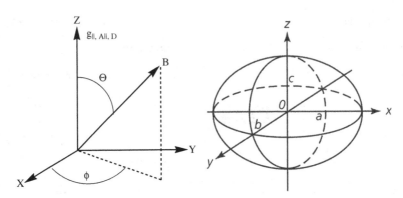

Fig. 3.37 Symbols used in formulae for axial (3.2.1) and rhombic (3.2.2) symmetry. Parameters $g_{||}$, $A_{||}$ and D occur along the Z axis, while $g_\perp$, and $A_\perp$, are in the XY plane. Due to the axial symmetry angle Φ does not enter the equations. The g-, A- and D-tensors with rhombic symmetry can be thought of as the lengths and the directions of the x, y, and z axes of an ellipsoid. The principal axes of the different tensors need not coincide with each other or with the X, Y, Z axes

A more general spin Hamiltonian of the form (3.9) is applied when zero-field splittings ($S > \frac{1}{2}$), anisotropic hyperfine interactions ($I \neq 0$), and nuclear quadrupole couplings ($I \geq 1$) occur.

$$H = \mu_B \mathbf{B} \cdot \mathbf{g} \cdot \mathbf{S} + \mathbf{S} \cdot \mathbf{D} \cdot \mathbf{S} + \mathbf{S} \cdot \mathbf{A} \cdot \mathbf{I} - \mu_N g_N \mathbf{B} \cdot \mathbf{I} + \mathbf{I} \cdot \mathbf{Q} \cdot \mathbf{I} \qquad (3.9)$$

The first three terms are usually the ones of relevance for the ESR analysis, where **D** and **A** are the zero-field (or fine structure) and hyperfine coupling tensors. They are represented by 3·3 symmetric matrices and specified by three principal values and three principal directions as for the g-tensor. The remaining nuclear Zeeman and quadrupole ($I > \frac{1}{2}$) terms do not affect the ESR spectra, unless they are of comparable magnitude to the hyperfine coupling, but must be taken into account in the analysis of ENDOR and ESEEM spectra. The spin Hamiltonian formalism introduced by M.H.L. Pryce and A. Abragam [79] is used explicitly or implicitly in the ESR literature as a convenient way to summarise resonance parameters.

A3.2 Resonance Conditions for ESR and ENDOR in Single Crystals

Axial symmetry was usually assumed for the transition metal ions studied in early ESR work. In the low-symmetry complexes now commonly investigated axial symmetry cannot be assumed, and the g-, hyperfine and zero-field splitting tensors need not have parallel principal axes.

A3.2.1 ESR with Axial Symmetry

The formulas apply for systems with axial symmetry about Z in Fig. 3.37. The equations [80] were obtained under the conditions that the electronic Zeeman interaction dominates and the nuclear Zeeman and quadrupole couplings are negligible.

(1) $S = \frac{1}{2}$:

$$B_0[\text{mT}] = \frac{h \cdot \nu}{g \cdot \mu_B} = \frac{21419.49}{g \cdot \lambda[\text{cm}]} = 0.07144775 \frac{\nu[\text{MHz}]}{g}$$
$$g^2 = g_{\parallel}^2 \cdot \cos^2 \theta + g_{\perp}^2 \cdot \sin^2 \theta$$

(2) $S = \frac{1}{2}, I \neq 0$:

$$B = B_0 - Km_I - \frac{A_{\perp}^2}{4B_0} \left[\frac{A_{\parallel}^2 + K^2}{K^2} \right] \cdot \left[I(I+1) - m_I^2 \right]$$
$$- \frac{1}{2B_0} \left(\frac{A_{\parallel}^2 - A_{\perp}^2}{K} \right)^2 \left(\frac{g_{\parallel} g_{\perp}}{g^2} \right) \cdot m_I^2 \sin^2 \theta \cdot \cos^2 \theta$$
$$K^2 g^2 = A_{\parallel}^2 g_{\parallel}^2 \cdot \cos^2 \theta + A_{\perp}^2 g_{\perp}^2 \cdot \sin^2 \theta$$

(3) $S > \frac{1}{2}$:

$$B = B_0 - D(m_S - 1/2) \cdot \left[3\frac{g_{\parallel}^2}{g^2} \cdot \cos^2 \theta - 1 \right]$$
$$+ \frac{D^2}{2B_0} \left(\frac{g_{\parallel} g_{\perp}}{g^2} \right)^2 [4S(S+1) - 24m_S(m_S - 1) - 9] \cdot \sin^2 \theta \cos^2 \theta$$
$$- \frac{D^2}{8B_0} \cdot \left(\frac{g_{\perp}}{g} \right)^4 \cdot [2S(S+1) - 6m_S(m_S - 1) - 3] \cdot \sin^4 \theta$$

A3.2.2 ESR and ENDOR with Rhombic Symmetry

Similar formulae that are applicable when the principal axes of the coupling tensors do not coincide have been given by several authors [4, 81].

The equations below follow as a rule the notation of Iwasaki [19]. Note however, the use of M and m in place of M_S (m_S) and M_I (m_I) quantum numbers to simplify the notations in the formulae. Resonance parameters for paramagnetic species of low symmetry are usually expressed in a laboratory axes system, for instance the crystallographic axes of a crystalline sample, or a molecule-fixed coordinate system. The **g**, hyperfine (**A**), zero-field (**D**), and nuclear quadrupole (**Q**) interactions are specified by symmetric tensors in the laboratory system, while **E** is the unit tensor. The orientation of the applied magnetic field is given by a unit vector **l** with direction cosines in terms of the polar angles (θ, φ) in Fig. 3.37 as:

$$\ell_X = \sin\theta \cos\Phi, \; \ell_Y = \sin\theta \sin\Phi, \; \ell_Z = \cos\theta.$$

Energies are given to 1st and 2nd order assuming that the electronic Zeeman energy is the dominating one, *i.e.* the high field approximation holds. The principal axes of the coupling tensors need not coincide with the (X, Y, Z) axes. The principal axes of the different tensors need not coincide.

(1) Zero order energy

$$E^0(M) = g\mu_B B \cdot M = \Delta \cdot M \tag{3.10}$$

$$g^2 = \mathbf{l} \cdot \mathbf{g}^2 \cdot \mathbf{l} = \sum_{i,j=1}^{3} (\mathbf{g}^2)_{ij} \, \ell_i \ell_j \tag{3.11}$$

$$\mathbf{u} = \mathbf{g} \cdot \mathbf{l}/g \tag{3.12}$$

(2) First order energy

$$E'(M,m) = \frac{1}{2} F \left[3M^2 - S(S+1) \right] + G_M\, m + \frac{1}{2} P_M \left[3m^2 - I(I+1) \right] \tag{3.13}$$

The first and third terms are to be included when $S > \frac{1}{2}$ and $I > \frac{1}{2}$, respectively.

$$F = \mathbf{u} \cdot \mathbf{D} \cdot \mathbf{u} \tag{3.14}$$

$$G_M^2 = \mathbf{l} \cdot \mathbf{T}_M^T \cdot \mathbf{T}_M \cdot \mathbf{l} \tag{3.15}$$

$$\mathbf{T}_M = \left(\frac{M}{g} \mathbf{Ag} - h\nu_N \mathbf{E} \right), \qquad \mathbf{T}_M^T = \left(\frac{M}{g} \mathbf{gA} - h\nu_N \mathbf{E} \right) \tag{3.16}$$

$$\nu_N = g_N \mu_N B/h \tag{3.17}$$

$$\mathbf{k}_M = \mathbf{T}_M \cdot \mathbf{l}/G_M \tag{3.18}$$

$$P_M = \mathbf{k}_M \cdot \mathbf{Q} \cdot \mathbf{k}_M \tag{3.19}$$

Equation (3.19) is valid when $|G_M| \gg |P_M|$.
(3) Second order energy

$$E''(M,m) = E_d''(M) + E_a''(M,m) + E_{ad}''(M,m) + E_q''(M,m)$$

The subscripts d, a, and q denote zero-field, hyperfine coupling and nuclear quadrupole coupling energies, respectively, while ad stands for a cross-term.

$$E''_d(M, m) = -d_1 M \left[4 \cdot S(S+1) - 8M^2 - 1 \right] + d_2 M \left[2 \cdot S(S+1) - 2 \cdot M^2 - 1 \right]$$
$$(3.20)$$

$$d_1 = \frac{\mathbf{u} \cdot \mathbf{D}^2 \cdot \mathbf{u} - (\mathbf{u} \cdot \mathbf{D} \cdot \mathbf{u})^2}{2 \cdot \Delta} \tag{3.21}$$

$$d_2 = \frac{2 \cdot Tr\mathbf{D}^2 + (\mathbf{u} \cdot \mathbf{D} \cdot \mathbf{u})^2 - 4 \cdot \mathbf{u} \cdot \mathbf{D}^2 \cdot \mathbf{u}}{8 \cdot \Delta} \tag{3.22}$$

$$E''_q(M, m) = -q_1 m \left[4 \cdot I(I+1) - 8m^2 - 1 \right] + q_2 m \left[2 \cdot I(I+1) - 2 \cdot m^2 - 1 \right]$$
$$(3.23)$$

$$q_1 = \frac{\mathbf{k}_M \cdot \mathbf{Q}^2 \cdot \mathbf{k}_M - (\mathbf{k}_M \cdot \mathbf{Q} \cdot \mathbf{k}_M)^2}{2 \cdot G_M} \tag{3.24}$$

$$q_2 = \frac{2 \cdot Tr\mathbf{Q}^2 + (\mathbf{k}_M \cdot \mathbf{Q} \cdot \mathbf{k}_M)^2 - 4 \cdot \mathbf{k}_M \cdot \mathbf{Q}^2 \cdot \mathbf{k}_M}{8 \cdot G_M} \tag{3.25}$$

$$E''_a(M, m) = a_1 Mm^2 - a_2 m \left[S(S+1) - M^2 \right] + a_3 M \left[I(I+1) - m^2 \right] \tag{3.26}$$

$$a_1 = \frac{\mathbf{k}_M \cdot \mathbf{A}^2 \cdot \mathbf{k}_M - (\mathbf{k}_M \cdot \mathbf{A} \cdot \mathbf{u})^2}{2\Delta} \tag{3.27}$$

$$a_2 = \frac{(\mathbf{u} \cdot \mathbf{A}^{-1} \cdot \mathbf{k}_M) \cdot \det(\mathbf{A})}{2\Delta} \tag{3.28}$$

$$a_3 = \frac{Tr\mathbf{A}^2 - \mathbf{u} \cdot \mathbf{A}^2 \cdot \mathbf{u} - \mathbf{k}_M \cdot \mathbf{A}^2 \cdot \mathbf{k}_M + (\mathbf{k}_M \cdot \mathbf{A} \cdot \mathbf{u})^2}{4 \cdot \Delta} \tag{3.29}$$

$$E''_{ad}(M, m) = -d_a \left[S(S+1) - 3M^2 \right] m \tag{3.30}$$

$$d_a = \frac{\mathbf{u} \cdot \mathbf{D} \cdot \mathbf{A} \cdot \mathbf{k}_M - (\mathbf{u} \cdot \mathbf{D} \cdot \mathbf{u})(\mathbf{k}_M \cdot \mathbf{A} \cdot \mathbf{u})}{\Delta} \tag{3.31}$$

The line positions of an ESR spectrum can accordingly be calculated to 2nd order. The different coupling tensors need not have parallel principal axes. We refer to [19] for procedures to calculate the intensity of the ESR lines under the same assumptions. The formulae have been incorporated in software for the simulation of powder ESR spectra, see Section 3.4.1.7.

A3.3   Resonance Conditions for Powder ENDOR

A3.3.1   S = ½, I = ½ and S = ½, I > ½ with *nqc* < *hfc*

The ENDOR frequencies for the nuclear transition $(M,m) \leftrightarrow (M,m+1)$ can be obtained from the 1st, $E'(M,m)$, and 2nd, $E''(M,m)$, order equations for the energies in Appendix A3.2.2 [19].

$$v(M,m \leftrightarrow m+1) = \left| G_M + 3P_M\,(m+1/2) + \left[E''(M,m+1) - E''(M,m) \right] \right| \quad (3.32)$$

The quantity G_M contains contributions from the hyperfine tensor **A** and the nuclear Zeeman term v_N. A quadrupole energy term P_M contributes when $I > ½$. By applying 2nd order corrections frequencies can be obtained with better accuracy when the condition $|G_M| \gg |P_M|$ does not strictly apply. Equation (3.33) for the ENDOR transition probability first given by Toriyama et al. [45b] applies for species with small *g* anisotopy.

$$W^2(M,m \leftrightarrow m+1) \propto \left[\boldsymbol{r} \cdot \mathbf{T}_M^T \cdot \mathbf{T}_M \cdot \boldsymbol{r} - (\mathbf{k}_M \cdot \mathbf{T}_M \cdot \boldsymbol{r})^2 \right] \quad (3.33)$$

The direction of the radiofrequency field (RF) is specified by the unit vector $\boldsymbol{r}$. An exhaustive analysis of ENDOR transition probabilities including cases with anisotropic *g* and with circularly polarised RF fields has been presented by Schweiger and Günthard [45c].

The formulas, (3.32) and (3.33), apply to a single crystal, with the tensors and field directions specified in a suitably chosen coordinate system, e.g. the crystallographic axes. Simulation of powder spectra is usually performed by a numerical integration over the polar angles (θ, φ) for the magnetic field direction as in Eq. (3.34) below.

A3.3.2   S = ½, I = ½ and S = ½, I > ½ with *nqc* ≈ *hfc*

The powder ENDOR spectra in this book were calculated with software developed by Erickson [53]. The main equations employed are reproduced, in slightly different notation from that used previously [12, 53]. The powder line shape at the ENDOR frequency v and static magnetic field B is thus given as:

$$Y'(B,v) = \int_\theta \sin\theta\, d\theta \int_\varphi \sum_{ij} s(B - B_{ij}) V_{ij}^2 \left\{ \sum_k t(v - v_{ik}) \overline{W}_{ik}^2 + \sum_l t(v - v_{jl}) \overline{W}_{jl}^2 \right\} d\varphi$$

$$(3.34)$$

The ENDOR signal is assumed to be proportional to the ESR absorption with the field locked at the ESR transition $|-½, i\rangle \leftrightarrow |½, j\rangle$, *i.e.* to $s(B - B_{ij}) V_{ij}^2$ as indicated in Fig. 3.38 [53]. This factor affects the shape of the calculated ENDOR spectra, and gives rise to angular selection even when the *g*- or other anisotropy is not completely resolved in the corresponding ESR spectra.

Fig. 3.38 Mechanism of
angular selection in powder
ENDOR. The ENDOR signal
for the transition between the
states |M, j> ↔ |M, k>, M =
±½ with the field locked at
the ESR transition |−½, i> ↔
|½, j> depends on the strength
$s(B − B_{ij})V_{ij}^2$ of the ESR
absorption as suggested
in [53]

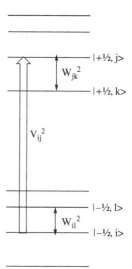

The ENDOR intensity to first order for the transition between the states |M, j>
↔ |M, k> in Eq. (3.34) is

$$\overline{W}_{jk}^2(\theta, \varphi) = \frac{1}{2}\left(\frac{B_2}{B}\right)^2 (\alpha^*\mathbf{T}^2\alpha - \alpha^*\mathbf{T}\mathbf{l} \cdot \alpha\mathbf{T}\mathbf{l}), \qquad (3.35)$$

where the bar indicates it applies to the powder. **l** is the unit vector of the static
magnetic field (not the radiofrequency field), $\mathbf{T} = (\frac{M}{g}\mathbf{Ag} - \nu_N\mathbf{E})$ is the "ENDOR
tensor" appearing also in the energy expressions in Appendix A3.3.2 and α is a
complex vector with the components $< j \mid I_x \mid k >$, $< j \mid I_y \mid k >$ and $< j \mid I_z \mid k >$.

The ENDOR frequencies are obtained from the energy eigenvalues $E(M, j)$ by
diagonalization of the Hamiltonian H′:

$$H' = \mathbf{I}\mathbf{A}\mathbf{u}S_u + \mathbf{I}\mathbf{Q}\mathbf{I} - g_N\beta_N\mathbf{B} \cdot \mathbf{I} \qquad (3.36)$$

S_u is the component of S along the effective Zeeman field direction $\mathbf{u} = \mathbf{g}\mathbf{l}/g$.

$$\nu_{jk} = (E(M,j) - E(M,k))/h \qquad (3.37)$$

Here j and k denote states belonging to the same electron quantum number, M i.e.
either $m_S = ½$ or $m_S = -½$. In general the states are mixtures of nuclear spin
states. Thus, the ordinary selection rule $\Delta m_I = 1$ does not apply for systems with
appreciable nuclear quadrupole couplings.

ENDOR signals due to nuclei with arbitrary values of the nuclear spin I can be
analyzed by this unconventional perturbation method, where the electronic Zeeman
term is the dominating one, while the hyperfine, quadrupole and nuclear Zeeman

interactions are treated, simultaneously, as a joint perturbation. Thus, no assumptions need to be made regarding their relative magnitude or the relative orientation of the principal axes of the tensors.

A3.3.3 S > 1/2

The formulae by Iwasaki [19] reproduced in Appendix A3.2.2 are applicable also for $S > \frac{1}{2}$, when the electron Zeeman term dominates. Procedures based on diagonalisation of the spin Hamiltonian [59] apply generally.

A3.4 Resonance Conditions for Powder ESEEM

Simulation of powder ESEEM spectra is usually performed by a numerical integration over the magnetic field directions. Frequently used Equations [54, 57, 61, 82–85] are reproduced below. Angular selection can be taken into account in a manner analogous to that used in powder ENDOR simulations.

A3.4.1 S = I = 1/2

The echo signal can be expressed in terms of the frequencies ν_α and ν_β and the amplitudes I_α and I_β, defined in the ESR stick diagram schematically shown in Fig. 3.39.

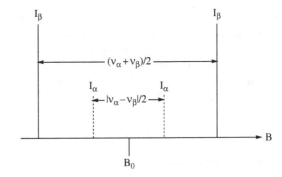

Fig. 3.39 $S = \frac{1}{2}$ ESR line pattern for an $I = \frac{1}{2}$ nucleus with an anisotropic hyperfine coupling and a nuclear Zeeman term of comparable magnitude. The intensities I_α and I_β depend on the values of ν_α and ν_β and the nuclear Zeeman energy ν_N in frequency units according to Eqs. (3.38) and (3.39)

In case of an isotropic g-factor one has for the frequencies and amplitudes [57]:

$$\nu_\alpha^2 = 1 \cdot \left(+\frac{1}{2}A - \nu_N E \right)^2 \cdot 1 \qquad I_\alpha = \frac{\left| \nu_n^2 - (1/4)\left(\nu_\alpha + \nu_\beta \right)^2 \right|}{\nu_\alpha \nu_\beta} \tag{3.38}$$

$$\nu_\beta^2 = 1 \cdot \left(-\frac{1}{2}A - \nu_N E \right)^2 \cdot 1 \qquad I_\beta = \frac{\left| \nu_n^2 - (1/4)\left(\nu_\alpha - \nu_\beta \right)^2 \right|}{\nu_\alpha \nu_\beta} \tag{3.39}$$

It is customary to introduce a modulation depth factor $k = 4 \cdot I_\alpha \cdot I_\beta$ and to use the angular frequency $\omega = 2\pi\nu$ in the equations for the time dependence of the ESEEM signal.

Two-pulse [54]:

$$V(\tau) = 1 - 2k \sin^2\left(\frac{\omega_\alpha \tau}{2}\right) \sin^2\left(\frac{\omega_\beta \tau}{2}\right) \tag{3.40}$$

Three-pulse [54]:

$$V(\tau, T) = 1 - k\left[\sin^2\left(\frac{\omega_\alpha \tau}{2}\right)\sin^2\left(\frac{\omega_\beta(\tau+T)}{2}\right) + \sin^2\left(\frac{\omega_\beta \tau}{2}\right)\sin^2\left(\frac{\omega_\alpha(\tau+T)}{2}\right)\right] \tag{3.41}$$

HYSCORE [84, 85]:

$$V(t_1, t_2) = 1 - I_\alpha \cdot I_\beta \left[C_0 + C_1 \cdot \phi_1 + C_2 \cdot \phi_2 + C_3 \cdot \phi_3\right] \tag{3.42}$$

$$C_0 = 3 - \cos(\omega_\beta \tau) - \cos(\omega_\alpha \tau) - I_\alpha \cos(\omega_+ \tau) - I_\beta \cos(\omega_- \tau) \tag{3.42a}$$

$$C_1 = I_\beta \cos\left(\omega_\beta \tau - \frac{\omega_\alpha \tau}{2}\right) + I_\alpha \cos\left(\omega_\beta \tau + \frac{\omega_\alpha \tau}{2}\right) - \cos\left(\frac{\omega_\alpha \tau}{2}\right) \tag{3.42b}$$

$$C_2 = I_\beta \cos\left(\omega_\alpha \tau - \frac{\omega_\beta \tau}{2}\right) + I_\alpha \cos\left(\omega_\alpha \tau + \frac{\omega_\beta \tau}{2}\right) - \cos\left(\frac{\omega_\beta \tau}{2}\right) \tag{3.42c}$$

$$C_3 = -2\sin\frac{\omega_\alpha \tau}{2} \cdot \sin\frac{\omega_\beta \tau}{2} \tag{3.42d}$$

$$\phi_1 = \cos\left(\omega_\alpha t_1 + \frac{\omega_\alpha \tau}{2}\right) + \cos\left(\omega_\alpha t_2 + \frac{\omega_\alpha \tau}{2}\right) \tag{3.42e}$$

$$\phi_2 = \cos\left(\omega_\beta t_1 + \frac{\omega_\beta \tau}{2}\right) + \cos\left(\omega_\beta t_2 + \frac{\omega_\beta \tau}{2}\right) \tag{3.42f}$$

$$\phi_3 = I_\beta \left\{\cos\left(\omega_\alpha t_1 + \omega_\beta t_2 + \frac{\omega_+ \tau}{2}\right) + \cos\left(\omega_\beta t_1 + \omega_\alpha t_2 + \frac{\omega_+ \tau}{2}\right)\right\}$$

$$+ I_\alpha \left\{\cos\left(\omega_\alpha t_1 - \omega_\beta t_2 + \frac{\omega_- \tau}{2}\right) + \cos\left(\omega_\beta t_1 - \omega_\alpha t_2 + \frac{\omega_- \tau}{2}\right)\right\} \tag{3.42g}$$

A3.4.2   S = ½, I = 1, nqc = 0

Two-pulse [54]:

$$V(\tau) = 1 - \frac{16k}{3}\sin^2\left(\frac{\omega_\alpha \tau}{2}\right)\sin^2\left(\frac{\omega_\beta \tau}{2}\right) + \frac{16k^2}{3}\sin^4\left(\frac{\omega_\alpha \tau}{2}\right)\sin^4\left(\frac{\omega_\beta \tau}{2}\right) \tag{3.43}$$

Three-pulse [54]:

$$V(\tau, T)$$

$$= 1 - \frac{8k}{3}\left[\sin^2\left(\frac{\omega_\alpha \tau}{2}\right)\sin^2\left(\frac{\omega_\beta(\tau+T)}{2}\right) + \sin^2\left(\frac{\omega_\beta \tau}{2}\right)\sin^2\left(\frac{\omega_\alpha(\tau+T)}{2}\right)\right]$$

$$+ \frac{8k^2}{3}\left[\sin^4\left(\frac{\omega_\alpha \tau}{2}\right)\sin^4\left(\frac{\omega_\beta(\tau+T)}{2}\right) + \sin^4\left(\frac{\omega_\beta \tau}{2}\right)\sin^4\left(\frac{\omega_\alpha(\tau+T)}{2}\right)\right] \tag{3.44}$$

A3.4.3 $S = \frac{1}{2}, I = 1$ with $nqc \neq 0$

An analytical treatment applicable when the quadrupole interaction is small compared to the nuclear Zeeman term is given in [57]. A numerical method due to Hoffman et al. [58] is valid irrespective of the relative magnitudes of the hyperfine, nuclear Zeeman and quadrupole terms.

A3.4.4 $S = \frac{1}{2}, I > 1$ with $nqc \neq 0$

A numerical method for arbitrary nuclear spin values has been described [54] using a similar procedure as applied in the original treatment by Mims [82]. The method is thus applicable for all values of the hyperfine, nuclear Zeeman and quadrupole terms smaller than the electron Zeeman energy. The frequencies are obtained as for ENDOR in Appendix A3.3.2. The modulation amplitudes depend on the elements of a matrix denoted M by Mims. The elements are calculated as the dot product $M_{ik} = \mathbf{F}_i^{(\alpha)} \cdot \mathbf{F}_k^{(\beta)}$ [83] between the nuclear state eigenvectors i for electronic level $m_S = \frac{1}{2}$ (α) and k for $m_S = -\frac{1}{2}$ (β). The M-matrix thus replaces the modulation depth factor k employed for nuclei with $I = \frac{1}{2}$.

Two-pulse: The modulation pattern due to a nucleus with spin I can be calculated as

$$V_I = \frac{1}{2I+1}\mathrm{Re}\sum_{i,j}\sum_{l,n}M_{il}^*M_{in}M_{jn}^*M_{jl}e^{-i\left(\omega_i^{(\alpha)}-\omega_j^{(\alpha)}+\omega_l^{(\beta)}-\omega_n^{(\beta)}\right)\tau} \tag{3.45}$$

Indices i and j refer to $m_S = \frac{1}{2}$, l and n to $-\frac{1}{2}$. A form suitable for numeric calculation is given in [54].

Three-pulse: Two terms corresponding to $m_S = \frac{1}{2}$ (α) and $-\frac{1}{2}$ (β) electronic states are added [54].

$$V_I(\tau, T) = \frac{1}{2(2I+1)}\left[V_I^{(\alpha)}(\tau, T) + V_I^{(\beta)}(\tau, T)\right] \tag{3.46}$$

$$V_I^{(\alpha)} = \mathrm{Re}\sum_{i,j}\sum_{l,n}M_{il}M_{in}^*M_{jn}M_{jl}^*e^{-i\left(\omega_i^{(\alpha)}-\omega_j^{(\alpha)}\right)(\tau+T)-i\left(\omega_l^{(\beta)}-\omega_n^{(\beta)}\right)\tau} \tag{3.46a}$$

$$V_I^{(\beta)} = \mathrm{Re} \sum_{i,j} \sum_{l,n} M_{il}^* M_{in} M_{jn}^* M_{jl} e^{-i\left(\omega_i^{(\alpha)} - \omega_j^{(\alpha)}\right)\tau - i\left(\omega_l^{(\beta)} - \omega_n^{(\beta)}\right)(\tau+T)} \tag{3.46b}$$

Equations suitable for programming are given in [54].

Hyscore: The signal is given by [71, 73]

$$V(t_1, t_2) = V^\alpha(t_1, t_2) + V^\beta(t_1, t_2) \tag{3.47}$$

$$V^\alpha(t_1, t_2) = \sum_{j,k,l} \sum_{m,n,p} M_{jk} \cdot M_{lk}^* \cdot M_{lm} \cdot M_{nm}^* \cdot M_{np} \cdot M_{jp}^* \mathrm{T}^\alpha(t_1, t_2) \tag{3.47a}$$

with

$$\mathrm{T}^\alpha(t_1, t_2) = e^{-i\left[\left(\omega_l^{(\alpha)} - \omega_j^{(\alpha)} + \omega_k^{(\beta)} - \omega_m^{(\beta)}\right)\tau + \left(\omega_l^{(\alpha)} - \omega_n^{(\alpha)}\right)t_1 + \left(\omega_k^{(\beta)} - \omega_p^{(\beta)}\right)t_2\right]}$$

$V^\beta(t_1, t_2)$ is obtained by interchanging α and β in the equations.

A3.4.5 S ≥ 1

ESEEM signals for $S > \frac{1}{2}$ species are as a rule affected by angular selection due to the zero-field splitting in addition to that caused by g-tensor anisotropy for $S = \frac{1}{2}$ species. The zero-field splitting usually also exceeds the maximum spectral width that can be excited with pulsed ESR instruments presently in use. We refer to the literature for general analysis procedures of ESEEM for $S > \frac{1}{2}$ with approximative and exact methods [54, 71, 82]. In the special case of $S = 1$ the authors employed the ESR line shape as a weighting function for the ESEEM signal in analogy to the procedure employed by Erickson for the angular selection in ENDOR [53]. The effect of the limited bandwidth was approximated by neglecting signals outside the window of the estimated bandwidth of the instrument. This treatment does not take into account e.g. electron spin flip angles different from 90° or 180°, and was introduced as a first approximation to simulate ESEEM spectra of a $(NO)_2$–Na^+ triplet state adsorption complex with a wide ESR spectrum, see Section 3.4.3.8 and Chapter 6.

References

1. A. Carrington, A.D. McLachlan: '*Introduction to Magnetic Resonance*', Harper & Row, New York, NY (1967).
2. H. Kurreck, B. Kirste, W. Lubitz: '*Electron Nuclear Double Resonance Spectroscopy of Radicals in Solution*', VCH Publishers, New York, NY (1988).
3. N.M. Atherton: '*Principles of Electron Spin Resonance*', Ellis Horwood, Chichester (1993).
4. J.A. Weil, J.R. Bolton: '*Electron Paramagnetic Resonance: Elementary Theory and Practical Applications*', 2nd Edition, Wiley, Hoboken, NJ (2007).
5. R.W. Fessenden, R.H Schuler: J. Chem. Phys. **43**, 2704 (1965).
6. J. Heinzer: QCPE **11**, 197 (1970).
7. (a) B. Kirste: Anal. Chim. Acta **265**, 191 (1992). (b) http://www.chemie.fu-berlin.de/chemistry/epr/eprft.html.
8. B.L. Bales, E. Wajnberg, O.R. Nascimento: J. Magn. Reson. A **118**, 227 (1996).

9. D.S. Schonland: Proc. Phys. Soc. (Lond.) **73**, 788 (1959).
10. J.A. Weil, J.E. Clapp, T. Buch: Adv. Magn. Reson. **6**, 183 (1973).
11. W.H. Nelson: J. Magn. Reson. **38**, 71 (1980).
12. A. Lund, M. Shiotani (eds.): '*EPR of Free Radicals in Solids, Ttrends in Methods and Applications*', Kluwer Academic Publishers, Dordrecht (2003).
13. C.P. Poole, Jr., H.A. Farach: J. Magn. Reson. **4**, 312 (1971).
14. A.R. Sørnes, E. Sagstuen, A. Lund: J. Phys. Chem. **99**, 16867 (1995).
15. A. Lund, T. Vänngård: J. Chem. Phys. **42**, 2979 (1965).
16. K. Minakata, M. Iwasaki: Mol. Phys. **23**, 1115 (1972).
17. H. Vrielinck, H. De Cooman, M.A. Tarpan, E. Sagstuen, M. Waroquier, F. Callens: J. Magn. Reson. **195**, 196 (2008).
18. J.R. Byberg, S.J.K. Jensen, L.T. Muus: J. Chem. Phys. **46**, 131 (1967).
19. M. Iwasaki: J. Magn. Reson. **16**, 417 (1974).
20. F.K. Kneubühl: J. Chem. Phys. **33**, 1074 (1960).
21. M. Ikeya: '*New Applications of Electron Paramagnetic Resonance, Dating, Dosimetry and Microscopy*', World Scientific, Singapore (1993).
22. H.M. McConnell, J. Strathdee: Mol. Phys. **2**, 129 (1959).
23. J.A. Weil, J.H. Anderson: J. Chem. Phys. **35**, 1410 (1961).
24. R. Lefebvre, J. Maruani: J. Chem. Phys. **42**, 1480 (1965).
25. T. Vänngård, R. Aasa: '*Proceedings of the First International Conference on Paramagnetic Resonance (1962)*', volume II, Academic Press, New York, NY (1963), p. 509.
26. S. Kokatam, K. Ray, J. Pap, E. Bill, W.E. Geiger, R.J. LeSuer, P.H. Rieger, T. Weyhermüller, F. Neese, K. Wieghardt: Inorg. Chem. **46**, 1100 (2007).
27. F.G. Herring, C.A. McDowell, J.C. Tait: J. Chem. Phys. **57**, 4564 (1972).
28. E. Sagstuen, E.O. Hole: In '*Electron Paramagnetic Resonance: A Practitioner's Toolkit*', ed. by M. Brustolon and E. Giamello, Wiley, Hoboken, NJ (2009), Chapter 9.
29. O. Edlund, A. Lund, M. Shiotani, J. Sohma: Mol. Phys. **32**, 49 (1976).
30. P.H. Kasai: Chem. Mater. **6**, 1581 (1994).
31. (a) J. Maruani, C.A. McDowell, H. Nakajima, P. Raghunathan: Mol. Phys. **14**, 349 (1968). (b) J. Maruani, J.A.R. Coope, C.A. McDowell: Mol. Phys. **18**, 165 (1970).
32. J.R. Pilbrow: '*Transition Ion Electron Paramagnetic Resonance*', Clarendon Press, Oxford (1990).
33. Simfonia EPR simulation programme, www.bruker-biospin.com/epr_software.html
34. G.R. Hanson, K.E.Gates, Ch.J. Noble, M. Griffin, A. Mitchell, S. Benson: (a) J. Inorg. Biochem. **98**, 903 (2004–2005); (b) In '*EPR of Free Radicals in Solids, Trends in Methods and Applications*' ed. by A. Lund, M. Shiotani, Kluwer Academic Publishers, Dordrecht (2003), Chapter 5.
35. S. Stoll, A. Schweiger: J. Magn. Reson. **178**, 42 (2006).
36. A. Lund, R. Erickson: Acta Chem. Scand. **52**, 261 (1998).
37. (a) J. Eloranta, M. Vuolle: Magn. Reson. Chem. **36**, 98 (1998). (b) J. Eloranta: EPR Newsl. (International EPR society) **10**, 3 (1999).
38. (a) R.B. Clarkson, R.L. Belford, K.S. Rothenberger, H.C. Crookham: J. Catal. **106**, 500 (1987). (b) R.B. Clarkson, K. Mattson, W. Shi, W. Wang, R.L. Belford: In '*Radicals on Surfaces*' ed. by A. Lund, C. Rhodes, Kluwer, Dordrecht (1995), Chapter II. (c) A. Lund, C. Rhodes (guest eds.): Mol. Eng. **4**, 3 (1994).
39. (a) F. Gerson: Acc. Chem. Res. **27**, 63 (1994). (b) A. Lund, M. Shiotani (eds.): '*Radical Ionic Systems*', Kluwer Academic Publishers, Dordrecht (1991).
40. (a) B.M. Hoffmann, J. Martinsen, R.A. Venters: J. Magn. Reson. **59**, 110 (1984). (b) B.M. Hoffman, R.J. Gurbiel: J. Magn. Reson. **82**, 309 (1989).
41. A.J. Hoff (ed.): '*Advanced EPR: Applications in Biology and Biochemistry*', Elsevier, Amsterdam (1989).
42. R.J. Gurbiel, C.J. Batie, M. Sivaraja, A.E. True, J.A. Fee, B.M. Hoffman, D.P. Ballou: Biochemistry **28**, 4861 (1989).

43. C.J. Bender, M. Sahlin, G.T. Babcock, B.A. Barry, T.K. Chandrasekar, S.P. Salowe, J. Stubbe, B. Lindström, L. Pettersson, A. Ehrenberg, B.-M. Sjöberg: J. Am. Chem. Soc. **111**, 8076 (1989).
44. C. Tommos, X.-S. Tang, K. Warncke, C.W. Hoganson, S. Styring, J. McCracken, B.A. Diner, G.T. Babcock: J. Am. Chem. Soc. **117**, 10325 (1995).
45. (a) D.H. Whiffen: Mol. Phys. **10**, 595 (1966). (b) K. Toriyama, K. Nunome, M. Iwasaki: J. Chem. Phys. **64**, 2020 (1976). (c) A. Schweiger, Hs.H. Günthard: Chem. Phys. **70**, 1 (1982).
46. G. Feher: Phys. Rev. **103**, 834 (1956).
47. G.H. Rist, J.S. Hyde: J. Chem. Phys. **52**, 4633 (1970).
48. B.M. Hoffman: (a) Acc. Chem. Res. **24**, 164 (1991); (b) ibid. **36**, 522 (2003).
49. A. Kreiter, J. Hüttermann: J. Magn. Reson. **93**, 12 (1991).
50. A. Pöppl, T. Rudolf, P. Manikandan, D. Goldfarb: J. Am. Chem. Soc. **122**, 10194 (2000).
51. (a) M. Iwamoto, H.Yahiro: Catal. Today **22**, 5 (1994). (b) H. Yahiro, M. Iwamoto: Appl. Catal. A **222**, 163 (2001).
52. (a) J. Niklas, T. Schulte, S. Prakash, M. van Gastel, E. Hofmann, W. Lubitz: J. Am. Chem. Soc. **129**, 15442 (2007). (b) W. Lubitz, F. Lendzian, R. Bittl: Acc. Chem. Res. **35**, 313 (2002).
53. R. Erickson: (a) *Electron magnetic resonance of free radicals. Theoretical and experimental EPR, ENDOR and ESEEM studies of radicals in single crystal and disordered solids*, Ph. D thesis, Linköping studies in science and technology. Dissertation No. 391 (1995); (b) Chem. Phys. **202**, 263 (1996).
54. S.A. Dikanov, Yu.D. Tsvetkov, *Electron Spin Echo Envelope Modulation (ESEEM) Spectroscopy*, CRC Press, Boca Raton, FL (1992).
55. J. McCracken: In *Applications of Physical Methods to Inorganic and Bioinorganic Chemistry* ed. by R.A. Scott, C.M. Lukehart, Wiley, Chichester (2007).
56. L. Kevan: In *Handbook of Zeolite Science and Technology* ed. by S.M. Auerbach, K.A. Carrado, P.K. Dutta, Marcel Dekker, Inc., New York, NY (2003), Chapter 7.
57. M. Iwasaki, K. Toriyama: J. Chem. Phys. **82**, 5415 (1985).
58. H.-I. Lee, P.E. Doan, B.M. Hoffman: J. Magn. Reson. **140**, 91 (1999).
59. C.P. Keijzers, E.J. Reijerse, P. Stam, M.F. Dumont, M.C.M. Gribnau: J. Chem. Soc. Faraday Trans. 1 **83**, 3493 (1987).
60. N.P. Benetis, A.R. Sørnes: Concepts Magn. Reson. **12**, 410 (2000).
61. A. Schweiger, G. Jeschke: *Principles of Pulse Electron Paramagnetic Resonance*, Oxford University Press, Oxford (2001).
62. N.P. Benetis, U.E. Nordh: Chem. Phys. **200**, 107 (1995).
63. S.A. Dikanov, M.K. Bowman: J. Magn. Reson. A **116**, 125 (1995).
64. Y. Deligiannakis, M. Louloudi, N. Hadjiliadis: Coord. Chem. Rev. **204**, 1 (2000).
65. A.G. Maryasov, M.K. Bowman: J. Magn. Reson. **179**, 120 (2006).
66. J. Woodworth, M. Bowman, S. Larsen: J. Phys. Chem. B **108**, 16128 (2004).
67. S.A. Dikanov, D.R.J. Kolling, B. Endeward, R.I. Samoilova, T.F. Prisner, S.K. Nair, A.R. Crofts: J. Biol. Chem. **281**, 27416 (2006).
68. R.J. Hulsebosch, I.V. Borovykh, S.V. Paschenko, P. Gast, A.J. Hoff: J. Phys. Chem. B **103**, 6815 (1999).
69. A.V. Astashkin, A.M. Raitsimring: J. Chem. Phys. **117**, 6121 (2002).
70. Z.L. Madi, S. Van Doorslaer, A. Schweiger: J. Magn. Reson. **154**, 181 (2002).
71. N.P. Benetis, P.C. Dave, D. Goldfarb: J. Magn. Reson. **158**, 126 (2002).
72. W. Liu, A. Lund, M. Shiotani, J. Michalik: Appl. Magn. Reson. **24**, 285 (2003).
73. N.P. Benetis, private communication.
74. M. Gutjahr, R. Bottcher, A. Pöppl: Appl. Magn. Reson. **22**, 401 (2002).
75. R. Szosenfogel, D. Goldfarb: Mol. Phys. **95**, 1295 (1998).
76. L. Kevan, R.N. Schwartz, (eds.): *Time Domain Electron Spin Resonance*, Wiley, New York, NY (1979).
77. L. Sun, J. Hernandez-Guzman, K. Warncke: (a) J. Magn. Reson. **200**, 21 (2009); (b) http://www.physics.emory.edu/faculty/warncke/optesim/index.php

78. G.E. Pake: '*Paramagnetic Resonance*', W. A. Benjamin, Inc., New York, NY (1962).
79. M.H.L. Pryce, A.Abragam: Proc. R. Soc. A **205**, 135 (1951).
80. D.J.E. Ingram: '*Spectroscopy at Radio and Microwave Frequencies*', 2nd Edition Butterworths, London (1967).
81. A. Rockenbauer, P. Simon: (a) Mol. Phys. **28**, 1113 (1974); (b) J. Magn. Reson. **22**, 243 (1976).
82. W.B. Mims: Proc. R. Soc. London, A **283**, 452 (1965).
83. M.K. Bowman, R.J. Massath: In '*Electron Magnetic Resonance of the Solid State*' ed. by J.A. Weil, The Canadian Society for Chemistry, Ottawa (1987).
84. P. Höfer, A. Grupp, H. Nebenfuehr, M. Mehring: Chem. Phys. Lett. **132**, 279 (1986).
85. C. Gemperle, G. Aebli, A. Schweiger, R.R. Ernst: J. Magn. Reson. **88**, 241 (1990).

Exercises

E3.1 Apply a stick plot analysis to the ESR spectrum of CH_3-substituted galvi-noxyl radical using the parameters quoted in Fig. 3.3. The spectra were originally analysed by this means but also with simulations [C. Besev et al.: Acta Chem. Scand. **17**, 2281 (1963)].

E3.2 (a) Construct the ESR stickplot of two equivalent ^{14}N ($I = 1$) by succes-sively drawing the stickplot due to the first nucleus and the stickplot of the second on top of each line of the first. What is the intensity ratio of the five hyperfine lines of the two equivalent ^{14}N?

(b) The ESR spectrum and stickplot pattern due to the $N_2H_4^+$ radical in solution shown below is rather complex due to similar coupling con-stants of four equivalent ^1H ($a_H = 11.0$ G) and two equivalent ^{14}N ($a_N = 11.5$ G). Predict the shape of the ENDOR spectrum at X-band, assuming that signals are obtained both from ^1H and ^{14}N.

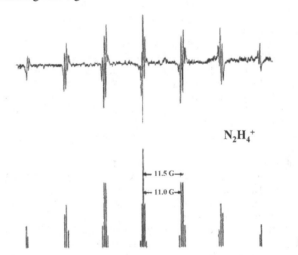

$N_2H_4^+$

←11.5 G→

←11.0 G→

X-band ESR spectrum for $N_2H_4^+$ radical with lines due to four equivalent ^1H and two equivalent ^{14}N nuclei. The figure is reproduced from [J.Q. Adams, J.R. Thomas: J. Chem. Phys. **39**, 1904 (1963)] with permission from the American Institute of Physics.

E3.3 The ENDOR lines of the Wurster blue cation discussed in the text (Section 3.2.3.1) are expected to appear at the frequencies $\nu_{\pm} = \left| \pm \frac{A}{2} - \nu_0 \right|$, where ν_0 is the nuclear frequency at the applied magnetic field of ca. 3390 G. Group the lines 1–6 due to hyperfine couplings with ^{1}H (two groups of inequivalent nuclei) and ^{14}N (two equivalent nuclei) in the figure below in pairs $\nu_{\pm}$ corresponding to the electron spin quantum number $m_S = \pm\frac{1}{2}$.

(a) Support the assignment by comparing the measured values of ν_H and ν_N with the theoretically expected frequencies, $\nu_H = 14.4$ MHz, $\nu_N = 1.04$ MHz.

(b) Then, obtain the ^{1}H and ^{14}N hyperfine couplings.

Note: (a) The ENDOR lines of the protons are centred about ν_H and separated by the respective hyperfine couplings, while the ^{14}N lines are separated by $2 \cdot \nu_N$ and centred at $a_N/2$. This is because $|A_H/2| < \nu_H$ while $|A_N/2| > \nu_N$. (b) The schematic X-band ENDOR spectrum due to the hyperfine structure of two groups of inequivalent ^{1}H and two equivalent ^{14}N nuclei shows amplitude differences due to the hyperfine enhancement effect. In an experimental study [G. Grampp et al.: Monatshefte für Chemie **136**, 519 (2005)] hyperfine couplings due to ^{1}H were obtained from ENDOR, while signals due to ^{14}N did not appear. This behaviour is quite common in CW-ENDOR studies.

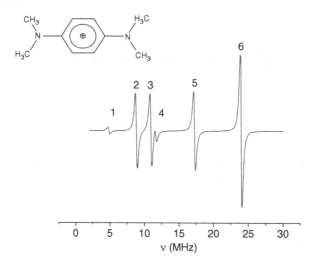

Schematic X-band ENDOR spectrum for Wurster blue cation in solution with lines due to two inequivalent ^{1}H and two equivalent ^{14}N nuclei

E3.4 (a) Measure the hyperfine coupling of the X-band ESR spectrum of trapped hydrogen atoms shown below. The positions of the hyperfine lines are shifted due to 2nd and 3rd order effects, see Section 5.2.1.1 in Chapter 5.

(b) Calculate an approximate value of the shift, given by the 3rd term in the 2nd order expression, $B(m_I) = B_0 - A \cdot m_I - (1/2)A^2/B_0 \left[I(I+1) - m_I^2 \right]$, where m_I $(= \pm\frac{1}{2})$ is the ^1H nuclear quantum number. B_0 may be roughly estimated from the experimental spectrum.

(c) Note that the g-factor, $g = h\nu/(\mu_B B_0)$, is not obtained exactly halfway between the hyperfine lines.

(d) Calculate the g-factor, using the experimental values $B(m_I = \frac{1}{2}) = 3,117.2$ G, $B(m_I = -\frac{1}{2}) = 3,624.5$ G and the calculated value of the 2nd order shift.

Note: According to the literature g = 2.00228 [J.S. Tiedeman and H.G. Robinson: Phys. Rev. Lett. **39**, 602 (1977)]. A deviation from the obtained value may in part be ascribed to experimental errors, in part to inaccuracies of the 2nd order approximation. More accurate values for the line positions can be obtained by the Breit-Rabi equation [G. Breit and I.I. Rabi: Phys. Rev. **38**, 2082 (1931)] originally employed to obtain the nuclear spins of Cs and Rb.

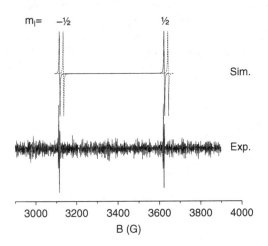

X-band ($\nu = 9.5000$ GHz) ESR spectrum of the hydrogen atom. The displacement of the "1st order" simulated spectrum (*dashed*) compared to the "exact" simulated and experimental spectra is due to a second order shift in the line positions

E3.5 (a) How many lines are theoretically expected in the spectra of the coronene cations (C^+) and $(C_2)^+$ in Fig. 3.7?

(b) The relative line intensities (R) for 12 equivalent ^1H in (C^+) are given in Table below. How many lines can reasonably be detected with a signal/noise ratio of 100?

(c) Estimate the intensity of the sharp lines flanked by satellite lines in figure below. Discuss evidence for attributing the lines to $(C_2)^+$. 24 equivalent ^1H give 25 lines with intensity ratios for lines 1–13 equal to 1 : 24 : 276 : 2024 : 10626 : 42504 : 134596 : 346104 : 735471 : 1307504 : 1961256 : 2496144 : 2704156.

Table: Relative intensities (R) and positions (B) of 12 equivalent ^{1}H ($I = $ ½) by the Pascal triangle method:

R = 1	B = 6*A	m = −6
R = 12	B = 5*A	m = −5
R = 66	B = 4*A	m = −4
R = 220	B = 3*A	m = −3
R = 495	B = 2*A	m = −2
R = 792	B = 1*A	m = −1
R = 924	B = 0*A	m = 0
R = 792	B = −1*A	m = 1
R = 495	B = −2*A	m = 2
R = 220	B = −3*A	m = 3
R = 66	B = −4*A	m = 4
R = 12	B = −5*A	m = 5
R = 1	B = −6*A	m = 6

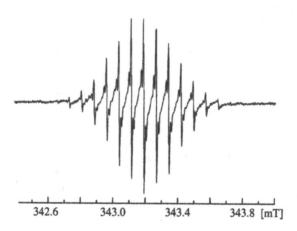

E3.6 The vector addition of two $I = $ ½ nuclear spins I_1 and I_2 symbolically shown in the figure gives rise to effective spins I_{eff} in the range $|I_1–I_2|$, $|I_1–I_2|+1$, ... $I_1 + I_2$ in steps of one. Two equivalent $I = $ ½ nuclei therefore give $I_{eff} = 0$ and $I_{eff} = 1$, each appearing once (Mult. = 1). The procedure can be iterated with the calculated I_{eff} as new I_1 values and adding an $I = $ ½ spin to obtain new I_{eff} spins. Use this method to obtain effective spin values for the radical CF_3 with three equivalent ^{19}F nuclei.

Note: an effective spin I_{eff} can appear more than once (Mult. > 1). The procedure is used to analyse spectra of radicals with large hyperfine couplings, as exemplified in [5] and in Chapter 5. The method can be extended to an arbitrary number of equivalent nuclei with nuclear spin $I \geq \frac{1}{2}$.

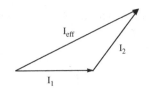

I_1	I_2	Ieff	Mult.
½	½	1	1
		0	1

E3.7 (a) Calculate the fields corresponding to $g_{ref} = 1.9$, $g = 2.1$ and the field difference $\Delta B = B_{ref} - B_0$ at X-band (9.5 GHz).
   (b) Recalculate g with the commonly employed formula: $g = g_{ref}/(1 + \Delta B/B_{ref})$.
   (c) Try to identify the approximation in the equation that causes the deviation from the assumed g-factor.

E3.8 Single crystal spectra with the field at a specific orientation and in the opposite direction are identical.

   (a) Show that the Schonland equation (3.1) has a period of 180° consistent with this fact.
   (b) Rotation of a sample in the wrong sense is in several cases difficult to note experimentally. Geometric [9] and numeric calculations [17] show, however, that only two cases with different principal values need to be considered. This can be deduced by considering the system of equations used to determine these values. The system applies for each of the three principal values (t) and the corresponding three sets of direction cosines (R_1, R_2, R_3) of e.g. $\mathbf{T} = \mathbf{g}^2$:

$$T_{11} \cdot R_1 + T_{12} \cdot R_2 + T_{13} \cdot R_3 = t \cdot R_1$$
$$T_{21} \cdot R_1 + T_{22} \cdot R_2 + T_{33} \cdot R_3 = t \cdot R_2$$
$$T_{31} \cdot R_1 + T_{32} \cdot R_2 + T_{33} \cdot R_3 = t \cdot R_3$$

Assume that the tensor is symmetric $T_{ij} = T_{ji}$ and that T_{13} and T_{23} were obtained with wrong sign:

$$T_{11} \cdot R_1' + T_{12} \cdot R_2' - T_{13} \cdot R_3' = t' \cdot R_1'$$

$$T_{21} \cdot R_1' + T_{22} \cdot R_2' - T_{23} \cdot R_3' = t' \cdot R_2'$$
$$-T_{31} \cdot R_1' - T_{32} \cdot R_2' + T_{33} \cdot R_3' = t' \cdot R_3'$$

The original equations are regained by the substitution $R_1' = R_1$, $R_2' = R_1$, $R_3' = -R_3$. The principal values are therefore equal for the two sets, $t' = t$. The direction cosines have the same magnitude, but differ in the sign of R_3. Analyse in a similar way the situation when T_{12} and T_{13} are obtained with the wrong sign.

One can show in a similar way that a wrong sign of a single element, e.g. T_{13} or wrong signs of T_{12}, T_{13} and T_{23} all give the same principal values. The principal values in general differ from the correct ones. Uncertainty in the sense of rotation of the crystal in the magnetic field accordingly gives rise to just two sets of principal values. Procedures to resolve this "Schonland ambiguity" in ESR [9] and ENDOR single crystal studies [17] have been developed.

E3.9 Expressions of the type below for the coupling constant K are often employed to analyse the orientation dependence of the hyperfine coupling in ESR single crystal studies. Give conditions when this analysis is adequate, and examples where it fails.

$$K^2 = (\mathbf{A}^2)_{xx} \cos^2 \theta + (\mathbf{A}^2)_{yy} \sin^2 \theta + 2(\mathbf{A}^2)_{xy} \sin \theta \cos \theta$$

E3.10 The case in which the nuclear Zeeman energy is much larger than the hyperfine coupling is rare at X-band and lower frequencies, but is of interest for measurements at W- and higher frequency bands that are now frequently employed. The variation of the hyperfine coupling can be analyzed using:

$$Kg^2 = (\mathbf{gAg})_{xx} \cos^2 \theta + (\mathbf{gAg})_{yy} \sin^2 \theta + 2(\mathbf{gAg})_{xy} \sin \theta \cos \theta$$

for the xy-plane and analogous equations for the yz and zx planes.

Simplify to the case with isotropic g to discuss the possibility to obtain the relative signs of the principal values of the hyperfine coupling tensor. Consider the angular dependences of the two ^{1}H hfc lines in the xy-plane of rotation for the cases a) $A_{xx} > 0$, $A_{yy} > 0$, b) $A_{xx} > 0$, $A_{yy} < 0$. For simplicity assume that x and y are principal axes for the tensor.

E3.11 Theoretically the sum of the principal values zero-field splitting should be zero, $D_X + D_Y + D_Z = 0$, i.e. the zero-field splitting has no isotropic component. Show that the usual zero-field splitting parameter D is therefore given by $D = (3/2) D_Z$.

E3.12 The ^{14}N hyperfine and quadrupole coupling tensors for the $N_2D_4^+$ cation formed by X-irradiation of $Li(N_2D_5)SO_4$ were obtained by single crystal ENDOR measurements.[Y. Itagaki et al.: J. Phys. Chem. A **106**, 2617 (2002)].

Tensor	Principal values (MHz)			Eigenvectors		
	A	a_{iso}	B	$\langle x \rangle$	$\langle y \rangle$	$\langle z \rangle$
^{14}N hfc	81.56		49.70	0.210	−0.781	−0.588
	7.50	31.86	−24.36	0.055	0.610	−0.791
	6.53		−25.33	−0.976	−0.134	−0.171
^{14}N nqc	−1.228			0.406	−0.595	−0.693
	0.357			−0.830	−0.557	−0.008
	0.871			0.381	−0.579	0.721

The directions of the principal axes are given with respect to the crystallographic axes. $Li(N_2D_5)SO_4$ is orthorhombic with space group $Pna2_1$, with lattice positions [Space Group Diagrams and Tables, http://img.chem.ucl.ac.uk/sgp/mainmenu.htm]:

Site	Lattice position		
1	x	y	z
2	½ − x	½ + y	½ + z
3	½ + x	½ − y	z
4	−x	−y	½ + z

(a) How many sites can at most be observed in the ENDOR measurements?

(b) How many sites can be observed with the magnetic field in the (yz-), (xz-), and (xy-) planes?

(c) How many sites can be observed experimentally in the ENDOR data shown in E2.8? Is the result consistent with the numbers predicted in (b)?

(d) $N_2D_4^+$ is probably planar like $N_2H_4^+$ discussed in Chapter 1. The ^{14}N hyperfine coupling is then at a maximum perpendicular to the radical plane. Determine the orientation of this axis with respect to the crystal axes using the ENDOR data in the table.

E3.13 Would the asymmetry in Fig. 3.13 be the same in the W-band spectrum?

E3.14 The g-tensor for the $\dot{C}O_2^-$ radical anion trapped in calcite has been determined by single crystal measurements as $g_x = 2.00320$, $g_y = 1.99727$, $g_z = 2.00161$. Estimate if lines corresponding to these g-factors would be resolved in the powder spectra at X- and W-bands? Assume a line-width of 0.3 mT.

E3.15 The direct field model introduced at an early stage [23] provides a pictorial explanation for the occurrence of forbidden ($\Delta m_I = 1$) hyperfine lines. The nucleus is exposed to the hyperfine ($\pm\mathbf{B_A}$) and the external ($\mathbf{B}$) fields. The nuclear spin is oriented in the effective fields ($\mathbf{B_\pm}$) which are not parallel when the hyperfine coupling is anisotropic. A closer examination shows that $\Delta m_I = 1$ lines then occur in the ESR spectrum. (a) How would the effective fields ($\pm\mathbf{B_A}$) be oriented in the limits $B \ll B_A$ and $B \gg B_A$? ($B \ll B_A$ applies at X-band for nuclei with small g_N, e.g. for ^{14}N hyperfine structure at X-band. $B \gg B_A$ might occur in high-field experiments). (b) Can the model explain why forbidden lines do not appear for isotropic couplings? Consult the original paper and/or [1] for more details.

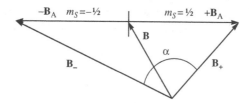

E3.16 g-factors are normally calculated from the centre position of the spectrum. Why is this unsatisfactory for the ^{13}C hyperfine lines in Fig. 3.19?

E3.17 In the original single crystal measurements of triplet state naphtahalene in a durene single crystal at K-band (22.7 GHz) one obtained $g = 2.0030 \pm 0.0004$, $D = 0.1003 \pm 0.0006$ cm^{-1}, and $E = -0.0137 \pm 0.0002$ cm^{-1} [C.A. Hutchison, B.W. Mangum: J. Chem. Phys. **34**, 908 (1961)].

(a) Calculate the D and E parameters in mT and cm^{-1} by measurements on the powder spectrum at X-band (9.5 GHz) below.

(b) Consider possible errors in this procedure.

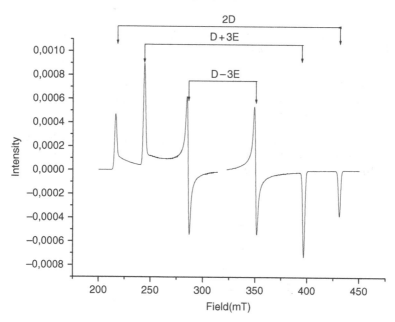

ESR powder spectrum of triplet naphthalene with $D = 0.1003$ cm^{-1}, $E = -0.0137$ cm^{-1}

E3.18 Suggest a reason why the $\Delta m_S = 2$ line is narrower than the $\Delta m_S = 1$ spectrum in Fig. 3.24 by considering the form of the 1st order energies $E'(m_S, m_I)$ in Eq. (3.13) for a triplet state molecule. Would the same argument hold for the $\Delta m_S = 2$ line for $S = 3/2$?

E3.19 Distances in the range 5–10 Å between two electron spins can be measured
        by the splitting D due to magnetic dipole coupling.

(a) Calculate the radical-radical distance from the spectrum below.
(b) Compare the results with literature values for radical pairs trapped in
      dimethyl glyoxime [Y. Kurita: J. Chem. Phys. **41**, 3926 (1964)].

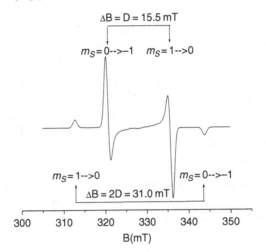

Simulated spectrum of a pair of dimethyl glyoxime radicals separated by 0.56 nm. The hyperfine
structure observed experimentally has been suppressed for clarity.

E3.20 The procedure to obtain the anisotropic hyperfine couplings with rhombic
        symmetry is indicated in the idealized ENDOR spectrum below obtained at
        X-band ($v = 9.5$ GHz).

(a) Assume $0 < A_x < A_y < A_z < 2 \cdot v_N$, estimate the nuclear frequency v_N
      and calculate the isotropic and anisotropic (dipole) couplings.
(b) Which element could give rise to the spectrum?

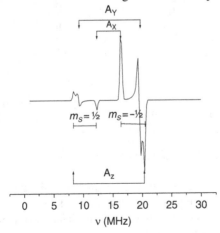

Chapter 4
Multi-Frequency and High Field ESR

Abstract ESR measurements at high microwave frequency and corresponding high field increase the resolution between species with different g-factors and of g-anisotropies in a single species. Field-independent spectral features due to hyperfine couplings (*hfc*) and zero-field splittings (*zfs*) can be separated from g-factor effects by multi-frequency measurements. The *zfs* for $S \geq 1$ species is obtained directly when the microwave frequency is much larger than the *zfs*. Very large *zfs* (larger than the microwave energy available) can be indirectly determined by multi-frequency measurements of the effective g-factor. The sign of the *zfs* (D) can be deduced by high field ESR measurements of the relative intensities of the lines due to *zfs*. The conditions for obtaining the Heisenberg exchange energy (J) between two species forming a coupled system are considered. Examples of applied studies in the solid state are presented. The influence of forbidden transitions and nuclear spin flip lines frequently complicating the analysis of anisotropic *hfc* due to ^{1}H is discussed.

4.1 Introduction

Measurements with multi-frequency and high-field methods [1] are modern trends in ESR both in basic studies and for applications, although X-band spectrometers are still the most common for general use. Commercial equipment is at present available up to a microwave frequency of 263 GHz, while spectrometers operating at even higher frequencies have been constructed in specialised laboratories. The progress is discussed in more detail in a recent review by Riedl and Smith [2]. The use of very high frequency ESR in studies of radicals and metal sites in proteins and small inorganic models has also been reviewed [3].

As mentioned in Chapter 1 high field ESR gives a better spectral resolution of anisotropic g-factors, and allows simpler analysis than is possible at X-band of spectra with large zero-field (another term is fine structure) and hyperfine splittings. These features, (1)–(3) below, are the most important in general applications, but some other characteristics of high field ESR are also of relevance [2]:

A. Lund et al., *Principles and Applications of ESR Spectroscopy*,
DOI 10.1007/978-1-4020-5344-3_4, © Springer Science+Business Media B.V. 2011

(1) *Higher g-resolution: increase in spectral resolution due to the higher magnetic field*: Small g-anisotropies can be resolved and species with differing g-factors can be recognized. Theoretical analysis of g-anisotropies can give an insight into the local electronic structure, e.g. regarding the influence of hydrogen bonding on the g-tensor of radicals in biological systems, see [4] for a review.

(2) *First order spectra*: First order spectra are obtained when the microwave energy $h\nu \gg D$ (the zero field splitting, *zfs*, for $S \geq 1$) or $h\nu \gg a$ (the hyperfine coupling, *hfc*, for $I \neq 0$). The zero-field or hyperfine couplings can be read directly from the spectra and the g-factor can be measured at the centre of the spectrum.

(3) *Accessibility of spin systems with large zero field splitting*: Spectra may not even be observable at lower frequencies when $D \gg h\nu$.

(4) *Less dynamic effects*: The tumbling rate of paramagnetic species leads to an averaging of g-anisotropy. The averaging is reduced relative to low fields.

(5) *Powder ENDOR*: Hyperfine couplings obtained by the angular selection method with the field set at anisotropic g_x, g_y and g_z features can give single-crystal-like ENDOR spectra from randomly oriented samples. The enhanced resolution of g-anisotropy at high magnetic field increases orientation selectivity of ENDOR spectra in amorphous systems.

(6) *Higher sensitivity*: For samples of limited quantity the sensitivity increases at high microwave frequency, thus spectra can be resolved with better signal/noise ratio.

(7) *Spread of g-factors, g-strain*: Measurement of g-strain and D-strain are indicative of the spread of sites of the paramagnetic centre

The frequency bands are often denoted by one-letter symbols as reported in Table 4.1.

Typical frequencies (bands) for multi-frequency measurements with commercial instruments are 1.0 (L-), 3.0 (S-), 9 (X-), 24 (K-), 34 (Q-), and 95 GHz (W-band). L-, S-, X-, K- and Q-band microwave bridges may be employed as accessories to a single ESR spectrometer. Regular electromagnets can be used up to Q-band frequencies and corresponding magnetic field, while at W-band and higher superconducting magnets are employed. Measurements at several frequencies are of interest for a number of reasons:

Table 4.1 Notation for frequency bands in multi-frequency ESR

Band	L	S	X	K	Q	W	D
Wavelength (cm)	30	10	3.0	1.25	0.88	0.32	0.14
Frequency (GHz)	1	3	9	24	34	95	220
Field at $g = 2$ (mT)	35.7	10.7	321.5	857.3	1,215	3,394	7,859
$T_\nu = h\nu/k$ (K)[a]	0.048	0.14	0.43	1.15	1.63	4.56	10.6

[a] See Section 4.4.1 for the determination of the sign of *zfs* at $T \approx T_\nu$.

(a) Origin of spectral features: A classical application is to clarify if an observed spectral feature depends on g-anisotropy. Its position is field dependent while line separations due e.g. to *zfs* and hyperfine structure are, to first order, field independent.

(b) Microwave frequency dependence of g-factor: This may signal the presence of a *zfs* $D > h\nu$. The frequency dependence can be used to obtain D and the asymmetry of the *zfs* (the *E/D* ratio).

(c) Hyperfine couplings by ESEEM: The modulation amplitude in ESEEM depends on the relative magnitudes of the (anisotropic) hyperfine coupling and the field-dependent nuclear Zeeman frequency. A microwave frequency lower than X-band can in some cases amplify the amplitude.

(d) Sample type: The sample size depends on the wavelength, large at L- and small at W-band. L- and S-band spectrometers are therefore suited for bulky samples, e.g. for ESR imaging, while measurements at W-band and higher are of interest when the amount of sample is limited, for instance in biochemical applications.

At present, multi-frequency and high field studies are mostly carried out in CW-mode. However, ongoing development may also make high-field pulsed ESR more generally accessible with the benefits of increased sensitivity and g-factor resolution available at high fields, see e.g. [5] for recent advances in modern ESR spectroscopy. Pulsed ENDOR above X-band is also of current interest. Developments of optical detection methods to lower the detection limit and to obtain time-resolution are of interest in semiconductor materials research. High frequency (95–140 GHz) ODMR has been developed in several research laboratories worldwide. These developments are, nonetheless, beyond the scope of this chapter, and we refer to previous reviews [2, 6]. The rest of the chapter is concerned with examples of CW ESR applications, mostly in rigid amorphous matrices, e.g. frozen solutions, multiphase systems and in polycrystalline samples where the additional information from multifrequency and/or high field measurements are the most crucial for accurate characterisation. The term 'powder spectrum' is frequently used in the following.

4.2 g-Factor Resolution

The development of high frequency/high field methods, initiated by the construction of a CW instrument for the 2 mm band (140 GHz) has the advantage to increase the resolution of samples with only slight g-factor anisotropy and to separate spectra of paramagnetic species with different g-factors [1].

The resonances of two species with different g-factors, g_1 and g_2 are separated by:

$$\Delta B = B_2 - B_1 = \frac{h}{\mu_B} \cdot \frac{g_1 - g_2}{g_1 \cdot g_2} \cdot \nu$$

The field separation (ΔB) and thus the g-factor resolution increases with the microwave frequency ν. Anisotropy of the g-factor occurring in solid materials is also better resolved; g_1 and g_2 could correspond to the axial ($g_{\parallel}$) and perpendicular ($g_{\perp}$) components in an axially symmetric case, or two of the three g-factors of a g-tensor with rhombic symmetry.

Highly symmetric species show rigorous axial symmetry, but the g-anisotropy $\Delta g = g_{\perp} - g_{\parallel}$ of organic radicals is often small. The anisotropy is therefore only partially or not at all resolved in measurement performed at X-band as discussed in a following section, see also Exercise E4.2 at the end of the Chapter. In other cases g- anisotropy was not observed even in extremely well resolved spectra. One example is illustrated in Chapter 5 for an ESR study of matrix isolated methyl radicals affected by quantum rotation effects at low temperature. The g-anisotropies of transition metal ions and of inorganic radicals are in general more pronounced and can often be resolved even at X-band [7, 8]. The following examples concerned with g-resolution in powder spectra at high magnetic field are limited to species with small g-anisotropies, typical of free radicals. Systems with an apparent isotropic or axially symmetric g-tensor observed at low frequency have in many cases been shown to have three distinct g-factors, i.e. possess rhombic symmetry

4.2.1 Resolution of Apparent Isotropic or Axially Symmetric g-Tensor

In initial work principal g-values in the range 2.0024–2.0072 of phenoxyl radicals were resolved with an uncertainty of $\pm 5 \cdot 10^{-5}$ [9]. Such small anisotropy is practically impossible to determine at low frequency, but is clearly resolved at the 2 mm band in Fig. 4.1 because of the 15 times better g-factor resolution as compared to that of the X-band. The g-factor resolution is correspondingly higher in recently constructed instruments with frequencies up to 500 GHz [10].

The ESR spectrum of radicals formed by illumination of photosystem II (PS II) at low temperature exhibits a single line at X-band. The D-band (130 GHz) spectrum of deuterated Mn-depleted PS II shown in Fig. 4.2 was attributed to carotenoid and chlorofyll cation radicals with slightly rhombic g-tensor components of $g_{xx} = 2.00335$, $g_{yy} = 2.00251$, $g_{zz} = 2.00227$ and $g_{xx} = 2.00312$, $g_{yy} = 2.00263$, $g_{zz} = 2.00202$, respectively [11].

The g-tensors of the individual species were determined by spectral simulations, and the assignments were made by comparisons of the g-tensors with data for related systems. Previously unresolvable cofactor signals in PS II could thereby be separated. It was accordingly concluded that the carotenoid and chlorophyll radicals were both generated in PS II by illumination at low temperature.

The g-resolution can be further increased by measurements at even higher microwave frequencies and correspondingly higher magnetic fields as shown by the following recent applications.

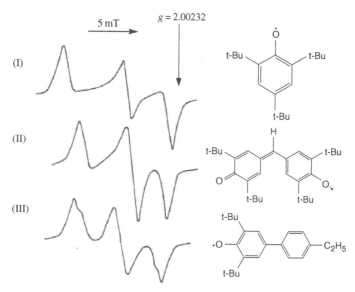

Fig. 4.1 2 mm band ESR spectra of phenoxyl radicals in frozen solution at 145 K. The principal g-values were determined to be **g**(I) = (2.00718, 2.00428, 2.00239), **g**(II) = (2.00636, 2.00404, 2.00251), **g**(III) = (2.00645, 2.00442, 2.00261), respectively. The spectra are adapted from [9] with permission from Taylor and Francis

Flavin radicals, present as intermediates in flavo-enzymes have been investigated by ESR over a period of more than 40 years, see e.g. [12] for an early study of the anion radical in aqueous solution. Hyperfine couplings due to hydrogen and nitrogen were derived and employed to interpret the electronic structure in terms of spin densities. The ESR spectra of frozen solutions are unresolved at X-band. However, taking advantage from high-field EPR studies, the g-anisotropy of flavin adenine dinucleotide (FAD) radicals could be partly resolved already at W-band (93.9 GHz) and completely resolved at 360 GHz in a recent multi-frequency ESR study [13]. Two forms of the radical were present, the FAD$^-$ anion and the neutral FADH depending on the experimental conditions, Scheme 4.1.

The single line spectrum at X-band of the anion radical was attributed to unresolved *hfc* due to hydrogen and nitrogen. The g-anisotropy is small and is smeared out. In addition to the partly resolved g-anisotropy, features attributed to *hfc* due to nitrogen were observed on the high-field wing of the W-band spectrum. At 360 GHz the principal g-values were resolved to such extent that they could be directly read from the spectrum, see Fig. 4.3.

Spectral fittings to the experimental data were made with the non-linear least-squares method to obtain the principal g-values $g_x = 2.00429(3)$, $g_y = 2.00389(3)$, $g_z = 2.00216(3)$ yielding $g_{iso} = 2.00345(3)$. These values differed only slightly between FAD$^-$ and the protonated FADH in several investigated protein-bound

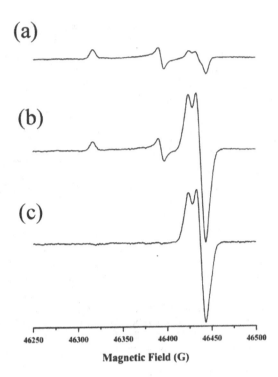

Fig. 4.2 D-band (130 GHz) ESR spectra of deuterated Mn-depleted PS II. (**a**) Background signals recorded in the dark from stable tyrosine radicals. (**b**) Signals recorded after illumination at 30 K. (**c**) Light-minus-dark difference spectrum (**b−a**) showing signals from carotenoid and chlorofyll radicals. The spectra are reproduced from [11] with permission from the American Chemical Society

flavin radicals. The *hfc* of the anion and the protonated form measured by pulsed ENDOR differed, however. We refer to the literature for data regarding the *hfc* of FAD$^-$ and FADH. It was concluded that the principal *g*-values obtained by high field ESR are well suited to identify flavin radicals, even in cases where the optical absorption spectra show large band shifts. In future work, high-field ESR in conjunction with ENDOR spectroscopy might play an important role in identifying the protonation state of flavin radicals and their overall binding situation to the protein surroundings.

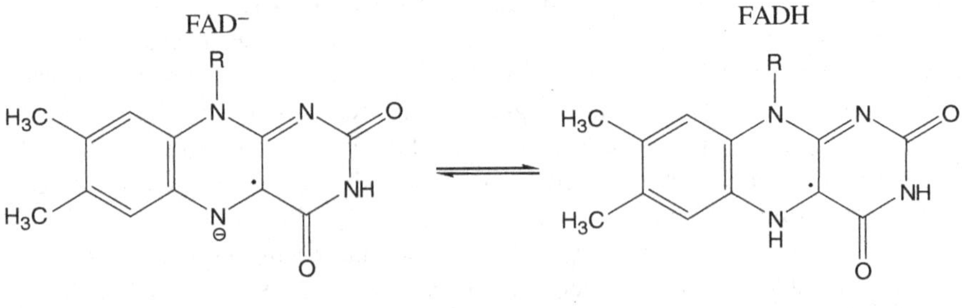

Scheme 4.1 Molecular structures of anionic (FAD$^-$) and neutral (FADH) flavin radicals. R denotes theribityl adenosine diphosphate side chain. The scheme is adapted from [13] with permission from the American Chemical Society

Fig. 4.3 X-band, W-band, and 360-GHz first derivative ESR spectra of FAD⁻, in a frozen solution at pH 10. Simulations using g- and hfc-parameters obtained by least-squares fittings are shown with dashed lines. The spectra are reproduced from [13] with permission from the American Chemical Society

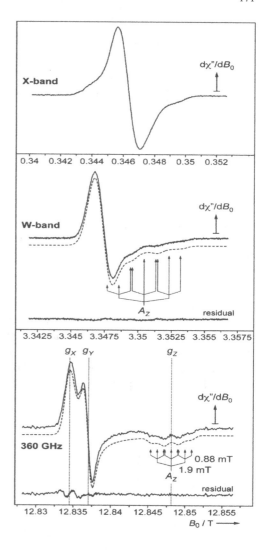

Stable organic radicals present in humic acids may be involved in processes taking place in soil and in water, for instance in the degradation of organic material. Those radicals cannot be well characterised by ESR at X-band, however, because of the small deviation (Δg) of the g-factor from that of the free electron value in the liquid and the correspondingly slight anisotropy of the g-tensor in the solid, typical for most organic radicals. The difference between the principal values of a single species or between the g-factors of different radicals is usually of the order $\Delta g \approx 10^{-3}$ or smaller. Those limitations were overcome by high-field ESR measurements at 285 GHz which can resolve the g-anisotropy [14]. In spectra obtained at X-band a single line is observed, which however was resolved to yield the structure at high field, Fig. 4.4.

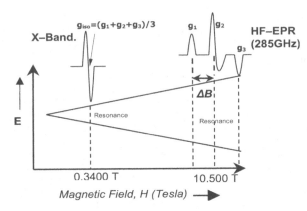

Scheme 4.2 Schematic representation of *g*-factor resolution by high-field ESR spectroscopy applied to radicals in humic acids. The scheme is reproduced from [14] with permission from the American Chemical Society

This is attributed to the increased field separation between the features corresponding to the principal *g*-values, schematically indicated in Scheme 4.2. The experimental spectra of samples isolated from different sources, e.g. from coal or soil, or from commercially obtained peat suggested that two types of radicals are abundant, (I) at acidic pH with $g_x = 2.0032$, $g_y = 2.0032$, and $g_z = 2.0023$, Fig. 4.4. The other (II) prevailing at alkaline pH, is characterized by $g_x = 2.0057$,

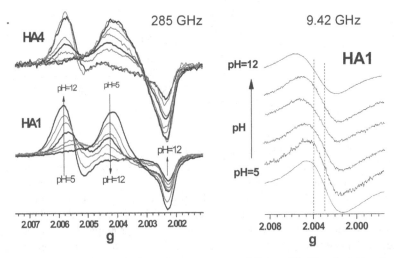

Fig. 4.4 ESR spectra at 285 GHz of frozen solutions at different pH of humic acid radicals in samples obtained from coal (HA1) and peat (HA4) (*Left panel*), and at 9.52 GHz of HA1 samples (*Right panel*). The figure is reproduced from [14] with permission from the American Chemical Society

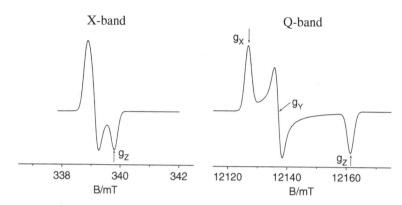

Fig. 4.5 Schematic microwave frequency dependence of the powder ESR spectrum of a radical with $g_x = 2.0032$, $g_y = 2.0014$, $g_z = 1.9975$ obtained by simulation using experimental ESR data for the carbon dioxide anion radical, $\dot{C}O_2^-$ [15]

$g_y = 2.0055$, and $g_z = 2.0023$. Both types of radical centres are consistent with π-type radicals. They persist not only in liquid solutions but also in humic acid powders.

The resolution of rhombic g-factors in powder spectra of apparent axial symmetry spectra at low frequency can be more easily obtained for species having larger g-anisotropy as schematically shown in Fig. 4.5 using parameters for the carbon dioxide radical, CO_2^-, with $g_x = 2.0032$, $g_y = 2.0014$, $g_z = 1.9975$ obtained from single crystal measurements. The powder spectrum at X-band shows an apparent axial symmetric spectrum shape. With a typical line-width of ca 0.2–0.4 mT, the g_x and g_y features are resolved already at Q-band. The anisotropic g-factors can be directly measured at the absorption-(g_x) and emission-(g_z) like lines and at the centre (g_y) of the 1st derivative Q-band spectrum.

Measurements at this band can be made using regular magnets. At higher frequencies equipment with a superconducting magnet is required, which is often not available in standard laboratories.

4.2.2 Resolution of g-Tensor in Presence of Hyperfine Structure (hfs)

Further splittings of the main resonance line, due to hyperfine interactions, are (to first order) independent of the microwave frequency. Measurements at different frequencies can therefore clarify if an observed spectrum is split by Zeeman interactions (different g-factors) or other reasons. The example reported in Fig. 4.6 due to Tempone (2,2,6,6- tetramethylpiperidine-N-oxyl) spin label illustrates how spectral features caused by ^{14}N hyperfine structure and g-anisotropy can be differentiated by measurements at X and W-bands [16].

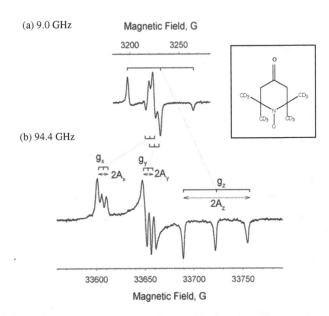

Fig. 4.6 9.0 (**a**) and 94.4 GHz (**b**) ESR spectra from 0.05 mM solution of perdeuterated 2,2′,6,6′-tetramethyl- 4-piperidone-1-nitroxide (TEMPONE) in toluene-d$_8$ at 130 K. Approximate principal g- and ^{14}N hyperfine (A)-tensors are indicated. The figure is reproduced from [16] with permission from the Royal Society of Chemistry

Features due to g-anisotropy are in many cases difficult to differentiate from those due to *hfs* in X-band spectra. The advantage of multi-frequency measurements in this case is demonstrated in Fig. 4.7 for a semiquinone complex of monovalent copper in a frozen toluene solution. The g-anisotropy is clearly resolved at the 2 mm band because of the better g-factor resolution. The *hfs* due to Cu and the two P atoms are not resolved at this band. The analysis of the *hfs* observed in the X-band spectrum is, however, simplified when the principal values of the g-tensor are known. A simulation based on the analysis is shown at the bottom of the figure with parameters given in the legend.

The resolution of **g**-anisotropy is a major reason for the use of very high frequency ESR in applications such as studies of radicals and metal sites in proteins [17]. The new experimental methods to determine accurate g-tensors with high field EPR have also caused a revival of theoretical research to predict these quantities following early seminal work on free radical systems [18]. The subject has been summarised [4]. Several commercial and free-ware programs are available. The performance of various methods and software to compute g-tensors and other ESR properties were compared in a recent article by Neese [19].

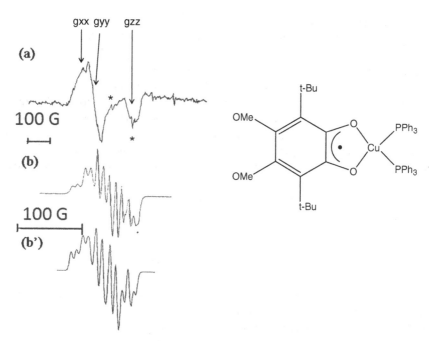

Fig. 4.7 Anisotropic ESR spectra of monovalent copper complex in frozen toluene solution. (**a**) 2 mm band ESR (140 GHz). (**b**) 3 cm band ESR (9.6 GHz): experimental spectrum, (**b'**) simulated spectrum with **g** = (2.0078, 2.0059, 1.9991), **A**(P*) = (20, 23, 20) G, **A**(P) = (8, 11, 8) G, and **A**(Cu) = (20, 0, 17) G for the principal values of the *g*-tensor and hyperfine coupling tensors of two inequivalent phosphorous and one Cu atom. The spectra are adapted from [R.R. Rakhimov et al.: Chem. Phys. Letters **255**, 156 (1996)] with permission from Elsevier

4.3 Zero-Field Splitting (*zfs*)

The second main advantage with high-field ESR is that spectra with large fine structure and/or hyperfine coupling constants are more easily interpreted when the Zeeman term dominates at high field, i.e. when $h\nu \gg D$ and/or $h\nu \gg a$. The spectra are then first order, with equal spacing between hyperfine lines ($I \geq 1$) or fine structure lines ($S \geq 1$). In the following Section 4.3.2 the effects of large *zfs* for species with $S \geq 1$ are considered.

Splittings that occur even at zero magnetic field are due to two physically different effects:

- Magnetic coupling between the orbital and spin magnetic moments in an atom or ion. This spin-orbit coupling is the source of the *zfs* of transition metal ions with electron spin $S \geq 1$.
- Magnetic coupling between unpaired electron spins. This magnetic dipole coupling occurs in triplet state molecules. A similar coupling occurs between the

electron spins in radical pairs and biradicals, and can also occur in complexes containing two transition metal ions or a metal ion and a radical.

The theory is complicated for transition metal ions where the zero-field splitting originates from spin-orbit coupling within the metal complex. The theory for anisotropic g-factors is based on similar theory, making it possible to establish relations between the g-tensor and the zfs. A detailed presentation is given in a text-book by Atherton [20]. The magnetic dipolar coupling typical of dimeric transition metal ion complexes, organic triplet state molecules, biradicals and radical pairs is conceptually simpler, and is of current interest in applications ranging from biological complexes to artificially made molecular magnet systems.

The two interactions contribute to the magnetic energy in formally the same way, mathematically described by the Hamiltonian H_f, where the subscript refers to the alternative term fine structure:

$$H_f = D_X S_X^2 + D_Y S_Y^2 + D_Z S_Z^2 \equiv D[S_Z^2 - 1/3S(S+1)] + E(S_X^2 - S_Y^2)$$

In the first form one has $D_X + D_Y + D_Z = 0$, signifying that the zfs is a completely anisotropic interaction. The zfs is therefore averaged to zero by rapid tumbling in liquid phase. The second form is the usual one employed in classical ESR studies of transition metal ions. The two notations are connected by the relations:

$$D_X = E - D/3$$
$$D_Y = -E - D/3$$
$$D_Z = \frac{2D}{3}$$

The Z-component is by convention taken as the numerically largest. The system is axially symmetric when $E = 0$, and deviates the maximum amount from axial when $|E| = 1/3 \cdot |D|$, therefore $0 \le |E| \le 1/3 \cdot |D|$. The sign of D can be determined by employing multi-frequency ESR measurements described later in this chapter. Occasionally, measurements at lower fields can be employed to obtain the relative signs of the zfs and an anisotropic hfc (Section 4.4.2). The sign of the zfs is of interest for zero or numerically small values of E, and becomes immaterial in the limit of maximum asymmetry.

In transition metal complexes and triplet state molecules the Z-axis for maximum zfs as well as the X and Y-axes are often determined by symmetry and agree with the axes for the g-tensor when the latter is anisotropic. The orientation of the axes for the g-tensor and the zero field tensor need not have an intuitive relation to molecular geometry in complexes of low symmetry, however. In this case the zfs tensor is usually specified by the principal values D_X, D_Y, and D_Z, together with the orientation of the X-, Y-, and Z-axes. The orientation is given by their direction cosines in a suitably chosen system, relating to the crystallographic axes of a crystalline sample or to an assumed geometry of the complex in a disordered matrix. The directions can also be given, more compactly, by the three Euler angles. The orientation can be obtained relatively straightforwardly in single crystals by a variant of the Schonland procedure (Chapter 3). For disordered systems multi-frequency ESR in conjunction with simulations can in favourable cases yield the *relative* orientation of the *g*- and *D*-tensors, while information about the orientation with respect to the molecular structure has to rely on complementary data obtained e.g. by quantum chemistry calculations.

At present, software that can extract theoretically the *zfs* term is not as easily available as the well established methods for the calculation of the *hfc*- and *g*-tensors [4, 19, 21].

4.3.1 Zero-Field Splitting in Powder Spectra

Single crystal measurements are not always applicable to determine the *zfs* in complex systems, e.g. of enzymes containing transition metal ions. ESR studies of such systems are commonly made on frozen solutions [17, 22, 23]. The "first order" "powder" ESR absorption spectrum with axially symmetric *zfs* ($E = 0$) increases step-wise at the magnetic fields corresponding to the $m_S = 1 \leftrightarrow 0$ and $0 \leftrightarrow -1$ transitions with the magnetic field along Z. This is the orientation for maximum zero-field coupling. Transitions cannot occur at lower or higher fields, and the absorption is zero. Because of the step-wise rise of the absorption, the transitions corresponding to $B||Z$ appear as absorptions and emissions in the derivative spectrum. Two lines separated by D, referred to as the perpendicular features occur in the plane perpendicular to Z (Fig. 4.8).

4.3.2 Deviations from First Order Appearance

Deviations from 1st order spectra occur when the high field condition $D \ll h\nu$ does not strictly apply:

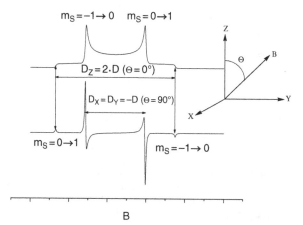

Fig. 4.8 Schematic 1st order powder spectrum showing fine structure ($D > 0$) with axial symmetry ($E = 0$) for an $S = 1$ species. The absorption spectrum increases step-wise at the magnetic fields corresponding to the $m_S = 1$ $\leftrightarrow 0$ and $0 \leftrightarrow -1$ transitions with the magnetic field along Z corresponding to maximum dipolar coupling. Two lines separated by D, referred to as the perpendicular features occur in the XY plane ($\theta = 90°$)

(1) "Forbidden" $\Delta m_S = 2$ transitions appear. Lines occur at approximately half field of the allowed $\Delta m_S = 1$ ones.

(2) The positions of $\Delta m_S = 1$ lines are shifted. The splitting between adjacent zero-field lines for $S \geq 1$ species is not constant. The centre of the spectrum is displaced from the field $B_0 = \frac{h\nu}{g\mu_B}$ at which the g-factor is measured.

(3) A strong dependence of spectral shapes occurs in multifrequency measurements when $D \approx h\nu$ at a particular band, see Fig. 4.11 below.

4.3.2.1 The $\Delta m_S = 2$ Transition

An influence of a large *zfs* that is frequently observed experimentally is the breakdown of the $\Delta m_S = 1$ transition rule. The strength of the transitions induced by a microwave magnetic field B_1 perpendicular to the applied field B is normally zero unless $\Delta m_S = 1$. But when D is appreciable the wave-functions are not the pure $|m_S\rangle$ states but are linear combinations of functions with different values of the magnetic spin quantum number m_S. This is often expressed in the way that m_S is not a good quantum number, making the $\Delta m_S = 2$ transition slightly allowed. The breakdown is observable experimentally by a line at approximately half the field of the $\Delta m_S = 1$ transitions, even for the relatively small *zfs* indicated in Fig. 4.9 with $D < h\nu$. The condition that $h\nu \gg D$ does not hold rigorously was taken into account and accurately analysed in an early ESR study at X-band of the naphthalene triplet in an amorphous matrix [24]. The $\Delta m_S = 2$ line can be suppressed by measurements at high frequency but is often a useful feature to confirm the presence particularly of $S = 1$ species in disordered matrices.

The spectrum schematically shown in Fig. 4.10 is of the type first observed [24] for triplet state naphthalene molecules at low temperature. In this case the $\Delta m_S = 2$

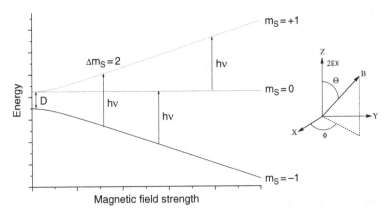

Fig. 4.9 Breakdown of the $\Delta m_S = 1$ transition rule when $h\nu \gg D$ does not hold rigorously for a triplet state molecule trapped in a single crystal oriented for B perpendicular to Z. The breakdown is observable experimentally as the $\Delta m_S = 2$ transition, and indicated theoretically by the deviation from straight lines at low field of the curves marked by $m_S = \pm 1$ at high field

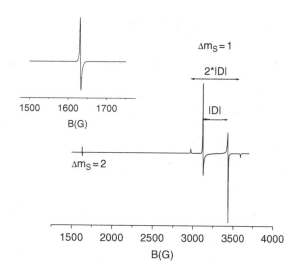

Fig. 4.10 Schematic appearance of the powder spectrum of a triplet state. Axial symmetry is assumed so that the $\Delta m_S = 1$ perpendicular and parallel features are separated by $|D|$ and $2 \cdot |D|$, respectively. The $m_S = 1 \leftrightarrow -1$ $\Delta m_S = 2$ transition is a single, *narrow line*, unaffected by anisotropy of the *zfs*

transition is a single, narrow line. The reason is that the line position is unaffected by the anisotropy caused by *zfs*.

The half field line usually saturates less readily than the $\Delta m_S = 1$ lines and can therefore be recorded at higher microwave power to enhance the intensity. The reason is that a weakly allowed transition is less efficient to equalize the populations of the magnetic levels than an allowed one so that saturation does not occur. Specially designed cavities employing parallel B_1 and B fields make it even easier to observe the $\Delta m_S = 2$ transitions [25]. The observation of $\Delta m_S = 2$ lines is as mentioned of value experimentally particularly to verify the assignment of $S = 1$ species from powder spectra.

4.3.2.2   D < hν, Second Order Corrections

Second order corrections can be relatively easily applied to systems with axial symmetry when 1st order treatment is insufficient, i.e. when $D \ll$ hν does not strictly apply. One can note from the equations for axial symmetry in Chapter 3 that, for the $S = 1$ case, 2nd order energies contribute equally to the $m_S = 1 \leftrightarrow 0$ and $0 \leftrightarrow 1$ transitions so that the *zfs* measured between the two lines is correct to 2nd order. For $S > 1$ the splitting between adjacent lines is not exactly constant, but the separation between the $m_S \leftrightarrow (m_S - 1)$ and $-(m_S - 1) \leftrightarrow -m_S$ lines is correct to 2nd order. The expressions for the 2nd order energies are complicated in the general case with non-coincident principal axes of g and D [26, 27] but one can note from the formulae in Chapter 3 that the 2nd order energies are smaller at high field in proportion to the factor $\frac{1}{g\mu_B B}$. This factor also enters in the expressions for the 2nd order hyperfine energy and in a cross-term involving both hyperfine- and fine-structure terms.

4.3.2.3 Zero-Field Splitting of Intermediate Size

Measurements at high field are advantageous to obtain g-, D- and A-tensors by the first order procedure when $D \approx h\nu$ applies at lower frequency, e.g. at X-band. Large zfs is common for transition metal ions with effective spin $S \geq 1$ [7] and also for artificially made high-spin molecules [28, 29].

The spectra in Fig. 4.11 show a case when the influence of the microwave frequency on the spectral shape is quite strong, with on one hand a not readily interpreted spectrum at X-band, and a 1st order spectrum at W-band on the other hand. For even larger zfs a multi-frequency method to determine the magnitude of D is applicable as discussed in the next section.

4.3.2.4 Large Values of the Zero-Field Splitting, $D \geq h\nu$

A zero-field splitting D that is larger than the microwave energy of presently available spectrometers ($\nu \leq 500$ GHz) can in some cases be measured indirectly using multi-frequency ESR. The technique has so far mostly been applied to systems with half-integer spin S, while systems with integer spin values tend to be "ESR silent" for large values of D. For half-integer S an ESR transition can occur even with $D \gg h\nu$ (Fig. 4.12).

The levels involved in the transition are referred to as a Kramers' doublet having effective spin $S' = 1/2$ and an effective g-factor that depends on the zfs. The splitting can be obtained by measurements at two microwave frequencies, applying an analytic procedure developed by Aasa et al. several years ago [30]. Values of D for the Fe(III) ESR spectrum of lipoxygenase estimated by measurements of the effective g-values at X- and Q-band frequencies and by analysis of the temperature variation of the signal strength in the range $0.6 < D$ (cm^{-1}) < 3 were reported. The asymmetry

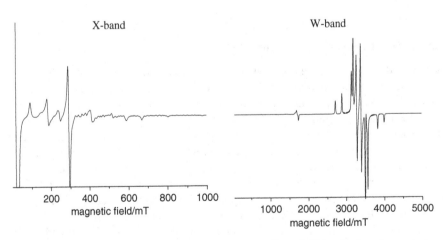

Fig. 4.11 Simulated $S = 2$ ESR powder spectra at X- and W-bands for zfs parameters $D = 0.200$ cm^{-1} and $E = 0.0314$ cm^{-1}. Spectra are reproduced from [29] with permission from Springer

Fig. 4.12 Energy of an
$S = 3/2$ system with *zfs*, $D \gg$
$h\nu$. A transition occurs
yielding a spectrum
characterized by an effective
g-factor of an effective $S' = \frac{1}{2}$
doublet. The diagram and the
spectrum are drawn for $B \parallel z$

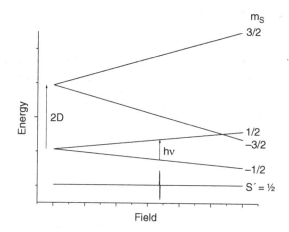

of the *zfs* tensor, i.e. the E/D ratio can be determined even when $D \gg h\nu$. In this
case the effective *g*-factor does not depend on the exact value of D, however.

The method developed in [30] is applicable to systems with $S = 5/2$. An
alternative method for arbitrary values of S employed more recently is based on
simulations of experimental spectra obtained at several microwave frequencies [22].
An observed shift of the effective *g*-factor with microwave frequency was used to
adjust the D-, E-, and *g*-values for best experimental agreement with experiment
for a Fe(II)-EDTA complex incubated with NO. The spectra shown in Fig. 4.13 are
similar to those calculated by Andersson and Barra [22], but have been recalculated
and redrawn for better printing quality.

A species of this type is probably best described as a $[Fe^{3+}\cdots NO^-]$ complex, with
a resultant spin $S = 3/2$ characteristic of a strongly magnetically coupled complex.
Magnetically coupled systems exhibiting the combined influence of *zfs* and an
electrostatic interaction termed Heisenberg exchange are discussed in Section 4.5.

Fig. 4.13 Shift of g_{eff} with
microwave frequency of an
$S = 3/2$ system with isotropic
$g = 2.0$, $D = 11.5$ cm^{-1},
$E = 0.1$ cm^{-1}. D-, E- and
g-values were adjusted in [22]
for best agreement with
experiment on a Fe(II)-EDTA
complex incubated with NO.
The spectra are adapted from
[22] with permission from
Elsevier

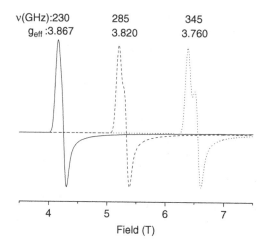

4.3.3 Sign of Zero-Field Splitting

The magnitude of the *zfs* can be obtained directly from the powder spectrum when 1st order conditions apply ($D << g\mu_B B$). These conditions are approximately satisfied for many triplet state organic molecules, such as biradicals and radical pairs and for several transition metal ions even at X-band. On the other hand the sign of the *zfs* can only in exceptional cases (see Section 4.5) be determined at X-band. This can, however, be done more generally by measurements at high field by comparing the relative intensities of the different transitions, e.g. $m_S = 1 \leftrightarrow 0$ and $m_S = 0 \leftrightarrow -1$ for the triplet state.

4.3.3.1 Triplet States

The probability for an electron spin to be in a specific electronic state with energy $E(m_S)$ is according to the Boltzmann distribution [31] given by:

$$p(m_S) = \frac{e^{-E(m_S)/kT}}{\sum\limits_{m'_S=-S}^{+S} e^{-E(m'_S)/kT}}$$

The intensities for the transitions corresponding to $m_S = 1 \leftrightarrow 0$ and $m_S = 0 \leftrightarrow -1$ are proportional to $p(0) - p(1)$ and $p(-1) - p(0)$ [32]. The electron Zeeman energy is conventionally expressed in terms of a temperature $T_\nu = g\mu_B B/k$ [22] so that

$$\frac{I(1 \leftrightarrow 0)}{I(0 \leftrightarrow -1)} = e^{-\frac{g\mu_B B}{kT}} = e^{-\frac{T_\nu}{T}}.$$

The intensity ratio depends strongly on the T/T_ν ratio (Fig. 4.14).

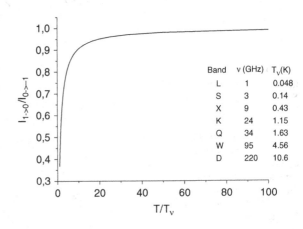

Fig. 4.14 Temperature dependence of the $\frac{I(1\leftrightarrow 0)}{I(0\leftrightarrow -1)}$ intensity ratio for the $m_S = 1 \leftrightarrow 0$ and $m_S = 0 \leftrightarrow -1$ transitions. The values of $T_\nu = g\mu_B B/k$ for different frequency bands are given in the table

Band	ν (GHz)	T_ν(K)
L	1	0.048
S	3	0.14
X	9	0.43
K	24	1.15
Q	34	1.63
W	95	4.56
D	220	10.6

At high temperature the intensities are equal, as usually observed at X-band. The $m_S = 1 \leftrightarrow 0$ transition becomes the weaker when the temperature is lowered towards T_v. This is due to the preferential electron spin population in the lowest $|m_S = -1\rangle$ spin level, an effect which is enhanced by the increased level separation at higher magnetic field. The term spin polarisation is sometimes used. In a temperature range about T_v the intensity ratio deviates significantly from unity. This range is at higher temperatures for high microwave frequency, see Table 4.1 and Fig. 4.14.

For axial symmetry ($E = 0$) the line positions in a crystal vary as:

$$m_S = 1 \leftrightarrow 0 \quad B = B_0 - (1/2)D\left(3\cos^2\theta - 1\right)$$

$$m_S = 0 \leftrightarrow -1 \quad B = B_0 + (1/2)D\left(3\cos^2\theta - 1\right)$$

The "parallel" ($\theta = 0°$) $m_S = 1 \leftrightarrow 0$ line of the powder spectrum therefore occurs at low field and the $m_S = 0 \leftrightarrow -1$ line at high field, for $D > 0$. The high field parallel line is stronger than that of the low field in the 94 GHz spectrum as schematically shown in Fig. 4.15. The opposite case applies for the perpendicular features ($\theta = 90°$).

To obtain the sign of D it is thus preferable to make measurements at high field for two reasons:

(1) The required magnetic polarisation can be achieved at a temperature that is not inconveniently low.
(2) The intensity ratio between the lines is obtained under the condition that the high field condition $|D| \ll g\mu_B B$ applies.

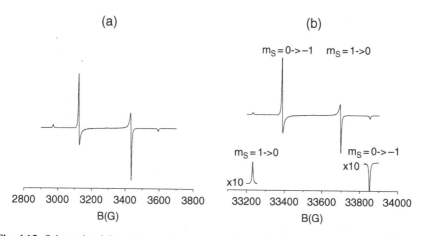

Fig. 4.15 Schematic triplet state powder spectra showing *zfs* at 10 K. (**a**) The intensities of the lines at X-band are equal. The sign of D cannot be determined. (**b**) The intensities at 94 GHz differ. The lines can be assigned as shown to $m_S = 1 \leftrightarrow 0$ and $m_S = 0 \leftrightarrow -1$ transitions. This permits determination of the sign of D, in this case $D > 0$

In theory the relative signs of D and g are determined by this method. In practice the electronic g-factors are always positive and the absolute sign of D is accordingly obtained. The sign of the *zfs* is of particular interest when it is close to axially symmetric, $|D| \gg |E|$, or alternatively $|D_Z| \gg |D_X-D_Y|$ for the principal values, where $D_X + D_Y + D_Z = 0$. Many systems show both zero-field and hyperfine splittings. A procedure to assign the signs of the principal values of the hyperfine coupling tensor after the sign of D has been established and is described in the next section.

4.3.3.2 Transition Metal Ions with $S \geq 1$

The change in intensity of the different transitions as a response to temperature changes can in a similar way be employed to determine the sign of D through high field measurements of transition metal ion complexes. If the sign of this parameter can be established it may also be possible to determine the absolute signs of the anisotropic *hfc* due to ligand groups. The $[MnF_6]^{2-}$ complex with electron spin $S = 5/2$ [33] provides an example. The ESR powder spectra of K_2MnF_6 in K_2GeF_6 at 230 GHz showed a temperature dependence in the range 5–25 K that could only be accounted for by a negative value of D (−0.362 cm^{-1}). In addition *hfc* due to the fluorine ligands was observed. The absolute signs of the axially symmetric couplings with $A_{||}$ negative, $A_{\perp}$ positive was the only possibility that could be fitted by spectrum simulations to the experimental data. By analogy a negative value of $A_{||}$ and a positive for $A_{\perp}$ for the ^{19}F couplings is anticipated for the Cs_2MnF_6 diluted in a Cs_2GeF_6 crystal although this cannot be confirmed in relation to the D parameter as this was zero for the cubic complex studied experimentally. The six-line spectrum in Fig. 4.16 is due to the *hfc* of ^{55}Mn with additional structure due to the *hfc*

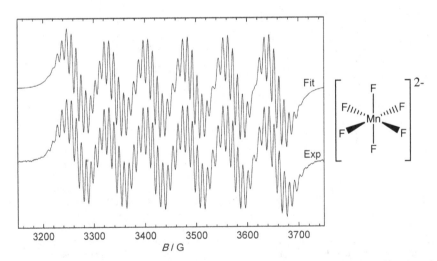

Fig. 4.16 Experimental and fitted ESR spectra of diluted Cs_2MnF_6 single crystal with magnetic field along the 3-fold crystal axis at 25 K, $\nu = 9.6396$ GHz. The figure was adapted from [33] with permission from the American Chemical Society

of the ^{19}F ligands. The number of hyperfine lines exceeded the seven lines for six equivalent ^{19}F, that was expected from the usual $\Delta m_I = 0$ selection rule, however. Fitting procedures could in this case verify that $A_{||}$ and $A_\perp$ are of opposite signs, The method is applicable when the hyperfine coupling is anisotropic with principal values of comparable magnitudes to the nuclear Zeeman energy. This condition applies at X-band for the Cs_2MnF_6 system. The direct field model described in the following sections provides a pictorial model.

4.4 Hyperfine Couplings (*hfc*)

To first order the hyperfine couplings (*hfc*) are directly obtained from the corresponding splitting between the lines. The splitting is in general independent of the microwave frequency. Multi-frequency measurements are therefore useful for distinguishing features due to anisotropic g-factors and *hfc* (Section 4.2.2). Differences in the hyperfine structure of spectra obtained at different frequencies can, however, occur in two cases:

- The high field condition with the electron Zeeman energy $g\mu_B B \gg hfc$ does not apply.
- The *hfc* is anisotropic and the nuclear Zeeman energy $g_N\mu_N B \approx hfc$.

The first effect may be eliminated experimentally by measurements at a higher frequency band. It is also relatively easy at least in the case of axial symmetry to evaluate the couplings from the measured splitting obtained at a lower frequency by applying the higher order corrections exemplified in Chapter 3 and Chapter 5. The second effect is more important in applied studies.

Hyperfine splittings in solids are in some cases dependent on the magnetic field strength, due to the influence of the nuclear Zeeman energy, rather than the 2nd order effect caused by large hyperfine couplings. This so-called direct field effect may be less familiar to the non-specialist, although the effect was taken into account in early works [34, 35]. The influence on the shape of the ESR spectra at different frequency bands can be pronounced for species with anisotropic hyperfine couplings of moderate size due to nuclei with large nuclear g-factors like ^{1}H or ^{19}F. The effect is of relevance in several applications, e.g. in studies of radicals with anisotropic ^{1}H hyperfine coupling, of transition metal ions with *hfc* due to fluorine ligands and of triplet state species featuring asymmetric ^{1}H hyperfine patterns between the $m_S = -1 \leftrightarrow 0$ and $m_S = 0 \leftrightarrow +1$ branches. To avoid confusion with the 2nd order effect caused by large hyperfine couplings the term direct field effect employed in initial work [36] is used in the following. This causes an admixture of nuclear-spin states, and the appearance of a complicated pattern of hyperfine lines, which is conventionally although somewhat misleadingly attributed to the breakdown of the $\Delta m_I = 0$ selection rule. The direct field effect is suppressed when $g_N\mu_N B \ll hfc$ which is the normal condition to obtain hyperfine coupling tensors by ESR of transition metal ions. The opposite case with $g_N\mu_N B \gg hfc$ can occasionally occur even at X-band, particularly for weak anisotropic *hfc* due to ^{1}H. The splitting between the inner lines is anisotropic. The outer lines are weak and nearly isotropic with a splitting of *ca.* $2g_N\mu_N B/g\mu_B$, see Fig. 4.19 for examples.

The influence of the direct field effect on the hyperfine structure is considered in more detail in the following section. Some technical properties are summarized in Appendix A4.1.

4.4.1 The Direct Field Effect

Nuclei with large magnetic moments like ^{1}H ($g_N = 5.586$) and ^{19}F ($g_N = 5.525$) give rise to a correspondingly large nuclear Zeeman energy in an applied field. When this energy is comparable to the energy due to an anisotropic hyperfine coupling with the nucleus, more transitions occur than the $2 \cdot I + 1$ lines predicted by the $\Delta m_I = 0$ selection rule. An $I = \frac{1}{2}$ nucleus like ^{1}H and ^{19}F gives rise to four lines instead of two. Generally $(2 \cdot I + 1)^2$ transitions can occur. The phenomenon can be explained by the effective field acting on the ^{1}H nucleus shown in Fig. 4.17.

The field due to the hyperfine interaction, $(\pm B_A)$ with the unpaired electron can take two opposite directions depending on the electron spin quantum number m_s ($m_s = \pm\frac{1}{2}$ for $S = \frac{1}{2}$). The direction of the applied magnetic field B is along B_N and does not change. B_N has the magnitude $B_N = \frac{g_N\mu_N}{g\mu_B}B$, *ca.* 0.5 mT at X-band. Organic radicals of the common π-electron type H—C$_\alpha\langle$ have anisotropic hyperfine couplings of magnitudes in the same range, i.e. 1.0–3.0 mT. The reason for the appearance of a four line spectrum is that that the effective fields B_+ and B_- acting on the ^{1}H nucleus for $m_s = \pm\frac{1}{2}$ are not parallel (or antiparallel), implying different quantization axes for the nuclear spin depending on the orientation of the electron spin. In quantum mechanical terms the levels are not pure nuclear states but are linear combinations of the $m_I = \pm\frac{1}{2}$ states, allowing ESR transitions involving all nuclear states as in

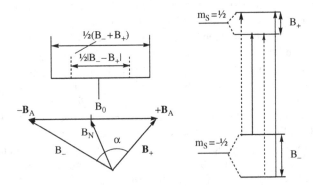

Fig. 4.17 Effective field model for anisotropic hyperfine structure. $+B_A$ and $-B_A$ are the fields at the nucleus from the hyperfine coupling with an unpaired electron for $m_s = +\frac{1}{2}$ and $m_s = -\frac{1}{2}$, respectively. B_N is in the direction of the applied field with a magnitude of $B_N = \frac{g_N\mu_N}{g\mu_B}B$. B_+ and B_- are the effective fields acting on the nucleus when $m_s = +\frac{1}{2}$ and $m_s = -\frac{1}{2}$, respectively

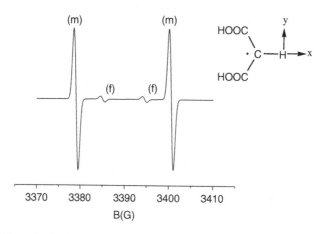

Fig. 4.18 Schematic single crystal X-band ESR spectrum of a π-electron radical H_α—Ċ⟨ with the magnetic field at an angle of 45° to the x-axis in the xy-plane. The main (m) and forbidden (f) lines occur as an outer and an inner doublet. The spectrum is calculated with $A_x = -90.25$, $A_y = -56.52$, and $A_z = -27.42$ MHz obtained from an ENDOR study [E. Sagstuen et al.: J. Phys. Chem. A **104**, 6362 (2000)].

Fig. 4.17. The spectrum for $I = \frac{1}{2}$ is composed of an inner and an outer doublet as in Fig. 4.18. The inner doublet of the spectrum in Fig. 4.18 is in this case the weakest and these lines are sometimes referred to as "forbidden". In other cases like for the ^{19}F *hfc* of the MnF_6^{2-} complex in Fig. 4.16 the "allowed" and "forbidden" lines can have comparable intensities, making the distinction immaterial.

This type of spectra are frequently observed at X-band for π-electron radicals of the common type H—$Ċ_\alpha$⟨. The appearance of the spectra is different for weakly coupled nuclei, in which case so-called spin flip lines appear.

Nuclear spin flip ($\Delta m_I = 1$) lines occur in systems with anisotropic hyperfine couplings of smaller magnitude than the nuclear Zeeman energy. In this case satellite lines appear on each side of an often unresolved doublet of central lines. At X-band the satellites are separated approximately by $B_N = \frac{g_N \mu_N}{g \mu_B} B \approx \pm 0.5$ mT from the centre. The separation increases and the intensity decreases at Q-band and higher frequencies, Fig. 4.19.

The intensity ratio between the main and satellite lines has been used to estimate the distance between a paramagnetic centre and protons at surrounding molecules. The method is based on the assumptions that the point dipole approximation applies for the anisotropic *hfc*, and that this coupling is much smaller than the nuclear Zeeman energy of the proton. It was found advisable to measure this ratio at as high microwave frequency as possible to achieve the latter condition, i.e. $|B_A| \ll B_N$ in terms of the direct field model. The procedure was adequate at Q-band but not at X-band to obtain the distance between a proton and the PO_3^{2-} radical in a single crystal of $Na_2HPO_3 \cdot 5H_2O$, see [37] for details.

In other applications the presence of forbidden and spin flip lines makes interpretations more difficult, particularly when they occur in the same spectrum as for the malonic acid radical in Fig. 4.20. The early work by McConnell and co-workers to obtain the first ESR single crystal data of this radical, including an interpretation of the forbidden transitions must therefore be considered as an extra-ordinary

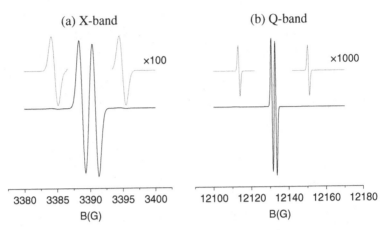

(a) X-band   (b) Q-band

×100   ×1000

3380 3385 3390 3395 3400   12100 12120 12140 12160 12180
B(G)   B(G)

Fig. 4.19 Schematic X- and Q-band spectra with an anisotropic ^{1}H hyperfine coupling of smaller magnitude than the nuclear Zeeman energy. Weak spin flip ($\Delta m_I = 1$) lines separated by $2B_N = \frac{2g_N\mu_N}{g\mu_B}B$, *ca.* 1 (X) and 3.5 (Q) mT appear on the wings of the two main lines. The intensity of the spin-flip lines are suppressed at higher field

achievement [34]. The ^{1}H hyperfine coupling tensor of this radical has turned out to be typical for a whole class of π-electron radicals of the general type H$_\alpha$—$\dot{\text{C}}\langle$, and the data have served as a reference in numerous later studies of similar species, both experimentally and theoretically.

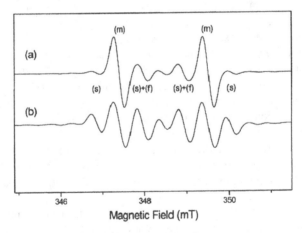

(m)   (m)

(a)

(s)   (s)+(f)   (s)+(f)   (s)

(b)

346   348   350

Magnetic Field (mT)

Fig. 4.20 First-derivative ESR spectra of the radical H$_\alpha\dot{\text{C}}$(COOH)$_2$ from an X-irradiated single crystal of malonic acid recorded at 295 K at (**a**) low (0.2 mW) and (**b**) high (100 mW) microwave power. The various resonance lines have been designated according to their categorization as main lines (m), forbidden transitions (f) due to hyperfine coupling with H$_\alpha$, and spin-flip lines (s) due to hyperfine coupling with distant hydrogen atoms at neighbour molecules. The figure is reproduced from [E. Sagstuen et al.: J. Phys. Chem. A **104**, 6362 (2000)] with permission from the American Chemical Society

The H_α-atom of the malonic acid radical (Fig. 4.18) has an anisotropic hyperfine coupling that gives rise to allowed and forbidden lines, while spin flip lines due to distant hydrogen atoms at neighbour molecules become apparent particularly at increased microwave power. Those lines have a lower transition probability than the main lines and a correspondingly lower tendency to equalize the populations of the energy levels. Saturation therefore occurs less readily. By varying the power as shown in Fig. 4.20 such lines can be experimentally recognized, which otherwise could be erroneously interpreted as an additional *hfc*. Due to the limited resolution of the ESR spectra recent studies have almost exclusively been made by ENDOR, however, which eliminates the problems with allowed and forbidden transitions and offers a higher resolution as discussed in Chapter 2.

The hyperfine pattern of a sample may differ at different microwave frequencies due to the difference in the nuclear Zeeman energy, with $B_N \approx 0.5$ mT at X-band, 1.8 mT at Q-band for the ^{1}H- *hfc* in a free radical ($g \approx 2.00$). An example is shown in Fig. 4.21(a) for the hydrazine cation radical discussed in Chapter 1. For the magnetic field oriented along the N—N (Y) bond the four hydrogen atoms are accidentally equivalent giving a normal quintet of lines at X-band. The *hfc* due to the two

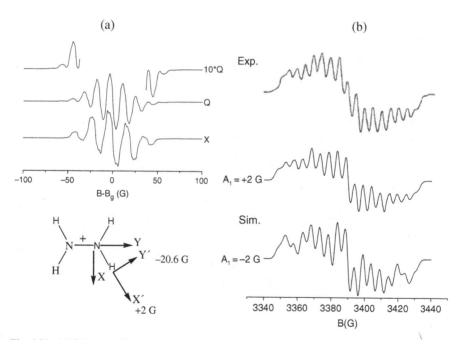

Fig. 4.21 (a) Schematic X- and Q-band single crystal spectra of the hydrazine cation radical, with the magnetic field B‖Y. The hyperfine structure due to accidentally equivalent 4 H is the expected quintet at X-band, while "spin-flip" lines occur on the Q-band spectrum. (b) Experimental and simulated X-band spectra with the magnetic field at an angle of 25° with X in the XY-plane. The spectra were adapted from [O. Edlund et al.: J. Chem. Phys. **49**, 749 (1968)] with permission from the American Institute of Physics

equivalent ^{14}N atoms is unresolved in the radical plane, and has its maximum along the perpendicular to the plane. Additional outer lines occur in the Q-band spectrum in which case the nuclear Zeeman and hyperfine terms are of comparable magnitude, i.e. $|B_A| \approx B_N$ in the direct field model. The features in the figure are accounted for by a simulation taking simultaneously account of the nuclear Zeeman and *hfc* terms using the method by Lefebvre and Maruani [38]. The line positions agree with those obtained by the direct-field model, while the intensities are calculated by a quantum mechanical method.

In conclusion hyperfine structure affected by direct field effects are difficult to interpret because of the admixture of nuclear-spin states leading to the appearance of a complicated pattern of hyperfine lines. Systems featuring such effects are in general better studied by ENDOR. The effect can, however, be employed to determine the relative signs of the principal values of *hfc* tensors particularly for ^{1}H and ^{19}F nuclei in cases where the ENDOR method is not applicable or is not available. The subject is discussed in the following section.

4.4.2 Sign of Anisotropic Hyperfine Couplings

The sign of an isotropic hyperfine coupling can as a rule not be determined from the ESR spectrum, and the same applies in general to the signs of the principal values of the hyperfine coupling tensor. The microwave frequency dependence on hyperfine patterns due to the direct field effect can sometimes be of advantage for determining the *relative signs* of the principal values of the hyperfine coupling tensor. This method is applicable for an anisotropic *hfc* of the same magnitude as the nuclear Zeeman energy, in other words for comparable strengths of the direct and hyperfine fields on the nucleus in the direct field model. In this special case with a non-negligible nuclear Zeeman term the relative signs can be deduced, by comparing the relative intensities of the "allowed" and "forbidden" lines of simulated spectra with the experimental values. This case is common at X- and Q-bands for the *hfc* due to ^{1}H and occurs occasionally for ^{19}F hyperfine couplings of modest magnitude as in the $MnF_6{}^{2-}$ complex [33] referred to in Section 4.3.3.2. At higher frequency bands direct field effects may occur also for nuclei with small g_N. The nuclear Zeeman energy of ^{14}N at W band is for instance of the same magnitude as for ^{1}H at X-band. The ESR line-shape can deviate significantly from that at X-band, as schematically shown in Fig. 3.17 of Chapter 3.

4.4.2.1 S = ½ Species

The analysis of hyperfine structure for the hydrazine cation radical discussed in Chapter 1 provides an example of the use of simulations to obtain the relative signs of ^{1}H principal hyperfine values by comparison between experimental and simulated spectra. The experimental X-band spectrum in Fig. 4.21(b) was simulated by the method developed in [38] to predict the positions and intensities for the "allowed" and "forbidden" hyperfine lines due to the four ^{1}H nuclei ($I = \frac{1}{2}$). Equations applicable to the $I = \frac{1}{2}$ case are reproduced in Appendix A4.1. Comparison with the experimental spectrum gives a positive sign of A'_X assuming a negative sign for A'_Y. The difference in shape between the corresponding Q-band spectra is less pronounced for this system.

4.4.2.2   S ≥ 1 Species

An asymmetry in the ^{1}H hyperfine pattern between the $m_S = 1 \leftrightarrow 0$ and $m_S = 0 \leftrightarrow -1$ electronic transitions, first observed in the ESR spectra of radical pairs in a single crystal of monofluoroacetamide, can be attributed to the direct field effect causing a mixing of the nuclear spin states. The sign of D was deduced based on assumed signs of the hyperfine coupling, see the work by Minakata and Iwasaki for details [39]. We refer to Section 3.3.6 for a similar application to an organic triplet state molecule, trimethylene methane, trapped in a rigid crystal matrix, and to Section 4.3.3.2 for a metal complex, $[MnF_6]^{2-}$, with electron spin S = 5/2 [33]. The absolute signs of the principal values of the *hfc* due to ^{19}F could be established by simulation of the spectrum with an exact method after the sign of the *zfs* had been determined by high-field measurements at low temperature.

4.5   Magnetically Coupled Systems

Paramagnetic systems that contain two or more subunits are quite common. Classical examples are transition metal ion dimers, biradicals and radical pairs formed by irradiation or chemical treatment [7, 40, 42]. Naturally occurring species like enzyme systems often contain several transition metal ions [3, 17, 22], while high spin molecules of both inorganic and organic origin have been synthesized and characterized by ESR [28, 29]. Systems containing two S = ½ subunits are suitable model compounds to exemplify high-field and multifrequency ESR procedures to obtain magnetic parameters of magnetically coupled systems.

4.5.1   S = ½ Dimers

The interaction between two $S = \frac{1}{2}$ species can be of two kinds, magnetic dipolar coupling and exchange coupling. The first interaction gives rise to an anisotropic coupling and can be explained classically as due to the coupling between the magnetic moments of the two electron spins. It causes a splitting in two lines similar to the *zfs* for $S = 1$ and is accordingly only visible in solid-state samples. The splitting between the lines depends on the distance r between the electron spins. A procedure to obtain distances by ESR measurements of the electron-electron dipolar coupling was first described by Kurita in a work where the spin-spin distance within a radical pair in an irradiated single crystal was estimated [40]. The point dipole approximation was applied under the condition that the magnitude of the exchange energy is much larger than the magnetic dipole coupling, $|J| \gg D$:

$$D(\text{mT}) = \frac{\mu_0}{4\pi} \frac{3 \, g\mu_B}{2r^3} = \frac{2780}{r^3} \text{ with } r \text{ in Å} \quad \text{and with } g = 2.00232$$

At shorter distances occurring for instance in triplet state molecules a quantum mechanical averaging is necessary [35]. The opposite case with $|J| \ll D$ occurs for instance in distance measurements with pulsed ESR discussed in Chapter 2.

First order analysis has been applied to obtain the magnitude of the coupling even from measurements at X-band [40] except for the short distances appearing in e.g. some Cu(II) complexes, where distances between the Cu(II)···Cu(II) pair were in the range 3.5–4 Å. Exact (rather than first order) analyses of the X-band spectra were then employed to deduce the magnitudes of the dipolar coupling even in early works, see Exercise E4.3. Measurements at higher frequency are, however, employed to experimentally obtain the *sign* of the coupling in analogy to the procedure for the *zfs* in Section 4.3.

4.5.1.1 The Exchange Coupling (*J*)

The second interaction originates from the electrostatic Heisenberg exchange term in quantum mechanics. The magnitude of the interaction is conventionally denoted by the symbol J. The conventions for the sign of J differ between treatises [7]. In this work the sum of the zero-field and exchange interactions is expressed with a Hamiltonian of the type:

$$H_{12} = \mathbf{S}_1 \mathbf{D}_{12} \mathbf{S}_2 + J_{12} \mathbf{S}_1 \mathbf{S}_2$$

With this convention the ground state is a triplet ($S = 1$) when $J < 0$ in which case the coupling is ferromagnetic. In the opposite case with $J > 0$ the system is anti-ferromagnetic. The magnitude of the exchange coupling J occurring between the unpaired electron spins of the subunits is of interest to obtain experimentally for a magnetically coupled system.

It is convenient to use a resultant effective spin when the $S = \frac{1}{2}$ subunits are identical. It is then assumed that the complex can be treated as a species with $S = 1$ or 0. The mathematical procedure given in several text-books [20, 27, 35] to obtain the *zfs* under those conditions is technical and is not reproduced here. An expression for the exchange interaction can, however, easily be calculated using the relation for the addition of the spin angular moment vectors in quantum mechanics:

Fig. 4.22 The singlet-triplet energy gap for S = 1 system when $J > 0$

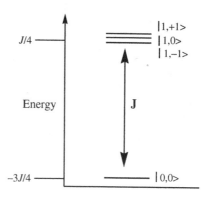

$$S(S+1) = \mathbf{S}^2 = (\mathbf{S}_1 + \mathbf{S}_2)^2 = S_1(S_1+1) + S_2(S_2+1) + 2 \cdot \mathbf{S}_1 \cdot \mathbf{S}_2 = \frac{3}{4} + \frac{3}{4} + 2 \cdot \mathbf{S}_1 \cdot \mathbf{S}_2$$

This gives:

$$S = 1 : \mathbf{S}_1 \cdot \mathbf{S}_2 = 1/2 \left(2 - \frac{3}{2}\right) = \frac{1}{4} \qquad S = 0 : \mathbf{S}_1 \cdot \mathbf{S}_2 = -\frac{3}{4}.$$

The exchange energy is $E_J^0 = -\frac{3J}{4}$ for the singlet (S = 0), $E_J^1 = \frac{J}{4}$ for the triplet state (S = 1). In the latter case there are three degenerate states, Fig. 4.22. The degeneracy is attributed to the different quantum numbers $m_S = 1, 0$, and -1.

According to this treatment the exchange coupling contributes only a constant energy shift to the energy levels of the triplet state and has therefore no influence on the ESR spectrum. The subunits of the pair may, however, be different and/or have different orientations of the principal axes of the tensors describing the magnetic interactions. An analysis of several possibilities for the interaction in dipolar coupled transition metal dimer complexes has been given by Pilbrow [7]. The treatment shows that the Heisenberg term cannot always be ignored in the analysis of magnetically coupled systems. The conditions when the value of J can be obtained from ESR measurements are considered in the following section.

4.5.1.2 Estimate of J by Multi-Frequency ESR

Multi-frequency measurements can be applied to obtain the magnitude of the exchange coupling when the subunits of the pair have different g-factors with $\Delta g = g_1 - g_2$. Following the treatment given by Fournel et al. [41], one finds that the spectrum shape depends on the microwave frequency as schematically shown in Fig. 4.23 for the case of two S = ½ spins with different g-factors. For a large value of the ratio $\Delta g \mu_B B / J$ (high microwave frequency) a spectrum of the isolated species centred at their different g-values is obtained. For small ratios the

Fig. 4.23 Schematic shapes of the ESR spectrum of two $S = \frac{1}{2}$ centres with different g-factors g_1 and g_2 coupled by an exchange interaction at different microwave frequencies

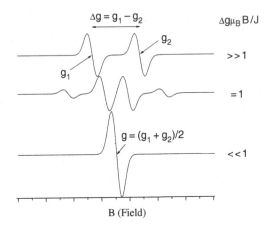

spectrum is centred about the average g, while intermediate spectra with shapes different from the two extreme cases occur when $\Delta g \mu_B B / J \approx 1$, allowing an estimate of the magnitude of the exchange term.

The work by Fournel et al. [41] provides an instructional example of the use of multi-frequency ESR to obtain detailed data for a complex biochemical system (a bacterial enzyme), in particular the exchange interaction between a transition metal ion and a radical and the determination of the magnitude and sign of the zero-field coupling. Procedures are described for the less complex biradical systems in the following section.

4.5.1.3 Biradicals

The magnetic properties of biradicals in solution have been thoroughly investigated for several decades. Of particular interest in the context of this chapter are high field experiments in solid matrices performed in complement to usual X band [42]. The g-anisotropy has a predominant influence on the spectral shape at 245 GHz of the 1,4-Bis(4″,4″-dimethyloxazolidine-3″-oxyl)-cyclohexane biradical.

At 245 GHz and at low temperatures (< 50 K), the differences between the populations of triplet energy levels have a perceptible influence on the intensities of the $\Delta m_S = 1$ lines. These differences result essentially from the dominating electron Zeeman term. The outermost peaks in Fig. 4.24, corresponding to maximum *zfs* have an intensity ratio opposite to that in Fig. 4.15, and therefore $D < 0$, as confirmed by the spectral simulations in the figure. The value of J was not determined in this case, however.

The exchange coupling can in certain cases be determined by an analysis of the hyperfine pattern due to nitrogen observed for instance in nitroxide biradicals in liquid solution. A procedure to extract the value of J was developed many years ago [43]. This procedure does not depend on the microwave frequency and measurements were usually made at X-band. Three cases may appear, (1) $|J| \ll |a_N|$, (2) $|J| \sim |a_N|$, (3) $|J| \gg |a_N|$, where a_N is the isotropic *hfc* of the mono-radical. In (1) the spectrum of the mono-radical with $a = a_N$ is obtained, in (3) the nitrogen atoms are equivalent with $a = a_N/2$. The value of J can be extracted when case (2) applies. The

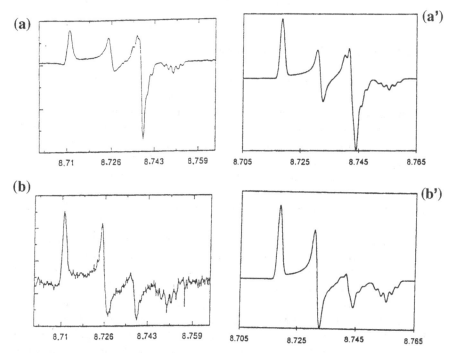

Fig. 4.24 *Left*: (**a**) 245 GHz, (**b**) 9 GHz spectra of the 1,4-Bis(4″,4″-dimethyloxazolidine-3″-oxyl)cyclohexane biradical. *Right*: (**a′**) 245 GHz, (**b′**) 9 GHz spectra computed with the parameters $D = -122$ G, $E = 3$ G, $A_{xx} = A_{zz} = 3.5$ G, $A_{yy} = 17$ G, $g_{xx} = 2.008$, $g_{yy} = 2.00235$, $g_{zz} = 2.0067$. The X-ray determination of the structure indicated that X and x as well as Z and y make an angle of 23° in the XZ plane. This has been taken into account in the fitting of the experimental spectra. The figure is reproduced from [http://www.esr-spectsim-softw.fr/triplets.htm] with permission from Dr. C. Chachaty

analysis presented e.g. in [27, 43] is technical and is not reproduced here. Analysis by simulations provides a modern alternative. We are not aware of dedicated software, but the general programs in Table 3.5 of Chapter 3 marked "Exact" would be applicable. Typical applications involve studies on the dependence of J in relation to the structure and on the temperature variation. The applications are outside the scope of the chapter, however.

4.5.2 Other Coupled Systems

Complexes containing two or more paramagnetic species that take part in biochemical processes or are artificially made high-spin molecules have been extensively studied by high-field and multi-frequency ESR. We refer to the literature for interesting examples of the use of high field and multi-frequency ESR to obtain detailed data for complex biochemical systems [3, 17, 22, 41] concerning the interactions between transition metal ions or between a radical and a transition metal ion. Applications to high-spin molecules have recently been reviewed [28, 29].

4.6 Multifrequency ENDOR and ESEEM

ENDOR and ESEEM studies at high magnetic field have until recently been performed at specialized laboratories with home-made instruments. Commercial equipment at W-band is now available, however. The greater g-resolution at high microwave frequency increases the angular selection in measurements by ENDOR and ESEEM on disordered systems [2]. An application concerned with the structure of the sodium-nitric oxide adsorption complex in NaA zeolites, is described in Chapter 6. The *hfc* due to ^{14}N along the axes of the g-tensor (Fig. 6.4) could be measured by pulsed ENDOR. This was attributed to the resolution of the g-features at W-band. The angular selection method described in Section 2.2.3.3 could accordingly be applied.

At X-band microwave frequency traditionally employed in ESEEM studies, the envelope modulation patterns are in certain cases not well developed. This applies particularly for weakly coupled nuclei. The intensity of the forbidden transitions are then suppressed and consequently also the modulation depth. The use of a *lower* microwave frequency than X-band is accordingly expected to be particularly useful for enhancement of the ESEEM for weakly coupled nuclei. Dramatic enhancement of the ESEEM depth was observed for weakly coupled nuclei in initial studies with a pulsed EPR spectrometer constructed to operate at S-band. Nuclear quadrupole resonance frequencies were observed for ^{14}N in some nitroaromatic systems under cancellation conditions for the *hfc*, i.e. when $|\frac{1}{2}A - \nu_N| \approx 0$ in terms of the direct field model [44].

We refer to several recent specialist reports for details and additional applications of multi-frequency and high field studies by ENDOR and ESEEM [2, 45–47].

4.7 Summary

Multi-frequency and high field studies are at present mostly carried out in CW-mode. The chapter is concerned with examples of CW ESR applications, mostly in rigid amorphous matrices, e.g. frozen solutions, multiphase systems and in polycrystalline samples where the additional information from multifrequency and/or high field measurements are the most crucial for accurate characterisation.

CW-ESR measurements at high microwave frequency and corresponding high field increase the resolution between species with different g-factors and of small g-anisotropies in a single species. Spectral features due to other interactions, e.g. hyperfine coupling (*hfc*) and zero field splitting (*zfs*) that to 1st order are field independent, can be separated from Zeeman effects by multi-frequency measurements. The *zfs* is obtained directly as the line separation in the ESR spectra of species with $S \geq 1$ when the microwave frequency is much larger than the *zfs*. ESR spectra can be obtained at high field from samples that are "ESR-silent" at lower microwave frequencies due to large *zfs*. Very high *zfs* (larger than the microwave energy available) can in some cases be obtained by measurement of an effective g-factor at different microwave frequencies. The g-factor is in this case frequency-dependent, and gives information about the magnitude and the asymmetry of the *zfs* (the E/D ratio). The intensity ratio of electronic transitions $m_S = -1 \leftrightarrow 0$ and $m_S = 0 \leftrightarrow +1$ can be used to obtain the sign of the *zfs* (D) for a triplet species. An analogous procedure applies for systems with $S > 1$. The effect depends on the different populations of the electronic Zeeman levels. The difference increases with decreasing temperature, and is more pronounced at high field due to the increased level separations. Measurements can therefore be made at the relatively high temperatures of *ca.* 6–50 K that are conveniently reached with standard cryostats. The Heisenberg exchange interaction term between two species forming a coupled system affects the ESR spectra to different extents, as it depends on the difference in the splitting due to different g-factors between the units. The interaction can be estimated by measurements at a microwave frequency where the Heisenberg term and g-factor splitting are of comparable magnitude. The *hfc*, although to first order field-independent, shows a microwave frequency dependence due to the direct field effect particularly for atoms like ^{1}H and ^{19}F present in many free radical, transition metal ion, and triplet state systems. The occurrence of forbidden transitions and nuclear spin flip lines is due to the mixing of nuclear states, when the nuclear Zeeman energy and hyperfine energy are of comparable magnitudes. This well-known complication in X-band studies of free radicals in solids can be employed to obtain the relative signs of the principal hyperfine couplings in special cases when ENDOR is not applicable. Due to the high g-resolution at high field angular selection can be employed to obtain single-crystal-like ENDOR spectra of disordered systems, even for moderate g-anisotropy. ESEEM measurements at high frequency bands also increase orientation selectivity, while the modulations due to weakly coupled nuclei are enhanced at low frequency, e.g. at S-band.

Appendix

A4.1 Hyperfine Splittings by the Direct Field Model for S = ½

The magnetic fields B_+ and B_- acting on a magnetic nucleus is in the direct field model obtained by vector addition of B_N due to the externally applied magnetic field, and $\pm B_A$ due to the magnetic moment of the electron.

The magnitude b of $\mathbf{B}_N$ corresponds to the nuclear Zeeman energy. The direction is specified by the unit vector $\mathbf{l}$ along the applied field:

$$\mathbf{B}_N = -\frac{g_N \mu_N B}{g \mu_B} \cdot \mathbf{l} = b \cdot \mathbf{l} \tag{4.1}$$

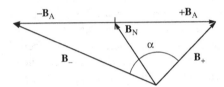

Fig. 4.25 Effective fields $\mathbf{B}_+$ ($m_S = +\frac{1}{2}$) and $\mathbf{B}_-$ ($m_S = -\frac{1}{2}$) acting on a nucleus with an anisotropic *hfc* and a nuclear Zeeman term of comparable magnitude. $\mathbf{B}_A$ is due to the hyperfine coupling, $\mathbf{B}_N$ to the applied magnetic field

The field $\mathbf{B}_A$ can be oriented two ways depending on the value of the electron spin quantum number ($m_S = \pm\frac{1}{2}$). The field is according to the direct field model [36] given by (4.2) when the electronic g-factor is isotropic:

$$\pm \mathbf{B}_A = \pm 1/2 \frac{\mathbf{A}}{g \mu_B} \cdot \mathbf{l} = \pm\frac{1}{2} \cdot \left(\mathbf{A}' \cdot \mathbf{l}\right) \tag{4.2}$$

The hyperfine coupling is specified in energy and field units by the tensors $\mathbf{A}$ and A', respectively. The field $\mathbf{B}_A$ is not along the applied field for anisotropic *hfc*. The strengths of the total fields are therefore given by (4.3):

$$B_{\pm}^2 = (\pm \mathbf{B}_A + \mathbf{B}_N)^2 = B_A^2 + B_N^2 \pm 2 \cdot \mathbf{B}_A \cdot \mathbf{B}_N \tag{4.3}$$

$$B_{\pm}^2 = (\mathbf{T}_{\pm} \cdot \mathbf{l})^2 \quad \mathbf{T}_{\pm} = \pm\frac{1}{2}\mathbf{A}' + b\mathbf{E} \tag{4.4}$$

The result (4.4) follows from (4.1, 4.2, and 4.3) by algebra that is not reproduced here. One can note, however, that $\mathbf{T}_{\pm}$ is the tensor commonly employed in the analysis of ENDOR data (Section 3.4.4) with $\mathbf{E}$ equal to the unit tensor. The final result (4.4) agrees with that derived more strictly by quantum mechanics methods [26, 36].

The ESR line positions B_o and B_i relative to the centre for the outer and inner doublets of an $I = \frac{1}{2}$ nucleus can be expressed in terms of $B_{\pm}$ i.e.:

$$B_0 = \pm\tfrac{1}{2}(B_+ + B_-) \qquad\qquad B_i = \pm\tfrac{1}{2}|B_+ - B_-| \qquad (4.5)$$

The intensities of the lines t can be expressed in terms of the angle α between the effective fields for $m_S = \pm\tfrac{1}{2}$ by a quantum mechanical treatment, see e.g. [38] for details. The intensities I_i and I_o for the inner and outer doublets of an $I = \tfrac{1}{2}$ nucleus are given by (4.6):

$$I_i = \cos^2\frac{\alpha}{2} \qquad I_o = \sin^2\frac{\alpha}{2} \qquad (4.6)$$

The angle α is calculated by quantum mechanics [38] or in the direct field model by applying the cosine theorem (Fig. 4.25) according to equation (4.7):

$$\cos\alpha = \frac{B_N^2 - B_A^2}{B_+ \cdot B_-} = \frac{b^2 - (\tfrac{1}{2}\cdot \mathbf{A}' \cdot \mathbf{1})^2}{B_+ \cdot B_-} \qquad (4.7)$$

$(2\cdot I + 1)^2$ lines are obtained for an arbitrary value of I with intensities and positions given in [38]. In the limits $B_N \ll B_A$ and $B_N \gg B_A$ or when the hyperfine coupling is isotropic the normal selection rule $\Delta m_I = 0$ applies, yielding $2\cdot I + 1$ hyperfine lines of the same intensity. Two equivalent formulations taking account of the combined effects of hyperfine coupling and nuclear Zeeman interaction in case of dominant electron Zeeman energy are commonly employed.

Weil and Anderson formulation [36]: Each electronic level is split in $2\cdot I + 1$ sublevels. The splittings in field units for the electronic states $m_S = \pm\tfrac{1}{2}$ are given by equation (4.8a):

$$G_\pm = \sqrt{\left\{\sum_{i=1}^{3}\left[\left(\pm\tfrac{1}{2}\sum_{j=1}^{3}A_{ij}u_j\right) + b_i\right]^2\right\}} \qquad (4.8a)$$

The quantities $G_\pm$ contain contributions from the hyperfine tensor $\mathbf{A}$ and from the nuclear Zeeman term $b_i = -B\frac{g_N\mu_N}{g\mu_B}\ell_i$ where ℓ_x, ℓ_y and ℓ_z are the direction cosines of the applied magnetic field with respect to a suitably chosen coordinate system of a crystal or molecule. In the general case with non-negligible g-anisotropy the effective direction of the field is $\mathbf{u} = \frac{\mathbf{g}\cdot\mathbf{l}}{g}$ following the notation of Weil and Anderson [36]. The matrix notation (4.8b) is also employed, e.g. in [26] that also provides the 2nd order corrections reproduced in Chapter 3.

$$G_{m_S}^2 = \mathbf{1}\left(\frac{m_S}{g}g\mathbf{A} + b\mathbf{E}\right)\left(\frac{m_S}{g}\mathbf{A}g + b\mathbf{E}\right)\mathbf{1} \qquad (4.8b)$$

The hyperfine energy for the transition $(m_S, m_I) \leftrightarrow (m_S +1, m_I')$ is $|G_{m_S}\cdot m_I - G_{m_S+1}\cdot m_I'|$, where the nuclear quantum numbers m_I and m_I' need not be the same. Each nucleus may therefore give $(2 \cdot I + 1)^2$ rather than the $2 \cdot I + 1$ hyperfine lines when $\Delta m_I = 0$ applies. A spectrum with $S = I = \frac{1}{2}$ has the schematic appearance shown in Fig. 4.26.

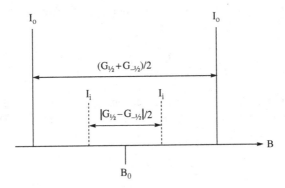

Fig. 4.26 $S = \frac{1}{2}$ ESR line pattern for an $I = \frac{1}{2}$ nucleus with an anisotropic hyperfine coupling and a nuclear Zeeman term of comparable magnitude. The inner (i) and outer (o) lines occur at $B_0 \pm \frac{|G_{1/2} - G_{-1/2}|}{2}$ and $B_0 \pm \frac{|G_{1/2} + G_{-1/2}|}{2}$ with intensities $I_i = \cos^2 \frac{\alpha}{2}$ and $I_o = \sin^2 \frac{\alpha}{2}$, respectively. The angle α is between the effective field directions in Fig. 4.25

Poole and Farach formulation [48]: Equations obtained for the special case with $S = I = \frac{1}{2}$ derived by Poole and Farach [48] are useful when both the inner and outer doublet splittings T_i and T_o are observed. In this case the hyperfine coupling tensor can directly be obtained in single crystal measurements by the Schonland method from the relation:

$$\mathbf{1} \cdot A^2 \cdot \mathbf{1} = T_0^2 + T_i^2 - (2B_N)^2 \tag{4.9}$$

It is also of interest that the intensities of the outer and inner doublets can be determined, from measurements of those doublets, using the equations:

$$I_0 = \frac{T_o^2 - (2g_N \mu_B B)^2}{T_0^2 - T_i^2} \qquad I_i = \frac{(2g_N \mu_B B)^2 - T_i^2}{T_0^2 - T_i^2} \tag{4.10}$$

The hyperfine coupling tensor is not needed to predict the intensities under the condition that the inner and outer doublet splittings are experimentally observed.

Depending on the relative magnitudes of the hyperfine and nuclear Zeeman terms, different cases occur.

(1) *Anisotropic **g**- and **A**-tensors, negligible b*: When the nuclear Zeeman energy is small compared to the hyperfine coupling as is usually the case for transition metal ions, the spectrum is independent of microwave frequency. The hyperfine coupling K depends on the crystal orientation according to the equation (4.11):

$$(g \cdot K)^2 = \mathbf{l} \cdot \mathbf{g}A^2\mathbf{g} \cdot \mathbf{l} = \sum_{i,j=1}^{3} (\mathbf{g}A^2\mathbf{g})_{ij}\ell_i\ell_j \tag{4.11}$$

The effective field acting on the nucleus is caused by the hyperfine coupling, so that $\alpha \approx 180°$ and the normal selection rule $\Delta m_I = 0$ applies. The tensor $\mathbf{A}^2$ is accordingly obtained by a modified Schonland procedure described in Chapter 3.

(2) *Isotropic* **g**, *anisotropic* **A**, *non-negligible b:* This case occurs particularly for proton hyperfine couplings in organic radicals, but has also been observed for transition metal ions with fluorine ligands [33]. The ^{1}H and ^{19}F nuclei have unusually large g_N-factors, and correspondingly large B_N fields in the direct field model. The four line *hfc* pattern for a single $I = \frac{1}{2}$ nucleus is schematically shown in Fig. 4.26. Formulas that take into account the nuclear Zeeman term on the line positions and intensities were first implemented in the classical works by Lefebvre and Maruani in spectrum simulations of organic radicals with hyperfine structure due to ^{1}H in amorphous samples. We refer to the literature for the treatment of $I = 1$ and 3/2 nuclei [38].

(3) *Anisotropic* **A**, *with principal values* $\ll$ *b*: Apart from the spin flip lines due to weak dipolar coupling with distant H atoms of the matrix this case is rare at X-band and lower frequencies, but may become of interest for measurements at high microwave frequency bands. The situation is most likely to occur for species with anisotropic hyperfine couplings of moderate size due to nuclei with large nuclear g-factors like ^{1}H or ^{19}F. The effective field acting on the nucleus is dominated by the applied field, so that $\alpha \approx 0$ and the normal selection rule $\Delta m_I = 0$ applies. The splitting between the lines is $K = \mathbf{l} \cdot A \cdot \mathbf{l}$ and $K \cdot g = \mathbf{l} \cdot \mathbf{A}\mathbf{g} \cdot \mathbf{l}$ for isotropic and anisotropic g-factors, respectively.

(4) *Anisotropic* **A** *and* **g**, *and non-negligible b*: The general case with appreciable g- and hyperfine anisotropy and a nuclear Zeeman term that is comparable with the hyperfine coupling is rarely considered in practical applications.

References

1. A.A. Galkin, O.Ya. Grinberg, A.A. Dubinskii, N.N. Kabdin, V.N. Krimov, V.I. Kurochkin, Ya.S. Lebedev, L.G. Oranskii, V.F. Shovalov: Instrum. Exp. Tech. (Eng. Transl.) **20**, 284 (1977).
2. P.C. Riedi, G.M. Smith: In '*Electron Paramagnetic Resonance*' (Vol. **18**) Royal Society of Chemistry Specialist Periodical Reports. Thomas Graham House, Cambridge (2002), pp. 254–303.
3. A.-L. Barra, A. Gräslund, K.K. Andersson: In '*Very High Frequency (VHF) ESR/EPR, Biological Magnetic Resonance*' (Vol. 22) ed. by O. Grinberg, L.J. Berliner, Kluwer Academic Publishers, Dordrecht (2004), Chapter 5.
4. M. Kaupp: In '*EPR of Free Radicals in Solids, Trends in Methods and Applications*' ed. by A. Lund, M. Shiotani, Kluwer Academic Publishers, Dordrecht (2003), Chapter 7.
5. D. Goldfarb (guest editor): '*Modern EPR spectroscopy*', Themed Issue, PCCP **11**, 6537 (2009).

6. W.M. Chen: In '*EPR of Free Radicals in Solids, Trends in Methods and Applications*' ed. by A. Lund, M. Shiotani, Kluwer Academic Publishers, Dordrecht (2003), Chapter 15.

7. J.R. Pilbrow: '*Transition Ion Electron Paramagnetic Resonance*', Clarendon Press, Oxford (1990).

8. P.W. Atkins, M.C.R. Symons: '*The Structure of Inorganic Radicals. An Application of ESR to the Study of Molecular Structure*', Elsevier Publishing Company, Amsterdam (1967).

9. A.Yu. Bresgunov, A.A. Dubinsky, O.G. Poluektov, Ya.S. Lebedev, A.I. Prokof'ev: Mol. Phys. **75**, 1123 (1992).

10. A.L. Barra: Appl. Magn. Reson. **21**, 619 (2001).

11. K.V. Lakshmi, M.J. Reifler, G.W. Brudvig, O.G. Poluektov, A.M. Wagner, M.C. Thurnauer: J. Phys. Chem. B **104**, 10445 (2000).

12. L.E.G. Eriksson, A. Ehrenberg: Acta Chem. Scand. **18**, 1437 (1964).

13. A. Okafuji, A. Schnegg, E. Schleicher, K. Möbius, S. Weber: J. Phys. Chem. B **112**, 3568 (2008).

14. K.C. Christoforidis, S. Un, Y. Deligiannakis: J. Phys. Chem. A **111**, 11860 (2007).

15. D.W. Ovenall, D.H. Whiffen: Mol. Phys. **4**, 135 (1961).

16. A.I. Smirnov: In '*Electron Spin Resonance Specialist Periodical Reports*' (Vol. **18**), ed. by M. Davies, B. Gilbert, N. Atherton, The Royal Society of Chemistry, Cambridge(2002) pp. 109–136.

17. K.K. Andersson, P.P. Schmidt, B. Katterle, K.R. Strand, A.E. Palmer, S-K Lee, E.I. Solomon, A Gräslund, A-L Barra: J. Biol. Inorg. Chem. **8**, 235 (2003).

18. A.J. Stone: Proc. R. Soc. A. **271**, 424 (1963).

19. F. Neese: EPR Newsl. **18**(4), 10, International EPR (ESR) Society (2009).

20. N.M. Atherton: '*Principles of Electron Spin Resonance*', Ellis Horwood, New York, NY (1993).

21. F. Ban, J.W. Gauld, S.D. Wetmore, R.J. Boyd: In '*EPR of Free Radicals in Solids, Trends in Methods and Applications*' ed. by A. Lund, M. Shiotani, Kluwer Academic Publishers, Dordrecht (2003), Chapter 6.

22. K.K. Andersson, A.-L. Barra: Spectrochim. Acta A **58**, 1101 (2002).

23. A. Kawamori: In '*EPR of Free Radicals in Solids, Trends in Methods and Applications*' ed. by A. Lund, M. Shiotani, Kluwer Academic Publishers, Dordrecht (2003), Chapter 13.

24. H. van der Waals, M.S. de Groot: Mol. Phys. **2**, 333 (1959).

25. http://www.bruker-biospin.com/epr.html

26. M. Iwasaki: J. Magn. Reson. **16**, 417 (1974).

27. J.A. Weil, J.R. Bolton: '*Electron Paramagnetic Resonance: Elementary Theory and Practical Applications*', 2nd Ed., Wiley, New York, NY (2007).

28. T. Takui, H. Matsuoka, K Furukawa, S. Nakazawa, K. Sato, D. Shiomi: In '*EPR of Free Radicals in Solids, Trends in Methods and Applications*' ed. by A. Lund, M. Shiotani, Kluwer Academic Publishers, Dordrecht (2003), Chapter 11.

29. M. Baumgarten: In '*EPR of Free Radicals in Solids, Trends in Methods and Applications*' ed. by A. Lund, M. Shiotani, Kluwer Academic Publishers, Dordrecht (2003), Chapter 12.

30. S. Slappendel, G.A.Veldink, J.F.G. Vliegenthart, R. Aasa, B.G. Malmström: Biochim. Biophys. Acta **624**, 30 (1980).

31. P. Atkins, J. de Paula: '*Atkins' Physical Chemistry*', 7th Ed., Oxford University Press, Oxford (2002), p. 632.

32. C. Chachaty: http://www.esr-spectsim-softw.fr/triplets.htm

33. C.A. Thuesen, A-L. Barra, J. Glerup: Inorg. Chem. **48**, 3198 (2009).

34. H.M. McConnell, C. Heller, T. Cole, R.W. Fessenden: J. Am. Chem. Soc. **82**, 766 (1960).
35. A. Carrington, A.D. McLachlan: '*Introduction to Magnetic Resonance with Applications to Chemistry and Chemical Physics*', Harper & Row, New York, NY (1967).
36. J.A. Weil, J.H. Anderson: J. Chem. Phys. **35**, 1410 (1961).
37. S. Schlick, L. Kevan: J. Magn. Reson. **21**, 129 (1976).
38. R. Lefebvre, J. Maruani: J. Chem. Phys. **42**, 1480 (1965).
39. K. Minakata, M. Iwasaki: Mol. Phys. **23**, 1115 (1972).
40. Y. Kurita: J. Chem. Phys. **41**, 3926 (1964).
41. A. Fournel, S. Gambarelli, B. Guigliarelli, C. More, M. Asso, G. Chouteau, R. Hille, P. Bertrand: J. Chem. Phys. **109**, 10905 (1998).
42. S. Gambarelli, D. Jaouen, A. Rassat, L.C. Brunel, C. Chachaty: J. Phys. Chem. **100**, 9605 (1996).
43. R. Briere, R.M. Dupeyre, H. Lemaire, C. Morat, A. Rassat, P. Rey: Bull. Soc. Chim. France **11**, 3290 (1965).
44. (a) R.B. Clarkson, M D. Timken, D.R. Brown, H.C. Crookham, R.L. Belford: Chem. Phys. Lett. **163**, 277 (1989). (b) R.B. Clarkson, D.R. Brown, J.B. Cornelius, H.C. Crookham, W.-J. Shi, R.L. Belford: Pure Appl. Chem. **64**, 893 (1992).
45. Grinberg, L.J. Berliner (eds.): '*Very High Frequency (VHF) ESR/EPR, Biological Magnetic Resonance*', Vol. 22, Kluwer Academic Publishers, Dordrecht (2004).
46. Y. Deligiannakis, M. Louloudi, N. Hadjiliadis: Coord. Chem. Rev. **204**, 1 (2000).
47. K. Möbius, A. Savitsky: '*High-Field EPR Spectroscopy on Proteins and Their Model Systems*', Royal Society of Chemistry, Cambridge (2009).
48. C.P. Poole, Jr., H.A. Farach: J. Magn. Reson. **4**, 312 (1971).

Exercises

E4.1 (a) Organic radicals typically have g-factors close to the free-electron value $g = 2.00232$, with only slight anisotropy. In an ESR study of carotenoid and chlorophyll cation radicals in photosystem II at D-band (130 GHz), Lakshmi et al measured $g_x = 2.00335$, $g_y = 2.00251$, $g_z = 2.00227$ [11]. The line-width was 4–5 G. Are the g_x, g_y, and g_z features completely resolved under those conditions? (The spectra are reproduced in Fig. 4.2) Would it be possible to resolve the features corresponding to these g-factors at 385 GHz?

   (b) Inorganic radicals can have larger g-anisotropies. Can one expect to resolve the $g_x = 2.0029$, $g_y = 2.0225$, and $g_z = 2.0384$ features of the ClSS· radical [F.G. Herring, C.A. McDowell, C. Tait: J. Chem. Phys. **57**, 4564 (1972)] at X-band? (The spectrum is reproduced in Chapter 3, Fig. 3.20).

E4.2 ESR spectra obtained of methyl radicals trapped in a frozen matrix of CO were interpreted by assuming that the radicals rotated rapidly about the three-fold axis of $\dot{C}H_3$, as shown in the figure. The g-tensor and the *hfc*-tensor of the three equivalent ^1H atoms are then axially symmetric about this axis. The parameters $A_\perp = 23.4$ G, $A_\parallel = 22.3$ G, $g_\perp = 2.0027$, $g_\parallel = 2.0022$, and line-width $= 0.43$ G were obtained by comparison of experimental and simulated spectra.

(a) Examine the shape of the experimental spectrum to explain why it was necessary to consider both *g*- and *hfc*-anisotropy in the analysis of the experimental spectrum.

(b) Would the observed spectrum look different at Q-band, and if so would the parameters obtained at Q-band be more accurately determined? (Only X-band data were experimentally available).

(c) Why can "forbidden" lines due to *hfc* from ^{1}H be ignored in the analysis of the experimental spectrum?

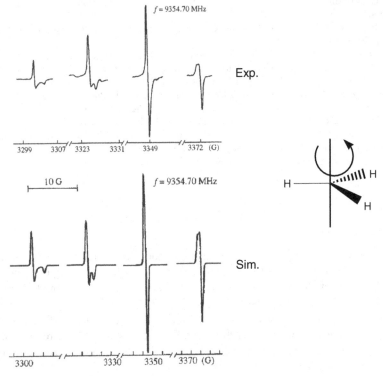

Experimental and simulated X-band spectra of methyl radicals in a frozen matrix of CO. The figure is reproduced from [Yu.A. Dmitriev, R.A. Zhitnikov: J. Low Temp. Phys. **122**, 163 (2001)] with permission from Springer

E4.3 In the first experimental study of radical pairs by Kurita [40], ESR spectra of the type shown below were observed in single crystal measurements of γ-irradiated dimethylglyoxime at X-band. Suggest an interpretation of the splitting D in the spectrum with H||c′ and a reason why the splitting appears smaller with H||a′. What kind of interaction could cause the additional splittings in the spectra?

ESR spectra of irradiated single crystals of dimethylglyoxime at 77 K with the magnetic field directed along the c' and a' axis. The curves represent 2nd derivatives of the actual absorption spectra. The diagram is reproduced from [40] with permission from the American Institute of Physics

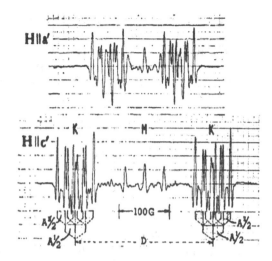

E4.4 The naturally occurring copper-containing complex, turacin, exhibits ESR spectra of frozen solutions that were interpreted to have a dimer structure containing two Cu^{2+} ions, separated by 3.5 Å. A suggested structure of two turacin molecules forming a sandwich complex obtained from X-band ESR measurements of frozen solutions of samples extracted from touraco feather components is as shown. Magnetic dipolar coupling gives a splitting in two lines separated by $F(T) = \frac{\mu_0}{4\pi} \frac{3\,g\mu_B}{2r^3}(3\cos^2\theta - 1)$. Is first order analysis of the X-band spectra recommendable i.e. is $F << B_0 = h\nu/g\mu_B \approx 0.33$ T?

Note: The actual analysis by W.E. Blumberg, J. Peisach: J. Biol. Chem., **240**, 870 (1965) was exact, i.e. spectra were calculated by diagonalizing the Hamiltonian matrix, and the *zfs* and the anisotropic Cu hyperfine structure were adjusted to fit the experimental data.

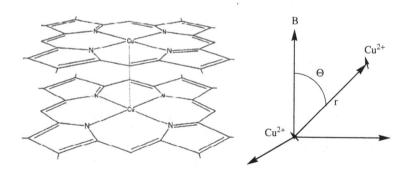

E4.5 In the paper by Fournel et al. [41] the exchange coupling between a radical and a transition metal complex was found to be positive $J = 0.72$ cm^{-1}.

(a) How would an energy diagram analogous to that in Fig. 4.21 look like for this *antiferromagnetic* coupling?

(b) Estimate a range of magnitudes of J that would be possible to measure by observing the effect on the spectrum shapes at X-, Q- and W-bands for a coupled system with $S_1 = S_2 = \frac{1}{2}$, $g_1 = 1.9$, $g_2 = 2.1$.

E4.6 The intensities of the two sets of lines for the radical pair of dimethylglyoxime in the spectra shown in E4.3 are approximately equal at 77 K. At what temperature range would it be possible to determine the sign of D by observing different intensities of the two sets in the ESR spectra at X and W band? Assume that a difference in intensities of *ca.* 20% is sufficient for the sign assignment.

E4.7 Suggest a reason why the direct field effect causes complications in ESR but not in ENDOR.

E4.8 The orientations of the effective fields $B_\pm$ in Fig. 4.17 determine the quantization axes of the nuclear spin for $m_S = \pm\frac{1}{2}$.

(a) What are the values for the angle α between the axes for the extreme cases $B_N \ll B_A$ and $B_N \gg B_A$? Which doublet is observed in each case?

(b) What angle applies for isotropic hyperfine couplings? Why do "forbidden" lines not occur in this case?

Note: Answers can be formally obtained by using formulae in Appendix A4.1. The formulae apply also in the quantum mechanics treatment employed in the first general program for the simulation of ESR spectra of free radicals in amorphous solids [38].

(c) Suggest a reason why the spin flip lines tend to become less intense at Q-band than at X-band in the spectra of Fig. 4.19.

E4.9 The relative signs of the *zfs* (D) and the principal values of an anisotropic *hfs* due to ^1H and ^{19}F of radical pairs in a single crystal of monofluoroacetamide were determined by analysing the asymmetry of the hyperfine patterns for the $m_S = 1 \rightarrow 0$ and $m_S = 0 \rightarrow -1$ transitions [39]. The asymmetry is pronounced when the principal *hfs* couplings are of comparable magnitudes to the nuclear frequencies of ^1H and ^{19}F. A similar analysis can be applied to triplet state species.

Would it be more difficult to determine the relative signs of zero-field and hyperfine couplings of triplet state trimethylenemethane by single crystal ESR measurements at W-band than at X-band? The principal values of the six ^1H nuclei are equal and given in the figure.

H   H

H   •   •   H

H

H   H

↗ −37.8 MHz

−26.3 MHz

↘ −14.0 MHz

E4.10 How many hyperfine lines due to ^{14}N are expected in liquid solution from
the "biradical 2" when $|J| \ll |a_N|$ and $|J| \gg |a_N|$, respectively? For details
see [K. Komaguchi et al.: Chem. Phys. Lett. **387**, 327 (2004)].

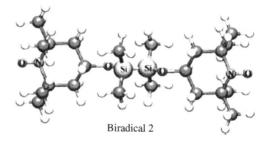

Biradical 2

E4.11 For most nuclei except ^1H and occasionally ^{19}F the hyperfine coupling is
usually much larger than the nuclear Zeeman energy at X-band.

(a) Estimate the magnitude of B_N (Fig. 4.17) for ^1H and ^{14}N at W-band and
discuss possible occurrence of forbidden lines, see Section 3.4.1.2.

(b) Estimate the value of B_N for the ^{14}N nucleus at X-band. Is there any
risk of "forbidden" lines due to ^{14}N *hfs* for the $N_2H_4^+$ ion discussed in
Chapter 1? Is there such risk for the lines due to ^1H?

Part II
Applications of ESR

Chapter 5
Applications to Molecular Science

Abstract CW-ESR spectroscopy combined with matrix isolation method and ionizing radiation is applied to investigate structure and reactions of intermediate radicals in low temperature solid matrices. ESR parameters are predicted with considerable precision by computations, affording a valuable bridge between experiment and theory. *Cyclic*-$C_nF_{2n}^-$ radicals ($n = 3$–5) have a planar structure with an entirely delocalized singly occupied MO. Unsaturated $C_nF_{2n-2}^-$ radicals have a distorted pyramidal structure occurring by mixing the π^* and σ^* orbitals. The acetylene anion has a trans-bent structure. Structural distortion occurs in methane cation from original T_d to C_{2v} symmetry due to the Jahn-Teller effect. A similar symmetry lowering from D_{3h} to C_{2v} is discussed for the trimethylenemethane cation. High-resolution ESR spectra of D-labelled methyl radicals in solid Ar are discussed in terms of nuclear spin-rotation couplings. The formation of "H---CH_3" radical pairs and of a "H---H_2" complex" in solid Ar is presented. The structure of $[H_2(H_2)H_2]^+$ formed in *para*-H_2 is discussed based on the ESR spectra of D-substituted samples and on computations.

5.1 Introduction

The ESR method potentially can provide not only an unambiguous assignment of radicals, but also experimental information about their geometrical and electronic structures in the ground and excited states as well as about their reactions. For example, molecular radical anions are formed by the addition of one electron to the lowest unoccupied molecular orbital (LUMO) of neutral (diamagnetic) molecules, while molecular radical cations are formed by the loss of one electron from the highest occupied molecular orbital (HOMO). Both the radical anions and cations are in a doublet electronic state possessing one unpaired electron ($S = 1/2$) in a singly occupied molecular orbital (SOMO) and are accordingly ESR active chemical species. In this chapter we deal with applications of CW-ESR spectroscopy combined with matrix isolation method to investigate structures and reactions of reactive neutral and ionic radicals generated by ionizing radiation and trapped in solid inert matrices at low temperatures.

A. Lund et al., *Principles and Applications of ESR Spectroscopy*,
DOI 10.1007/978-1-4020-5344-3_5, © Springer Science+Business Media B.V. 2011

Matrix isolation is an experimental technique used in chemistry and physics for trapping a material in a large excess of an unreactive host matrix by rapid condensation at low temperature. Originally this technique has been developed to study unstable reactive chemical species and is nowadays widely used for various types of research problems [1, 2]. A typical matrix isolation experiment involves a guest (solute) atom or molecule of interest being diluted in the gas phase with the host matrix (solvent), usually a noble gas cooled to below 20 K. The samples can then be studied using various spectroscopic methods including ESR. Individual guest atoms or molecules are trapped and isolated from one another in a low temperature solid matrix. Under these conditions, the guest atoms or molecules do not interact with each other, but interact very weakly with the surrounding matrix, thereby simulating the gas phase. A sample thus prepared can be preserved as long as the matrix is maintained. Thus, the matrix isolation technique is particularly well-suited to the study of highly reactive chemical species including small organic and inorganic radicals which are difficult to be generated and maintained in appreciable abundance in the gas phase. Reactive species are usually generated either in the gas phase prior to deposition in the matrix, or after deposition by in-situ photolysis, radiolysis, or other methods of an appropriate matrix-isolated precursor (Fig. 5.1).

The matrix isolation method combined with ionizing radiation (i.e., γ-rays, X-rays, etc.) at low temperature is versatile and has been extensively developed for ESR study of radical ions [3–13]. The procedure consists of dissolution of the solute molecule (atom) of interest in an appropriate solvent (matrix), freezing at low temperature (in general below 77 K), irradiation (or illumination), and ESR measurements before and after thermal treatment. The solvent molecules (atoms) are ionized by the irradiation to yield an electron (e^-) and a positive hole (h^+). The electron is transferred in general to a solute molecule with higher electron affinity than that of the solvent molecule to form a solute molecular radical anion (Fig. 5.2). On the other hand, the positive hole is transferred to a solute molecule with first ionization energy (potential) lower than that of the matrix molecule, resulting in the

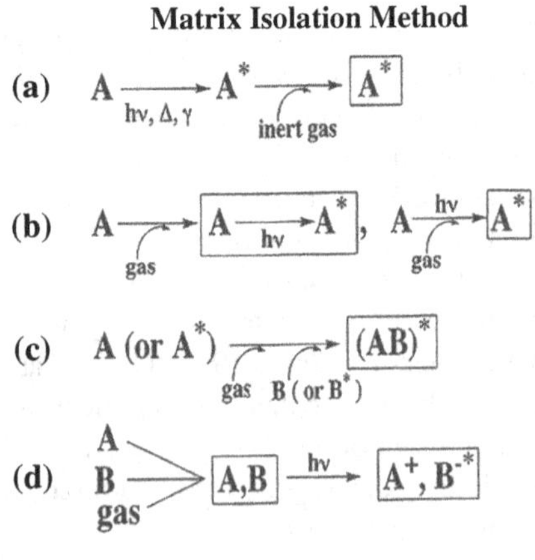

Matrix Isolation Method

Fig. 5.1 Various generation methods of reactive chemical species using a matrix. "A", "B" and "gas" stand for solute molecules or atoms and "inert gas" stands for matrix. The terms "hν", "Δ", and "γ" imply generation methods by photolysis, thermolysis and radiolysis by γ-rays, respectively

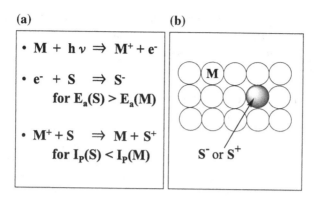

Fig. 5.2 (a) Generation and (b) trapping of solute molecular radical anions or cations (S$^-$ or S$^+$) in matrix (M). M is ionized by radiolysis or photolysis (hν), and then the generated hole (M$^+$) or electron (e$^-$) is transferred to a solute (S) to generate S$^-$ or S$^+$. E$_a$ and I$_p$ stand for electron affinity and first ionization energy, respectively

formation of a solute molecular radical cation (Fig. 5.2). Thus formed molecular radical anions and cations can undergo a wide variety of reactions by illuminating the samples with visible or UV light, or by annealing them at elevated temperatures.

ESR parameters, especially hyperfine (*hf*) and *g*-tensors, can potentially provide detailed information about geometrical and electronic structures of radicals. The *hf*-tensor provides information about the interactions between electronic spin density and certain nuclei within a given radical, whereas the *g*-tensor reflects properties of the entire molecule, i.e., a general spin density distribution and bonding features, and interactions with the environment, etc. Due to the relatively complicated nature of the *hf*- and *g*-tensors their analysis by quantum chemical means is required for correctly relating them to structures and bonding. Furthermore, the correct analysis of experimental ESR spectra is often difficult based on the ordinary trial method for ESR spectral simulation, especially in powder samples, because of a number of adjustable parameters that are needed; in addition to three principal values of the *hf* tensors, their directions (three parameters) must be specified for each of the atoms with nuclear spins. In addition the *g*-tensor is specified by another six adjustable parameters in which the principal directions are in general fixed to the molecular coordinate differing from those in the *hf*-tensor. Fortunately quantum chemical computations can also provide an independent theoretical approach to this difficult spectral analysis and radical identification problems as will be presented in Sections 5.2 and 5.3 (radical anions and radical cations). Thus quantum computations have now become an indispensable tool for the experimentalists who can assess the validity of a proposed radical model by comparing its calculated ESR parameters with experimental ones.

During the last two decades remarkable progress has been made in the quantum chemical calculations of the *hf*- and *g*-tensors [14, 15]. The calculations on organic radicals used to be essentially restricted to semi-empirical methods such as (extended) Hückel, INDO (Intermediate Neglect of Differential Overlap), MINDO

(Modified Intermediate Neglect of Differential Overlap), MNDO (Modified Neglect of Differential Overlap), etc. [16–18]. Today, a variety of ab initio and density functional theory (DFT) methods are available; a brief summary on quantum chemical computation methods is given in Appendix A5.1. Accurate hf couplings can be predicted by the highest levels of electron correlation theory for small radical species, comparable results can often be obtained from the methods based on DFT such as B3LYP [18–21]. The DFT methods require considerably less computational time and memory than the conventional correlated ab initio procedures and have now become powerful alternative methods for the computations of hf-couplings and g-tensors. The usefulness and accuracy of DFT methods to examine the structure and ESR parameters of radicals are extensively discussed in the literature (for example, [14, 22–29]), especially for organic radicals. ESR spectra are measured in generally at a temperature above 4 K. The excitation of vibrational motions at higher temperatures may have effects on the molecular structure of the radical, and therefore on hf couplings. The theoretical methods incorporating temperature or vibrational effects by means of vibronic coupling, vibrational averaging, and molecular dynamics have been discussed by Eriksson [25], Adamo [27] and others [24, 26]. The performance of various levels of theory for the calculations of hf- and g-tensors has been reviewed in recent articles by Ban et al. [14] and Kaupp [15], respectively.

The first topic in this chapter is concerned with the geometrical and electronic structure of radical anions such as perfluorocycloalkane anion (c-$C_nF_{2n}^-$) and perfluoroalkene anion ($C_nF_{2n-2}^-$). Based on the detailed comparisons between experimental and theoretical results planar molecular structures with a high symmetry and pyramidally distorted structures are proposed for the former and latter radical anions, respectively. Furthermore, a trans bent structure of acetylene radical anion is presented. The second topic is concerned with Jahn-Teller (J-T) distortion in the radical cations of organic molecules with a doubly degenerate HOMO such as methane and trimethylenemethane. A lowering of symmetry from original T_d to C_{2v} symmetry upon one-electron oxidation is presented based on the ESR results of D(^{2}H)-labelled methane radical cations combined with theoretical computations. The third topic is concerned with high resolution ESR spectra of methyl radical, hydrogen atom – hydrogen molecule complex, and H_6^+ radical cation generated in a solid Ar or a quantum solid of $para$-H_2 ($I = 0$) matrix. The ESR results are discussed in terms of quantum effects at cryogenic temperature.

5.2 Radical Anions

A large number of radical ions have been studied by ESR since its applications to free radical chemistry started in the middle of the 1950s. They were generally prepared in the liquid phase by redox (reduction-oxidation) reactions, electrolysis or photolysis, and rather stable radical ions possessing conjugation system such as aromatics have been studied. The results have been summarized, for example, in a book "Radical Ions" by Kaiser and Kevan [30]. Since the middle of the 1960s a number of radical anions were generated by the matrix-isolation method combined with ionizing radiation and their structures and reactions have been extensively studied by ESR spectroscopy. Some selected radical anions reported until the end of the 1980s have been listed in CRC Handbook of Radiation Chemistry [6], which includes fluorinated hydrocarbon radical anions and σ^* radical anions with a trigonal pyramidal (C_{3v}) structure such as CF_3X^- (X: Cl, Br, I). In the studies the following molecules have been adopted as suitable matrices to generate and isolate solute radical anions: ethanol (ETOH), 2-methylhydrofuran (2-MTHF), neopentane, tetramethylsilane (TMS), SF_6, and adamantane. MTHF and ETOH form a rigid glassy matrix by rapid cooling the samples to 77 K [6b, 10]. The others form a plastically crystalline matrix and, even above the solid-solid phase transition (i.e., plastic phase), the solute radical anions have been found to be stable, which allowed to observe well resolved ESR spectra and make the spectral analysis easier.

We do not intend to show full coverage of the ESR studies of radical anions reported so far, but to show some typical examples and to demonstrate general trends. Thus, here we present our recent ESR studies combined with quantum chemical computations on electron delocalization in perfluoroalkane radical anions [3, 31, 32] and on structure distortion in perfluoroalkene radical anions [33] and in the acetylene radical anion [34].

5.2.1   Perfluorocycloalkane Radical Anions: Electron Delocalization

Perfluorocarbons have remarkable properties such as chemical inertness, thermal stability, high hydrophobicity, low dielectric constant, and large electronegativity. These properties have led to their widespread use in many industrial products which include blood substitutes, lubricants, surface modifiers, oxygen carriers, etchants for semiconductor fabrication, etc. [32, 35a]. The structure and stability of perfluorocycloalkane (c-C_nF_{2n}) negative ions are of interest because of the unusual nature of these species and their significant role in numerous potential applications of the parent compounds. The detailed geometrical and electronic structures of perfluorocycloalkane radical anions (c-$C_nF_{2n}^-$) have thus been subjected to ESR spectroscopic study. The ESR parameters, especially the hf couplings, can be predicted with considerable precision by recent advances in computational methods involving density functional theory (DFT) [18–21], thereby affording a valuable bridge between experiment and theory at a most fundamental level.

5.2.1.1   Experimental Results

Radical anions of c-$C_nF_{2n}^-$ ($n = 3$–5) were generated and stabilized in γ-irradiated (plastically crystalline) tetramethylsilane (TMS) and (rigid) 2-methyltetrahydrofuran (2-MTHF) matrices. The identification of radical anions was confirmed by generating the identical ESR spectra in photoionization experiments using N,N,N',N'-tetramethyl-p-phenylenediamine (TMPD) [3, 31, 32]. By recording their temperature-dependent ESR spectra in TMS, both isotropic and anisotropic spectra were observed. The isotropic ESR spectra of c-$C_3F_6^-$, c-$C_4F_8^-$, and c-$C_5F_{10}^-$ in TMS matrix show the second-order hyperfine structures [36, 37] (see below) characteristic of six, eight and ten equivalent ^{19}F atoms with their isotropic hf couplings of 19.8, 14.85, and 11.6 mT, respectively (see Table 5.1). The spectrum of c-$C_4F_8^-$ is shown in Fig. 5.3 as an example. The values of the ^{19}F hf couplings are in inverse ratio to the total number of fluorine atoms per anion; the total coupling being approximately the same value (117.0 ± 2.0 mT) in each case. This indicates that the unpaired electron is delocalized over the entire molecular framework in a SOMO of high symmetry. Furthermore, the isotropic ^{19}F hf couplings observed in TMS are in excellent agreement with those obtained for these radical anions in the neopentane [31] and hexamethylethane (HME) [35] matrices, suggesting a negligible influence of matrix effects on the electronic structure of these radical anions. This makes it possible to compare the experimental ^{19}F hf couplings with the theoretical ones computed on the basis of the isolated molecule approximation.

Higher-order hyperfine splitting: ESR transitions with isotropic hyperfine (hf) and g values can be expressed as follows [36b]:

$$\nu = g\beta h^{-1}B + m_1 a + C_2(a^2/D) + C_3(a^3/D^2) + \cdots \qquad (5.1)$$

where, $D = (g-g_I)\mu_B h^{-1}B$, $C_2 = (1/2)[I(I+1) - m_I^2]$, $C_3 = (m_I/2)\,[m_I^2 + (1/2) - I(I+1)]$. Here ν is the microwave frequency and B the magnetic field at the particular line. The terms g, μ_B and h all have the usual meaning (see Tables G1 and G2 in General Appendix G). The quantity I is the total angular momentum of the equivalent nuclei (nuclear spin quantum number) in the group and m_I is its z component. The nuclear g *factor* (g_I) is expressed in terms of the Bohr magneton (μ_B). The third and forth terms of $C_2(a^2/D)$ and $C_3(a^3/D^2)$ correspond to the second-order and the third-order hyperfine splittings. These "higher-order" splittings become visible in the ESR spectrum with narrow line-width when the isotropic hf splitting (a) is not negligibly small compared to B.

Here we briefly apply the above equation to the spectrum of c-$C_4F_8^-$ radical anion possessing eight equivalent ^{19}F nuclei ($I = 8 \times 1/2$) with $a = 14.85$ mT and $g = 2.0021$. The theory predicts, for example, the second-order shifts of ($4a^2/H$, $3a^2/H$, $2a^2/H$) at the $m_I = \pm 1$ components. The values exactly agree with the experimentally observed four lines as seen in Fig. 5.3. The hf lines are selectively broadened depending on the (m_I, I) values to the extent that the weak outermost (± 4, 4) components are barely detectable in the experimental spectrum. The hf splittings (a) calculated independently from the (± 1, 1), (± 2, 2), and (± 3, 3) components are also in excellent agreement with each other. Readers can refer to a book of Atherton [37a] for more details about "higher-order hyperfine splittings" which include the anisotropic hf- and g-values.

Table 5.1 Experimental ^{19}F and ^{13}C hyperfine splittings and g-values for perfluorocycloalkane, $c\text{-}C_nF_{2n}{}^-$ ($n = 3\text{–}5$), perfluorocycloalkene, $c\text{-}C_nF_{2n-2}{}^-$ ($n = 3\text{–}5$), and related radical anions from matrix ESR studies

Radical anion	Matrix[a]	T/K	g value	$a(^{19}F)$ and $a(^{13}C)/mT$[b]	Symmetry (State)[c]	References
$c\text{-}C_3F_6{}^-$	TMS	147	2.0028^d (g$_\parallel$ = 2.0040, g$_\perp$ = 2.0022)	$a(6\ ^{19}F) = 19.8^d$ (A$_\parallel$ = 17.6, A$_\perp$ = 20.8)	D_{3h} ($^2A_2''$)	[32]
$c\text{-}C_4F_8{}^-$	TMS	113	2.0022^d (g$_\parallel$ = 2.0023, g$_\perp$ = 2.0022)	$a(8\ ^{19}F) = 14.85^d$ (A$_\parallel$ = 14.45, A$_\perp$ = 15.05)	D_{4h} ($^2A_{2u}$)	[32]
$c\text{-}C_5F_{10}{}^-$	TMS	167	2.0031^d (g$_\parallel$ = 2.0027, g$_\perp$ = 2.0034)	$a(10\ ^{19}F) = 11.6^d$ (A$_\parallel$ = 12.2, A$_\perp$ = 11.3)	D_{5h} ($^2A_2''$)	[32]
$cC_3F_4{}^-$	TMS-d_{12}	142	2.0047	$A_\parallel(2\ ^{19}F) = 20.1$, $A_\perp(2\ ^{19}F) = 18.4$ {a $(2\ ^{19}F) = 18.9$}	C_2 (^2A)	[33, M.B. Yim: private communication]
	2-MTFH	86	$g_\parallel$ = 2.0045, $g_\perp$ = 2.0058	$A_\parallel(2\ ^{19}F) = 26.5$, $A_\perp(2\ ^{19}F) = 14.3$ {a $(2\ ^{19}F) = 18.4$}		
$cC_4F_6{}^-$	TMS-d_{12}	167	2.0030	$a_1(2\ ^{19}F) = 15.2$, a_2 (2F) = 6.5, $a_3(2\ ^{19}F) = 1.1$	C_1 (^2A)	[33]
$cC_5F_8{}^-$	TMS-d_{12}	168	2.0027	$a_1(2\ ^{19}F) = 14.7$, a_2 (2 ^{19}F) = 7.4, $a_3(2\ ^{19}F) = 1.0$	C_1 (^2A)	[33]
$CF_2{=}CF_2{}^-$	TMS-d_{12} / 2-MTHF	120 / 83	2.0027	a (4F) = 9.43, a (2 ^{13}C) = 4.87, $A_\parallel(4\ ^{19}F) = 13.59$, $A_\perp(4\ ^{19}F) = 7.44$ {a $(4\ ^{19}F) = 9.49$}	C_{2h} (^2A_g)	[33, 38, 39]
$CF_3CF{=}CFCF_3{}^-$	TMS	134	2.0027	a (2 ^{19}F) = 16.93, a (6 ^{19}F) = 2.96	C_2 (^2B)	[33]

[a] TMS: Tetramethylsilane, Si(Me)$_4$; TMS-d_{12}: Si(Me-d_3)$_4$; 2-MTHF: 2-Methylhydrofuran.

[b] The isotropic and anisotropic ^{19}F hf-tensors computed for $c\text{-}C_4F_8{}^-$ are given in the text. The readers can refer to refs. [32, 33] for the ^{19}F hf-tensors computed for the other radical anions listed in the table.

[c] Molecular symmetry (electronic ground state).

[d] The isotropic g-value and ^{19}F hf-couplings were determined by averaging the measured parameters for a spectrum showing residual anisotropy.

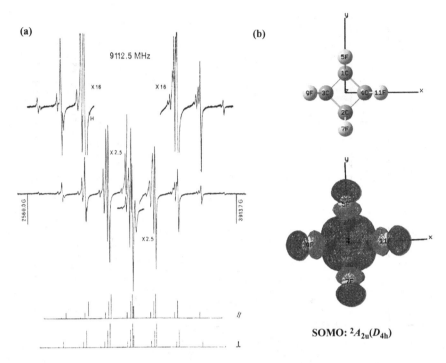

(a)

9112.5 MHz

(b)

SOMO: $^2A_{2u}(D_{4h})$

Fig. 5.3 (a) X-band ESR spectrum of c-$C_4F_8^-$ recorded at 113 K. The c-$C_4F_8^-$ radical was generated by γ-ray irradiation of the solid solution of 1 mol% c-C_4F_8 in TMS at 77 K. The line diagram shows the expected positions of the second-order ^{19}F-hyperfine lines calculated from $a(^{19}$F$) = 14.85$ mT for eight equivalent ^{19}F-nuclei of the c-$C_4F_8^-$ radical (10 gauss (G) = 1 mT). (b) Plots of the singly occupied molecular orbital (SOMO) projected to the molecular x-y plane computed for c-$C_4F_8^-$ with $^2A_{2u}$ ground electronic state in D_{4h} symmetry. The figure is adapted from [32] with permission from the American Chemical Society

5.2.1.2 Experimental vs. Computational Results

A comparison between the experimental and theoretical *hf* couplings was carried out not only for the isotropic ^{19}F *hf* couplings, but also for the anisotropic ^{19}F *hf* tensors to elucidate the geometrical and electronic structures of the anions [32]. The UHF MO computations with 6-311+G(d,p) basis set resulted in planar geometrical structures for the three radical anions; c-$C_3F_6^-$ with $^2A_2''$ (D_{3h} symmetry), c-$C_4F_8^-$ with $^2A_{2u}$ (D_{4h}), and c-$C_5F_{10}^-$ with $^2A_2''$ (D_{5h}) electronic state, in which the respective six, eight, and ten ^{19}F atoms are equivalent by symmetry. Here the geometrical structure (or molecular symmetry) is represented referring to the *point group* notation in *group theory* such as D_{3h} and D_{4h}. The electronic state is represented by the *irreducible representation* in the *character table* with large Roman letters such as A_2'' and A_{2u}, but the molecular orbital is represented with small italic letters such as a_2'' and a_{2u} orbitals for the orbitals of A_2'' and A_{2u} symmetry in a conventional way in the field: a brief summary on molecular symmetry is given in Appendix A5.2. The computations predict geometrical structures that are significantly altered by the electron attachment. For example, electron attachment to c-C_4F_8 results in a geometrical change from the puckered D_{2d} to the planar D_{4h} symmetrical structure. This remarkable increase in symmetry on negative ion formation is in sharp contrast to the lowering of symmetry in the acyclic perfluoro-compounds [35b], and is largely attributable to the stabilizing effect resulting from the complete delocalization of the added electron in the planar D_{4h} ring structure. The computations predict that, relative to the values in neutral c-C_4F_8, the C—C bond length decreases by ca. 0.1 Å, while the C—F bond increases by 0.1 Å. The changes in bond lengths arise from the nature of the high-symmetric SOMO (a_{2u} symmetric orbital) with its C—C bonding and C—F anti-bonding characteristics. Similar changes in bond lengths have been found to occur on electron attachment to c-C_3F_6 and c-C_5F_{10}. The isotropic ^{19}F hf couplings computed by the B3LYP method with the 6-311+G(2df,p) basis set for the optimized geometries are in almost perfect agreement with the experimental values: 19.8 (exp.) vs 19.78 (cal.) mT for c-$C_3F_6^-$; 14.85 (exp.) vs 14.84 (cal.) mT for c-$C_4F_8^-$; 11.6 (exp.) vs 11.65 (cal.) mT for c-$C_5F_{10}^-$.

5.2.1.3 Anisotropic ESR Spectra

The same computation method was employed to calculate the nearly axially symmetric anisotropic ^{19}F hf couplings for the magnetically equivalent ^{19}F atoms: for example, (B_{aa}: –3.54, B_{bb}: –3.48, B_{cc}: 7.02) mT for c-$C_4F_8^-$ [32]. These anisotropic ^{19}F hf splittings together with their direction cosines computed were then used to simulate an anisotropic "powder" spectrum of c-$C_nF_{2n}^-$ ($n = 3$–5). Since the principal values (three parameters) of the ^{19}F hf tensor and their directions (three parameters) must be specified for each of the ^{19}F nuclei of the radical anion, this is a much more direct procedure for ESR spectral simulation than the ordinary trial method which of necessity requires a large number of adjustable parameters. The overall ESR spectral features in the powder spectra of c-$C_nF_{2n}^-$ ($n = 3$–5) were reproduced quite well by the computations as shown for c-$C_4F_8^-$ in Fig. 5.4 as an example. Symmetry considerations indicate that in the carbon ring (x-y) plane, two different ^{19}F hf couplings with groups of four nuclei apply for c-$C_4F_8^-$. Thus, it was confirmed that the computed anisotropic ^{19}F hf coupling tensors were quite close to the actual values, paralleling the excellent agreement found for the isotropic ^{19}F hf couplings.

5.2.1.4 Photoexcitation Reaction

All the ESR spectral lines attributable to the c-$C_nF_{2n}^-$ radical anions are removed by exposure of the sample to light in the visible range 560–530 nm [3, 31, 32]. In addition experimental evidence for photoinduced electron transfer from c-$C_4F_8^-$ to SF_6 has been obtained by photobleaching experiments in γ-irradiated TMS solutions of both c-C_4F_8 and SF_6. These photobleaching studies provide direct evidence that the photoexcited states of these perfluorocycloalkane anions can undergo simple

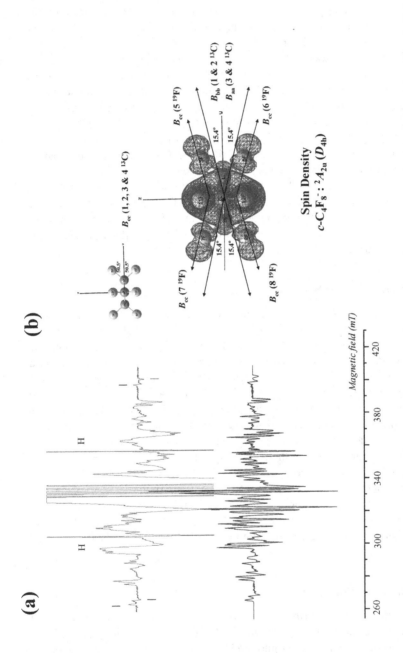

Fig. 5.4 (a) (*Upper*) Anisotropic ESR spectrum of $c\text{-}C_4F_8^-$. The radical anion was generated by γ-irradiation of solid solution of 1 mol% $c\text{-}C_4F_8$ in 2-methyltetrahydrofuran (2-MTHF) and recorded at 77 K. (*Lower*) Theoretical ESR spectrum of $c\text{-}C_4F_8^-$ computed using the isotropic and anisotropic ^{19}F *hf* couplings given in the text. (b) Plots of the spin density projected to the y–z plane computed for $c\text{-}C_4F_8^-$ with $^2A_{2u}$ symmetry (D_{4h}). The principal directions of the ^{19}F *hf* couplings at positions 5–8 are indicated in the figure. The figure is adapted from [32] with permission from the American Chemical Society

electron detachment. Photobleaching resulting from an alternate decay process by C—F bond dissociation to give a perfluorocycloalkyl radical and a fluoride anion (F⁻) is considered less likely in view of the highly delocalized and non-bonding character of the excess electron in these excited states. The electronic excitation energies and oscillator strengths were computed for the ground electronic states of c-$C_3F_6^-$ ($^2A_2''$ in D_{3h}), c-$C_4F_8^-$ ($^2A_{2u}$ in D_{4h}), and c-$C_5F_{10}^-$ ($^2A_2''$ in D_{5h}) by the TD (Time-Dependent) – DFT/Gaussian method [32]. The computations predict weak oscillator strengths of $f = 0.02$–0.03 in a visible range of 560–530 nm, which correspond to the electronic transitions from the SOMO to the LUMO of $5a_2'' \Rightarrow 7a_2'$ for c-$C_3F_6^-$, $5a_{2u} \Rightarrow 6a_{1g}$ for c-$C_4F_8^-$, and $6a_2'' \Rightarrow 8a_2'$ for c-$C_5F_{10}^-$.

5.2.2 Perfluoroalkene Radical Anions: Structural Distortion

Similar to the perfluoroalkane radical anions perfluoroalkene radical anions such as tetrafluoroethylene ($C_2F_4^-$), hexafluorocyclopropene (c-$C_3F_4^-$), hexafluorocyclobutene (c-$C_4F_6^-$), octafluorocyclopentene (c-$C_5F_8^-$) were generated by electron attachment to the solute molecule in a γ-irradiated solid matrix at 77 K and subjected to an ESR study [33]. ESR spectra observed using plastically crystalline neopentane, tetramethylsilane (TMS) or TMS-d_{12} as a matrix molecule are characterized by the isotropic ^{19}F hf splittings listed in Table 5.1. An isotropic ESR spectrum of $CF_3CF{=}CFCF_3^-$ was also observed in the TMS matrix. As an example the experimental spectrum of c-$C_4F_6^-$ in TMS-d_{12} is shown in Fig. 5.5.

5.2.2.1 Experimental vs. Computational Results

The isotropic ^{19}F hf splittings were computed by the B3LYP (DFT) method with 6-311+G(2df,p) basis set for the geometry optimized by the UHF and/or MP2 methods [33]. The computed theoretical splittings are within 6.0% error of the experimental values. Considering the large value of the atomic hyperfine constant for ^{19}F and its high sensitivity to the structural distortions, this agreement in the isotropic splittings is excellent. By comparison with the results of DFT computations, the large ^{19}F hf splittings of 9–19 mT observed for the perfluoroalkene radical anions are attributable to the two ^{19}F nuclei attached to the C=C bond. The computations predict that the geometrical structures of the perfluoroalkenes are strongly distorted by one electron reduction to form their radical anions; $C_2F_4^-$: D_{2h} symmetry (^1A_g electronic ground state) $\Rightarrow C_{2h}(^2A_g)$, c-$C_3F_4^-$: $C_{2v}(^1A_1) \Rightarrow C_2(^2A)$, c-$C_4F_6^-$: $C_{2v}(^1A_1) \Rightarrow C_1(^2A)$ and c-$C_5F_8^-$: $C_s(^1A') \Rightarrow C_1(^2A)$; see Fig. 5.6 for $C_2F_4^-$ and c-$C_4F_6^-$. The structural distortion arises from a mixing of the π^* and higher-lying σ^* orbitals so as to give a pyramidal structure at the C=C carbons similar to that for unsaturated alkyne and alkene radical anions including the acetylene and ethylene radical anions [34, 40, 41]. The SOMOs for c-$C_4F_6^-$ and $C_2F_4^-$ are given in Figs. 5.5(c) and 5.7(b), respectively. In each case the unpaired electron is primarily localized in the sp^3-like hybrid orbitals formed by

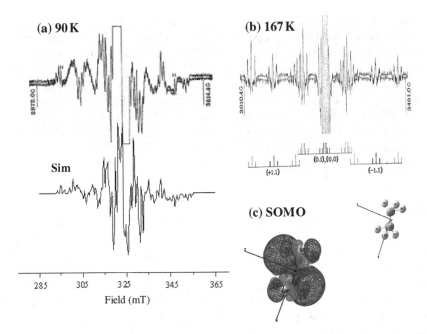

Fig. 5.5 (**a**) Anisotropic ESR spectra (first derivative) of c-$C_4F_6^-$ at 90 K. The c-$C_4F_6^-$ radical was generated by γ-ray irradiation of the solid solution of 1 mol% c-C_4F_6 in TMS-d_{12} at 77 K. Lower spectrum was calculated for c-$C_4F_6^-$ using the principal values and principal directions of ^{19}F hf splittings by the B3LYP/6-311+G(2df,p) // UHF/6-311+G(d,p) method, see ref. [33] for details. (**b**) Isotropic ESR spectrum (second derivative) of c-$C_4F_6^-$ at 167 K. The line diagram shows the expected positions of the second-order ^{19}F-hyperfine lines calculated from the parameters given in Table 5.1. The values in parentheses, (+1, 0) etc., correspond to (m_I, I), where I is the total nuclear spin quantum number and m_I is its z-component. (**c**) The singly occupied molecular orbital (SOMO) computed for c-$C_4F_6^-$ with ^2A ground electronic state in C_1 point group (symmetry). The figure is adapted from [33] with permission from the American Chemical Society

the π^* and σ^* orbital mixing and is transferred to the fluorine orbitals so as to give large hf splittings for the two (or four) ^{19}F nuclei at the original C=C bond.

5.2.2.2 Anisotropic Spectra

The anisotropic ^{19}F hf splittings can potentially provide more detailed experimental information about the electronic structure of the radical anions. Fully anisotropic ESR spectra were observed as shown in Fig. 5.5(a) for c-$C_4F_6^-$ in TMS-d_{12}. As mentioned already in the former sections, it is very difficult in general to analyze the experimental "powder" spectrum using an ordinary ESR spectral simulation method because of the large number of adjustable parameters which include the hf principal values (three parameters) and their directions (three parameters) for each of the ^{19}F nuclei of the radical anion. However, the experimental anisotropic spectra of c-$C_4F_6^-$, c-$C_5F_8^-$ and CF_2=CF_2^- were satisfactorily reproduced by the ESR spectral simulation method using the computed hf principal values and directions of ^{19}F

Fig. 5.6 Geometrical structure distortion in perfluoroalkene radical anions. **(a)** C_2F_4 and **(b)** c-C_4F_6 before and after one electron reduction together with ground electronic state and point group (symmetry) (in *parentheses*). The diagram is adapted from [33] with permission from the American Chemical Society

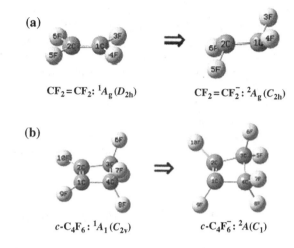

(a)

$CF_2 = CF_2$: ^1A_g (D_{2h}) $\Rightarrow$ $CF_2 = CF_2^-$: ^2A_g (C_{2h})

(b)

c-C_4F_6: $^1A_1$ (C_{2v}) $\Rightarrow$ c-$C_4F_6^-$: $^2A(C_1)$

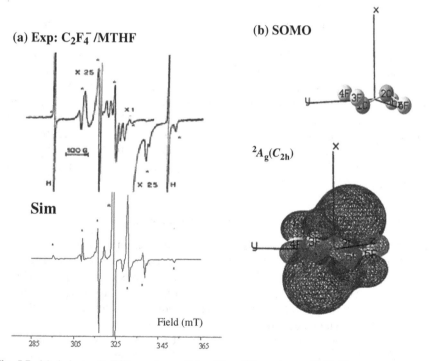

(a) Exp: $C_2F_4^-$ /MTHF

Sim

285   305   325   345   365

(b) SOMO

$^2A_g(C_{2h})$

Field (mT)

Fig. 5.7 **(a)** Anisotropic ESR spectrum of the $CF_2=CF_2^-$ anion at 83 K (*Upper*). The anion was generated by γ-ray irradiation of a solid solution of 1 mol% $CF_2=CF_2$ in 2-MTHF at 77 K. (*Lower*) Simulated spectrum for $CF_2=CF_2^-$ calculated using the principal values and directions of ^{19}F $(I=1/2)$ *hf* couplings by the B3LYP/6-311+G(2df,p) // UHF/6-311+G(d,p) method. **(b)** SOMO computed for $CF_2=CF_2^-$ with ^2A_g ground electronic state and C_{2h} point group (symmetry). The figure is adapted from [33] with permission from the American Chemical Society

nuclei; the computed anisotropic spectra being shown in Figs. 5.5 (c-$C_4F_6^-$) and 5.7 ($CF_2{=}CF_2^-$). For the simulation of $CF_2{=}CF_2^-$ the anisotropic ^{19}F hf couplings of (B_{aa}: -2.80, B_{bb}: -2.39, B_{cc}: 5.19) mT were used together with isotropic couplings of $a = 9.55$ mT for the magnetically equivalent four ^{19}F atoms. The readers can refer to [33] for the directions of ^{19}F nuclei used for the simulations.

Both the isotropic and anisotropic ^{19}F hf splittings to the two fluorines at the $C{=}C$ carbons of c-$C_3F_4^-$ are about two times larger than those to the four fluorines of $C_2F_4^-$, see Table 5.1. This suggests that both anions have almost the same type of hybrid orbital for the unpaired electron, but that the degree of unpaired electron delocalization to the "α"-fluorine orbitals is inversely proportional to the number of F-atoms.

5.2.3 Acetylene Radical Anions: Trans-Bent Structure

Acetylene is one of the simplest hydrocarbons and a fundamentally important chemical in organic chemistry. The electronic structure of acetylene and its related compounds have been extensively studied both experimentally and theoretically, and a number of interesting observations have been reported. For example, a trans-bent structure has been reported for acetylene in an excited state [42, 43] and also in a radical anion form [34, 44]. The interaction between acetylene and metal atoms such as Al and Li has been studied by inert gas matrix isolation (MI) ESR and IR methods [45–48]. Here we present our ESR study on the structural distortion in acetylene radical anion in the glassy 2-MTHF matrix, which is another example showing a mixing of the π^* and higher-lying σ^* orbitals at the $C{\equiv}C$ carbons similar to that at the $C{=}C$ carbons for the perfluoroalkene radical anions. In addition photo-induced isomerization reaction of the acetylene radical anion in the matrix is briefly presented.

5.2.3.1 ESR Identification of Acetylene Radical Anion

The acetylene radical anion was generated by an electron transfer to the solute acetylene ($^{12}CH{\equiv}^{12}CH$) in γ-ray irradiated 2-MTHF matrix at 77 K. The ESR spectrum observed immediately after γ-irradiation consists of a sharp central singlet at $g = 2.0023$ due to a trapped electron (e_t^-) and a broad septet with $ca.$ 1.9 mT hf splitting due to the matrix 2-MTHF radical. By illumination with $\lambda \geq$ 580 nm light the singlet due to e_t^- disappeared and concomitantly a spectrum due to the solute radical became visible (Fig. 5.8(a)). Upon successive illumination with $\lambda \geq 330$ nm the solute radical spectrum completely disappeared and only the matrix radical spectrum was observed (Fig. 5.8(b)). By subtracting spectrum (b) from (a) the solute radical spectrum was separately observed as shown in Fig. 5.8(c). Thus an anisotropic triplet spectrum was clearly revealed and attributed to the hf splitting due to two ^1H atoms of the solute acetylene radical anion with a trans-bent structure ($trans$-$^{12}CH{=}^{12}CH^-$) based on the comparison of the experimental ESR spectrum with the theoretical one as will be described below.

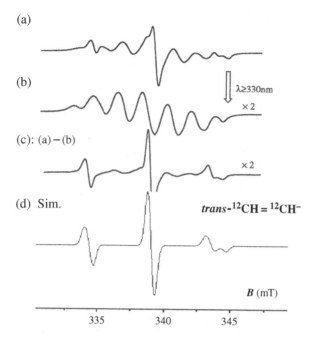

Fig. 5.8 ESR spectra of a solid solution of 1 mol% of $^{(12)}CH{\equiv}^{(12)}CH$ in glassy 2-MTHF matrix irradiated by γ-rays at 77 K; (**a**) and (**b**) were recorded after light illumination with $\lambda \geq 580$ nm and $\lambda \geq 330$ nm, respectively; (**c**) was obtained by subtracting (**b**) from (**a**) and identified as *trans*-$^{12}CH{\equiv}^{12}CH^-$; (**d**) is a spectrum simulated to (**c**). The simulated spectrum was calculated using the theoretical principal values and principal directions of ^1H *hf* A-, and g-tensors computed for *trans*-CH≡CH⁻ radical anion (see Table 5.2 and Fig. 5.9), and a Gaussian line width of $\Delta B_G = 0.45$ mT. The spectra are adapted from [34] with permission from the American Chemical Society

When $^{12}CD{\equiv}^{12}CD$ was used instead of $^{12}CH{\equiv}^{12}CH$, the triplet was changed into an anisotropic quintet due to two magnetically equivalent D atoms, as shown in Fig. 5.9(a); the *hf* coupling decreased, as expected, by a factor of 6.514, the ratio of the ^1H and $^2H(D)$ nuclear g-factors. Thus, the triplet observed for the irradiated CH≡CH/2-MTHF system was attributable to the two magnetically equivalent protons belonging to the solute molecule. Three isomer radical anions of acetylene, *trans*-CH≡CH⁻, *cis*-CH≡CH⁻, and CH₂=C⁻, are candidate species responsible for the triplet because they have two magnetically equivalent H atoms.

The observation of ^{13}C-*hf* couplings is very important to clarify the electronic structure of $^{12}CH{\equiv}^{12}CH^-$ because a large portion of the unpaired electron is expected in the orbitals of the skeletal carbons. Thus, similar experiments were carried out using ^{13}C-enriched acetylenes, $^{13}CH{\equiv}^{13}CH$, instead of $^{12}CH{\equiv}^{12}CH$ as a solute. Fig. 5.9(c) is the experimentally obtained spectrum of the $^{13}CH{\equiv}^{13}CH^-$ radical anion which shows a complicated anisotropic line shape and is difficult to be simulated based on the ordinary trial method for ESR spectral simulation. An independent approach using the results of quantum chemical computations is potentially capable of overcoming the difficult spectral simulation problem as stated in

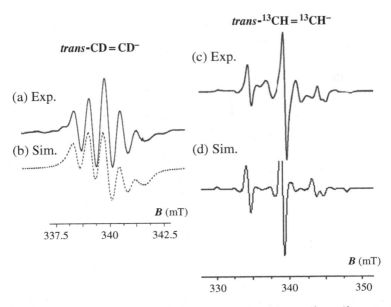

Fig. 5.9 (**a**) and (**c**) Experimental spectra of *trans*-CD≡CD⁻ and *trans*-¹³CH≡¹³CH⁻ in the 2-MTHF matrix at 77 K, respectively. (**b**) and (**d**) Spectra simulated to (**a**) and (**c**), which were calculated using the theoretical principal values and principal directions of ¹H and ¹³C *hf* A-, and g-tensors computed for *trans*-CH≡CH⁻ radical anion (see Table 5.2; Fig. 5.8). In the spectral simulation of *trans*-CD≡CD⁻ (**b**) the *hf* principal values reduced by the ratio of the ¹H and ²H(D) nuclear g-factors, $g_H/g_D = 6.514$, were used for the two D-atoms. The spectra are adapted from [34] with permission from the American Chemical Society

the foregoing sections. In the original paper [34] the anisotropic spectra of both ¹²CH≡¹²CH⁻ and ¹³CH≡¹³CH⁻ were analyzed using the ¹H and ¹³C *hf* principal values and directions computed by the semi-empirical INDO method and analytical dipolar (ANADIP) calculation method for isotropic and anisotropic *hf* couplings, respectively, and by the the AM1 method for g-tensor, for the geometrical structure optimized by ab initio method (uhf/6-31++G**).

5.2.3.2 Experimental vs. Computational Results

Here we present our recent results by more advanced DFT computational methods. The principal values and directions of the ¹H and ¹³C *hf*- and g-tensors were computed for three candidate radical anions possessing two equivalent ¹H-atoms, *trans*-CH≡CH⁻, *cis*-CH≡CH⁻, and CH₂=C⁻, by the B3LYP/6-311+G(2D,P) method for the geometries optimized by the B3LYP/6-31+G(d,p) method. The results of the computations are summarized in Table 5.2. The DFT computations predicted that the geometrical structure of acetylene is distorted by one electron reduction to form its radical anion: C_s symmetry ($^1A''$ electronic ground state) ⇒ $C_{2h}(^2A_g)$; thus the computations fully supported the original identification of the experimental spectra to the *trans*-CH=CH⁻ radical anion. The structural distortion

Table 5.2 ESR parameters for *trans*-CH≡CH− and related radical anions with two magnetically equivalent H-atoms computed by the DFT method; ESR (method/basis sets) // Geometry (method/basis sets): B3LYP/6-311+G(2DF,P) // B3LYP/6-31+G(D,P) [M. Shiotani, A. Lund: unpublished results]

	Trans-CH≡CH− (^2A_g state in C_{2h} symmetry)	*Cis*-CH≡CH− ($^2B_2$ state in C_{2v} symmetry)	CH$_2$=C− ($^2B_2$ state in C_{2v} symmetry)
A(^{13}C) tensor[a, b] (in mT)	Two equivalent ^{13}C-atoms A_{aa}: 1.21 (B_{aa}: −1.14) A_{bb}: 1.23 (B_{bb}: −1.14) A_{cc}: 4.35 (B_{cc}: 2.28) (a_{iso}: 1.97)	Two equivalent ^{13}C-atoms A_{aa}: 11.59 (B_{aa}: −0.86) A_{bb}: 11.62 (B_{bb}: −0.83) A_{cc}: 14.14 (B_{cc}: 1.69) (a_{iso}: 12.45)	^{13}C(CH$_2$=)-atom A_{aa}: −2.05 (B_{aa}: −0.38) A_{bb}: −1.67 (B_{bb}: 0.00) A_{cc}: −1.29 (B_{cc}: 0.38) (a_{iso}: −1.67) ^{13}C(=C)-atom A_{aa}: −0.39 (B_{aa}: −2.56) A_{bb}: −0.05 (B_{bb}: −2.22) A_{cc}: 6.95 (B_{cc}: 4.78) (a_{iso}: 2.17)
A(^{1}H) tensor[a, b] (in mT)	Two equivalent ^{1}H-atoms A_{aa}: 4.23 (B_{aa}: −0.52) A_{bb}: 4.61 (B_{bb}: −0.13) A_{cc}: 5.40 (B_{cc}: 0.65) (a_{iso}: 4.75)	Two equivalent ^{1}H-atoms A_{aa}: 5.79 (B_{aa}: −0.39) A_{bb}: 6.08 (B_{bb}: −0.10) A_{cc}: 6.67 (B_{cc}: 0.49) (a_{iso}: 6.18)	Two equivalent ^{1}H-atoms A_{aa}: 4.81 (B_{aa}: −0.16) A_{bb}: 4.82 (B_{bb}: −0.15) A_{cc}: 5.28 (B_{cc}: 0.31) (a_{iso}: 4.97)
g tensor[c]	g_{xx}: 2.0001(3) g_{yy}: 2.0022(4) g_{zz}: 2.0027(8)	g_{xx}: 2.0005(1) g_{yy}: 2.0023(1) g_{zz}: 2.0027(4)	g_{xx}: 2.0009(8) g_{yy}: 2.0021(9) g_{zz}: 2.0032(3)

[a] Hyperfine (*hf*) principal values: A_{ii} (i = x, y, or z) = a_{iso} (isotropic part) + B_{ii} (anisotropic part).
[b] For the principal directions of the A-tensor see the inserts in Fig. 5.10.
[c] The principal directions of the g-tensor coincide with the molecular x-y-z coordinate system.

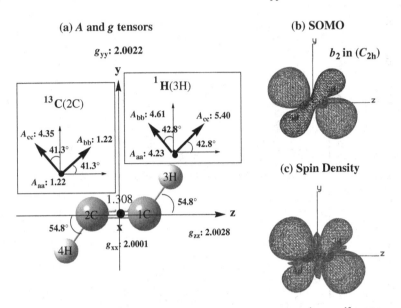

(a) A and g tensors   **(b) SOMO**

(c) Spin Density

Fig. 5.10 (a) Principal values (in mT) and directions (in degree) of the ^{1}H and ^{13}C *hf* A-tensors and the *g*-tensor computed for *trans*-CH≡CH$^-$ radical anion (in C_{2h} symmetry). The C—C bond length is in Å. The *A*- and *g*-tensors were computed by the B3LYP/6-311+G(2d,p) (ESR) // B3LYP/6-31+G(d,p) (Geometry) method. **(b)** and **(c)**: Plots of the SOMO (b_2 orbital) and spin density projected to the y-z plane computed for *trans*-CH≡CH$^-$ with $^2B_2$ ground state (in C_{2h} symmetry). The principal directions of ^{1}H(4H) and ^{13}C(1C) can be obtained from those of ^{1}H(3H) and ^{13}C(2C) by a C_2 symmetry operation about the *x*-axis [M. Shiotani, A. Lund: unpublished results]

arises from a mixing of the π* and higher-lying σ* orbitals at the C≡C carbons similar to the case for perfluoroalkene radical anions described in Section 5.2.2. Plots of the SOMO and spin density (SD) for *trans*-CH=CH$^-$ are given in Fig. 5.10, in which the unpaired electron is primarily localized in the sp^2-like hybrid orbitals in the skeletal carbons and is transferred to the hydrogen $1s$ orbital so as to give a positive isotropic *hf* splitting for the two ^{1}H-atoms in the molecular plane.

The theoretical ESR spectra, which were calculated for *trans*-^{12}CH≡^{12}CH$^-$, *trans*-^{12}CD≡^{12}CD$^-$, and *trans*-^{13}CH≡^{13}CH$^-$ using the computational results, agree almost perfectly with the experimental ones as shown in Figs. 5.8 and 5.9. The agreement leads us to conclude that the computed principal values and directions of the A(^{1}H and ^{13}C)- and *g*-tensors are quite close to the experimental ones. The ^{13}C A-principal values are perfectly axially symmetric: $(A_{aa}, A_{bb}, A_{cc}) = (1.12, 1.12, 4.35)$ (mT). The principal direction of A_{cc} (or B_{cc}), which corresponds to the maximum ^{13}C anisotropic coupling, occurs along the symmetry axis defining the cylindrical cross section of the SD plots; the direction lying in the y-z plane with an angle of 41.3° (anti-clockwise direction) from the *y*-axis for the 2C atom, see Fig. 5.10. It should be noted that the SD plots are very similar to those of the SOMO as shown in the figure as observed for a series of perfluorocycloalkane and perfluorocycloalkene radical anions [32, 33].

Fig. 5.11 (a)–(d):
Successive change in the ESR
spectra by rearrangement of
trans-^{12}CH≡^{12}CH⁻ (peaks
marked as "*a*") to
^{12}CH₂=^{12}C⁻ (peaks marked
as "*b*") upon illumination
with light of λ ≥ 430 nm.
(e) Spectrum simulated to **(d)**
^{12}CH₂=^{12}C⁻ using the
A(^{1}H and *hf*)- and *g*-tensors
in the text. The spectra are
adapted from [34] with
permission from the
American Chemical Society

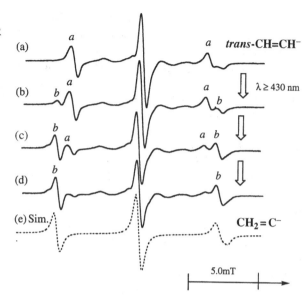

5.2.3.3 Photo-Isomerization of *trans*-CH≡CH⁻

By illuminating with λ ≥ 430 nm light, the anisotropic triplet due to *trans*-^{12}CH≡^{12}CH⁻ was irreversibly changed into a new triplet with slightly different A- and *g*-tensors, as shown in Fig. 5.11. The new triplet was characterized by the anisotropic *hf* couplings of (A_{aa}, A_{bb}, A_{cc}: 5.4, 6.0, 5.7) mT with two equivalent ^{1}H-atoms and the *g*-tensor, (g_{aa}, g_{bb}, g_{cc}: 2.0026, 2.0008, 2.0026), based on the assumption that the A- and *g*-tensors had co-axial principal directions. Either *cis*-CH≡CH⁻ or H₂C=C⁻ can be a candidate radical because both species have two magnetically equivalent ^{1}H-atoms with a slightly larger coupling predicted theoretically than the *trans*-CH≡CH⁻ radical. The new triplet was finally attributed to the CH₂=C⁻ radical by observing the *hf* couplings due to non-equivalent two ^{13}C-atoms in the experiments using ^{13}C enriched samples [34], although the DFT computations do not perfectly reproduce the experimental A- and *g*-tensors. Furthermore, the electron absorption (*EA*) band was observed at λ_{max} = 374 nm for CH₂=C⁻ and was attributable to the electronic transition from the SOMO ($^2B_2$ ground state) to the LUMO ($^2A_1$ excited state) in C_{2v} symmetry. Readers can refer to [34, 44, 45] for further details of the ESR studies on the *trans*-CH≡CH⁻ radical anion.

5.3 Radical Cations

Except for conjugated systems such as aromatics, chemically important radical cations of alkanes, alkenes and other small organic and inorganic molecules have never been observed by ESR spectroscopy until the development of new matrix-isolation methods by Shida et al. and Knight et al. because of either their short life time, high ionization potential, or both. In 1978 Shida [49] first demonstrated the usefulness of halocarbon matrices, especially $CFCl_3$, for ESR studies of solute radical cations generated by ionizing radiation. Since then a variety of halocarbon matrices have been newly developed for the studies of radical cations and a large number of radical cations have been subjected to ESR studies which include some n-alkane cations, c-alkane cations, branched-alkane cations, alkene and alkyne cations, aromatic and hetero-aromatic cations, ester cations, ether and some other cations [4–9, 12, 13]. In the studies halocarbons (such as $CFCl_3$, $CFCl_2CF_2Cl$, c-C_6F_{12}, c-$C_6F_{11}CF_3$), SF_6 and zeolites with high ionization potential (energy) have been used as matrices for ESR studies of radiolytically generated radical cations; the ionization potentials of some halocarbons used are listed in Table 5.3. An additional advantage of the halocarbons is that the ESR spectra of the matrix radicals widely spread out due to large anisotropic hyperfine coupling of ^{19}F nuclei, so that the matrix radicals give a relatively weak background signal. It is sometimes sufficient to work at 77 K or higher for rather stable radical cations. For highly reactive small radical cations, however, lower temperatures available by using liquid helium (or liquid hydrogen) cryostats are essential.

In 1982 Knight reported another method for the ESR observation of radical cations in an inert-gas matrix at cryogenic temperatures [50]. The method involves radical generation techniques such as fast atom bombardment, electron bombardment or photoionization from discharged neon gas during deposition on a cold finger surface at 4 K in an ESR matrix isolation apparatus [50–52]. By employing this method a number of radical cations were successfully generated and were subjected to ESR studies, which include a variety of small radical cations such as H_2O^+ [42] and CH_4^+ [53, 54] in a neon matrix which have never been detected by the halocarbon matrix isolation method. ESR studies on Jahn-Teller (J-T) distortion of $^2H(D)$-labelled methane radical cations are presented in some detail as an example in the following section.

Review articles by Shida et al. [55] and Symons [56] cover earlier studies in this field until 1984. For more recent studies readers can refer to the review papers by Shiotani [4], Lund [5], Knight [51], and to the books (book chapters) [6–9, 12, 13].

Table 5.3 First vertical ionization potential (IP_1) for some molecules/atoms used as matrices for generation of radical cations by ionizing radiation

Molecule or atom (abbreviation)	IP_1(eV)	References
$CFCl_3$ (F-11)	11.78	[49]
CCl_4	11.69	[57]
CF_3CCl_3 (F-113a)	11.73	[4]
$CFCl_2CF_2Cl$ (F-113)	11.98	[4]
CF_2ClCF_2Cl (F-114)	12.66	[4]
Cyclic-C_6F_{12}	12.90	[4]
Cyclic-$C_6F_{11}CF_3$	13.06	[4]
SF_6	15.69	[58]
Ne	24.59	[59]
Ar	15.76	[59]

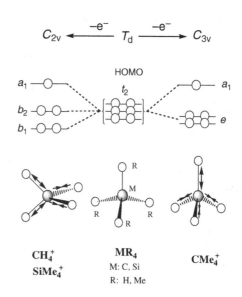

Fig. 5.12 Schematics showing how three-fold degeneracy of the t_2 HOMO is lifted when tetrahedral (T_d) molecules release an electron to yield the radical cations with a lower symmetry of either C_{3v} or C_{2v}. The *arrows* ⟷ and ⟶⟵ indicate an elongated and a compressed σ bond, respectively. The diagram is reproduced from [60] with permission from the Royal Society of Chemistry

5.3.1 Jahn-Teller Distortion of T_d and D_{3h} Molecules

Tetrahedral molecules in T_d symmetry have a three-fold degenerate t_2 highest occupied molecular orbital (HOMO). When they release one electron, the degeneracy is possibly lifted and the associated radical cations are formed having a lower symmetry of either C_{3v}, C_{2v} or D_{2d} due to the so called static Jahn-Teller (*J-T*) distortion (Fig. 5.12) [4, 7, 13, 41]. A C_{2v} structure distortion has been reported for the radical cations of methane [53, 54] and tetramethylsilane (TMS) [60, 61]. The neopentane radical cation, $C(CH_3)_4{}^+$, is a typical example for a C_{3v} distortion from T_d in which the unpaired electron is mainly located in one of the C—C bonds. Large couplings of 4.2 mT are due to the three H atoms at trans positions of the other methyl groups [62]. Here we present ESR studies of selectively deuterated methane radical cations by Knight [53, 54], which exemplify the C_{2v} distortion of $CH_4{}^+$.

Furthermore, we deal with a structural distortion of the trimethylenemethane radical cation (TMM$^+$) [63]. The neutral TMM has a high symmetry of D_{3h} and is in a triplet state with a doubly degenerate e'' HOMO in the ground electronic state [64]. The D_{3h} structure of TMM is expected to be distorted to a C_{2v} structure with a lower symmetry in TMM$^+$ due to the *J-T* effect, so as to split the original degenerate e'' orbitals into a_2 and b_1 orbitals.

5.3.2 D-Labelled Methane Radical Cations: $CH_4{}^+$, $CDH_3{}^+$, $CD_2H_2{}^+$, CD_3H^+ and $CD_4{}^+$

Knight et al. observed ESR spectra of a series of selectively deuterated methane radical cations, $CH_4{}^+$, $CDH_3{}^+$, $CD_2H_2{}^+$, CD_3H^+ and $CD_4{}^+$, generated and stabilized in a neon matrix at 4 K [53, 54]. The observed isotropic *hf* coupling constants

Table 5.4 Experimental isotropic *hf* splittings, a_{iso} (mT) for ^{1}H and D atoms in D-labelled methane radical cations in neon matrix at 4 K. The table is reproduced from [54] by permission of the American Institute of Physics

	CH$_4^+$	CDH$_3^+$	CD$_2$H$_2^+$	CD$_3$H$^+$	CD$_4^+$
a_{iso}(H)	5.43	7.64	12.2	12.5	–
a_{iso}(D)^a: a_{iso}(H)/(g_H/g_D)	0.83	1.2	1.9	1.9	–
a_{iso}(D)	–	(–)0.22	(–)0.22	0.45	0.81
a_{iso}(H) : a_{iso}(D) · (g_H/g_D)	–	(–)1.4	(–)1.4	2.9	5.3

^{a} The experimental a_{iso}(H) and a_{iso}(D) *hf* values were converted to a_{iso}(D) and a_{iso}(H) *hf* "scales" by dividing and multiplying by the ratio of the H and D nuclear *g*-factors, $g_H/g_D = 6.514$, respectively.

(a_{iso}) are summarized in Table 5.4. The value of a_{iso}(H) = 5.43 mT for ^{1}H in CH$_4^+$, divided by the ratio of the ^{1}H and ^{2}H(D) nuclear *g* factors, $g_H/g_D = 6.514$, yields an equivalent value of a_{iso}(D) = 0.83 mT for deuterium. This value agrees well with the observed value of a_{iso}(D) = 0.81 mT in CD$_4^+$, in which all the hydrogens are replaced with D atoms. No unusual characteristics are evident for CD$_4^+$. However, for the methane radical cations containing a mixture of ^{1}H and D atoms, i.e. CDH$_3^+$, CD$_2$H$_2^+$ and CD$_3$H$^+$, the simple conversion from the ^{1}H *hf* scale to the D *hf* scale using the appropriate ratio of the nuclear *g*-factors does not account for the observed *hf* values. For example, in CDH$_3^+$ the observed value of a_{iso}(H) = 7.64 mT for three ^{1}H atoms would predict a deuterium *hf* value of a_{iso}(D) = 1.2 mT, which is compared to a directly observed value of a_{iso}(D) = (–)0.22 mT for one D atom.

In order to account for the observed a_{iso} values of ^{1}H and D atoms the following rules were proposed for the location (or site) of ^{1}H and D atoms in the radical cation by Knight [54]. (a) Methane radical cation possesses a C_{2v} type geometrical structure with two distinctly different electronic sites, i.e. apical "*a*" site and equatorial "*e*" site. (b) Site "*a*" can accommodate two ^{1}H or D atoms and is coplanar with the carbon *p*-orbital with an unpaired electron; site "*e*" can accommodate two atoms and lies in the nodal plane of this same carbon *p*-orbital, see Fig. 5.13. (c) D atoms

Fig. 5.13 Schematics of the C_{2v} geometrical structure in the CH$_4^+$ cation with the $^2B_1$ electronic state. Sites marked as "*a*" are coplanar with the carbon *p*-orbital with an unpaired electron and sites "*e*" lie in the nodal plane of the *p*-orbital. ^{1}H atoms prefer the co-planar "*a*" site and D atoms favor the nodal plane site "*e*" [13, 54]

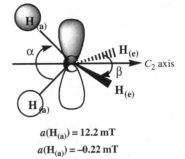

$a(H_{(a)}) = 12.2$ mT
$a(H_{(a)}) = -0.22$ mT

prefer the nodal plane site "*e*" and ^{1}H atoms prefer the co-planar site "*a*". (d) ^{1}H atoms exchange with other ^{1}H atoms even between the different sites, "*a*" and "*e*", but not with D atoms, and D atoms likewise exchange with other D atoms but not with ^{1}H atoms in a different electronic environment. The orbitals of the atoms that occupy the nodal plane site "*e*" cannot efficiently mix with the *p*-orbital. The *hf* interaction is dominated by a spin polarization mechanism so as to produce a small negative a_{iso} value; refer to Chapter 1 for the negative a_{iso} value. The 1*s* orbitals of the ^{1}H and D atoms in the co-planar site "*a*" can mix with the carbon *p*-orbital and yield a large positive a_{iso} value.

The a_{iso} *hf* values observed for D atoms in CDH_3^+ and $CD_2H_2^+$ are almost the same and equal to (–)0.22 mT. There are two equivalent positions in the nodal plane site "*e*", which are strongly preferred by the D atoms. Likewise, for $CD_2H_2^+$ and CD_3H^+ the first two ^{1}H atoms can occupy their preferred co-planar site "*a*" and exhibit a large *hf* coupling of 12.2 and 12.5 mT, respectively. Thus one can anticipate that a_{iso} for site "*a*" has a value of *ca.* 12.2 mT and a_{iso} for site "*e*" has a hydrogen equivalent value of –1.4 mT (–0.22 mT for D). A weighted average of these two basic a_{iso} values can account for the observed *hf* splittings. For CH_4^+, where the H atoms exchange between two different "*a*" and "*e*" sites, we have 5.39 mT [(2 · 12.2 – 2 · 1.43)/4] *vs.* the experimental value of 5.43 mT. Furthermore, for CDH_3^+ such H atom exchanges also occur and we have 7.66 mT [(2 · 12.2 – 1.43)/3] *vs.* the experimental value of 7.64 mT. However, for $CD_2H_2^+$ and CD_3H^+, the H atoms in site "*a*" do not exchange their positions with the D atoms in site "*e*" and the observed hydrogen a_{iso} values are almost the same (12.2 and 12.5 mT). The C_{2v} structure of the CH_4^+ cation has been theoretically supported by calculations with several different methods [65–68]. For example the DFT/DZP method resulted in the two long C—C bonds (1.208 Å) and the two short C—H bond (1.104 Å) lying in the mutually orthogonal planes: angle "α" between the long bonds and angle "β" between the short bonds are 58.6° and 124.5°, respectively; for the angles see Fig. 5.13 [68]. The experimental ^{1}H *hf* splittings were also well reproduced by the theoretical calculations: for 2H at "*a*" sites, 12.2 (exp) vs. 12.11 mT (cal) and for 2H at "*e*" sites, –1.43 (exp) vs. –1.63 (cal).

The observed deuterium isotope effects on the preferences of occupation site can be interpreted in terms of a zero point vibrational energy (ZPVE) of the C—H bond stretching vibration. The singly occupied molecular orbital (SOMO) is delocalised from the centre carbon to the hydrogens making the corresponding C—H bond weaker and longer. To a first approximation the ZPVE is proportional to $(k/m)^{1/2}$, where *k* is the force constant and *m* is the mass of the hydrogen nucleus. The decrease in ZPVE upon deuteration is greater for bonds having larger force constants. Thus, the deuterium atoms can preferentially occupy positions with a larger force constant, i.e. the nodal plane positions with a short bond distance. A similar D-isotope effect has been reported for the methyl group conformation of D-labeled dimethyl ether radical cations studied by ESR spectroscopy and by ab initio and DFT quantum chemical methods [26].

5.3.3 Trimethylenemethane Radical Cation

Trimethylenemethane (TMM) is one of the most attractive organic molecules for investigating the electronic structure because of the simplest known 4π-electron system of D_{3h} symmetry [69]. In TMM a pair of electrons occupies a bonding π-MO (BMO), and the remaining two electrons are accommodated by a doubly degenerate nonbonding π-MO (NBMO or HOMO). Several ESR studies have shown that the ground electronic state of TMM is a triplet with the two NBMOs (e'' orbital in D_{3h}) singly occupied by the two unpaired π-electrons [64, 70]. The associated radical cation (TMM$^+$) is a 3π-electron system, in which the BMO is doubly occupied and the unpaired electron is shared by the NBMOs. Thus, TMM$^+$ has a degenerate ground state subject to *J-T* distortion and its structural distortion was studied by ESR spectroscopy combined with ab initio MO calculations [63].

5.3.3.1 Experimental vs. Computational Results

The TMM$^+$ radical cation was generated by γ-ray irradiation of solid solutions containing methylenecyclopropane (MCP) in a CF_2ClCF_2Cl matrix. The 4.2 K ESR spectrum consists of a septet of *hf* lines with the relative intensity close to a binomial one, 1:6:15:20:15:6:1, and is successfully reproduced by employing six magnetically equivalent ^1H atoms with $a_{iso}(H) = 0.93$ mT splitting. The ESR lineshape did not change appreciably between 4 and 125 K. The observed *hf* splitting due to six equivalent protons can not be attributed to any radicals with the original ring-closed MCP geometrical structure, but to a ring-opened TMM$^+$ radical cation with an apparent D_{3h} structure as described below.

For TMM$^+$ the original D_{3h} symmetry of TMM is expected to be reduced to a lower one of C_{2v} due to *J-T* effect so as to split the degenerate NBMOs of e'' (in the D_{3h} point group) into a_2 and b_1 MOs in the C_{2v} point group:

$$a_2 = (\phi_1 - \phi_5)/\sqrt{2}$$
$$b_1 = (\phi_1 + \phi_5 - 2\phi_8)/\sqrt{6} \qquad (5.2)$$

where ϕ_1, ϕ_5, and ϕ_8 stand for the $2p_z$-AOs at the three peripheral C atoms of TMM, see Figs. 5.14 and 5.15 for the numbering of C atoms. The a_2 and b_1 orbitals are symmetric and antisymmetric with respect to the vertical mirror plane containing the C_2 symmetry axis, respectively. In order to see which of the orbitals that is responsible for the observed ESR result of TMM$^+$, DFT quantum chemical computations were performed. Fig. 5.15 shows two electronic states of $^2A_2$ and $^2B_1$ in the C_{2v} point group, which were optimized by the B3LYP method with 6-31+G(d,p) basis set; the associated SOMO (or NBMO) of a_2 or b_1 being depicted in Fig. 5.14. The $^2A_2$ structure has one shorter C—C bond of 1.394 Å along the C_2 symmetry axis, the other two C—C bonds having a longer bond length of 1.428 Å. On the other hand, in the $^2B_1$ structure the unique C—C bond becomes longer (1.431 Å) than the other two C—C bonds (1.408 Å). The $^2A_2$ structure was calculated to be more stable by *ca.* 0.07 eV in the total energy than the $^2B_1$ structure. Thus, it is predicted by

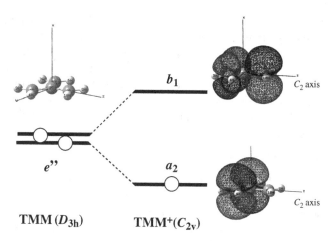

Fig. 5.14 Schematic representation showing that the original D_{3h} symmetry of TMM is reduced to a lower one of C_{2v} due to *J-T* effect for TMM$^+$ so as to split the doubly degenerate NBMOs of e'' into a_2 and b_1 orbitals, together with plots of the latter two orbitals projected to the x-z plane. The computations of a_2 and b_1 MOs were performed using the B3LYP method with 6-31+G(d,p) basis set

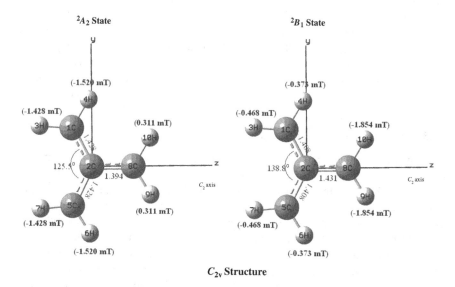

Fig. 5.15 Optimized geometrical structures corresponding to the $^2A_2$ and $^2B_1$ states of TMM$^+$ with C_{2v} symmetry calculated by the B3LYP method with 6-31+G(d,p) basis set. The bond lengths are in Å. The isotropic ^{1}H *hf* splittings (in parentheses) were computed by the B3LYP method with 6-311+G(2d,p) basis set for the two geometry optimized structures

the computations that the $^2A_2$ structure is preferable as the ground state of TMM$^+$ with the distorted C_{2v} structure. The ^{1}H hf splittings were computed by the B3LYP method with 6-311+G(2d,p) basis set for the two geometry optimized structures, see the values in parentheses in Fig. 5.15. The hf splittings consist of three different pairs of protons for both structures and can not directly be compared to the experimental one with six equivalent protons. Thus we failed to directly detect a distorted C_{2v} structure of TMM$^+$ even at low temperature of 4.2 K in the matrix.

5.3.3.2 Dynamic Models

There are two possible dynamic models to explain the reason why not the distorted C_{2v} structure but the apparent D_{3h} structure of TMM$^+$ was experimentally observed. One is the intra-molecular dynamics of TMM$^+$ in the ground $^2A_2$ state. There are three energetically equivalent C_{2v} structures depending on which of the three C—C bonds is the shorter bond along the C_2 symmetry axis, see Fig. 5.16. An intra-molecular dynamic conversion among the three C_{2v} structures can average out the structural distortion to give an apparent D_{3h} structure with six equivalent H-atoms (due to dynamic J-T effect) as observed. In this case the total ^{1}H hf splittings of $(-)5.27$ mT in the static $^2A_2$ state compares well with the experimental value of 5.58 mT ($0.93 \cdot 6$ mT). Similar intra-molecular dynamics have been discussed for the radical cations of cyclohexane, benzene and their derivatives in low temperature solid matrices [4, 8, 13, 71–73]. The other dynamic model that may account for the six equivalent protons is a state mixing between the ground and lowest excited states, $^2A_2$ and $^2B_1$, as observed for benzene radical ions in solution. The population of the excited $^2B_1$ state is, however, negligibly small even at a relatively high temperature of 77 K based on the energy difference of 0.07 eV between the two states evaluated by the DFT computations and the latter model may be ruled out.

5.3.3.3 TMM-Me$_4$$^+$ radical cation

The degeneracy of the NBMOs (e'' in D_{3h}) can be removed by perturbing the π-MO system of TMM. Bally et al. recently reported an ESR and ENDOR study of the radical cation of 1,1,2,2-tetramethyltrimethylenemethane (TMM-Me$_4$) [74]. The radical cation was generated by ring opening of 2,2,3,3-tetramethylmethylenecyclopropane (MCP-Me$_4$) upon ionization by γ-irradiation in frozen CFCl$_3$, or by X-irradiation in an Ar matrix. The experimental ESR spectrum exhibited a hyperfine splitting due to the two methylene α-protons with $|a_{iso}| = 1.99$ mT, while the ^{1}H ENDOR signals were associated with coupling constants $|a_{iso}| = 0.53$ and 0.19 mT arising from two sets, each of six protons in two methyl substituents (β-protons in ESR nomenclature), see Scheme 5.1. These hyperfine data are fully consistent with those expected for the radical cation of TMM-Me$_4$$^+$, in which the shape of the SOMO closely resembles that of b_1 in TMM$^+$, and the ground state should therefore have $^2B_1$ symmetry in the C_{2v} point group. This ^{1}H hf assignment was corroborated by B3LYP/6-31G* calculations, which yielded $a_{iso} = -2.05$ mT for the two methylene α-protons and a_{iso}(average) $= 0.47$ (*endo*)/0.25 mT (*exo*) for the two sets of

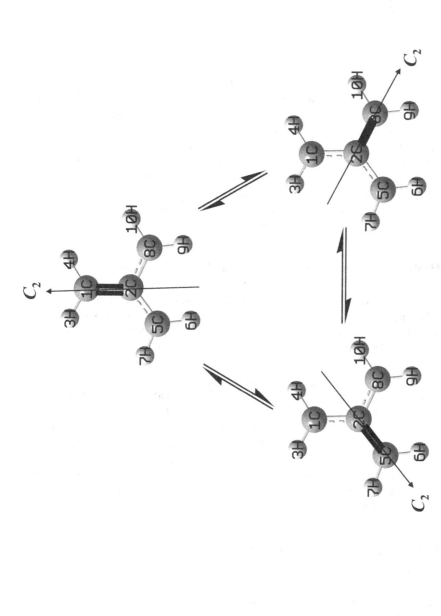

Fig. 5.16 Intra-molecular dynamics model for the interchange H and C atoms of TMM$^+$ between three equivalently distorted C_{2v} structures with the ground $^2A_2$ state

Scheme 5.1 The orbital degeneracy in the ground state of TMM⁺ is removed by the methyl substitution to yield TMM-Me₄⁺. The diagram is adapted from [74] with permission from the American Chemical Society

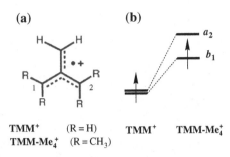

six methyl β-protons. These values hold for TMM-Me$_4$⁺ at a computed equilibrium geometry of C_2 symmetry: the C_{2v} symmetry is lowered due to steric congestion by the two endo-methyl groups. Methyl substitution of the π-radical cation TMM⁺ to yield TMM-Me$_4$⁺ thus removes effectively the degeneracy of the ground state as illustrated in Scheme 5.1. An analogous removal of the degeneracy by lowering of the symmetry from D_{3d} to C_{2h} has been observed for the σ-radical cation of cyclohexane upon 1,4-dimethyl substitution [71].

Readers can refer to the literature [4, 7, 8, 41, 60] for further details of *J-T* distortion of other T_d and D_{3h} molecules.

5.4 High Resolution ESR Spectra and Quantum Effects

In this section we deal with high resolution ESR studies on deuterium (D or ^{2}H) labelled small radicals generated in solid Ar or *para*-H$_2$ matrix. Emphasis is put on quantum effects at cryogenic temperatures [13]. We start with the high-resolution ESR spectra of D-labelled methyl radicals generated in irradiated solid Ar matrix. The observed H/D-isotope effects on the ESR lineshapes are discussed in terms of nuclear spin-rotation couplings [75]. Next the high-resolution spectra of hydrogen atom – methyl radical pairs in solid Ar matrix are presented. Their formation reactions and trapping sites in the matrix are discussed based on the ESR results [76]. Then, we move on to the ESR studies of the formation of hydrogen atom – hydrogen molecule complex in solid Ar. The results are discussed in terms of quantum effects on H/D isotope reactions [77, 78]. In the end we deal with recent ESR studies on the generation and structure of "H$_2$⁺-core" H$_6$⁺ (D_{2d} symmetry) complex radical cation and its D-substitutions, H$_{6-n}$D$_n$⁺ ($n = 1, 2$), in quantum solid *para*-H$_2$ matrix at cryogenic temperatures by Kumagai et al. [79].

5.4.1 D-Labelled Methyl Radicals: Nuclear Spin-Rotation Couplings

The methyl radical has attracted much attention since it is the most simple and fundamentally important alkyl radical. ESR spectra have been observed for methyl radicals generated in liquids [36c], isolated in inert gas matrices [46, 75], adsorbed on solid surfaces [80, 81], and trapped in organic media [82]. In this section we present high resolution ESR spectra of selectively D-labelled methyl radicals, CH_3, CH_2D, CHD_2, and CD_3, which were generated by X-ray irradiation and isolated in a solid Argon (Ar) matrix at cryogenic temperatures [75]. Ar is one of the most appropriate matrices among the inert gases to observe high resolution ESR spectra because all its stable isotopes lack nuclear spin. The observed H/D-isotope effects on the ESR lineshapes and their temperature dependencies are discussed in terms of nuclear spin-rotation couplings using a three-dimensional free quantum-rotor model.

5.4.1.1 CH_3 Radical

Figure 5.17 shows ESR spectra of the CH_3 radical trapped in the solid Ar matrix at various temperatures from 6 K to 40 K. Highly resolved quartet hyperfine (*hf*) lines with an equal-intensity (1:1:1:1) were observed at 6 K and attributed to the CH_3 radical affected by nuclear spin-rotation coupling. The observed quartet with ^1H *hf* splitting of 2.315 mT was assigned to A_1-lines in D_3 symmetry (point group) as will be more fully described below. Upon increasing the temperature to 12 K, a new doublet, so called *E*-lines, with the same splitting became clearly visible at the $m_I = \pm 1/2$ positions. The *E*-line positions are at 0.024 mT higher fields than

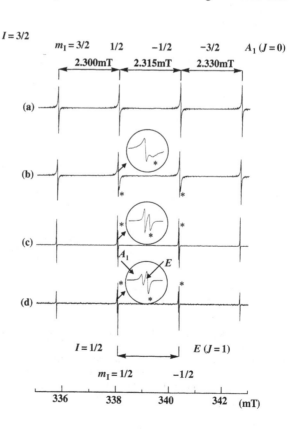

Fig. 5.17
Temperature-dependent ESR spectra of the CH_3 radical generated by X-ray irradiation of Ar matrix containing 0.2 mol% CH_4 at 4.2 K. The spectra were recorded at (**a**) 6 K, (**b**) 12.0 K, (**c**) 20.0 K, and (**d**) 40.0 K. Only the *hf* quartet due to $I = 3/2$ (A_1 lines) is visible at 6 K. The doublet due to $I = 1/2$ (*E* lines) marked as a *star* (*) increases in intensity with increasing temperature. Because of the high resolution spectra the A_1 and *E* lines are resolved by a small second-order shift (0.024 mT) of the ^1H *hf* splitting. The figure is adapted from [75] with permission from the American Chemical Society

the inner $m_I = \pm 1/2$ lines of A_1-lines. The difference in the resonance lines comes from the second-order hf shift [36a, 36b] (see Section 5.2.1.1) between the total nuclear spins of $I = 1/2$ (E-lines) and $I = 3/2$ (A_1-lines). The intensity of the E-lines increased with temperature and reached a value two times stronger than that of $m_I = \pm 1/2$ (A_1-lines) at 40 K. The result suggests that higher rotational energy levels (with the angular momentum quantum number $J \geq 1$) are populated at the elevated temperature as described below. The successful observation of the high resolution ESR spectra suggests that the methyl radicals are well isolated from the host Ar atoms and they behave as if in the gas phase.

The observed temperature dependent ESR spectra can be explained in terms of nuclear spin-rotation couplings using a three-dimensional free quantum-rotor model (Fig. 5.18). The Pauli principle [83] requires that, when the labels of any two identical fermions (particles with half integer value of spin) [84] are interchanged, the total wave function changes sign. The protons (^{1}H) in CH$_3$ are fermions with nuclear spin $I = 1/2$. The total wave function (ψ) of methyl radical system is a product of four wave functions, ψ_E (electron wave function), ψ_V (vibrational wave function), ψ_R (rotational wave function) and ψ_I (nuclear spin function). The wave functions of ψ_E and ψ_V are anti-symmetric (change sign) and symmetric (retain the same sign) when the labels a and b are interchanged in the methyl radical system (refer to Fig. 5.18). Thus, the Pauli principle requires that the product of $\psi_R \otimes \psi_I$ is symmetric for CH$_3$.

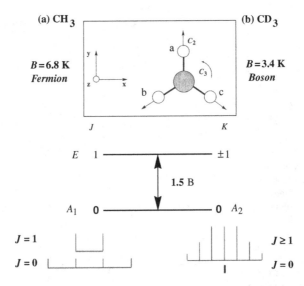

Fig. 5.18 A three-dimensional free quantum-rotor model and possible nuclear spin-rotational state couplings ($\psi_R \otimes \psi_I$) for (**a**) CH$_3$ and (**b**) CD$_3$ radicals in the D_3 symmetry. The observed 1:1:1:1 quartet and 2:2 doublet of CH$_3$ are attributed to four A_1 and two doubly degenerated E nuclear spin states coupled with rotational ground $J = 0$ (even) and excited $J = 1$ (odd) states, respectively. For CD$_3$ only one A_2 nuclear spin state is possible in the $J = 0$ state. "B" stands for the theoretical rotational constant

Applying the Pauli principle the observed 1:1:1:1 quartet of CH_3 is attributable to the four totally symmetric A_1 nuclear spin states coupled with the rotational ground state, $J = 0$ (even), in the D_3 symmetry (point group) [13, 75, 85]:

$$\Psi_{a,b,c}^{A_1}(0,0,0) = |0,0,0\rangle |n,n,n\rangle$$

$$\Psi_{a,b,c}^{A_1}(0,0,0) = \frac{1}{\sqrt{3}} |0,0,0\rangle \left[|n',n,n\rangle + |n,n',n\rangle + |n,n,n'\rangle \right]$$

$$(5.3)$$

In the above and following expressions a nuclear spin-rotation ($R \otimes I$) basis function is designated by $\Psi_{a,b,c}^{\Gamma}(J,K,M)$, where Γ is an irreducible representation of the point group which is either A_1 or A_2 for the overall symmetry of the methyl protons in the D_3 point group (see the character table in Appendix A5.2). The terms J, K and M stand for the angular momentum quantum number, the quantum number used to signify a component on the unique axis of the molecule (or z-component of J) with $K = 0, \pm 1, \pm 2, \ldots, \pm J$, and a component on an externally defined axis, respectively [75, 86a]; $(J,K,M) = (0,0,0)$ in Eq. (5.3) corresponds to the ground rotational level of CH_3 radical with $J = 0$, $K = 0$, and $M = 0$. Suffixes a, b, and c stand for the labelling of the methyl protons, see the model of the CH_3 radical rotation in Fig. 5.18. In Eq. (5.3) the projections (or z-components) of the nuclear spins, n and n', are given by $n = -n' = \pm 1/2$ for the CH_3 protons with $I = 1/2$.

On the other hand, the E-lines are attributable to the nuclear spin states coupled with the $J = 1$ (odd) rotational state. For the present D_3 rotor the E-lines are almost superimposed on the A_1-lines at $m_I = \pm 1/2$, however. It should be noted that, in an earlier study of β-protons with the C_{3v} symmetry ($\cdot CRR'\text{-}CH_3$ type of radical), the isotropic hf splitting of a quadratic cosine form couples the degenerate rotational states with $K = \pm 1$ and splits them, giving the characteristic triplet called E-lines [87].

5.4.1.2 CD$_3$ Radical

The CD_3 spectra behaved much differently from those of CH_3. A strong singlet was superimposed on the central peak of an underling weak septet (Fig. 5.19). The septet showed a line splitting of 0.359 mT, which is 6.45 times smaller (close to the ratio of the ^1H and D nuclear g-factors, $g_H/g_D = 6.514$) than in the quartet of CH_3 (2.315 mT), and is attributable to the CD_3 radical. The central singlet rapidly decreased with increasing temperature from 4 K. Concomitantly, the septet increased and the relative intensity reached to the expected ratio of 1:3:6:7:6:3:1 for three equivalent D ($I = 1$), characteristic of a high temperature type of "classical" spectrum due to a freely rotating CD_3, already at a lower temperature of 10 K. This indicates that the $J = 1$ level can be populated more rapidly in CD_3 than in CH_3 upon warming because of its two times smaller rotational constant (B), see Fig. 5.18.

The deuterium (D or ^2H) contained in CD_3 has nuclear spin $I = 1$ and belongs to the class of bosons (particles with integer value of spin) [88]. The Pauli principle

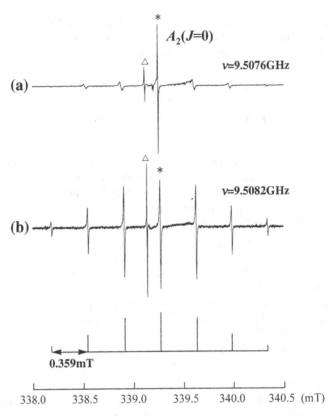

Fig. 5.19 High-resolution ESR spectra of the CD_3 radical in Ar matrix at (a) 4.2 K and (b) 25.0 K. The strong central singlet in (a), marked by a *star* (*), has a relative intensity of 105:1 to the outermost left peak. The line close to the centre, marked by *triangle* (Δ), is due to the D-atom ($m_I = 0$ transition), not relevant for this study. The figure is adapted from [75] with permission from the American Chemical Society

requires that, when the labels of any two identical bosons are interchanged, the total wave function retains the same sign. The wave functions of ψ_E and ψ_V are anti-symmetric and symmetric, respectively, for interchanging any two D-atoms in the CD_3 radical similar to the case of the CH_3 radical. Thus, the product of $\psi_R \otimes \psi_I$ should be anti-symmetric for CD_3. There is only one spin-rotation state available to meet the Pauli principle at the lowest rotational level, $J = 0$ (even), with the following anti-symmetric nuclear spin function of A_2 symmetry in the D_3 point group [75]:

$$
\Psi_{a,b,c}^{A_2}(0,0,0) = \frac{1}{\sqrt{6}} |0,0,0\rangle \{[|-1,0,1\rangle + |1,-1,0\rangle + |0,1,-1\rangle]
$$
$$
- [|1,0,-1\rangle + |-1,1,0\rangle + |0,-1,1\rangle]\}
$$

$$(5.4)$$

Here the values of -1, 0, or 1 in the nuclear spin function $\{[|-1,0,1\rangle + ...]\}$ stand for the m_I of nuclear spin $I = 1$ for the suffixes (a, b, c) of the CD_3 rotor (see Fig. 5.18). It is noted that, on the contrary, the CD_3 quantum-rotor in the C_{3v} symmetry (i.e., R_2C-CD_3 or β-CD_3) is expected to show an intensity ratio of $1:2:2:3:2:2:1$ for the A-lines (corresponding to A_2-lines for the D_3 symmetry) at the $J = 0$ level.

The observed seven hf lines of CD_3 at 25 K are successfully attributable to the allowed nuclear spin functions for the $J = 1$ (odd) level in the D_3 symmetry and the readers can refer to [75] for details.

5.4.1.3 CHD_2 and CH_2D Radicals

The 4.2 K CHD_2 spectrum consists of a double quintet due to one ^1H atom and two identical D atoms with the hf splittings of 2.315 and 0.359 mT, respectively. The central triplet of each quintet is much stronger than the outer lines and has an intensity ratio of 1:1:1, i.e., the quintet intensity ratio being far from 1:2:3:2:1, expected for the high temperature spectrum, see Fig. 5.20. This indicates that the CHD_2 spectrum at 4.2 K is a superposition of the strong triplet transitions at $m_I = 1$, 0, -1 on a weak quintet spectrum due to a freely rotating CHD_2 radical. The central triplet has a narrow line width of ca. 0.001 mT, whereas the outer two lines are about four times broader. The triplet intensity rapidly decreased with temperature.

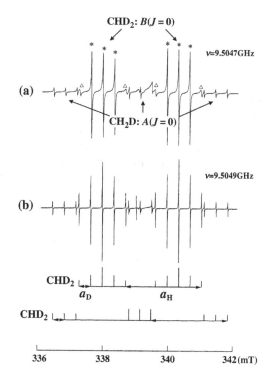

Fig. 5.20 High-resolution ESR spectra of the CH_2D and CHD_2 radicals in an Ar matrix at (**a**) 4.2 K and (**b**) 25.0 K, respectively. The strong triplet of CHD_2 at 4.2 K (*) has a relative intensity of ca. 10:1 to the associated outer two lines ($\triangle$). The figure is adapted from [75] with permission from the American Chemical Society

A doublet of quintets with relative intensities expected for the high temperature spectrum was observed at 25 K as shown in the figure.

The strong triplet *hf* lines can be attributed to the following three deuteron nuclear spin fuctions (corresponding to $m_I = -1, 1, 0$ resonance positions) [75]:

$$\Psi^B_{-1} = [|-1,0\rangle - |0,-1\rangle]/\sqrt{2}$$
$$\Psi^B_1 = [|0,1\rangle - |1,0\rangle]/\sqrt{2} \qquad (5.5)$$
$$\Psi^B_0 = [|1,-1\rangle - |-1,1\rangle]/\sqrt{2}$$

Here, only the deuteron spin projections (z-components) of the nuclear spin fuction (ψ_I) are given in the expression for simplicity. These three anti-symmetric nuclear spin functions belong to B irreducible representation in the C_2 symmetry (point group) of CHD_2 and are only allowed to couple with the rotational function (ψ_R) at the lowest level of $J = 0$ (even) by the Pauli principle.

For the rotational $J = 1$ (odd) level of CHD_2 the Pauli principle requires the following six possible deuteron nuclear spin functions with A irreducible representation (total symmetry) [75]:

$$\Psi^A_{-2} = [|-1,-1\rangle]$$
$$\Psi^A_0 = [|0,0\rangle]$$
$$\Psi^A_2 = [|1,1\rangle]$$
$$\Psi^A_{-1} = [|-1,0\rangle + |0,-1\rangle]/\sqrt{2} \qquad (5.6)$$
$$\Psi^A_1 = [|0,1\rangle + |1,0\rangle]/\sqrt{2}$$
$$\Psi^A_0 = [|1,-1\rangle + |-1,1\rangle]/\sqrt{2}$$

The A lines comprise an approximate 1:1:2:1:1 D *hf* pattern (corresponding to $m_I = -2, -1, 0, 1, 2$ resonance positions) at the $J = 1$ (odd) level. When they are superimposed on the three B lines for the $J = 0$ level the lines reach a relative intensity of 1:2:3:2:1 as observed at 25 K.

On the other hand, the 4.2 K CH_2D spectrum shows a triple triplet due to two equivalent ^1H atoms and one D atom. The triplet *hf* lines due to two ^1H atoms are of equal intensity, 1:1:1, but not a binomial one of 1:2:1. Upon warming above 15 K a new line became visible at the higher field side of each of the central triplet by 0.024 mT and the relative intensity of the new line to the old one reached a limiting value of 1:1 at 40 K. The triplet of CH_2D with equal intensity at 4.2 K can be attributed to three totally symmetric A nuclear spin functions (in the C_2 symmetry) of two ^1H-atoms which are coupled with the $J = 0$ rotational state [75]:

$$\Psi^A_{n,n} = |n,n\rangle$$

$$\Psi^A_{n,n'} = \frac{1}{\sqrt{2}}\left[|n,n'\rangle + |n',n\rangle\right] \qquad (5.7)$$

where $n = -n' = \pm 1/2$. It is worthy to note that Knight et al. [52] have observed a similar nuclear spin state for the H_2O^+ radical cation isolated in Ne matrix at low temperature close to 4 K.

When the $J = 1$ rotational level is populated at elevated temperatures one nuclear spin state with an anti-symmetric B irreducible representation appears in consistency with the experimental spectra:

$$\Psi_{n,n'}^{B} = \frac{1}{\sqrt{2}} \left[|n, n'\rangle - |n', n\rangle \right] \quad (n = -n' = \pm 1/2) \tag{5.8}$$

The reader can refer to [75, 85] for more details on the nuclear spin-rotation couplings of the selectively D-labelled methyl radicals.

5.4.2 Hydrogen Atom – Methyl Radical Pairs

A "singlet" molecule (with a paired electron spin arrangement) can be decomposed into a pair of "doublet" species (each with an unpaired electron, i.e. a radical) by chemical and physical methods including ionizing radiation and photolysis. A pair-wise trapping of radicals is therefore often inherent in radiation damage processes, especially at low temperature. Due to their importance as primary unstable species in chemical reactions, "triplet" state radical pairs (with two unpaired electrons) in irradiated organic crystals have attracted much attention and have been extensively studied by ESR spectroscopy [37b, 89–92]; the studies include a pioneer work on a dimethylglyoxime single crystal by Kurita [89]. The radical pairs are characterized by weak signals at the forbidden $\Delta m_s = 2$ transition ($g \approx 4$), and a fine structure (fs) or zero-field splitting (zfs) due to the electron dipole-dipole interaction at the allowed $\Delta m_s = 1$ transition ($g \approx 2$) in cw-ESR as mentioned in Chapter 4. The ESR parameters such as fs and hyperfine (hf) structures can potentially provide detailed information about the orientation and trapping sites of radical pairs as well as their inter-spin distance and electronic structures.

Here we present high-resolution ESR spectra of the radical pairs of a hydrogen atom coupled with a methyl radical, such as $H\cdots CH_3$, $H\cdots CHD_2$, $D\cdots CH_2D$, and $D\cdots CD_3$, observed at both the allowed $\Delta m_s = 1$ and forbidden $\Delta m_s = 2$ transitions. They were generated by X-ray irradiation of a solid Ar matrix containing selectively D-labelled methanes, CH_4, CH_2D_2, or CD_4, at 4.2 K. The ESR spectra of the radical pairs were observed together with the isolated hydrogen-atoms, methyl radicals and methane radical cation [13, 76]. The unique combination of guest methane and matrix Ar resulted in the observation by ESR of the radical pairs at high-resolution, as for the isolated methyl radicals discussed in the forgoing section (Section 5.4.1) [75], although with the different spectrometer settings to obtain high microwave power and gain.

5.4.2.1 H⋯CH₃ Radical Pairs

Figure 5.21(a) shows the ESR spectrum at the $\Delta m_s = 2$ transition observed at 4 K for a CH_4/Ar sample immediately after X-ray irradiation. The double-quartet ^{1}H *hf* splittings of *ca.* 26 and 1.16 mT (species "A") are attributed to H⋯CH₃ radical pairs. Upon annealing the sample to 10 K three sets of double quartets, pairs I, II, and III, were separately observed as seen in spectra (b) to (d) of the same figure; the three pairs consisting of a constant isotropic *hf* value of 1.16 mT for the quartet (due to three equivalent H-atoms), but of slightly different doublet splittings 25.1–26.3 mT. The former *hf* value is exactly one-half of the ^{1}H *hf* splitting of the isolated CH_3 radical (2.32 mT), and the latter *hf* values are very close to one-half of those of the H-atoms trapped in the interstitial tetrahedral and octahedral sites in the Ar crystal, 51.26 and 51.14 mT, respectively [92]. The result suggests that the observed

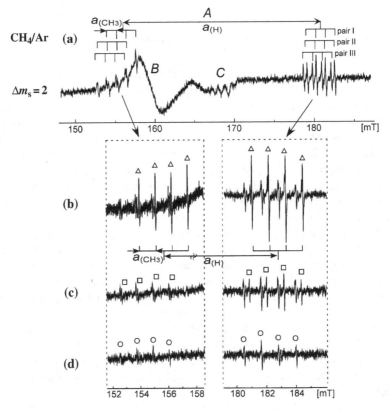

Fig. 5.21 ESR spectra observed at the $\Delta m_s = 2$ transition for the CH_4 (0.05 mol%)/Ar system at (a) 4 K, (b) 5.8 K, (c) 8.0 K, and (d) 10 K after X-ray irradiation at 4 K. The lines marked as *A* are attributable to the H⋯CH₃ radical pair. In spectra (b)–(d), each band of the doublet separated by ca. 26 mT is expanded. The peaks due to three sets of pairs are marked by △: pair I, □: pair II, o: pair III. See [76] for the lines marked as *B* and *C*. The figure is adapted from [76] with permission from the American Chemical Society

double quartets are attributable to the radical pairs of the H-atom coupled with the methyl radical, H···CH$_3$. The radical pairs are expected to be in the triplet electronic ground state which lies below the singlet state by $2J$ (the singlet triplet separation energy; see Chapter 4) with $|J| \gg a_{iso}$ (isotropic hf splitting) of the isolated H-atom ≈ 1.4 GHz (9.3×10^{-25} J). Because of the high resolution spectra, three different sets of the double quartets are separately observed and the ^{1}H-hf splittings at the $\Delta m_s = 2$ transition were evaluated with a high degree of accuracy.

The ^{1}H-hf splittings at the $\Delta m_s = 1$ transition were further split by the fine structure (fs) with axial symmetry due to the electron dipole-dipole coupling. Three different sets of the fs splitting, $d_\perp$, were resolved in the spectra of the H···CH$_3$ radical pairs and were evaluated by the ESR spectral simulation method to be $d_\perp = 4.2$, 4.9, and 5.1 mT for pairs I, II and III, respectively [76]. Applying the point dipole approximation, the $d_\perp$ value can be related to the separation distance, R, between two radicals by the following equation [37b, 89]:

$$d_\perp = (3/2)\mu_0\, g\mu_B/4\pi R^3 \qquad (5.9)$$

where μ_0, g, and μ_B are the magnetic constant, the g-factor of the radical (2.0022), and the Bohr magneton, respectively; for the numerical values of μ_0 and μ_B, see Table G2 in General Appendix G. Replacing $d_\perp$ by the experimental values, the distances are evaluated to be $R = 0.87$, 0.83 and 0.82 nm for pairs I, II and III, respectively.

5.4.2.2 D···CD$_3$, H···CHD$_2$ and D···CH$_2$D Radical Pairs

When a solid Ar sample containing an equimolar mixture of CH$_4$ and CD$_4$ (CD$_4$—CH$_4$/Ar system) was irradiated, the spectra of two radical pairs, D···CD$_3$ and H···CH$_3$, were observed at the $\Delta m_s = 2$ transition as shown in Fig. 5.22(a). The spectrum of D···CD$_3$ at $\Delta m_s = 2$ consists of a triple septet with the isotropic hf splittings of $ca.$ 3.9 and 0.18 mT, respectively; the former is one-half of the splitting of an isolated D-atom, 7.8 mT, and the latter one-half of an isolated CD$_3$ splitting, 0.36 mT. Furthermore, the formation of D···CD$_3$ radical pair was confirmed by observing exactly the same spectrum in the irradiated CD$_4$/Ar sample at 4 K. If the radical pairs were statistically generated in the CD$_4$—CH$_4$/Ar system, equal amounts of radical pairs should be observed for the four different sets, H···CH$_3$, H···CD$_3$, D···CH$_3$, and D···CD$_3$. Only two distinct pairs, H···CH$_3$ and D···CD$_3$, were observed, however, in the present system. When partially D-substituted methane, CH$_2$D$_2$, was used as a solute, the ESR spectra of two different radical pairs, H···CHD$_2$ and D···CH$_2$D, were observed as shown in Fig. 5.22(b). No radical pairs such as D···CHD$_2$ and H···CH$_2$D were observed.

The present ESR studies using the selectively D-labeled methanes, CD$_4$ and CH$_2$D$_2$, unambiguously revealed that the hydrogen atom and the methyl radical present in pairs originate from the same methane molecule, suggesting that the radical pair formation is caused by the homolytic C—H bond scission via the electronically excited methane molecule (CH$_4^*$). The H-atoms initially formed may

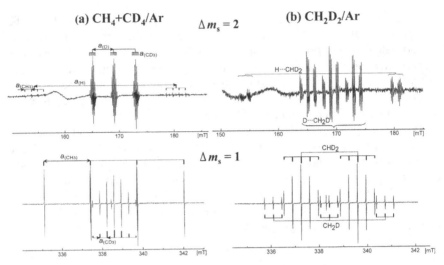

Fig. 5.22 (a) ESR spectra for an equimolar mixture of CH_4—CD_4 (0.25 mol% each)/Ar system irradiated by X-ray at 4 K. (*Upper*) 4 K spectra of D···CD_3 and H···CH_3 radical pairs at the $\Delta m_s = 2$ transition observed immediately after the irradiation. (*Lower*) 25 K spectra of isolated CD_3 and CH_3 radicals at the $\Delta m_s = 1$ transition observed after the radical pairs were thermally decayed out. (b) ESR spectra for CH_2D_2 (0.5 mol%)/Ar system. (*Upper*) 4 K spectra of D···CH_2D and H···CHD_2 at the $\Delta m_s = 2$ transition. (*Lower*) 25 K spectra of isolated CHD_2 and CH_2D radicals at the $\Delta m_s = 1$ transition. The figure is adapted from [76] with permission from the American Chemical Society

migrate in the Ar face-centered cubic (*fcc*) crystal lattice so as to be trapped in the interstitial tetrahedral (I_t) or octahedral (I_o) site, while the methyl radical may retain the substitutional site of the Ar crystal where the mother molecule originally occupied, see Fig. 5.23. In fact the I_t and the I_o sites are calculated to be at distances of 0.87 nm and 0.88 nm from the counter H atom, respectively; the distances are in very good agreement with the experimental value of $R = 0.87$ nm for pair I.

We close this section by noting that Knight et al. have reported ESR spectra of more simple triplet radical pairs of H···H, H···D, and D···D trapped in rare gas matrices at 4.2 K and developed a theoretical model of treating these spin-pairs as weakly interacting atoms [93].

5.4.3 Hydrogen Atom – Hydrogen Molecule Complex

In this section we deal with a complex formation between a hydrogen atom and a hydrogen molecule in the Ar matrix [77, 78], which demonstrates an important ESR spectroscopic application on elemental reaction processes between a simple atom and a simple molecule at cryogenic temperatures.

Fig. 5.23 Schematics showing proposed trapping sites for the H···CH$_3$ radical pair with $d_\perp = 4.2$ mT (i.e., $R = 0.87$ nm) in the solid Ar. The CH$_3$ radical occupies a substitutional site (S) in the Ar (fcc) crystalline lattice. The counter H-atom occupies (**a**) the interstitial tetrahedral site (I_t) at a distance of 0.87 nm from the CH$_3$ radical, and (**b**) the interstitial octahedral site (I_o) at a distance of 0.88 nm. The figure is adapted from [76] with permission from the American Chemical Society

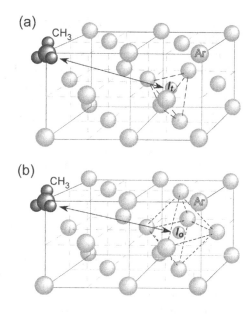

5.4.3.1 Formation and ESR Identification of H···HD, H···D$_2$ and D···D$_2$ Complexes

Figure 5.24(a) shows the low-field component of the doublet ESR *hf* lines due to three different hydrogen atoms generated in the Ar matrix by X-ray irradiation at 4.2 K. Immediately after irradiation the H atoms were found to be trapped in three

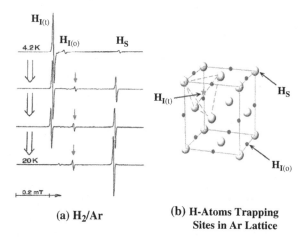

(a) H$_2$/Ar   (b) H-Atoms Trapping Sites in Ar Lattice

Fig. 5.24 (**a**) The low-field components of the doublet ESR spectra of hydrogen (H) atoms generated by ionizing radiation of an Ar sample containing 3 mol% H$_2$ molecules at 4.2 K and their spectral change upon annealing to 20 K. A new *hf* doublet with 51.2 mT is marked as an *arrow* (↓) in the spectra. (**b**) $I_{(t)}$, $I_{(o)}$ and S sites in solid Argon *fcc* lattice. The figure is adapted from [77] with permission from Elsevier

different trapping sites in an Ar *fcc* crystal lattice: interstitial tetrahedral site ($I_{(t)}$), interstitial octahedral site ($I_{(o)}$), and substitutional site (S), see Fig. 5.24(b). The H-atoms in these trapping sites, $H_{I(t)}$, $H_{I(o)}$ and H_S can be distinguished by three different isotropic ^1H *hf* couplings, 51.5, 51.4, and 50.7 mT, respectively. ^1H *hf* couplings larger than the theoretical value, 50.8 mT, have been explained in terms of the Pauli exclusion effect in preference to the van der Waals attraction effect [92]. The H atoms which are initially formed are predominately trapped in the narrowest $I_{(t)}$ site [77, 78]. Upon annealing the sample to 20 K $H_{I(t)}$ atoms start to migrate to S sites to form H_S atoms with concomitant decrease in the total amounts of H atoms due to the recombination reaction. At the same time a new doublet with a ^1H *hf* coupling of 51.2 mT appeared as seen in the figure.

Figure 5.25 shows the lowest-field component of the *hf* doublet or triplet at 20 K due to the newly appeared H or D atoms in solid Ar containing (a) H_2, (b) D_2 and (c) HD molecules. For the HD/Ar sample the doublet splits further into isotropic nine *hf* lines with relative intensity 1:1:2:1:2:1:2:1:1 in which the outer and inner peaks have 0.068 ($a_{2(D)}$) and 0.062 ($a_{3(D)}$) mT splittings, respectively, see Fig. 5.25 (c, left). The nine lines can be attributed to a superposition of two different spectra which consist of super-*hf* splittings due to the D_2 and HD molecules coupled with the H atom to form H$\cdots$D$_2$ (products) and H$\cdots$HD (reactants) complexes as will be mentioned below. Interestingly no further splittings were observed for the D atom lines in the same HD/Ar sample, see Fig. 5.25 (c, right). In contrast to the HD/Ar

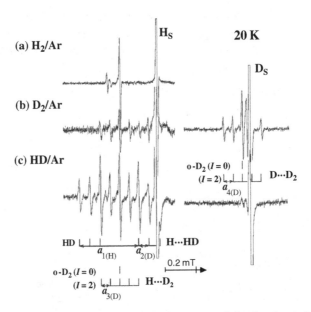

Fig. 5.25 Low-field components of the ESR spectra of H atoms (*left column*) and of D atoms (*right column*) which were newly formed upon annealing the irradiated samples to 20 K: **(a)** 1.0 mol% H$_2$/Ar, **(b)** 1.0 mol% D$_2$ (containing a trace of H$_2$ as an impurity)/Ar and **(c)** 3.3 mol% HD/Ar. The sticks show an analysis of (super) *hf* lines due to H$\cdots$HD, H$\cdots$D$_2$ and D$\cdots$D$_2$ complexes. The spectra are adapted from [77] with permission from Elsevier

system, for the H_2/Ar system only the simple doublet without further splittings was observed, see Figs. 5.24(a) and 5.25 (a, left). For the D_2/Ar system both the *hf* lines due to D and H atoms were further split into five lines with $a_{3(D)}$ with relative intensity 1:1:2:1:1 as seen in Fig. 5.25(b).

The experimental results on the HD/Ar system can be explained as follows. The H and D atoms initially formed in the irradiated HD/Ar sample may encounter an HD molecule located in a substitutional site of solid Ar in the course of their thermal migration. The D atoms may react with HD to form an H atom and a D_2 molecule in an exothermic process ($\Delta G < 0$). On the other hand the H atom reaction with HD is an endothermic process ($\Delta G > 0$) and is unlikely to occur [77, 94]:

$$\begin{array}{ccc} \text{D + HD} & \xrightarrow{} & \text{H + D}_2 \\ \text{(reactants)} & \Delta G < 0 & \text{(products)} \end{array}$$

$$\begin{array}{ccc} \text{H + HD} & \xrightarrow{}\!\!\!\!\!\times\!\!\!\!\!\xrightarrow{} & \text{D + H}_2 \\ \text{(reactants)} & \Delta G > 0 & \end{array}$$

The *hf* value of $a_{1(H)}$ (0.44 mT) observed for the HD/Ar system is larger by the ratio of the ^{1}H and D nuclear g-factors, $g_H/g_D = 6.514$, than the $a_{2(D)}$ value. Then, the double triplet splittings of $a_{1(H)}$ and $a_{2(D)}$ are attributable to the super-*hf* splittings due to the H and D nuclei of the reactant HD molecule, which interacts with the H atom to form the H$\cdots$HD complex, see the stick diagrams in Fig. 5.25 (c, left). On the other hand, the quintet with a splitting denoted $a_{3(D)}$ observed for the HD/Ar system is attributable to two D atoms of the product D_2 molecule, which interacts with an H-atom to form the H$\cdots$D$_2$ complex. The same quintet was observed for the D_2/Ar system and attributed to the D$\cdots$D$_2$ complex in which the D atom formed interacted with the reactant D_2 molecule. The relative peak intensity of the quintet, 1:1:2:1:1, originates from a superposition of a central singlet line due to *ortho*-D_2 (*o*-D_2) with $I = 0$ and five lines due to *o*-D_2 with $I = 2$, in which the intensity of each line is equal.

5.4.3.2 Evidence of Ortho-Para Conversions in Solid H_2 and D_2

The hydrogen nucleus (^{1}H) is a Fermi particle (*Fermion*) with nuclear spin $I = 1/2$. A sample of normal-H_2 (*n*-H_2) contains *ortho*-H_2 (*o*-H_2) with parallel nuclear spins ($I = 1$) and *para*-H_2 (*p*-H_2) with anti-parallel nuclear spins ($I = 0$) in a natural abundance of 25 and 75%, respectively. On the other hand the deuterium nucleus (D; ^{2}H) is a Bose particle (*Boson*) with $I = 1$ and D_2 consists of *o*-D_2 with $I = 0, 2$ and *p*-D_2 with $I = 1$. Applying the Pauli principle the allowed combinations of nuclear-spin states and rotational states for H_2 and D_2 molecules can be obtained. They are given in Table 5.5 together with their *ortho-para* designations. The lowest rotational states of *o*-H_2 and *p*-H_2 are characterized by $J = 1$ (odd) and $J = 0$ (even), respectively, while for *o*-D_2 by $J = 0$ and for *p*-D_2 by $J = 1$, in their ground rotational states. In the ground states *o*-D_2 ($J = 0$) is more stable by *ca.* 86 K than *p*-D_2 ($J = 1$), see Fig. 5.26 [96]. Paramagnetic H and/or D atoms can

Table 5.5 Allowed combinations of nuclear-spin states and rotational states for H_2 and D_2 molecules and their ortho-para designations. "Anti-symmetric" is abbreviated by AS and "symmetric" by S; I is the total molecular nuclear spin and J the rotational quantum number; Ψ_R and Ψ_I are the rotational and nuclear-spin wave functions, respectively. The table is adapted from [96] by permission of the American Physical Society

Hydrogen molecule		I	J	$\Psi_R \cdot \Psi_I$	Nuclear weight	Designation
H_2 (^{1}H with $I=1/2$)	State	0	Even			Para (p-H_2)
	Symmetry	AS	S	AS	1	
	State	1	Odd			Ortho (o-H_2)
	Symmetry	S	AS	AS	3	
D_2 (^{2}H with $I=1$)	State	1	Odd			Para (p-D_2)
	Symmetry	AS	AS	S	3	
	State	0, 2	Even			Ortho (o-D_2)
	Symmetry	S	S	S	6	

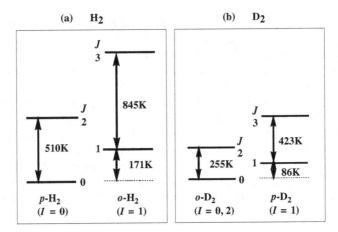

Fig. 5.26 The rotational energy levels for isolated free (**a**) H_2 and (**b**) D_2 molecules [96]. Similar diagrams apply for both H_2 and D_2, but for the latter the energy is scaled down by about a factor of two due to its two times larger moment of inertia

serve as magnetic perturbations to allow an adjacent p-D_2 molecule to convert into o-D_2 in the lowest rotational level, $J = 0$, at low temperature:

$$o\text{-}H_2 \xrightarrow{\text{H, D}} p\text{-}H_2$$
$$(I{=}1; J{=}1) \qquad\qquad (I{=}0; J{=}0)$$

$$p\text{-}D_2 \xrightarrow{\text{H, D}} o\text{-}D_2$$
$$(I{=}1; J{=}1) \qquad\qquad (I{=}0, 2; J{=}0)$$

In consistency with the HD/Ar system, a quintet split by $a_{3(D)} = 0.062$ mT with an 1:1:2:1:1 intensity ratio was observed for the D_2/Ar system and attributed to a D atom interacting with an o-D_2 ($I = 0, 2$ at $J = 0$) molecule in the D$\cdots$D$_2$ complex. Likewise, for the H_2/Ar system the H$\cdots$H$_2$ complex formation is expected. However, the observed new doublet does not show any super-hf couplings due to the H_2 molecule. This suggests that paramagnetic o-H_2 ($I = 1$) located near an H atom is converted into diamagnetic p-H_2 ($I = 0$) so as not to give any super-hf couplings. It has been suggested that the observed paired complexes are formed via a hydrogen atom tunnelling mechanism [77, 78, 95].

5.4.4 Hydrogen Molecular Complex Ions of $H_{6-n}D_n^+$ in para-H_2 Matrix

5.4.4.1 Solid para-H_2 Matrix and Hydrogen Molecular Ions

Solid hydrogen is called a quantum solid because of the large vibrational amplitude of the hydrogen molecule at zero-point energy level [97]. Furthermore, as mentioned in the foregoing section, n-H_2 molecules consist of 25% p-H_2 with $I = 0$ and 75% o-H_2 with $I = 1$ and the non-magnetic p-H_2 can be separately obtained from the magnetic o-H_2. Because of its non-magnetic property the solid p-H_2 is potentially useful as a matrix for high-resolution ESR spectroscopy. Miyazaki and his co-workers reported that H atoms (H_t) trapped in solid p-H_2 matrix gave rise to about three times narrower ESR line width (0.011 mT) than that of H_t in n-H_2 matrix (0.034 mT) at 4.2 K [98, 99]. The difference in the line width comes from the corresponding difference in the dipolar hf interaction between the two matrices. As the nuclear spin of p-H_2 is zero, the dipolar hf interaction between the electron spin (S) of H_t and the nuclear spins (I) of the surrounding p-H_2 molecules vanishes. Whereas, in the n-H_2 matrix the o-H_2 molecules with $I = 1$ that are accidentally located near H_t, cause a line width broadening due to the dipolar hf interaction.

Studies on the reactions and the structure of molecular ions of the simple hydrogen molecule, have a long history since Thomson [100] first discovered H_3^+ ions produced in a reaction in the gas phase:

$$H_2^+ + H_2 \rightarrow H_3^+ + H \qquad (5.10)$$

In recent years the associated hydrogen molecular ions have attracted much attention in relation to low temperature chemistry including the chemistry of interstellar clouds. Here we call attention to a recent ESR study by Kumagai et al. [79] on the "H_2^+-core" H_6^+ radical cation with D_{2d} symmetry generated and stabilized in irradiated solid p-H_2 matrix. It is demonstrated that ESR spectroscopy is a very useful experimental technique to observe and identify unstable intermediates attributed to hydrogen molecular ions. The assignment of H_6^+ radical ions was experimentally confirmed by generating deuterium substituted

$H_nD_{6-n}^+$ ($n = 1, 2$) ions in D_2/p-H_2 and HD/p-H_2 mixture systems. Furthermore, experimentally determined hf splittings compared well with those predicted theoretically.

5.4.4.2 ESR spectra and structures of H_6^+ and $H_{6-n}D_n^+$: experimental vs. computational results

Pure p-H_2 γ-irradiated at 4.2 K gave rise to a quartet of hf lines with slight anisotropy; see the lines marked by open circles (o) in Fig. 5.27(a). The quartet lines can be well simulated using the ^1H hf splitting due to two magnetic equivalent ^1H atoms given in the figure; the separation of the central two lines being attributable to the second-order hf splitting [36a, 36b, 37a] (see Section 5.2.1.1) between two different states of $[I = 1, m_I = 0]$ and $[I = 0, m_I = 0]$. Comparing these experimental results with the ab-initio calculations the quartet lines are attributed to the "H_2^+-core" H_6^+ radical cation with a D_{2d} symmetry abbreviated as $[H_2(H_2)H_2]^+$,

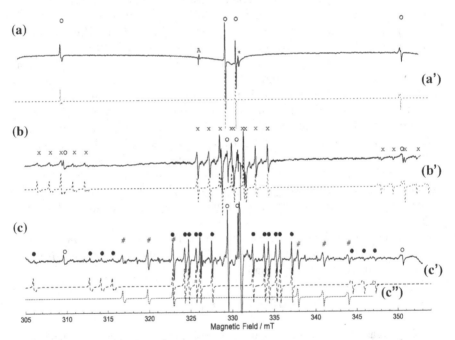

Fig. 5.27 ESR spectra of γ-ray irradiated solid (**a**) p-H_2(pure), (**b**) o-D_2(1 mol%)/p-H_2, (**c**) HD(1 mol%)/p-H_2 samples at 4.2 K. The simulated spectra of (**a'**) $[H_2(H_2)H_2]^+$, (**b'**) $[H_2(H_2)D_2]^+$, (**c'**) $[H_2(H_2)HD]^+$, and (**c''**) $[H_2(HD)H_2]^+$ are shown together with the assignment of the hf lines. For the spectral simulations hf splittings with a small anisotropy were assumed; the isotropic hf splittings used are given in Fig. 5.29. The anisotropic (axial symmetric) hf value used is, for example, (–)0.06 mT for the two equivalent H atoms of the H_2^+-core in $[H_2(H_2)H_2]^+$. The spectrum below 305 mT and above 354 mT could not be measured because of the overlapping of very intense signal due to trapped H-atoms [79]. The spectra were obtained from Dr. J. Kumagai

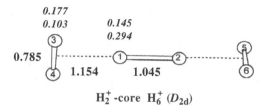

H_2^+-core H_6^+ (D_{2d})

Fig. 5.28 Schematic representation of optimized structure of H_2^+-core H_6^+ ion, $[H_2(H_2)H_2]^+$, with a D_{2d} symmetry at MP2/cc-pVTZ level of computations: bond lengths in Å. Numbers in italics give atomic net charges and spin densities in the upper and lower lines, respectively. The diagram is adapted from [101] with permission from the American Institute of Physics

where the H_2 in the parentheses is the H_2^+-core, and the two H_2 molecules at both ends correspond to side-on H_2 molecules, see Fig. 5.28.

No *hf* splittings due to the two side-on H_2 molecules were observed although a 9 mT splitting was predicted by the calculations. The reason for this can be explained as follows. The rotational constant of H_2 (*ca.* 8 meV) is much larger than the rotational barrier (1.4 meV) calculated for side-on H_2 along the main axis of H_6^+ and the thermal energy at 4 K (0.4 meV). Thus one can expect that almost all side-on H_2 molecules rotate freely and the $J = 0$ rotational state is only populated at 4.2 K. In this case, like for the *p*-H_2 molecule, the side-on H_2 at $J_{sid}(H_2) = 0$ is exclusively at $I_{sid}(H_2) = 0$ state due to the parity conservation rule on exchanging the ^{1}H fermion species, see Table 5.5. Furthermore, the experimental isotropic ^{1}H *hf* splitting (20.42 mT) obtained for the H_2^+-core is very close to the calculated one (23.31 mT). Thus, the $[H_2(H_2)H_2]^+$ radical cation was concluded to be preferentially generated in irradiated solid *p*-H_2 matrix.

D-substitution of the side-on H_2 molecules could be much informative, since the *hf* interaction with side-on D_2 or HD nuclei in $H_4D_2^+$ or H_5D^+ would be observed even if D_2 or HD are in free rotational states as will be explained below [102]. When *p*-H_2 containing *o*-D_2 (1 mol%) was irradiated at 4.2 K a number of resolved *hf* lines were newly observed in addition to those from irradiated pure *p*-H_2 as shown in Fig. 5.27(b). Interestingly, each of the quartet lines of $[H_2(H_2)H_2]^+$ was observed to be further split into a quintet of 1.44 mT with a relative intensity of *ca.* 1:1:2:1:1. The observed *hf* lines are attributable to $[H_2(H_2)D_2]^+$ in which the side-on H_2 and D_2 are freely rotating. Unlike H_2, D_2 in $J_{sid}(D_2)$ can be coupled with $I_{sid}(D_2) = 0$ and 2 states so as to give the quintet *hf* lines corresponding to $m_I = 2, 1, 0, -1$, and -2 with the relative intensity of 1:1:2:1:1; the *hf* lines due to the two states of the side-on D_2, $[I = 2, m_I = 0]$ and $[I = 0, m_I = 0]$ giving the double intensity at the centre.

The irradiated HD(1 mol%)/*p*-H2 sample gave rise to a spectrum with a large number of *hf* lines, see Fig. 5.27(c). The experimental spectrum was successfully analyzed by superimposing the spectra of three radical ions $[H_2(H_2)H_2]^+$, $[H_2(H_2)HD]^+$, and $[H_2(HD)H_2]^+$ as shown in the figure; the simulation spectra were calculated using the ^{1}H and D *hf* splittings given in Fig. 5.29.

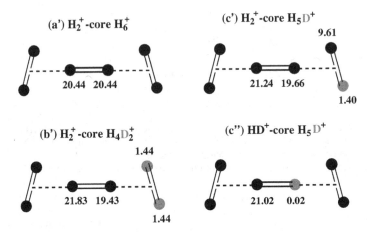

Fig. 5.29 Structures of "H_2^+-core" H_6^+ radical cation with a D_{2d} symmetry and its D-isotope substituents $H_{6-n}D_n^+$: (**a′**) $[H_2(H_2)H_2]^+$, (**b′**) $[H_2(H_2)D_2]^+$, (**c′**) $[H_2(H_2)HD]^+$, and (**c″**) $[H_2(HD)H_2]^+$ radical ions. The isotropic ^{1}H and ^{2}D hf splittings (in mT) used for the ESR spectral simulations are given in each structure. The diagram is adapted from [79] with permission from the American Institute of Physics

For the $[H_2(H_2)D_2]^+$ radical cation slightly non-equivalent ^{1}H hf splittings were observed for the core-H_2 atoms, see Figs. 5.27(b, b′) and 5.29(b′). This can be explained as follows. As H_2 is lighter than D_2, the amplitude of zero-point vibration near H_2 can be larger than that near D_2; the unharmonic term on vibrational levels near the H_2—H_2^+ bond can be larger than that near H_2^+—D_2 in $[H_2(H_2)D_2]^+$ even at ground vibrational state. Therefore, the averaged H_2—H_2^+ bond length in $[H_2(H_2)D_2]^+$ can be larger than the H_2^+—D_2 one. A similar explanation is possible for the observed non-equivalent ^{1}H hf splittings of $[H_2(H_2)HD]^+$.

5.4.4.3 D-Isotope Condensation in o-D_2/p-H_2 and HD/p-H_2

As shown in Fig. 5.27, even though only 1 mol% of o-D_2 or HD was introduced in the p-H_2 sample, the yields of $[H_2(H_2)D_2]^+$ or $[H_2(H_2)DH]^+$ ions were much higher than that of $[H_2(H_2)H_2]^+$. Furthermore, $[D_2(H_2)D_2]^+$ and $[HD(H_2)HD]^+$ cations were preferentially generated in both o-D_2(8 mol%)/ p-H_2 and HD(8 mol%)/p-H_2 systems rather than $[H_2(H_2)D_2]^+$ and $[H_2(H_2)HD]^+$ cations. The decay rate of $[H_2(H_2)H_2]^+$ ions in the o-D_2/p-H_2 and HD/p-H_2 systems was found to be faster than that in the pure p-H_2 system. Based on these experimental results the following D-isotope condensation (or enrichment) reactions were proposed for the irradiated o-D_2/p-H_2 and HD/p-H_2 systems:

$$[H_2(H_2)H_2]^+ + D_2 \rightarrow H_2 + [H_2(H_2)D_2]^+ + 17\,\text{meV} \tag{5.11}$$

$$[H_2(H_2)D_2]^+ + D_2 \rightarrow H_2 + [D_2(H_2)D_2]^+ + 18\,\text{meV} \tag{5.12}$$

$$[H_2(H_2)H_2]^+ + HD \rightarrow H_2 + [H_2(H_2)HD]^+ + 8\,meV \qquad (5.13)$$

$$[H_2(H_2)HD]^+ + HD \rightarrow H_2 + [HD(H_2)HD]^+ + 9\,meV \qquad (5.14)$$

The exothermic energies $[\Delta G\ (=-18 \sim -8\,meV) < 0]$ were theoretically evaluated from the difference in zero point vibration energies between the reactants and the products [79]. The exothermic energies are higher than the thermal energy at 4 K (0.4 meV) and no backward reactions are expected. Thus, D-isotope condensation reactions in the $H_6{}^+$ cation are rationalized. The results are very similar to the reported D-isotope enrichment reaction of $H_3{}^+$:

$$H_3^+ + HD \rightarrow H_2D^+ + H_2 \qquad (5.15)$$

in interstellar clouds whose temperatures are lower than the exothermic energy of the reaction [103].

5.5 Summary

ESR spectra can provide not only an unambiguous assignment of radicals, but also experimental information about their geometrical and electronic structures and reactions. CW-ESR spectroscopy combined with matrix isolation methods and ionizing radiation (γ-ray, X-ray, etc.) is applied to the studies on reactive intermediate radicals including anionic and cationic species trapped in low temperature solid matrices. ESR parameters, especially hyperfine (*hf*) couplings, are predicted with considerable precision by recent advances in computational methods such as density functional theory (DFT), which affords a valuable bridge between experiment and theory at a most fundamental level.

The first topic is concerned with geometrical and electronic structure of the radical anions of perfluorocycloalkane (c-$C_nF_{2n}{}^-$) and perfluoroalkene ($C_nF_{2n-2}{}^-$). On the basis of detailed comparisons between the experimental and theoretical results saturated c-$C_nF_{2n}{}^-$ ($n = 3$–5) anions are concluded to have planar molecular structures with a high symmetry, in which the singly occupied molecular orbital (SOMO) delocalizes over the entire molecule. Unsaturated $C_nF_{2n-2}{}^-$ anions such as $C_2F_4{}^-$ and c-$C_nF_{2n-2}{}^-$ ($n = 3$–5) have a pyramidally distorted structure at the C=C carbon which occurs by mixing of the π^* and higher-lying σ^* orbitals. The unpaired electron is primarily localized in the sp^3-like hybrid orbitals and is transferred to the fluorine orbitals to give rise to large *hf* splittings for the ^{19}F nuclei at the original double bond. Furthermore, an ESR study is demonstated for a structure distortion of acetylene upon one electron reduction to form CH≡CH$^-$ radical anion with a trans bent structure in a low temperature organic solid matrix.

The second topic is concerned with Jahn-Teller (*J-T*) distortion of organic radical cations whose mother molecules have either T_d or C_{3v} structure with a doubly degenerate HOMO. A series of D(^2H)-labelled methane radical cations ($CH_4{}^+$,

CDH_3^+, $CD_2H_2^+$, CD_3H^+ and CD_4^+) were generated in a neon matrix at 4 K and subjected to ESR studies. The experimentally observed isotropic hf values of ^1H and D atoms are fully discussed in terms of a lowering of the symmetry of methane from the original T_d to C_{2v} in the cation form. Two different electronic sites, apical "a" and equatorial "e", exist in the C_{2v} structure. The D atoms prefer to occupy the site "e", whereas the ^1H atoms prefer the site "a". The ^1H atoms exchange with other ^1H atoms even in a different electronic environment, but not with D atoms. The D atoms likewise exchange with other D atoms, but not with ^1H atoms. Furthermore, we discuss J-T structural distortion of trimethylenemethane radical cation (TMM$^+$) whose mother molecule has the simplest known 4π-electron system of D_{3h} symmetry. The DFT computations predict that in the TMM$^+$ ion the original D_{3h} symmetry is reduced to a lower one of C_{2v} so as to split an originally doubly degenerated e'' HOMO (corresponding to $^2E''$ state) in D_{3h} into a_2 and b_1 orbitals ($^2A_2$ and $^2B_1$ states) in C_{2v}. In contrast to prediction the TMM$^+$ cation generated in a fluorocarbon matrix shows a septet of hf lines due to six magnetically equivalent H atoms at 4 K, suggesting an apparent D_{3h} structure. The experimental spectrum is discussed in terms of an intramolecular dynamics of TMM$^+$ among the three equivalent C_{2v} structures with the $^2A_2$ ground state, depending on which of three C—C bonds is the shorter bond along the C_2 symmetry axis, so as to average out the structural distortion (dynamic J-T effect). The e'' orbital degeneracy of TMM can be removed by methyl substitution of the π-radical cation. The radical cation of 1,1,2,2-tetramethyltrimethylenemethane (TMM-Me$_4^+$; C_{2v} symmetry) gave rise to hf couplings characteristic of a SOMO that closely resembles the b_1 orbital in TMM$^+$.

The third topic is concerned with high resolution ESR spectra on D-labelled small radicals generated in a solid Ar or a quantum solid of *para*-H$_2$ ($I = 0$; nonmagnetic) matrix. Emphasis is put on quantum effects at cryogenic temperatures. High resolution ESR spectra are presented for selectively D-labelled methyl radicals (CH$_3$, CH$_2$D, CHD$_2$, and CD$_3$) generated in low temperature solid Ar matrices. The experimentally observed H/D-isotope effects on the ESR lineshape and their temperature dependencies are discussed in terms of nuclear spin-rotation couplings using a three-dimensional free quantum-rotor model. A pair-wise trapping of radicals is important as primary unstable species in chemical reactions. D-labelled hydrogen atom – methyl radical pairs such as H$\cdots$CH$_3$, H$\cdots$CHD$_2$, and D$\cdots$CH$_2$D, and D$\cdots$CD$_3$ were generated in irradiated Ar matrix and their high resolution ESR spectra were observed at both the allowed $\Delta m_s = 1$ and forbidden $\Delta m_s = 2$ transitions. Using CD$_4$ and CH$_2$D$_2$ as a solute it is concluded that a hydrogen atom and a methyl radical are generated by a homolytic C—H (or C—D) bond scission of methane and trapped nearby with an inter-spin distance of $R = 0.87$ nm to form the radical pair in the Ar matrix. This suggests that the H-atoms initially formed migrate in the Ar *fcc* crystal lattice so as to be trapped in the interstitial tetrahedral (I_t) or octahedral site (I_o), while the methyl radical retains the original substitutional (S) site.

A complex formation between a hydrogen atom and a hydrogen molecule in Ar matrix is demonstrated as an important example of ESR applications on elemental reaction processes between a simple atom and a simple molecule at cryogenic temperatures. An ESR study was carried out for H and D atoms generated in irradiated solid Ar containing H_2, HD and D_2 molecules at 4 K. Upon annealing the initially formed hydrogen atom starts to migrate in the Ar lattice and encounters a hydrogen molecule located in a S site so as to form the complex. ESR spectra of the H$\cdots$$D_2$ and H$\cdots$HD complexes, but not of the D$\cdots$HD complex were observed for the irradiated HD/Ar system. The results suggest that the D atom reacts with HD to form an H atom and a D_2 molecule in an exothermic process ($\Delta G < 0$), but the H atom reaction with HD is an endothermic process ($\Delta G > 0$) and is unlikely to occur. A super-hyperfine coupling of a quintet due to o-D_2 (with $I = 0$ and 2) of the D$\cdots$$D_2$ complex was observed for the D_2/Ar system. This suggests that H and/or D atoms can serve as magnetic perturbations to allow the adjacent p-D_2 to convert into o-D_2 in the lowest rotational level, $J = 0$, at low temperatures. Furthermore, we present ESR results showing that the paramagnetic o-H_2 ($I = 1$) near an H atom is converted into a diamagnetic p-H_2 ($I = 0$) so as to give no super-hf couplings due to an H_2 of the H$\cdots$$H_2$ complex in the H_2/Ar system.

In the end we deal with a recently reported high-resolution ESR spectrum of "H_2^+-core" H_6^+ radical cation with a D_{2d} symmetry, abbreviated as $[H_2(H_2)H_2]^+$, where the H_2 in the parentheses is the H_2^+-core and the two H_2 molecules in both ends are the side-on H_2 molecules. The 4.2 K γ-irradiated pure p-H_2 showed hyperfine structure due to two equivalent H-atoms of the "H_2^+-core". The identification and structure of $[H_2(H_2)H_2]^+$ were confirmed by comparing the experimentally obtained ESR spectra of deuterium substitutions such as $[H_2(H_2)HD]^+$, $[H_2(HD)H_2]^+$ and $[D_2(H_2)H_2]^+$ cations generated in irradiated D_2/p-H_2 and HD/p-H_2 mixture systems with ab-initio computations. D-substitution of the side-on H_2 molecules such as $[H_2(H_2)HD]^+$ and $[D_2(H_2)D_2]^+$ is informative because the hf interaction due to the side-on D_2 or HD nuclei is observable. For example, $[H_2(H_2)D_2]^+$ cation gave rise to a quintet of hf lines due to the o-D_2 with $I = 0$ and 2 states in addition to the quartet due to the "H_2^+-core". Thus, the $[H_2(H_2)H_2]^+$ radical cation is concluded to be preferentially generated in irradiated solid p-H_2 matrix.

Appendices

A5.1 A Brief Summary on Quantum Chemical Computation Methods

A5.1.1 Schrödinger Equation

The non-relativistic time-independent *Schrödinger equation* (SE) [104] for N nuclei and n electrons is:

$$H\Psi = E\Psi \tag{5.16}$$

where E is the energy of state described by the *wave function* (WF), Ψ, which depends both on the electronic and nuclear coordinates. Hamiltonian H can be written in atomic units as:

$$
H = -\sum_{i=1}^{n} \frac{1}{2} \nabla_i^2 - \sum_{A=1}^{N} \frac{1}{2M_A} \nabla_A^2 - \sum_{i=1}^{n} \sum_{A=1}^{N} \frac{Z_A}{r_{iA}}
$$
$$
+ \sum_{i=1}^{n} \sum_{j \geq i}^{n} \frac{1}{r_{ij}} + \sum_{A=1}^{N} \sum_{B \geq A}^{N} \frac{Z_A Z_B}{r_{AB}}
\tag{5.17}
$$

The first two terms in Eq. (5.17) are *kinetic energy* (KE) operators of the electrons and of the nuclei while the final three terms are the operators of electron-nuclear, electron-electron, and nuclear-nuclear *potential energies* (PEs). M_A is the mass of nucleus A divided by the mass of the electron and Z_A is the atomic number of nucleus A. Terms r_{ij}, r_{iA}, and r_{AB} are the distances between electrons i and j, between electron i and nucleus A, and between nuclei A and B, respectively.

The quantum chemical methods applied to the molecular science such as structural chemistry and chemical reactions described in this chapter are all based on the *Born-Oppenheimer approximation* [105]. Within the BO approximation the KE operators for nuclei are removed from the above Hamiltonian and the SE is solved for a fixed set of nuclei. The SE within the BO approximation is usually referred to as the electronic SE, and the resulting WFs as electronic states.

The electronic SE is impossible to solve analytically for systems containing more than one electron and further approximations are required. There are two methods available in modern quantum chemistry, i.e. *wave function theory* (WFT) and *density functional theory* (DFT) [18–21]. The WFT has been developed to compute an optimal WF for the system, whereas the DFT has been developed to find the optimal electron density (ED) for the system and has becomes an overwhelmingly popular method.

A5.1.2 Molecular Orbitals

A *molecular orbital* (MO) has a form of:

$$
\Phi_i(r) = \sum_{\mu=1}^{K} c_{i\mu} \chi_\mu(r)
\tag{5.18}
$$

where χ_μ is an *atomic orbital* (AO) of atom μ and the sum extends over all the *valence orbitals* of all atoms in a molecule. The coefficients of $c_{i\mu}$ can be evaluated by setting up the *secular equations* and the *secular determinant*. That is, the associated energies can be evaluated by solving the *secular determinant*, and then the coefficients are found by inserting these energies into the secular equations. The bond lengths and bond angles of the molecule can be predicted by calculating the total energy for a variety of nuclear positions and then identifying the conformation that corresponds to the lowest energy.

The π MO energies of conjugated molecules can be evaluated using a set of approximations suggested by E. Hückel [106]. In the *Hückel method* (HM) the π orbitals are treated separately from the σ orbitals, and the latter forms a rigid framework that determines the geometrical structure of the molecule. All the carbon (C) atoms are treated identically, so all the *Coulomb integrals* for the AOs that contribute to the π orbitals are set equal. The first level of sophistication is the *extended Hückel method* (EHM), which is a significant extension of the basic HM to include σ orbitals [107]. Current versions of EHM are easy to implement with a computer and a give useful qualitative pictures of MOs in molecules with known structure. Many of the difficulties associated with EHM have been overcome by more sophisticated theories that not only calculate the shapes and energies of MOs, but also predict with reasonable accuracy the structure and reactivity of molecules.

A5.1.3 Hartree-Fock Equations

The starting point for the WFT is provided by *Hartree-Fock (HF)* theory. In the HF theory the n-electron WF of a molecule is written as a determinant referred to as a single *Slater determinant* (SD), Ψ_{SD}, constructed from n occupied one-electron WFs, $\Phi_i(x)$:

$$\Psi_{SD} = \frac{1}{\sqrt{n!}} \begin{vmatrix} \Phi_1(x_1) & \Phi_2(x_1) & \cdots & \Phi_n(x_1) \\ \Phi_1(x_2) & \Phi_2(x_2) & \cdots & \Phi_n(x_2) \\ \cdot & \cdot & \cdot & \cdot \\ \cdot & \cdot & \cdot & \cdot \\ \cdot & \cdot & \cdot & \cdot \\ \Phi_1(x_n) & \Phi_2(x_n) & \cdots & \Phi_n(x_n) \end{vmatrix} \tag{5.19}$$

The initial term of $\frac{1}{\sqrt{n!}}$ is a normalization factor. The $\Phi_i(x)$ function is a product of a spatial molecular orbital, $\Phi_i(r)$, and a spin function, $\alpha(\omega)$ or $\beta(\omega)$. According to the *variation principle* the best WF is the one that minimizes the total energy $\langle \Psi | H | \Psi \rangle$ and the optimal MOs can be found by solving the set of n one-electron canonical *HF equations*:

$$f(i)\Phi(x_i) = \varepsilon\Phi(x_i), \quad i = 1 \ldots n, \tag{5.20}$$

where ε is the energy of the MO, Φ occupied by electron i, and $f(i)$ is the so-called *Fock operator*:

$$f(i) = h(i) + v^{HF}(i) \tag{5.21}$$

where, the first term, $h(i)$, is the *one-electron operator*:

$$h(i) = -\frac{1}{2}\nabla_i^2 - \sum_{I=1}^{N} \frac{Z_I}{r_{iI}} \tag{5.22}$$

which consists of the kinetic energy operator for electron i, and the potential energy between this electron and fixed nucleus I with atomic number Z_I. The second term, $v^{HF}(i)$, is the so called *HF potential operator* and can be written as:

$$v^{HF}(i) = \sum_j \{J_j(i) - K_j(i)\} \tag{5.23}$$

Two terms in the sum, $J_j(i)$ and $K_j(i)$, correspond to the *Coulomb interaction* between electron i and all other electrons j, and the *exchange interaction* between electrons with the same spins, respectively. The calculations of the HF potential energy for electron i require prior knowledge of all other occupied MOs. We have to guess the initial form of the MOs, use them in the definition of the Coulomb and exchange operators, and solve the HF equations iteratively in a process known as the *self-consistent field* (SCF) procedure.

The spin parts of the MOs in the HF equation, Eq. (5.20), are integrated out. This transform the HF equations to a set of equation involving only the spatial parts. When one forces a pair of α and β electrons to occupy the same spatial part, i.e. closed-shell species, the procedure is called the *restricted HF* (RHF). On the other hand, if the α and β electrons are associated with different spatial parts, the procedure leads to the *unrestricted HF* (UHF) equations, which are employed for radical species. The UHF has an advantage over the RHF to allow for a proper account of spin polarization, a feature which is very important for the study of, for example, hyperfine interactions of radicals. However, a disadvantage of UHF is that the corresponding WF is not an eigenfunction of the total spin operator for the electrons. This means that UHF WFs may be contaminated with spin states of higher multiplicity. One way of measuring the quality of an UHF calculation is to compare the computed *expectation value of* S^2 with the theoretical one.

A5.1.4   *Semi-Empirical* and *Ab Initio* Methods

There are two methods, *semi-empirical* and *ab initio* methods, for continuing the quantum chemical calculations. In the *semi-empirical* method, many of the integrals are estimated by using spectroscopic data or physical properties. On the other hand, in the *ab initio* method, all the integrals in the secular determinant are attempted to be directly calculated.

The Fock matrix has elements that consist of integrals of the following form:

$$\langle \chi_i \chi_j | \chi_k \chi_l \rangle = \int \chi_i^*(1)\chi_j(1)\frac{1}{r_{12}}\chi_k^*(2)\chi_l(2)d\tau_1 d\tau_2 \tag{5.24}$$

where $\chi_i^*(1)$, $\chi_j(1)$ and $\chi_k^*(2)$, $\chi_l(2)$ are AOs, which in general may be centered on different nuclei. One severe approximation is called CNDO (Complete

Neglect of Differential Overlap), in which all integrals are set to zero unless $\chi_i^*(1)$ and $\chi_j(1)$ are the same orbitals centered on the same nucleus, and likewise for $\chi_k^*(2)$ and $\chi_l(2)$. The more recent *semi-empirical* methods make less severe restrictions about which integrals are to be ignored, but they are all descendants of the CNDO method. These procedures are now readily available in commercial soft-ware packages such as MOPAC [16] and AMPAC [17] which enclose INDO (Intermediate Neglect Differential Overlap), MINDO (Modified Intermediate Neglect Differential Overlap), MNDO (Modified Neglect of Differential Overlap), MNDO/d, AM1 (Austin Model 1), PM3 (Parametric Method 3), SAM1 (Semi-Ab initio Method 1), etc. Here the objective is to use parameters to fit experimental heats of formation, dipole moments, ionization potentials, and geometries.

Commercial packages are also available for *ab-initio* calculations [21b]. The task to evaluate the integrals is generally facilitated by expressing the AOs used in the LCAOs as linear combinations of Gaussian orbitals. A Gaussian type orbital (GTO) is a function of the form $\exp(-\alpha r^2)$. The advantage of GTOs over the fundamentally more correct Slater orbitals (which are proportional to $\exp(-\alpha r)$) is that the product of two Gaussian functions is itself a Gaussian function that lies between the centers of the two contributing functions. In this way the four-centre integrals of $\langle \chi_i \chi_j | \chi_k \chi_l \rangle$ become two-centre integrals. Integrals of this form are much easier and faster to evaluate numerically than the four-centre integrals obtained with Slater orbitals.

A5.1.5 Density Functional Theory

The *density functional theory* (DFT) methods [18–21] focus on finding the optimal *electron density* (ED). The "functional" comes from the fact that the energy, and all other properties, of the system are computed as a function of the ED and the ED is itself a function of position (r), $\rho(r)$, and in mathematics a function of a function is called a "functional". Its advantages include less demanding computational effort, less computer time, and, in some cases (particularly *d*-metal complexes), better agreement with experimental values than is obtained from HF procedures.

Kohn and Sham (KS) [20] developed a scheme for optimizing the density given an approximate form of this functional, which is known as the KS scheme and underlies virtually all DFT methods used today. The *KS scheme* is orbital based and starts by introducing a non-interacting n electron system moving in an external potential, v_s. Such a system is described by a single SD and optimal orbitals are given by the following equations:

$$\left\{ -\frac{1}{2}\nabla^2 + v_s(r) \right\} \Psi_j(r) = \varepsilon_j \Psi_j(r) \tag{5.25}$$

The ED is constructed from the orbitals by:

$$\rho(r) = \sum_{j}^{n} \left| \Psi_j(r) \right|^2 \tag{5.26}$$

The KS theory proposed a separation of the exact unknown energy functional, $E[\rho]$, into four parts in terms of orbitals:

$$E[\rho(r)] = -\frac{1}{2}\sum_{j}^{n}\left\langle\psi_j(r)\left|\nabla^2\right|\psi_j(r)\right\rangle - \sum_{j}^{n}\int\sum_{A}^{N}\frac{Z_A}{|r-R_N|}\left|\psi_j(r)\right|^2 dr$$
$$+ \frac{1}{2}\sum_{j}^{n}\sum_{j'}^{n}\iint\left|\psi_j(r)\right|^2\frac{1}{|r-r'|}\left|\psi'_j(r')\right|^2 drdr' + E_{xc}[\rho(r)] \quad (5.27)$$

where the first term is the functional for the kinetic energy of the system of non-interacting electrons, the second and third terms are functionals for electron-nucleus and electron-electron Coulomb interactions, respectively. The final term, $E_{XC}[\rho]$, is the exchange-correlation (XC) functional and is defined as to contain all of what is unknown including the non-classical effects of both exchange and correlation. By applying the variation principle to Eq. (5.27) the set of orbitals that minimizes the energy has to fulfill the following equations:

$$\left[-\frac{1}{2}\nabla^2 - \sum_{A}^{N}\frac{Z_A}{|r-R_A|} + \int\frac{\rho(r')}{|r-r'|}dr' + v_{XC}(r)\right]\psi_j(r) = \varepsilon_j\psi_j(r) \quad (5.28)$$

where $v_{XC}(r)$ is the so called XC potential defined by the functional derivative of the $E_{XC}[\rho(r)]$:

$$v_{XC}(r) = \partial E_{XC}[\rho(r)]/\partial\rho(r) \quad (5.29)$$

The last three terms in the Hamiltonian in Eq. (5.28) define an effective one-body potential, $v_{eff}(r)$, which transforms the density of the non-interacting system into the real density. Then, by choosing $v_s(r) = v_{eff}(r)$ in Eq. (5.25) the effective potential is found. As was the case with HF, the one-electron Hamiltonian in the KS equations is solved iteratively and self-consistently. If the exact expression for $E_{XC}[\rho(r)]$ was known, the KS equations would provide the exact non-relativistic ground state solution within the space spanned by a given basis set, including all electron correlation effects. The latter are missing in HF. This is an important difference between HF and KS.

For details of the WFT and DFT the reader is referred to the books by Szabo and Ostlund [18], and Parr and Yang [19]. Furthermore, the book by Koch and Holthausen [20] is recommended as an introductory reading for chemists starting to carry out quantum chemical calculations. Ph.D theses by T. Fängström [108] and D. Norberg [109], and Atkins's physical chemistry book [86b] served as the main sources to the present appendix on Quantum Chemical Methods: Sections A5.1.1 and A5.1.5 [109], Section A5.1.3 [108, 109], and Sections A5.1.2 and A5.1.4 [86b].

A5.2 A Brief Summary on Molecular Symmetry

This appendix briefly summarizes "molecular symmetry" (symmetry elements, point groups and character tables) of a molecule so as to facilitate readers' understanding of the molecular structures and molecular orbitals presented in this chapter. We use some specific examples such as NH_3 with C_{3v} symmetry, CH_3 radical with D_3 symmetry (see Fig. 5.18) and c-$C_4F_8^-$ radical with D_{4h} symmetry (Fig. 5.3) to introduce and illustrate some important aspects of the molecular symmetry.

A5.2.1 Symmetry Elements and Symmetry Operations of Molecules

The methyl radical, CH_3, with a planar trigonal structure, looks the same if it is rotated by 120, 240, or 360° about an axis (C_3 axis in Fig. 5.18) perpendicular to the plane containing three equivalent carbon atoms. An operation that leaves a molecule (object) looking the same (or sending into itself or a position indistinguishable from the original) is a symmetry operation. Symmetry operations include seven symmetry elements in Table 5.6 which are commonly possessed by molecular systems.

A5.2.2 Point Groups

A *point group* consists of all possible *symmetry elements* possessed by a given molecule. There is always one point, at which an atom is not necessarily present, in the molecule that remains unchanged by the operation. In contrast to this, when we consider crystals, we meet a space group containing operations which displace the molecule to another position in space. Some points groups of chemical interest are listed in Table 5.7 with examples of molecules.

Table 5.6 Symmetry elements and symmetry operations

Symbol	Symmetry element	Symmetry operation
E	Identity	Doing nothing or leave a molecule alone
C_n	Proper symmetry axis	Rotate a molecule by $360/n$ degree around an n-fold axis of symmetry
σ_h	Horizontal plane	Reflect a molecule through a plane perpendicular to the major axis (principal axis)
σ_v	Vertical plane	Reflect a molecule through a plane containing the major axis
σ_d	Dihedral plane	Reflect a molecule through a plane bisecting two C_2 axes
S_n	Improper axis (n-fold axis of improper rotation)	Rotate a molecule by $360/n$ degrees around an improper axis and then reflect the molecule through a plane perpendicular to the improper axis
i	Inversion center	Inversion of a molecule through a center of symmetry

Table 5.7 Some point groups with examples of molecules

Point group	Symmetry element (h: order)	Example
C_1	E ($h = 1$)	CClFBrI
C_s	E, σ_h ($h = 2$)	Quinoline, $CH \equiv CH$
C_i ($= S_2$)	E, i ($h = 2$)	*Meso*-tartaric acid
C_2	E, C_2 ($h = 2$)	H_2O_2
C_{2v}	E, C_2, $\sigma_v \sigma'_v$ ($h = 4$)	H_2O, CH_2D, CH_4^+
C_{3v}	E, $2C_3$, $3\sigma_v$ ($h = 6$)	NH_3, CF_3Cl
C_{2h}	E, σ_h, i ($h = 3$)	$CF_2 = CF_2^-$
D_3	E, $2C_3$, $3C_2$ ($h = 6$)	CH_3, CH_3CH_3
D_{2h}	E, $3C_2$, i, σ_h, $2\sigma_v$ ($h = 8$)	$CF_2 = CF_2$, *trans*-$CH \equiv CH^-$
D_{3h}	E, σ_h, $2C_3$, $2S_3$, $3C'_2$, $3\sigma_v$ ($h = 12$)	c-C_3F_6, $C(CH_2)_3$ (Trimethylenemethane)
D_{4h}	E, $2C_4$, C_2, $2C'_2$, $2C''_2$, i, $2S_4$, σ_h, $2\sigma_v$, $2\sigma_d$ ($h = 16$)	c-$C_4F_8^-$
D_{2d}	E, C_2, $2C'_2$, $2\sigma_d$, $2S_4$ ($h = 8$)	$H_2C = C = CH_2$ (Allene), c-C_4F_8
T_d	E, $8C_3$, $3C_2$, $6\sigma_d$, $6S_4$ ($h = 24$)	CH_4, $C(CH)_4$
O_h	E, $8C_3$, $6C_2$, $6C'_2$, $3C''_2$, i, $6S_4$, $8S_6$, $3\sigma_h$, $6\sigma_d$ ($h = 48$)	SF_6

A molecule without any symmetry except the identity operation E, such as CHFClBr, belongs to the group C_1. The NH_3 molecule has one E element, two C_3 elements ($2C_3$) and three σ_v elements ($3\sigma_v$) which reflect the molecule through three vertical planes containing the C_3 axis and it belong to the group C_{3v}. A molecule possessing an n-fold principal (C_n) axis and n twofold axes perpendicular to the C_n axis belongs to the group D_n. For example, the CH_3 radical belongs to the group D_3 since it has the E, C_3, and $3C_2$ symmetry elements. If a molecule possesses a horizontal mirror plane (σ_h) in addition to the symmetry elements of the group D_n it belongs to the group D_{nh}. For example, c-$C_4F_8^-$ radical has the E, C_4, $4C_2$, and σ_h symmetry elements and it belongs to the D_{4h} group.

A5.2.3   Character Tables

A character table is a two-dimensional table whose columns correspond to *symmetry operations* of the group. Character tables for the point groups C_{2v}, C_{3v}, D_3 and D_{4h} are reproduced in Table 5.8. For example, the columns in the C_{3v} character table are headed by the E, C_3, and C_2 operations (Table 5.8b). The numbers multiplying each operation are those of members of each *class*. That is, two threefold rotations ($2C_3$; clockwise and counter-clockwise rotations by 360°/3) belong to the same class. The three reflections ($3\sigma_v$; one through each of the three vertical mirror planes) also belong to the same class. The two reflections (σ_v and σ'_v) of the group C_{2v}, however, fall into different classes, see Table 5.8a. Although one can not be transferred into the other by any symmetry operation of the group they belong to the same group by a rule of group theory: the product of any two columns of a character table must be

Table 5.8 Character tables for C_{2v}, D_3 and D_{4h} point groups are given as representative one. More character tables are to be found in the textbooks cited as references [86c, 110, 111]

(a) C_{2v} point group ($h = 4$)

	E	C_2	$\sigma_v(xz)$	$\sigma_v'(yz)$	III	IV
A_1	1	1	1	1	z	x^2, y^2, z^2
A_2	1	1	-1	-1	R_z	xy
B_1	1	-1	1	-1	x, R_y	xz
B_2	1	-1	-1	1	y, R_x	yz

(b) C_{3v} point group ($h = 6$)

	E	$2C_3$	$3\sigma_v$	III	IV
A_1	1	1	1	z	$z^2, x^2 + y^2$
A_2	1	1	-1	R_z	
E	2	-1	0	$(x, y)\ (R_x, R_y)$	$(x^2 - y^2, xy)\ (xz, yz)$

(c) D_3 point group ($h = 6$)

	E	$2C_3$	$2C_2$	III	IV
A_1	1	1	1		$z^2, x^2 + y^2$
A_2	1	1	-1	z, R_z	
E	2	-1	0	$(x, y)\ (R_z, R_y)$	$(x^2 - y^2, xy)\ (xz, yz)$

(d) D_{4h} point group ($h = 16$)

	E	$2C_4$	C_2	$2C_2'$	$2C_2''$	i	$2S_4$	σ_h	$2\sigma_v$	$2\sigma_d$	III	IV
A_{1g}	1	1	1	1	1	1	1	1	1	1		$x^2 + y^2, z^2$
A_{2g}	1	1	1	-1	-1	1	1	1	-1	-1	R_z	
B_{1g}	1	-1	1	1	-1	1	-1	1	1	-1		$x^2 - y^2$
B_{2g}	1	-1	1	-1	1	1	-1	1	-1	1		xy
E_g	2	0	-2	0	0	2	0	-2	0	0	(R_x, R_y)	(xz, yz)
A_{1u}	1	1	1	1	1	-1	-1	-1	-1	-1		
A_{2u}	1	1	1	-1	-1	-1	-1	-1	1	1	z	
B_{1u}	1	-1	1	1	-1	-1	1	-1	-1	1		
B_{2u}	1	-1	1	-1	1	-1	1	-1	1	-1		
E_u	2	0	-2	0	0	-2	0	2	0	0	(x, y)	

(The group D_{4h} is a direct product of the groups D_4 and C_i)

a column in that table. The number of symmetry operations in a group is called its *order* ("h" in Table 5.7).

Rows in character tables correspond to *symmetry properties of the orbitals*, more formally, *irreducible representations* of the group. The irreducible representations are labelled with large Roman letters such as A_1 and E, but the orbitals to which they apply are labelled with small italic equivalents: for example, an orbital of A_1 symmtry is called an a_1 orbital.

The entries consist of *characters*, the trace of the matrices representing group elements of the column's class in the given row's group representation. The *character tables of point groups* can be used to classify molecular orbitals that belong to the various atoms in a molecule by referring to the different symmetry types possible in the point group. For example, the characters in the rows labelled *A* and *B* and in the columns headed by symmetry operations other than the identity *E* indicate the behavior of an orbital under the corresponding operations: a "+1" indicates an orbital unchanged, and a "−1" indicates that it changes sign. Thus, one can identify the symmetry label of the orbital by comparing the result of changes that occurs to an orbital under each operation with the entries, "+1" or "−1", in a row of the character table for the point group concerned. The character of the *identity operation E* tells us the degeneracy of the orbitals. For example, the character in the row labelled *E* or *T* refers to the sets of doubly or triply degenerated orbitals, respectively.

In column III in character tables we see six symbols: x, y, z, R_x, R_y, R_z. The first three represent the coordinates x, y and z, while the R's stand for rotations about the axes specified in the subscripts. In column IV the squares and binary products of coordinates are classified according to their transformation properties. For example, the pair functions xz and yz in C_{2v} must have the same transformation properties as the pair x, y, since z goes into itself under all symmetry operations in the group.

The textbooks by Atkins and Paula [86c], McQuarrie and J.D. Simon [110], and Cotton [111] served as the main sources for the present appendix on molecular symmetry.

References

1. I.R. Dunkin: '*Matrix-Isolation Techniques – A Practical Approach*', Oxford University Press, Oxford (1998).
2. A.J. Barnes (ed.): '*Matrix Isolation Spectroscopy*', NATO advanced study institutes series. C (Mathematical and physical sciences) **76**, Reidel Publishing Company, Dordrecht (1981).
3. A. Hasegawa, M. Shiotani, F. Williams: Faraday Discuss. Chem. Soc. **63**, 157 (1977).
4. M. Shiotani: Mag. Res. Rev. **12**, 33 (1987).
5. A. Lund, M. Lindgren, S. Lunell, J. Maruani: In '*Molecules in Physics, Chemistry and Biology*' ed. by J. Maruani, Vol. **111**, Academic Publishers, Boston, MA (1988), p. 259.
6. (a) M. Shiotani: In '*CRC Handbook of Radiation Chemistry*' ed. by Y. Tabata, CRC Press, Boca Raton, FL (1991), Chapter III.B.7. (b) M. Shiotani, H. Yoshida: *ibid.*, Chapter VIII.C. (c) M. Shiotani: *ibid.*, Chapter IX.B.
7. A. Lund, M. Shiotani (eds.): '*Radical Ionic Systems – Properties in Condensed Phases*', Kluwer Academic Publisher, Dordrecht (1991).
8. M. Lindgren, M. Shiotani: In '*Radical Ionic Systems: Properties in Condensed Phases*' ed. by A. Lund and M. Shiotani, Kluwer Academic Publisher, Dordrecht (1991), Chapter I.5.
9. M. Shiotani, A. Lund: In '*Radical Ionic Systems: Properties in Condensed Phases*' ed. by A. Lund and M. Shiotani, Kluwer Academic Publisher, Dordrecht (1991), Chapter I.6.
10. A. Hasegawa: In '*Radical Ionic Systems: Properties in Condensed Phases*' ed. by A. Lund and M. Shiotani, Kluwer Academic Publisher, Dordrecht (1991), Chapter II.1.

11. M. Shiotani, M. Lindgren: In *'Radicals on Surfaces'* ed. by A. Lund, C. Rhode, Molecular Engineering **4** (1–3), Kluwer Academic Publisher, Dordrecht (1994), pp. 179–199.

12. A. Lund, M. Shiotani (eds.): *'EPR of Free Radicals in Solids: Trends in Method and Applications'*, Kluwer Academic Publisher, Dordrecht (2003).

13. M. Shiotani, K. Komaguchi: In *'EPR of Free Radicals in Solids: Trends in Method and Applications'* ed. by A. Lund, M. Shiotani, Kluwer Academic Publisher, Dordrecht (2003), Chapter 4.

14. F. Ban, J.W. Gauld, S.D. Wetmore, R.J. Boyd: In *'EPR of Free Radicals in Solids, Trend in Methods and Applications'* ed. by A. Lund, M. Shiotani, Kluwer Academic Publishers, Dordrecht (2003), Chapter 6.

15. M. Kaupp: In *'EPR of Free Radicals in Solids: Trend in Methods and Applications'* ed. by A. Lund, M. Shiotani, Kluwer Academic Publishers, Dordrecht (2003), Chapter 7.

16. MOPAC (http://openmopac.net/; http://en.wikipedia.org/wiki/MOPAC).

17. AMPAC (http://en.wikipedia.org/wiki/AMPAC).

18. A. Szabo, N.S. Ostlund: *'Modern Quantum Chemistry'*, 1st Edition, McGraw-Hill Publishing Company, New York, NY (1989).

19. R.G. Parr, W. Yang: *'Density-Functional Theory of Atoms and Molecules'*, Oxford University Press, New York, NY (1989).

20. W. Koch, M.C. Holthausen: *'A Chemist's Guide to Density Functional Theory'*, 2nd Ed., Wiley-VCH, Weinheim (2001).

21. (a) J.A. Pople et al.: *'Gaussian 03'* (Revision B.05), Gaussian Inc., Wallingford, CT (2004). (b) *Ab-initio* calculations (http://www.gaussian.com/g_prod/1.htm).

22. B. Engels, L. Eriksson, S. Lunell: Adv. Quantum Chem. **27**, 297 (1996).

23. T. Fängström, S. Lunell, B. Engels, L. Eriksson, M. Shiotani, K. Komaguchi: J. Chem. Phys. **107**, 297 (1997).

24. P. Wang, M. Shiotani, S. Lunell: Chem. Phys. Lett. **292**, 110 (1998).

25. L.A. Eriksson: In *'Encyclopedia of Computational Chemistry'*, Wiley, New York, NY (1998).

26. M. Shiotani, N. Isamoto, M. Hayashi, T. Fängström, S. Lunell: J. Am. Chem. Soc. **122**, 12281 (2000).

27. C. Adamo, M. Cossi, N. Rega, V. Barone: In *'Theoretical Biochemistry: Processes and Properties of Biological Systems'* ed. by L.A. Eriksson, Elsevier, New York, NY (2001), p. 467.

28. Z. Sojka, P. Pietrzyk: Spectorochim. Acta A **63**, 830 (2006).

29. (a) K. Komaguchi, D. Norberg, N. Nakazawa, M. Shiotani, P. Persson, S. Lunell: Chem. Phys. Lett. **410**, 1 (2005). (b) D. Norberg, M. Shiotani, S. Lunell: J. Phys. Chem. A **112**, 1330 (2008).

30. E.T. Kaiser, L. Kevan (eds.): *'Radical Ions'*, Interscience Publishers, New York, NY (1968).

31. M. Shiotani, F. Williams: J. Am. Chem. Soc. **98**, 4006 (1976).

32. M. Shiotani, A. Lund, S. Lunell, F. Williams: J. Phys. Chem. A **111**, 321 (2007).

33. M. Shiotani, P. Person, S. Lunell, A. Lund, F. Williams: J. Phys. Chem. A **110**, 6307 (2006).

34. Y. Itagaki, M. Shiotani: J. Phys. Chem. A **103**, 5189 (1999).

35. (a) A.M. ElSohly, G.S. Tschumper, R.A. Crocombe, J.T. Wang, F. Williams: J. Am. Chem. Soc. **127**, 10573 (2005). (b) A. Paul, C.S. Wannere, V. Kasalova, P. von R. Schleyer, H.F. Schaefer III: J. Am. Chem. Soc. **127**, 15457 (2005).

36. (a) A.W. Fessenden: J. Chem. Phys. **37**, 747 (1962). (b) R.W. Fessenden, R.H. Schuler: *ibid.* **43**, 2704 (1965). (c) R.W. Fessenden, R.H. Schuler: *ibid.* **39**, 2147 (1963).

37. N.M. Atherton: (a) *'Electron Spin Resonance: Theory and Applications'*, Wiley, New York, NY (1973), p. 108; (b) *ibid.*, p. 150.

38. R.I. McNeil, M. Shiotani, F. Williams, M.B. Yim: Chem. Phys. Lett. **51**, 433 (1977).

39. R.I. McNeil, M. Shiotani, F. Williams, M.B. Yim: Chem. Phys. Lett. **51**, 438 (1977).

40. M.N. Paddon-Row, N.G. Rondan, K.N. Houk, K.D. Jordan: J. Am. Chem. Soc. **104**, 1143 (1982).

41. (a) L.N. Shchegoleva, I.L. Bilkis, P.V. Schastnev: Zh. Strukt. Khim. (Russian) **25**, 19 (1984);
    (b) P.V. Schastnev, L.N. Shchegoleva: In *'Molecular Distortions in Ions and Excited States'*,
    CRC Press, Boca Raton, FL (1995), Chapter 3.
42. C.K. Ingold, G.W. King: J. Chem. Soc. **1953**, 2702 (1953); *ibid.* **1953**, 2704 (1953).
43. K.K. Innes: J. Chem. Phys. **22**, 863 (1954).
44. K. Matsuura, H. Muto: J. Chem. Phys. **94**, 4078 (1991); J. Phys. Chem. **97**, 8842 (1993).
45. (a) P.H. Kasai, D. McLeod Jr.: J. Am. Chem. Soc. **97**, 6602 (1975). (b) P.H. Kasai,
    D. McLeod Jr., T. Watanabe: *ibid.* **102**, 179 (1980). (c) P.H. Kasai: J. Phys. Chem. **86**, 4092
    (1982).
46. P.H. Kasai: J. Am. Chem. Soc. **104**, 1165 (1982); *ibid.* **105**, 6704 (1983); *ibid.* **114**, 3299
    (1992).
47. L. Manceron, L. Andrews: J. Am. Chem. Soc. **107**, 563 (1985); J. Phys. Chem. **89**, 4094
    (1985).
48. E.A. Piocos, D.W. Werst, A.D. Trifunac, L.A. Eriksson: J. Phys. Chem. **100**, 8408 (1996).
49. T. Shida, Y. Nosaka, T. Kato: J. Phys. Chem. **82**, 695 (1978).
50. L.B. Knight Jr., J. Steadman: J. Chem. Phys. **77**, 1750 (1982).
51. L.B. Knight Jr.: In *'Radical Ionic Systems: Properties in Condensed Phases'* ed. by A. Lund,
    M. Shiotani, Kluwer Academic Publisher, Dordrecht (1991), Chapter I.3.
52. L.B. Knight Jr., J. Steadman: J. Chem. Phys. **78**, 5940 (1983).
53. L.B. Knight Jr., J. Steadman, D. Feller, E.R. Davidson: J. Am. Chem. Soc. **106**, 3700 (1984).
54. L.B. Knight Jr., G.M. King, J.T. Petty, M. Matsushita, T. Momose, T. Shida: J. Chem. Phys.
    103, 3377 (1995).
55. T. Shida, E. Haselbach, T. Bally: Acc. Chem. Res. **17**, 180 (1984).
56. M.C.R. Symons: Chem. Soc. Rev. **13**, 393 (1984).
57. S. Katumata, K. Kimura: Bull. Chem. Soc. Jpn. **46**, 1342 (1973).
58. A.W. Pott, H.J. Lempka, D.G. Streets, W.C. Price: Trans. R. Soc. Lond. A **268**,
    59 (1970).
59. *'CRC Handbook of Chemistry and Physics,'* 83rd Edition, ed. by D.R. Lide, CRC Press,
    Baca Raton, FL (2002–2003), pp. 10–178.
60. K. Komaguchi, T. Marutani, M. Shiotani, A. Hasegawa: Phys. Chem. Chem. Phys. **3**, 3536
    (2001).
61. (a) L. Bonazzola, J.P. Michaut, J. Roncin: J. Phys. Chem. **95**, 3132 (1991); (b) L. Bonazzola,
    J.P. Michaut, J. Roncin: New J. Chem. **16**, 489 (1992).
62. M. Iwasaki, K. Toriyama, K. Nunome: J. Am. Chem. Soc. **103**, 3591 (1981).
63. K. Komaguchi, M. Shiotani, A. Lund: Chem. Phys. Lett. **265**, 217 (1997).
64. O. Claesson, A. Lund, T. Gillbro, T. Ichikawa, O. Edlund, H. Yoshida: J. Chem. Phys. **72**,
    463 (1980).
65. W. Meyer: J. Chem. Phys. **58**, 1017 (1973).
66. M.N. Paddon-Raw, D.J. Fox, J.A. Pople, K.N. Houk, D.W. Pratt: J. Am. Chem. Soc. **107**,
    7696 (1985).
67. R.F. Frey, E.R. Davidson: J. Chem. Phys. **88**, 1775 (1988).
68. L.A. Eriksson, S. Lunell, R.J. Boyd: J. Am. Chem. Soc. **115**, 6896 (1993).
69. W.T. Borden, H. Iwamura, J.A. Berson: Acc. Chem. Res. **27**, 109 (1994).
70. P. Dowd: Acc. Chem. Res. **5**, 242 (1972).
71. M. Shiotani, M. Lindgren, T. Ichikawa: J. Am. Chem. Soc. **112**, 967 (1990).
72. K. Komaguchi, M. Shiotani: J. Phys. Chem. **101**, 6983 (1997).
73. A. Hasagawa, M. Shiotani, Y. Hama: J. Phys. Chem. **98**, 1834 (1994).
74. T. Bally, A. Maltsev, F. Gerson, D. Frank, A. de Meijere: J. Am. Chem. Soc. **127**, 1983
    (2005).
75. T. Yamada, K. Komaguchi, M. Shiotani, N.P. Benetis, A.R. Sφrnes: J. Phys. Chem. A **103**,
    4823 (1999).
76. K. Komaguchi, K. Nomura, M. Shiotani: J. Phys. Chem. A **111**, 726 (2007).
77. K. Komaguchi, T. Kumada, Y. Aratono, T. Miyazaki: Chem. Phys. Lett. **268**, 493 (1997).

78. K. Komaguchi, T. Kumada, T. Takayanagi, Y. Aratono, M. Shiotani, T. Miyazaki: Chem. Phys. Lett. **300**, 257 (1999).
79. J. Kumagai, H. Inagaki, S. Kariya, T. Ushida, Y. Shimizu, T. Kumada: J. Chem. Phys. **127**, 024505 (2007).
80. M. Fujimoto, H.D. Gesser, B. Garbutt, A. Cohen: Science **154**, 381(1966).
81. M. Shiotani, F. Yuasa, J. Sohma: J. Phys. Chem. **79**, 2669 (1975).
82. A.R. Sϕrnes, N.P. Benetis, R. Erickson, A.S. Mahgoub, L. Eberson, A. Lund: J. Phys. Chem. A **101**, 898 (1997).
83. Pauli principle (http://en.wikipedia.org/wiki/Pauli_exclusion_principle).
84. Fermion (http://en.wikipedia.org/wiki/Fermion).
85. (a) N.P. Benetis: In '*EPR of Free Radicals in Solids*' ed. by A. Lund, M. Shiotani, Kluwer Academic Publisher, Dordrecht (2003), Chapter 3. (b) N.P. Benetis, Y. Dmitriev: J. Phys. Condens. Matter. **21**, 103201 (2009).
86. P. Atkins, J. de Paula: (a) '*Atkins' Physical Chemistry*', 7th Edition, Oxford University Press, New York, NY (2002), p. 501; (b) *ibid.*, Chapter 14; (c) *ibid.*, Chapter 15.
87. J. H. Freed: J. Chem. Phys. **43**, 1710 (1965).
88. Boson (http://en.wikipedia.org/wiki/Boson).
89. Y. Kurita: J. Chem. Phys. **41**, 3926 (1964).
90. J. Owen, E.A. Harris: In '*Electron Paramagnetic Resonance*' ed. by S. Geschwind, Plenum Press, New York, NY (1972), p. 427.
91. (a) W. Gordy, R. Morehouse: Phys. Rev. **151**, 207 (1966). (b) W. Gordy: '*Theory and Applications of Electron Spin Resonance*', Wiley, New York, NY (1980).
92. S.N. Foner, E.L. Cochran, V. A. Bowers, C.K. Jen: J. Chem. Phys. **32**, 963 (1960).
93. (a) L.B. Knight Jr., W.E. Rice, L. Moore, E.R. Davidson: J. Chem. Phys. **103**, 5275 (1995); (b) L.B. Knight Jr., W.E. Rice, L. Moore, E.R. Davidson, R.S. Dailey: J. Chem. Phys. **109**, 1409 (1997).
94. T. Takayanagi, S. Sato: J. Chem. Phys. **92**, 2862 (1990).
95. T. Miyazaki (ed.): '*Atom Tunneling Phenomena in Physics, Chemistry and Biology*', Springer, Berlin (2004).
96. I.F. Silvera: Rev. Mod. Phys. **52**, 393 (1980).
97. J. van Kranendonk: '*Solid Parahydrogen*', Plenum Press, New York, NY (1983).
98. T. Miyazaki, T. Hiraku, K. Fueki, Y. Tsuchihashi: J. Phys. Chem. **95**, 26 (1991).
99. T. Miyazaki, K. Yamamoto, J. Arai: Chem. Phys. Lett. **219**, 405 (1994).
100. J.J. Thomson: Philos. Mag. **24**, 209 (1912).
101. Y. Kurosaki, T. Takayanagi: J. Chem. Phys. **109**, 4327(1998).
102. T. Kumada, H. Tachikawa, T. Takayanagi: Phys. Chem. Chem. Phys. **7**, 776 (2005).
103. W.D. Watson: Rev. Mod. Phys. **48**, 513 (1976).
104. E. Schrödinger: Phys. Rev. **28**, 1049 (1926).
105. M. Born, J.R. Oppenheimer: Ann. Phys. **84**, 457 (1027).
106. E. Hückel: Z. Phys. **70**, 204 (1931).
107. R. Hoffmann: J. Chem. Phys. **39**, 1397 (1963).
108. T. Fängström: '*Quantum Chemical Studies of Radicals and Radical Reactions*', PhD thesis, Uppsala University (1997).
109. D. Norberg: '*Quantum Chemical Studies of Radical Cation Rearrangement, Radical Carbonylation, and Homolytic Substitution Reactions*', PhD thesis, Uppsala University (2007).
110. D.A. McQuarrie, J.D. Simon: '*Physical Chemistry – A Molecular Approaches*', University Science Books, Sausalito, CA (1997), Chapter 12.
111. F.A. Cotton: '*Chemical Applications of Group Theory*', 3rd Edition, Interscience, New York, NY (1990).

Chapter 6
Applications to Catalysis and Environmental Science

Abstract Electronic and geometrical structures of NO-Na$^+$ and Cu(I)-NO complexes formed in zeolites are discussed based on the g and the ^{14}N and ^{23}Na hf values evaluated by multi-frequency ESR, pulsed ENDOR and HYSCORE methods. The structure of (NO)$_2$ bi-radical formed in zeolites is discussed based on X- and Q-band ESR spectra. Microenvironment effects on the molecular dynamics and the thermal stability of triethyl- and tripropyl-amine radical cations as spin probes are presented referring to the CW-X-band ESR results and theoretical DFT calculations. X- and Q-band ESR studies on nitrogen-doped TiO$_2$ semiconductor reveal that the diamagnetic N$^-$ ion in the system absorbs visible light so as to excite an electron of N$^-$ to the conduction band. The photo-catalytic reactions of TiO$_2$ are modified by introducing O$_2$ molecules which scavenge a fraction of photoexcited electrons to generate O$_2^-$. ESR spectral characteristics of adsorbed O$_2^-$, g-tensor and hf structure of labeled ^{17}O ($I = 7/2$), are presented.

6.1 Introduction

To understand catalytic reactions it is indispensable to characterize catalytic materials and to clarify static and dynamic structures of reaction intermediates as well as active sites of reactions. Paramagnetic species are in general involved in many catalytic reactions as reaction intermediates and/or active sites, especially in heterogeneous catalytic reactions. Thus, the ESR method has played an important role to get valuable information on catalytic and/or surface reactions with high selectivity and high sensitivity, which has not been achieved by any other methods.

ESR applications to catalysis and solid surfaces have started in the beginning of the 1960s. The studies on Tigullar-Natta catalysis by Angelescu [1], and cromina-alumina catalysis by O'Reilly [2] are pioneer works in the field. Since then a large number of studies have been reported so far, including some important review papers and books cited as references [3–16]. They cover a broad range of research subjects: (a) ESR characterization of oxide supported transition metal ions/complexes relevant to catalysis and/or environmental pollutant control, (b) ESR identification and quantitative measurements of reactive organic and inorganic radical species formed

A. Lund et al., *Principles and Applications of ESR Spectroscopy*,
DOI 10.1007/978-1-4020-5344-3_6, © Springer Science+Business Media B.V. 2011

on catalytic surfaces, (c) catalytic and photo-catalytic reaction dynamics of radical species, (d) radicals on surfaces formed by ionizing radiation, (e) nature of surface centers and reactivity with adsorbed molecules, (f) chemical bonding or electronic structure of paramagnetic reaction intermediates like radical species, (g) diffusion and molecular dynamics of radicals on porous heterogeneous systems, etc.

With recent advancement in the measurement techniques and the data analysis methods ESR spectroscopy is an increasingly important tool in the studies on catalysis and solid surfaces. This chapter focuses on the following five specific subjects relevant to the ESR applications in catalysis and environmental science; (a) nitric oxide (NO) adsorbed on zeolites, (b) Cu(I)-NO complexes formed in zeolites, (c) structure and dynamics of organic radicals in zeolites, (d) titanium dioxide (TiO$_2$) semiconductor photo-catalysis, and (e) the superoxide (O$_2^-$) ion radical.

6.2 Surface Probing: Nitric Oxide Interactions with Metal Ions in Zeolites

Nitric oxide (NO) is an odd-electron molecule possessing one unpaired electron with electronic configuration, $[(K^2K^2)-(2s\sigma)^2(2s\sigma^*)^2(2p\pi)^4(2p\sigma)^2(2p\pi^*)^1]$. The reactions of NO molecule with metal ions are one of the major topics in catalysis and environmental research as well as in biochemistry and coordination chemistry [13, 17]. In catalysis and environmental studies a large number of researchers have been interested in the decomposition of NO into N$_2$ and O$_2$ over transition metal ion exchanged zeolites [18–22]. The NO molecule has been used also as a paramagnetic (spin) probe to characterize the catalytic activity, particularly the structure, concentration and acid strength of Lewis acid surface sites and metal ions of nanoporous materials including zeolites [3, 4, 8, 23–28]. Here ESR is the most appropriate spectroscopic method for the detection and identification of paramagnetic NO and can potentially provide valuable experimental information about the structure and dynamics of NO molecules interacting with metal ions and of the reactions involved.

Lunsford [3b] and Hoffman and Nelson [23] first reported the ESR spectra for adsorbed NO molecules. Then, Kasai [4b] revealed that ESR spectra of NO probe molecules are very sensitive to the interaction with metal ions and Lewis acid sites in zeolites. The earlier ESR studies of the NO/zeolite system have been summarized in several review papers [3a, 4a, 8]. A number of ESR studies have been also carried out for NO adsorbed on metal oxides such as MgO and ZnO as reviewed by Che and Giamello [5]. Modern ESR techniques such as pulsed ESR [25–27], ENDOR (Electron Nuclear Double Resonance) [26], and multi-frequency (X-, Q-, and W-band) ESR [28] are especially useful for an unambiguous identification of the ESR magnetic parameters (**g**, hyperfine *A*, and quadrupole tensors, etc.) and, consequently, for a detailed characterization of structural changes and motional dynamics involved. Some recent advancements in ESR studies on NO adsorbed on zeolites are presented in this section.

The decomposition of nitric oxide, a process in which NO is converted to harmless nitrogen and oxygen, deserves considerable practical attention, as the oxides

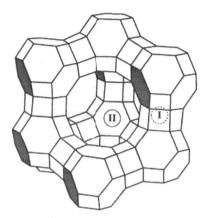

Fig. 6.1 A perspective view of part of Linde type Y (LTY) zeolite with the faujasite structure (Si/Al ratio of 2.73). Aluminum and silicium atoms lie on the corners, oxygen atoms near the midpoints of edges. *Site I* (S1) in hexagonal prisms and *site II* (S2) in supercage are indicated. The figure is adapted from [30] with permission from Elsevier. See Fig. 6.11 for other zeolites such as SAPO-37, SAPO-42 and Al-offretite

of nitrogen are regarded as the major cause of air pollution [18–22]. A number of transition metal ion exchanged zeolites have been reported to be active for NO decomposition. Among them, the copper exchanged high siliceous zeolites such as Cu-ZSM-5 have been observed to be highly active. The decomposition of NO has been reported to occur via the formation of a Cu(I)-(NO)$_2$ dimer, whose precursor is a Cu(I)-NO monomer [29]. This subject is presented separately in Section 6.3.

6.2.1 NO-Na$^+$ Complex Formed in Zeolites

Zeolites (Greek, ζέω (*zeō*), meaning "boil" and λίθος (*lithos*), meaning "stone") are microporous aluminosilicate with the general formula $\{[M^{n+}]_{x/n} \cdot [H_2O]_m\}\{[AlO_2]_x[SiO_2]_y\}^{x-}$, where M^{n+} stands for cations, such as H$^+$, Na$^+$, K$^+$, Ca^{2+}, Mg^{2+}, Cu^{2+}, and other ions (Fig. 6.1). The M^{n+} cations and H$_2$O molecules bind inside the cavities, or pores, of the Al-O-Si framework. The M^{n+} cations can be exchanged for other ions in a contact solution. Small molecules such as NO$_x$(x = 1, 2), CO$_2$, NH$_3$, and hydrocarbons including aromatic and amine compounds can be adsorbed to the internal surfaces and this partially accounts for the utility of zeolites as catalysts, we refer to Section 6.4 for some studies of organic radicals in zeolites.

6.2.1.1 *g*-Tensor Anisotropy of Adsorbed NO

The NO molecule exhibits the degenerate molecular orbitals, 2pπ*(x,y), in the ground electronic state, see Fig. 6.2. The orbital degeneracy can be lifted by the

Fig. 6.2 (a) Energy levels of NO interacting with M$^+$ located on a surface. (b) The x-y-z coordinate system used for NO adsorbed at metal ion (M$^+$) in zeolite. The term Δ stands for the energy splitting between 2pπ^*(x) and 2pπ^*(y) orbitals, and the term E for that between 2pσ^* and 2pπ^*(y) orbitals. The unpaired electron resides in the 2pπ^*(y) orbital in the absence of any spin-orbit coupling

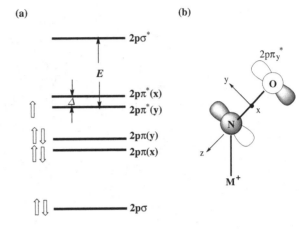

electrostatic field due to a metal ion where NO is adsorbed on inorganic solid matrices. The unpaired electron occupies the 2pπ^*(y) molecular orbital in the absence of any spin-orbit coupling, see Fig. 6.2. As a result of spin-orbit coupling, however, the perturbed wave function for the electron is a mixture of the 2pπ^*(y) and the excited 2pπ^*(x) orbital. The g-tensor for the NO adsorption complexes has been discussed by Lunsford [3], Kasai [4], and recently by Rudolf et al. [28] who proposed the following analytical expressions for the principal values g_{xx}, g_{yy}, and g_{zz} of the g-tensor:

$$g_{xx} = g_e \frac{\Delta}{\sqrt{\lambda^2 + \Delta^2}} - \frac{\lambda}{E}\left(\frac{\Delta - \lambda}{\sqrt{\lambda^2 + \Delta^2}} - 1\right) \tag{6.1}$$

$$g_{yy} = g_e \frac{\Delta}{\sqrt{\lambda^2 + \Delta^2}} - \frac{\lambda}{E}\left(\frac{\Delta - \lambda}{\sqrt{\lambda^2 + \Delta^2}} + 1\right) \tag{6.2}$$

$$g_{zz} = g_e - \frac{2\,l\lambda}{\sqrt{\lambda^2 + \Delta^2}} \tag{6.3}$$

Here "g_e" is the g-value for the "free" electron, "λ" the spin-orbit coupling constant of NO (123.16 cm^{-1}), "E" the energy splitting between 2pσ^* and 2pπ^*(y) orbitals, and "Δ" that between 2pπ^*(x) and 2pπ^*(y) orbitals, see Fig. 6.2. The quantity "l" stands for a covalency factor, which equals one for free NO or for purely ionic bonding and offers the opportunity to correct the g_{zz} principal value for possible spin delocalization ($l < 1$) [31]. The energy splittings, "E" and "Δ", are affected by the local electronic structure, leading to changes in the g tensors of NO adsorbed on different solids. The experimentally obtained principal values of the g tensor allow in principle to determine the three unknown parameters, "E", "Δ", and "l" using Eqs. (6.1), (6.2) and (6.3).

To evaluate accurate values of parameters E, Δ and l, three principal values of g_{xx}, g_{yy} and g_{zz} have to be determined with a high degree of accuracy. The orthorhombic distortion, $\Delta g = g_{xx} - g_{yy}$, is of the order of 10^{-3} or even less, but can be resolved by multi-frequency ESR measurements [28]. For the NO adsorption complexes at metal oxide surfaces, a distribution of the g tensor principal values and the corresponding splitting energies are expected because of the inhomogeneous properties of the surface and the varying orientations of the NO complexes. Thus, the linewidths of the ESR spectra can contain information about the g value distribution from which one can evaluate the distributions δE and δD of the splitting energies E and D.

6.2.1.2   Multi-Frequency ESR Spectra

An important advantage of using multi-frequency ESR spectroscopies is to separate the g and hyperfine (hf) A tensor components and to resolve the weak deviation ($g_{xx} - g_{yy}$) from the axially symmetry of the g tensor so as to evaluate accurate values of E, Δ and l according to Eqs. (6.1), (6.2) and (6.3).

Experimental X (9.3 GHz)-, Q (33.9 GHz)-, and W (93.9 GHz)-band ESR spectra of the NO/Na-LTA (Linde type A) zeolite system are shown in Fig. 6.3(a). The X-band spectrum shows a ^{14}N ($I = 1$) hf coupling with three transitions ($m_I = -1$, 0, 1) at the $g_{xx} \approx g_{yy}$ position. The g_{zz} position lies in the high field region and does not exhibit any hf splittings. The g_{xx} and g_{yy} values are very close to each other and cannot be resolved both at X- and Q-band. In the W-band spectrum, however, the g_{xx} spectral component is visibly separated from the ^{14}N hf triplet at the g_{yy} position. All three experimental spectra are satisfactorily simulated with an identical set of g and A principal values given in Table 6.1: see Fig. 6.3(a) vs (b). Thus, the frequency dependence of the powder ESR spectra with the overlapping g_{xx} and g_{yy} peaks at X- and Q-band and their successful separation at W-band were clearly demonstrated by T. Rudolf et al. [28]. The values of E, Δ and l for the Na-LTA/NO system, which were evaluated from the experimentally obtained g values, are given in Table 6.1 together with those for the Na-ZSM-5/NO zeolite system. The table also contains the distribution widths of δE, $\delta \Delta$ and δl, which are evaluated from the linewidth analysis at the g_{xx}, g_{yy} and g_{zz} peaks. The readers can refer to ref. [28] for further details.

The relative distribution of width, $\delta \Delta_{rel}$ ($\equiv \delta \Delta / \Delta$) $\approx 0.2\%$ for Na-LTA/NO is much narrower than that of $\delta \Delta_{rel} \approx 31\%$ for Na-ZSM-5/NO. This indicates that the LTA-type zeolite has a uniform structure, the local electric fields at the sodium ion adsorption sites do not significantly vary and the ion sites display uniform chemical properties in Na-LTA. In contrast, the relatively wide distribution of $\delta \Delta_{rel}$ for Na-ZSM-5/NO system suggests a variety of adsorption sites, which may originate from the more complicated structure for Na-ZSM-5 zeolites (possessing not only straight, but also zigzag channels) as well as the randomly distributed Al atoms in the Si-O-Al lattice. Thus, multifrequency ESR measurements with a NO probe molecule can be a very sensitive method to monitor the different site distributions in nanoporous materials.

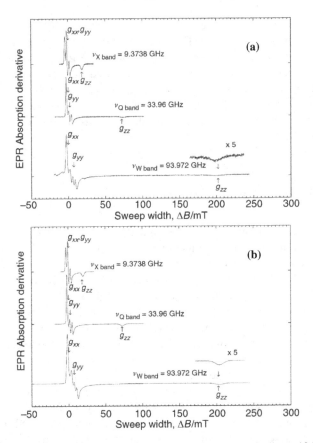

Fig. 6.3 (a) Experimental ESR spectra of NO adsorbed on Na-LTA zeolite at 10 K observed at X-, Q- and W-band microwave (MW) resonance frequencies. NO was adsorbed onto activated Na-LTA samples at room temperature with a gas pressure corresponding to approximately 10^{-1} molecules per unit cell. (b) ESR spectra simulated using an identical set of the A and g ESR parameters in Table 6.1. The figure is adapted from [28] with permission from the Royal Society of Chemistry

6.2.1.3 ^{14}N and ^{23}Na Hyperfine Couplings and Structure of NO-Na$^+$ Complex

For the NO-Na$^+$ complex in Na-LTA zeolite the ESR spectrum is characterised by a g-tensor with principal values of $g_{xx} = 1.999$, $g_{yy} = 1.993$, $g_{zz} = 1.884$, and a ^{14}N hyperfine coupling with $A_{xx} = A_{zz} \approx 0$, $A_{yy} = 91$ MHz (Table 6.1); refer to Section 3.4.2.3 (Surface complex structures) in Chapter 3. The deviation of the g tensor from axial symmetry, although small, is resolved at W-band with the g_{xx} spectral position distinguished from the g_{yy} region as mentioned above. Pöppl et al. [26] have successfully employed pulsed ENDOR spectroscopy at X- and W-band frequencies to precisely evaluate the ^{14}N($I = 1$) and ^{23}Na($I = 3/2$) *hf* couplings and to characterize the geometrical and electronic structure of NO-Na$^+$ complex in Na-LTA type zeolite at low temperature.

Table 6.1 Experimental and computed g, $A(^{14}N)$ and $A(^{23}Na)$, and $Q(^{23}Na)$ principal values of Na$^+$-NO complexes formed in Na-LTA and Na-ZSM-5 zeolites together with the energy splitting E, the crystal field parameters Δ, l and their distribution widths δE, $\delta\Delta$ and δl

NO-Na$^+$ complex	(a) Na-LTA Multi-band ESR (10 K)	(b) Na-ZSM-5 Multi-band ESR (10 K)	(c) Na-LTA Pulsed ENDOR	(d) Na-LTA DFT
g_{xx}	1.9993	1.9939	2.001	2.030
g_{yy}	1.9936	1.9914	1.996	1.999
g_{zz}	1.8842	1.8460	1.888	1.888
$A(^{14}N)$/MHz			X-band (5 K)	
A_{xx}	16.2	32.5	25.3 ± 0.2	16.9
A_{yy}	91.6	102.0	91.0 ± 0.5	82.0
A_{zz}	0.0	0.0	26.3 ± 0.2	16.9
$A(^{23}Na)$/MHz			W-band (4.3 K)	
A_{xx}			6.3 ± 0.2	5.16
A_{yy}			6.3 ± 0.2	5.16
A_{zz}			10.9 ± 0.2	11.56
$Q(^{23}Na)$/MHz			W-band (4.3 K)	
Q_{xx}			−0.41	−0.38
Q_{yy}			−0.23	−0.19
Q_{zz}			0.64	0.57
E/eV	5.67	13.45		
Δ/eV	0.272	0.165		
l	1.051	0.848		
δE/eV	0.28	9.25		
$\delta\Delta$/eV	0.005	0.051		
δl	0.032	0.019		
References	[28]	[28]	[26]	[32]

Notes: ESR g and $A(^{14}N)$ principal values in (a) and (b) were obtained from the simulation of CW ESR spectra observed at three different MW frequencies (multi-frequency ESR spectra). The E, Δ, l, δE, $\delta\Delta$ and δl values were evaluated from the g values and their distribution widths (δg). The computational DFT results correspond to the B3LYP/6-31+G(d) optimized geometry of the Na-NO complex in model 3A in [32].

The principal values and even the orientation of the principal axes of the ^{23}Na hyperfine coupling tensor with respect to axes of the g tensor could be determined from Mims' and Davies' pulsed ENDOR spectra, refer to Section 2.3.3 in Chapter 2. The values $A_{xx}(^{23}Na) = A_{yy}(^{23}Na) = 6.3$ and $A_{zz}(^{23}Na) = 10.9$ MHz were obtained by simulation taking angular selection into account. The so-called hyperfine enhancement of ENDOR intensities due to the interaction between the radio frequency field and the electron spin could lead to pronounced differences in the ENDOR intensities between signals from different m_s electron spin states in experiments at conventional MW frequencies such as in X-band, but also at the W-band. The ^{23}Na ($I = 3/2$) nuclear quadrupole tensor is almost coaxial to the A tensor, $Q_{zz} = 0.48$ MHz, $Q_{yy} = -0.07$ MHz, and $Q_{xx} = -0.41$ MHz. Simulation of orientation-selective ENDOR spectra as described in [26, 33] serves to refine the principal values of the hyperfine coupling tensors estimated from experiment. In addition the influence of the spectra on the orientation of the corresponding principal axes can be examined by simulation. The axes are specified in the principal axes system of the g-tensor.

In the X-band CW-ESR spectra of the NO-Na$^+$ complex, the ^{14}N ($I = 1$) hf splittings of $A_{xx}(^{14}\text{N})$ and $A_{zz}(^{14}\text{N})$ values were too small to be resolved and the third principal value of $A_{yy}(^{14}\text{N})$ was only detected. Orientation selective ENDOR spectroscopy was therefore applied to determine the couplings along the x and z axes of the g tensor, yielding the principal values $A_{xx}(^{14}\text{N}) = 25.3$ and $A_{zz}(^{14}\text{N}) = 26.3$ MHz.

The ^{14}N and the ^{23}Na hyperfine interactions were finally employed to obtain the spin densities in the molecular orbitals of the NO-Na$^+$ complex to give insight into the electronic structure of the adsorption complex. An isotropic hf coupling of $A_{iso}(^{23}\text{Na}) = 7.8$ MHz was evaluated from the above principal values of the $A(^{23}\text{Na})$ tensor. From the $A_{iso}(^{23}\text{Na})$ value an unpaired electron spin density in the Na 3s orbital is evaluated to be $\rho_{3s}(\text{Na}) = 0.9\%$ [34]. In addition, judging from the small anisotropic values of the $A(^{23}\text{Na})$ tensor, $B_{zz}(^{23}\text{Na}) \equiv A_{zz} - A_{iso} = 3.1 \pm 0.2$ MHz, the spin density in Na 3p orbitals is negligible. Thus, the unpaired electron in the Na$^+$-NO complex is concluded to be mainly localized at the NO molecule.

From the experimental hf tensor of $A(^{14}\text{N})$ the isotropic hf splitting of $A_{iso}(^{14}\text{N})$ and the principal values of the dipolar coupling tensor, $B(^{14}\text{N})$, are deduced: $A_{iso}(^{14}\text{N}) = 47.7$ and $(B_{xx}, B_{yy}, B_{zz})\,(^{14}\text{N}) = (-22.4, 43.8, -21.4)$ MHz. The value of $A_{iso}(^{14}\text{N})$ leads to a spin density of $\rho_{2s}(\text{N}) = 0.031$ in the nitrogen 2s orbital [34]. From the average dipolar ^{14}N hf coupling of $B_{xx}(\text{N})$ and $B_{zz}(\text{N})$ (i.e. -21.9 MHz) the spin density of $\rho_{2p}(\text{N}) = 0.458$ is deduced for the nitrogen 2p orbital which forms, together with the oxygen 2p orbital, an anti-bonding $2p\pi^*$ SOMO (singly occupied molecular orbital) for NO. Assuming the relation of $\rho_{total} = \rho_{2s}(\text{N}) + \rho_{2p}(\text{N}) + \rho_{2p}(\text{O}) + \rho_{3s}(\text{Na}) \equiv 1$ the spin density in the oxygen 2p orbital can be evaluated to be $\rho_{2p}(\text{O}) = 0.502$. The SOMO of the NO-Na$^+$ complex is then given by a linear combination of the following atomic orbitals (χ):

$$\psi = (0.03)^{1/2}\chi_{2s}(\text{N}) + (0.46)^{1/2}\chi_{2p}(\text{N}) + (0.50)^{1/2}\chi_{2p}(\text{O}) + (0.01)^{1/2}\chi_{3s}(\text{Na}) \quad (6.4)$$

The geometrical structure of the NO-Na$^+$ complex can be discussed based on the ^{23}Na hf (axially symmetrical) dipolar principal values $[(B_{xx}, B_{yy}, B_{zz})\,(^{23}\text{Na}) = (-1.5, -1.5, 3.0)$ MHz] and their orientation with respect to the g and $A(^{14}\text{N})$ principal coordinate systems. The ^{23}Na dipolar hf couplings can be related to the electron spin densities of $\rho_{2p\pi}(\text{N})$ and $\rho_{2p\pi}(\text{O})$ and the Na$^+$—N and Na$^+$····O distances, $r(\text{Na}^+$—N) and $r(\text{Na}^+$····O) by an analysis based on the point dipole approximation proposed by Hutchison and McCay [26, 35]. The bond angle of Na$^+$—N—O and the value of $r(\text{Na}^+$—N) were evaluated to be 142° and 0.21 nm, respectively. The structure proposed for the NO-Na$^+$ complex in Na-LTA zeolite is depicted in Fig. 6.4.

6.2.1.4 DFT Computations

Liu et al. [32] have recently reported a computational study on the adsorption site and the ESR magnetic parameters of NO adsorbed in Na-LTA zeolite employing

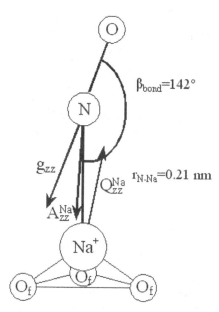

Fig. 6.4 Schematic drawing of the bent structure of the NO-Na$^+$ adsorption complex in Na-LTA zeolite. The z principal axes of the g, $A(^{23}$Na), and $Q(^{23}$Na) tensors lie within the Na$^+$-N-O complex plane and form an angle with the Na$^+$-N(O) bond direction of 38, 3, and 8°, respectively. The unpaired electron is localized mainly in the anti-bonding $2p\pi^*$ molecular orbital of the NO molecule (refer to Fig. 6.2), which is also within the plane of the complex. The cation at site S2 is coordinated to the framework oxygens, O$_f$, in the six-membered rings in a trigonal symmetry, refer to Fig. 6.1. Only the three oxygens in the first coordination sphere are shown. The figure is adapted from [26] with permission from the American Chemical Society

density functional theory (DFT); see Chapter 5 for "DFT". A rather good correspondence was obtained between the experimental and computed electronic g and $A(^{14}$N) tensors, and the $A(^{23}$Na) and $Q(^{23}$Na) tensors of the Na$^+$-NO complex, as summarized in Table 6.1. For the computations the following model of zeolite network were employed: a six-membered ring terminated by hydrogen atoms with one Na$^+$ ion above the ring, three additional Na$^+$ ions located at the centers of three imagined four-membered rings adjacent to the six-membered ring, and three additional four-membered rings adjacent to the six-membered ring. The optimized geometry of the complex agrees nicely with that estimated experimentally, except for the Na-N distance, where the computations resulted in $R(\text{N-Na}) = 0.266$ nm which is longer by as much as 0.05 nm than that deduced from the previous ENDOR experiments (0.21 nm) [26].

6.2.2 Triplet State of (NO)$_2$ Bi-Radical Formed in Zeolite

Kasai et al. [4c] have first reported that the CW X-band ESR spectrum of NO adsorbed on Na-LTA zeolite consists of two signals, one due to the NO-Na$^+$ complex (NO mono-radical) as described in the above section and the other due to an unusual

NO—NO dimer species with a triplet state (referred to as the $(NO)_2$ bi-radical or radical pair in the following). The ESR spectrum of $(NO)_2$ bi-radical shows the forbidden transition, $\Delta m_S = 2$, at *ca.* 170 mT ($g \approx 4$), when the corresponding allowed transitions, $\Delta m_S = 1$, are observed for the same sample at *ca.* 340 mT ($g \approx 2$). These verify the presence of the triplet electronic state. Thus the ESR study suggested that the zeolite can stabilize the $(NO)_2$ dimer as the triplet rather than the usual singlet state, indicating a great affinity of the NO molecule for the zeolites. The $(NO)_2$ bi-radical may play an important role as an intermediate species in the decomposition of NO [29]. ESR studies have accordingly continued to be of interest in recent decades.

6.2.2.1 X- and Q-Band ESR Spectra

The *X*-band spectrum of the NO/Na-LTA zeolite system is mainly due to the NO mono-radical when the pressure is low ($P_{NO} \leq 0.1$ kPa), while the $(NO)_2$ bi-radical becomes dominant at higher NO pressure ($P_{NO} \geq 10$ kPa) [24, 36]. The ESR signals due to the NO mono-radical (NO-Na^+ complex) and the $(NO)_2$ bi-radical are superimposed at intermediate pressures. The *Q*-band ESR spectrum helped very much to resolve the individual spectrum and to evaluate the accurate ESR parameters of the $(NO)_2$ bi-radical. As shown in Fig. 6.5, the *Q*-band spectral line-shape is well simulated using the following **g** tensor and the *D* and *E* parameters of the zero field splitting (ZFS) tensor for the $(NO)_2$ bi-radical: $(g_{xx}, g_{yy}, g_{zz}) = (1.9120, 2.0042,$

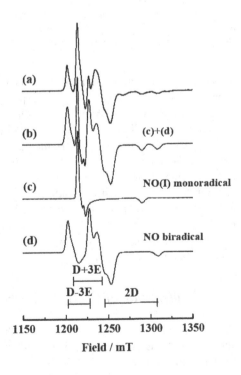

Fig. 6.5 (**a**) *Q*-band ESR spectra of NO ($P_{NO} = 13.2$ kPa) introduced in Na-LTA zeolite at 5 K. (**b**) ESR spectrum simulated to (**a**). The theoretical spectra (**c**) for NO(I) and (**d**) for the $(NO)_2$ bi-radical were calculated using Lorentzian lineshape with anisotropic line-width of 4.0, 7.5, and 6.0 mT for the *x*-, *y*-, and *z*-components, respectively. The simulation spectrum (**b**) is a superposition of (**c**) and (**d**) with the intensity ratio of 1:1. The figure is adapted from [36] with permission from the American Chemical Society

1.9770), $|D| = 33.1$ mT and $|E| = 2.8$ mT. The ESR spectra due to NO mono- and $(NO)_2$ bi-radicals were also observed for NO adsorbed on other ion-exchanged zeolites such as partially lithium ion-exchanged Na-LTA zeolite [37] and on sulfated zirconia [38].

ESR simulations revealed that the spectral line-shape of $(NO)_2$ bi-radical is very sensitive with respect to the relative orientations of the g and the D tensors and the principal axes of g_{ZZ} and D_{ZZ} cannot deviate by more than 30°(corresponding to angle "θ" in Scheme 6.1) from each other in this model. In the simulations the principal values of the g-tensors of each NO molecule were assumed to be the same as the experimental ones for the $(NO)_2$ biradical with the principal axes parallel and perpendicular to the N—O bond and along the common x-axis ($g = 2.0042$) and the direction of D_{ZZ} was taken along the line connecting the midpoints of the two N—O bonds as shown below [13, 36].

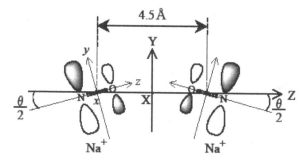

Scheme 6.1 A geometrical structure proposed for the $(NO)_2$ bi-radical in Na-LTA zeolite. The structure is adapted from [36] with permission from the American Chemical Society

The ZFS tensor of radical pairs generally originates from the dipolar-dipolar interaction between the two unpaired electrons of the radical units. As described in Chapter 4 the D-parameter can be related to the average distance between two radicals, R, by the following equation based on the point-dipole approximation,

$$D = \frac{\mu_0}{4\pi} \frac{3\, g\mu_B}{2R^3} \tag{6.5}$$

where g, μ_0, and μ_B are the spectroscopic g-factor, the magnetic constant, and the *Bohr* magneton, respectively; for the numerical values of μ_0, and μ_B, see Table G2 in General Appendix G, and refer to the H$\cdots$CH$_3$ radical pair in Section 5.4.2 where "$d_\perp$" is used in stead of "D". The values of R, evaluated from the experimental D values, are summarized in Table 6.2.

The NO molecule can be stabilized on a pair of adjacent cations (Na$^+$ ions in the present case) as shown in Scheme 6.2(a). However, a structure with the two NO molecules bonded to the same Na$^+$ ion and with nearly collinear N—O bonds (Scheme 6.2(b)) has also been proposed. The observation that the NO—NO distance became shorter from $R = 0.45$ nm in Na-LTA to 0.42 nm in Li-LTA (see Table 6.2)

Table 6.2 ESR g-tensor and zero-field splitting (ZFS) tensor for $(NO)_2$ bi-radical formed in Na-LTA zeolite and other matrices

Zeolite	(a) Na-LTA CW Q-band ESR	(b) Na-LTA *Pulsed* ESR	(c) Na-LTA X-band ESR	(d) Li-Na-LTA[a] X-band ESR	(e) Sulfated Zirconia X-band ESR
g-tensors					
g_{xx}	2.0042		1.976	1.989	1.993
g_{yy}	1.9770		1.976	1.989	1.993
g_{zz}	1.9120		1.912	1.900	1.994
ZFS (mT)					
$|D|$	33.1		28.8	41.0	19.5
$|E|$	2.8				
$A(^{23}Na)/MHz$					
A_{xx}		4.6			
A_{yy}		4.6			
A_{zz}		8.2			
$Q(^{23}Na)/MHz$		−0.3			
Q_{xx}		−0.3			
Q_{yy}		0.6			
Q_{zz}					
$R(nm)$[b]	0.45	0.45	0.45	0.42	0.52
References	[36]	[12]	[4c]	[37]	[38]

[a] The exchange level of lithium ion (Li$^+$) is 66%.
[b] NO—NO distance (Scheme 6.1).

provided support for the latter model taking into account the Na$^+$ ionic radius ($r = 0.095$ nm) in comparison with the Li$^+$ ion ($r = 0.060$ nm). Thus, the structure shown in Scheme 6.2(b) can not be completely ruled out. Further studies are required to clarify the details of the geometrical structure of the $(NO)_2$ bi-radical in the zeolite.

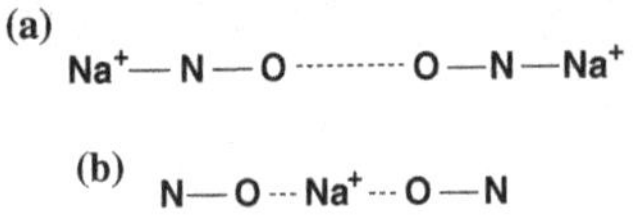

Scheme 6.2 Models for $(NO)_2$ bi-radical. **(a)** Two NO molecules stabilized on a pair of adjacent Na$^+$ ions. **(b)** Two NO molecules bounded to the same Na$^+$ ion.

(a) Na$^+$— N — O ·········· O — N —Na$^+$

(b) N — O ··· Na$^+$ ··· O — N

6.2.2.2 Pulsed ESR

Pulsed ESR was employed to study the $(NO)_2$ triplet-state bi-radical in Na-LTA type zeolite, with the purpose to resolve the interaction with surface groups, and to elucidate the role of the zeolite in stabilizing the triplet state rather than the usual singlet state [12]. Measurements performed at 5 K gave rise to FT (Fourier Transformation, see Chapter 2) spectra that were assigned to the $(NO)_2$ bi-radical interacting with one or two ^{23}Na nuclei (with $I = 3/2$), with $A(^{23}Na) = (4.6, 4.6, 8.2)$ MHz and

$Q(^{23}Na) = (-0.3, -0.3, 0.6)$ MHz for the hyperfine and nuclear quadrupole coupling tensors, respectively. The values are of similar magnitude as those determined for the NO-Na$^+$ complex (see Table 6.1).

6.2.3 Other Nitrogen Oxides as Spin Probes

Other nitrogen oxides such as nitrogen dioxide (NO_2) have been employed as a spin probe to characterize the zeolite structure, chemical properties of zeolites, and motional dynamics of small molecules in them by ESR [13].

NO$_2$ is a stable paramagnetic gaseous molecule at normal temperatures. The ESR parameters of NO$_2$ trapped in a solid matrix have been well established from single-crystal ESR measurements and have been related to the electronic structure by molecular orbital studies [39]. Thus, the NO$_2$ molecule has potential as a spin probe for the study of molecular dynamics at the gas-solid interface by ESR. More than two decade ago temperature-dependent ESR spectra of NO$_2$ adsorbed on porous Vycor quartz glass were observed [40]; Vycor® is the registered trademark of Corning, Inc. and more information is available at their website. The ESR spectral line-shapes were simulated using the slow-motional ESR theory for various rotational diffusion models developed by Freed and his collaborators [41]. The results show that the NO$_2$ adsorbed on Vycor displays predominantly an axial symmetrical rotation about the axis parallel to the O—O inter-nuclear axis below 77 K, but above this temperature the motion becomes close to an isotropic rotation probably due to a translational diffusion mechanism.

In contrast to the NO$_2$/Vycor glass system, translational diffusion (or Heisenberg type of spin exchange) of NO$_2$ in Na-MOR, Na-MFI and K-LTL types of zeolites were suggested by analysis of the temperature dependence of the ESR spectra using the slow motional theory [13, 42, 43]. Broadening at increased temperature is assigned to the spin exchange between the NO$_2$ molecules diffusing along zeolite channels. In Na-MFI the spin exchange rate increased rapidly with increasing Si/Al ratio of the zeolite, as expected if the hindrance against diffusion is caused by Na$^+$, the amount of which increases with a decreased Si/Al ratio [44]. The diffusion was also affected by the water content and by the channel structure, but not appreciably by replacing Na$^+$ with Li$^+$, Ca^{2+}, Sr^{2+}, K$^+$ or Cs$^+$ [45]. A detailed investigation of the NO$_2$/Na-MOR system has suggested that there exists a distribution of exchange rates at each temperature [46].

For the characterization of acid sites on solids such as silica-alumina, alumina, and zeolites, di-tert-butyl nitroxide (DTBN) , a stable organic nitroxide radical, was employed as a spin probe molecule by Hoffman et al. [47, 48] and others [49, 50]. Gutjahr et al. [50] studied the electron pair acceptor properties of the monovalent cations such as Li$^+$, Na$^+$, K$^+$, and Cs$^+$ in a faujacite (Y) type of zeolite by means of ESR using DTBN as a probe molecule. They reported a linear relationship between the nitrogen spin density estimated from the nitrogen hf constants of DTBN and the electro-negativity of the zeolite cations.

6.3 Cu(I)-NO Complexes Formed in Zeolites

Nitrogen oxides (NO_x) are hazardous pollutants formed as byproducts during the combustion processes in industrial boilers and vehicle engines and are responsible for smog formation, acid rain, and global warming [18–22]. Many metal ion exchanged zeolites have been reported to be active for catalytic decomposition and reduction of nitrogen monoxide. Among them, copper ion exchanged zeolites have received much attention due to their high activity and selectivity toward the decomposition of NO_x. High siliceous zeolites such as ZSM-5 and MCM-22 have been reported to be promising for the NO decomposition process [19, 29, 31, 51–55]. The Cu(I)-NO complexes have attracted special interest because they are important intermediates in the catalytic decomposition of nitric oxide over copper exchanged zeolites. The interaction of NO with Cu(I) is reported to be a complex redox (reduction-oxidation) process where the oxidation of the site is proposed to occur *via* the elimination of N_2O from dinitrosyl through the formation of Cu(I)-$(NO)_2$ from monomeric Cu(I)-NO complexes [29].

The Cu(I)-NO adsorption complexes formed on copper exchanged and auto-reduced Cu-ZSM-5 and Cu-MCM-22 and other zeolites were extensively studied by Giamello et al. [29] and by Pöppl et al. [31, 51–55] using multi frequency ESR and ENDOR spectroscopies. The increased spectral resolution and separation of the Cu(I)-NO signals from the Cu(II) signal at higher frequencies (Q- and W-bands) allowed an accurate determination of the ^{63,65}Cu and ^{14}N *hf* couplings and g-values from the powder ESR spectrum and successfully led to a detailed characterization of the Cu(I)-NO complexes [31]. Furthermore, pulsed electron nuclear double resonance (ENDOR), and hyperfine sublevel correlation spectroscopy (HYSCORE) were employed to characterize the local structure of Cu(I)-NO adsorption complexes formed over Cu-L and Cu-ZSM-5 type of zeolites [53]: see Chapter 2 for pulsed ENDOR and HYSCORE.

6.3.1 Multifrequency ESR Spectra

Multifrequency ESR spectra of Cu(I)-NO complexes formed on Cu-ZSM-5 were extensively studied by Pöppl et al. [31]. Figure 6.6 shows the X-, Q-, and W-band ESR spectra of Cu-ZSM-5 with adsorbed NO. Although an auto-reduction of Cu(II) to Cu(I) takes place to a large extent during the activation process at an elevated temperature in vacuum, the ESR signal due to Cu(II) is still visible in the spectrum. In the X-band spectrum, the Cu(II) spectrum is superimposed by the intense spectra due to the Cu(I)-NO complexes, which makes it difficult to determine accurate values of the ESR parameters. In the Q-band spectrum, the Cu(II) spectrum is almost separated from that of the Cu(I)-NO complexes. Finally, in the W-band spectrum complete separation between the Cu(II) and Cu(I)-NO complex spectra is achieved. The Cu(I)-NO complexes formed in the Cu-ZSM-5 system show an anisotropic spectrum in which the ^{63}Cu ($I = 3/2$) *hf* splitting with four lines is clearly visible in the g_{zz} and g_{xx} ($= g_{yy}$) spectral regions. With increasing NO pressure electronic dipole-dipole interactions among the Cu(I)-NO complexes resulted in an

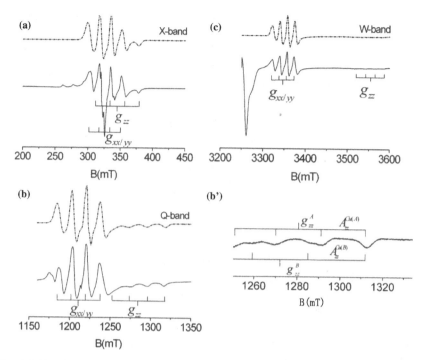

Fig. 6.6 Experimental (*solid lines*) and simulated (*dotted lines*) (**a**) X-, (**b**) Q-, and (**c**) W-band ESR spectra of Cu(I)-NO species formed in Cu-ZSM-5 zeolite after activation at 673°C in vacuum and subsequent NO adsorption at pressures of 5 Pa (X-, Q-bands) and 0.5 Pa (W-band) at 300 K. The stick diagrams indicate the ^{63}Cu *hf* splitting of the Cu(I)-NO ESR spectrum in the $g_{xx/yy}$ and g_{zz} spectral regions. (**b'**) Q-band spectra of the Cu(I)-NO complexes showing two different species A and B in the g_{zz} spectral region. The *hf* splittings of both ^{63}Cu (natural abundance 69.1 at.%) and ^{65}Cu (30.9 at.%) isotopes with nuclea spin $I = 3/2$ have been included in the spectral simulation, but the splittings in the stick diagrams only refer to the ^{63}Cu isotope. The ESR *hf* parameters and g-values used for the simulations are listed in Table 6.3. The spectra are adapted from [31] with permission from the American Chemical Society

ESR linewidth broadening so as to smear the ^{14}N ($I = 1$) *hf* triplet. In fact the *hf* triplet is not resolved in the *Q*-band ESR spectra of Cu(I)-NO complexes at NO pressure of 5 Pa, but could be observed at lower NO pressures of *ca.* 0.5 Pa as a splitting of *ca.* 3 mT resolved in the g_{xx} (= g_{yy}) spectral regions.

The ESR parameters of the Cu(I)-NO species are indicative of a N-centered complex with a bent geometry and a significant contribution of the Cu(I) 4s atomic orbital to the SOMO of the Cu(I)-NO complex. That is, the observed relation of $g_{zz} < g_e = 2.0023 < g_{xx} = g_{yy}$ suggests the presence of Cu 3d orbital contributions [56a] to the SOMO and the large isotropic Cu *hf* coupling shows that the structure of the complex is bent so as to allow an effective admixture of the 4s orbital of Cu(I) to the SOMO. Based on the experimental ESR parameters the geometrical structure shown in Fig. 6.7 was proposed for the Cu(I)-NO complex formed in the ZSM-5 zeolite.

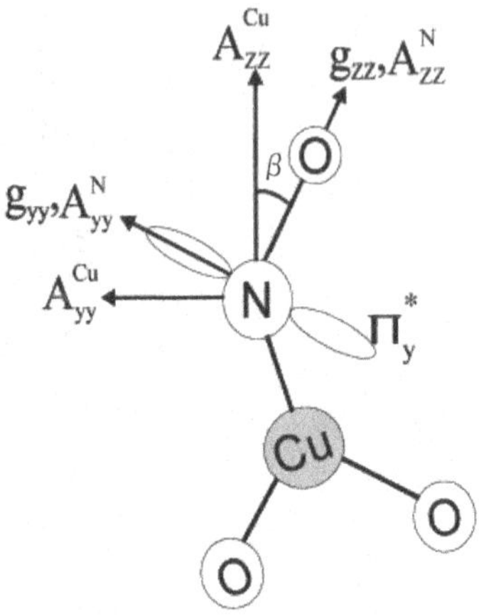

Fig. 6.7 Orientation of principal axes frames of the g, A(N), and A(Cu) tensors of the Cu(I)-NO moiety. The A(Cu) frame is rotated by the angle, β, about the common x principal axis, but the A_{zz}(Cu) principal axis is not necessarily parallel to the Cu—N bond. The y-z plane of the g tensor is spanned by the N—O bond and the symmetry axis of the $2p\pi_y^*$ orbital of the NO molecule. See Table 6.3 for ESR parameters and structural data. The figure is adapted from [31] with permission from the American Chemical Society

Table 6.3 ESR parameters (g and A hf tensors) and covalent parameter "l" of Cu(I)-NO complexes formed over Cu-ZSM-5 and Cu-MCM-22 zeolites[a]

Cu(I)-NO	Cu-ZSM-5		Cu-MCM-22	
Species	A	B	A	B
g_{xx}	2.0050	2.0050	2.0050	2.0050
g_{yy}	2.0050	2.0050	2.0050	2.0050
g_{zz}	1.8900	1.8999	1.8900	1.8999
$A(^{14}N)/10^{-4}$ cm^{-1} [b]				
A_{xx}	–	–	–	–
A_{yy}	29	29	33	33
A_{zz}	–	–	–	–
$A(^{63}Cu)/10^{-4}$ cm^{-1} [b]				
A_{xx}	158	158	157	157
A_{yy}	130	130	133	133
A_{zz}	216	239	200	220
A_{iso}	168	175	163	170
Covalent parameter				
"l" [c]	0.85	0.78	0.81	0.79

[a] The data taken from multi-band ESR results [31].
[b] 1 cm^{-1} = 4.6686 × 10^{-4} g mT.
[c] Refer to Table 6.1 for parameter "l".

The ESR signal intensity at low temperature of the Cu(I)-NO species in ZSM-5 zeolite decreased as a function of increased NO loading suggesting the formation of diamagnetic Cu(I)-(NO)$_2$ species at the expense of paramagnetic Cu(I)-NO. The Cu(I)-(NO)$_2$ is ESR silent with a singlet ($S = 0$) ground state as predicted by quantum chemical calculations [57]. This is in contrast to the triplet ($S = 1$) ground state that has been experimentally confirmed for Na$^+$-(NO)$_2$ complexes in Na-A zeolites as mentioned in Section 6.2. The ESR intensity due to the isolated Cu(II) cations was essentially independent of the NO loading at low temperatures, suggesting that the NO molecules do not form adsorption complexes with Cu(II) cations remaining after auto-reduction.

Two Cu(I)-NO complexes, A and B, were observed in Cu-ZSM-5 (and Cu-MCM-22) zeolites with similar ESR parameters, see Table 6.3. By comparing the experimental hf couplings of A_{iso}(^{63}Cu; A) < A_{iso}(^{63}Cu; B) with the results of theoretical computations [58] species A and B were tentatively assigned to Cu(I)-NO complexes with two and three oxygen neighbors, (RO)$_2$Cu(I)-NO and (RO)$_3$Cu(I)-NO, respectively. The assignment is qualitatively supported by the order of the covalent parameter of the two species, l(A) > l(B), derived from the corresponding g_{zz} values, indicating a larger transfer of unpaired electron spin density from the 2pπ_y* orbital of the NO to the Cu(I) ion for species B than for A.

6.3.2 Pulsed ENDOR and HYSCORE Studies

The local structure of Cu(I)-NO adsorption complexes formed over Cu-L and Cu-ZSM-5 zeolites were studied by pulsed ENDOR and HYSCORE methods by Umamaheswari et al. [53]. The ^{1}H ENDOR signals from residual distant protons were not detected in completely copper ion exchanged Cu-ZSM-5 zeolites. Such signals were, however, observed for the Cu-L zeolite, where the ^{1}H form of the zeolite was 30% ion exchanged with Cu(II) ions and subsequently dehydrated to (auto)reduce Cu(II) to Cu(I). For both systems, very broad ^{27}Al ENDOR spectra were observed. The ^{27}Al hf couplings were estimated using the point dipole approximation for the Cu(I)-NO center in Cu-L. The result shows that an aluminum framework atom is located in the third coordination sphere with respect to the NO molecule adsorbed on a Cu(I) cation site.

Less favorable experimental conditions were met for Cu(I)-NO complexes formed over Cu-ZSM-5 that prevented a determination of the ^{27}Al hf coupling data because of short electron spin relaxation times and larger distributions of ^{27}Al nuclear quadrupole couplings, probably due to an inhomogeneous distribution of Al framework atoms. Detailed local structures of the complexes in Cu-ZSM-5 zeolites, O$_2$-Al-O$_2$-Cu(I)-NO, were recently proposed on the basis of quantum chemical calculations [59]. To experimentally verify the theoretically proposed structural properties of the Cu(I)-NO species formed in ZSM-5, it is highly desirable to develop improved synthesis strategies for high siliceous zeolites that lead to a better statistical Al distribution in the crystallites.

6.4 Molecular Motion Probes: Radicals in Zeolites

In the foregoing sections the ESR studies of NO and other nitrogen oxides as a spin probe were shown to be useful for understanding the electronic and geometrical structure and the dynamical processes of molecules adsorbed on zeolites. The dynamics of NO_x (x = 1, 2) radicals are strongly dependent on properties of zeolites such as channel structure (multiple-channel or single-channel) and channel size. These observations indicate that the microenvironment plays an important role for the molecular dynamics of molecules incorporated in it. In this section applications are presented of X-band ESR spectroscopy to investigate microenvironment effects on molecular dynamics and thermal stability of relatively large organic radicals such as triethylamine $((CH_3CH_2)_3N^+; Et_3N^+)$, and tripropylamine $((CH_3CH_2CH_2)_3N^+; Pr_3N^+)$ cations used as spin probes [12, 60–62].

Amines have comparatively low ionization potentials ($Ip_1 = 7.82$ and 7.50 eV for Me_3N and Et_3N, respectively [63] and are used as electron donors [64]. Furthermore, amines are widely used as organic templates in synthesizing zeolites. Zeolites provide an appropriate microenvironment to retard back electron transfer and increase the lifetime of photoproduced radical ions, and long-living radicals could be observed even at room temperature in them [65]. Thus, zeolites incorporated with amines have attracted interest to investigate the molecular dynamics of radical cations such as Et_3N^+ and Pr_3N^+ formed inside the void structures by ionizing radiation. By analyzing the temperature-dependent X-band ESR spectral line-shapes experimental information about molecular dynamics, especially rotational motion as well as electronic and geometrical structures, were obtained.

6.4.1 Structure and Dynamics of Et_3N^+ and Pr_3N^+ in $AlPO_4$-5

$AlPO_4$-5, which is typical of the $AlPO_4$ family composed of AlO_4 and PO_4 tetrahedra [66], contains alternating Al and P atoms forming four- and six-membered rings. The Et_3N and Pr_3N encapsulated into $AlPO_4$-5 take on a tripod shape in the cylindrical channel. The Et_3N^+ radicals were generated inside the void structure of the $AlPO_4$-5 and gave well-resolved ESR spectra which allow to observe temperature-dependent ESR lineshapes over a wide temperature range, from 4 to 300 K [61]. The temperature-dependent spectra were successfully explained assuming a two-site exchange model of inequivalent β-methylene hydrogens with respect to the central ^{14}N $2p_z$ orbital with the unpaired electron.

6.4.1.1 High Temperature ESR Spectra

The X-band ESR spectrum of Et_3N^+ at 300 K consists of an intense and sharp hyperfine (hf) septet of 2.0 mT splitting with two weak additional lines at each wing which have almost the same splitting as the septet (Fig. 6.8(a)). The intense septet corresponds to the $m_I = 0$ band of the central $^{14}N(I = 1)$ atom and is attributed to the isotropic hf splittings due to six equivalent β-hydrogens of $^+N_{(\alpha)}$—$[C_{(\beta)}H_2$—$C_{(\gamma)}H_3]_3$. The weak wing lines have a spectral feature characteristic of hf anisotropy and are attributable to the parallel component, $A_{||}(^{14}N)$, at the $m_I = \pm 1$ lines of $^{14}N(I = 1)$. The corresponding perpendicular (x and y) features are hidden beneath the $m_I(^{14}N) = 0$ lines because the value of $A_\perp(^{14}N)$ is less than the line width. The experimental spectrum is successfully simulated using the following ESR parameters: $a_{iso}(^1H) = 2.0$ mT for six equivalent β-hydrogens, $A_{||}(^{14}N) = 4.4$ mT, $A_\perp(^{14}N) = 0 \pm 0.4$ mT, $g_{iso} = 2.0033$ and Gaussian line-width of 0.18 mT. The lines corresponding to $m_I(^{14}N) = \pm 1$ are hidden beneath the intense septet of $m_I(^{14}N) = 0$, only weak shoulders of the parallel features are detected in the outer wings of the spectrum.

6.4.1.2 Low Temperature ESR Spectra

The ESR spectrum of Et_3N^+ at 77 K corresponds to a rigid-limit (Fig. 6.8(b)). The spectrum consists of a broad *hf* sextet whose two outer-most lines have a 4.4 mT splitting, but the central quartet has a splitting of 3.6 mT. The quartet is attributed to three equivalent hydrogens, one from each β-methylene group, and the weak

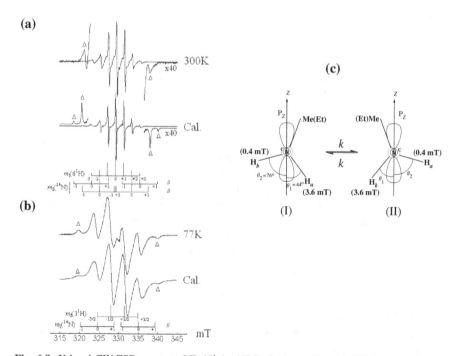

Fig. 6.8 X-band CW-ESR spectra of Et_3N^+ in $AlPO_4$-5 observed at (**a**) 300 K and (**b**) 77 K together with the theoretical spectra (Cal.) calculated by assuming a "fast limit" case and a "rigid limit" case, respectively. The ^1H and ^{14}N hyperfine splittings used in the simulations are given in the text. The weak wing lines marked as a *triangle* ($\triangle$) are attributable to $A_{||}(^{14}N)$ at $m_I(^{14}N) = \pm 1$ transitions. (**c**) Two-site exchange model employed for ESR spectral line-shape simulations. Two methylene hydrogens at $C_{(\beta)}$ labeled as H_a and H_b are interchanged by the hindered rotation about the $N_{(\alpha)}$—$C_{(\beta)}$ bond. Only one methylene group is labeled. The figure is adapted from [61] with permission from the Royal Society of Chemistry

wing lines to the parallel component of $m_I(^{14}N) = \pm 1$ transitions. The experimental spectrum is well simulated using the following ESR parameters: $a_{iso}(^1H) = 3.6$ mT for three equivalent β-hydrogens (neglecting the contribution of other hydrogens), $A_{\parallel}(^{14}N) = 4.4$ mT, $A_{\perp}(^{14}N) = 0 \pm 0.4$ mT, $g_{iso} = 2.003$ and Gaussian linewidth of 1.2 mT.

Assuming the hyperconjugative mechanism [56b] for the β-hydrogen hf splittings the twist angles in Fig. 6.8(c) can be evaluated to $\theta(1) = 44°$ and $\theta(2) = 76°$, which correspond to the 3.6 and 0.4 mT hf splittings, respectively, refer to Appendix A6.1.

6.4.1.3 Analysis of Temperature-Dependent ESR Line-Shapes

The ^1H hf splitting due to the β-hydrogens of Et$_3$N$^+$ was reversibly changed from 3.6 mT (3H: three magnetically equivalent hydrogens) at 77 K to 2.0 mT (6H: six hydrogens) at 300 K. The anisotropic ^{14}N hf splitting, however, remained constant and was not even partially averaged by the motion in the temperature range. This indicates that the temperature dependent spectral change is associated with a change in the number of magnetically equivalent β-hydrogens and their hf splitting. The temperature dependent ESR spectra were well simulated in terms of the two-site exchange model [67] in which the two inequivalent β-hydrogens of the Et or Pr groups interchange their positions with each other so as to become equivalent (Fig. 6.9). Two preferred conformations, (I) and (II) in Fig. 6.8(c), which are energetically equivalent with a mirror image structure, can undergo "exchange" with a temperature-dependent rate, $k(s^{-1})$.

A modified version of the Heinzer program [68] was employed for the line-shape simulations with the exchange rate constant, $k(s^{-1})$, as a variable parameter. Based on the good correlation between the experimental and calculated spectra as shown in Fig. 6.9, the exchange rate of Et$_3$N$^+$ was evaluated to increase by more than two orders of magnitude from 1.8×10^7 to 6.6×10^9 s^{-1} when the temperature increased from 110 to 270 K. In a similar manner the temperature dependent experimental ESR spectra of Pr$_3$N$^+$ were well reproduced by the line-shape simulations. Furthermore, from a linear Arrhenius plot of the exchange rates, activation energies of 9.1 and 11.4 kJ mol^{-1} were evaluated for Et$_3$N$^+$ and Pr$_3$N$^+$, respectively, as shown in Fig. 6.10. DFT calculations for the isolated cations resulted in energy barriers of 8.7 and 7.7 kJ mol^{-1} for the exchange of Et$_3$N$^+$ and Pr$_3$N$^+$, respectively. The reason why the experimentally determined energy barrier for Pr$_3$N$^+$ in AlPO$_4$-5 is higher than the calculated one is explained in terms of interaction with the zeolite wall. That is, in the calculations both Et$_3$N$^+$ and Pr$_3$N$^+$ are reasonably assumed to be located in the 12-ring channel (*ca.* 8 Å in diameter [69]) of the AlPO$_4$-5 framework (Fig. 6.10(a) and (b)). The diameters of Et$_3$N$^+$ and Pr$_3$N$^+$ are *ca.* 6.5 and 8.3 Å, respectively, based on the optimized structures predicted by the DFT calculations. The diameter of Et$_3$N$^+$ is significantly smaller than the ring channel, while Pr$_3$N$^+$ has almost the same diameter as the channel. On the one hand a comparatively weak interaction with the zeolite channel wall is expected for the Et$_3$N$^+$ system because there is still a significantly large vacancy in the channel. On the other hand, for

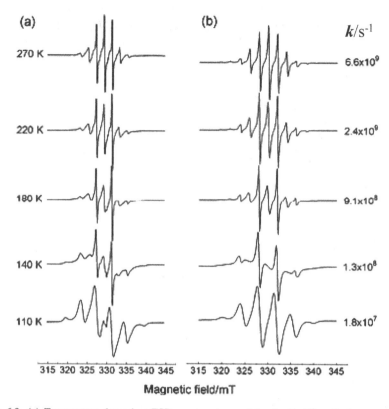

Fig. 6.9 (a) Temperature dependent ESR spectra observed for the Et_3N^+ radical generated in γ-ray irradiated $AlPO_4$-5 containing Et_3N. (b) Best-fit simulation spectra calculated using the two-site exchange model for the hindered rotation of two inequivalent β-hydrogens (see Fig. 6.8(c)). The exchange rates, $k(s^{-1})$, are given in the figure. The rigid limit ESR parameters used in the calculations are given in the text. The spectra are adapted from [61] with permission from the Royal Society of Chemistry

the Pr_3N^+ system there is essentially no vacancy in the channel so that Pr group can experience a stronger interaction with the wall. Thus, the reason why the experimentally determined energy barrier for Pr_3N^+ in $AlPO_4$-5 is higher than the calculated one is explained in terms of interaction with the zeolite wall, refer to [61] for more details.

6.4.2 Cage Effects on Stability and Molecular Dynamics of Amine Radicals in Zeolites

The ESR studies using amine radicals as a spin probe have been extended to investigate "cage effects" on stability and molecular dynamics of organic molecules in zeolites [62]. Three zeolites with different cages or channels were employed, *i.e.*,

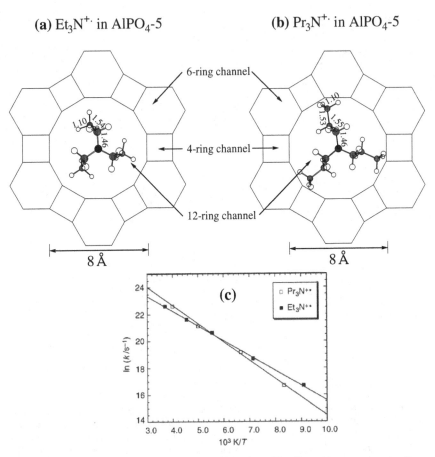

(a) $Et_3N^{+\cdot}$ in $AlPO_4$-5

(b) $Pr_3N^{+\cdot}$ in $AlPO_4$-5

Fig. 6.10 $AlPO_4$-5 framework viewed in the (001) plane. The Si or Al atoms occupy all corners and O atoms (not shown) are near the mid-points of edges. The incorporated **(a)** Et_3N^+ and **(b)** Pr_3N^+ are indicated by shaded and hatched circles. **(c)** The Arrhenius plots for the exchange rates, $\ln(k)$ vs. $1/T$ (K^{-1}), for Et_3N^+ and Pr_3N^+ in $AlPO_4$-5. The figure is adapted from [61] with permission from the Royal Society of Chemistry

SAPO-37 [70] with sodalite cages, Al-offretite with gmelinite cages and main channels and SAPO-42 [71] with α-cages, see Fig. 6.11. Two radical cations, $[(CH_3)_3N]^+$ and $[(CH_3)_3NCH_2]^+$, were generated by ionizing radiation in cages of the zeolites containing $[(CH_3)_4N]^+$ ions as an organic template. The $[(CH_3)_3N]^+$ radical cation is stable at room temperature when it is generated in relatively small size cages such as sodalite cages of SAPO-37 and β-cages of SAPO-42. On the other hand, the $[(CH_3)_3NCH_2]^+$ radical cation is stable not only in the sodalite cages of SAPO-37, but also in relatively large size cages such as gmelinite cages or main channels of Al-offretite [72] and α-cages of SAPO-42.

Strongly temperature dependent ESR spectra were observed for the $[(CH_3)_3NCH_2]^+$ radical cation stabilized in Al-offretite and SAPO-42, see

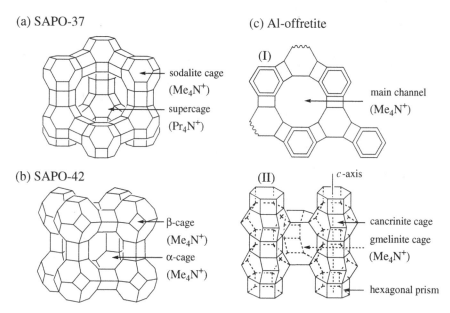

Fig. 6.11 Framework structures of (**a**) SAPO-37, (**b**) SAPO-42 shown as perspective view, and (**c**) Al-offretite viewed parallel (I) and approximately perpendicular (II) to the c axis. Si, Al or P atoms occupy all corners and O atoms (not shown) are near the midpoints of the edges. Possible locations of Me₄N⁺ [(CH₃)₄N⁺] and Pr₄N⁺ [(CH₃CH₂CH₂)₄N⁺] are shown in the structures. The figure is adapted from [62] with permission from the Royal Society of Chemistry

Fig. 6.12(a). The spectra are well simulated by assuming a three-site exchange model for the methyl protons (Fig. 6.12(b)). The three H-atoms labelled H_a, H_b and H_c of a specific methyl group in Fig. 6.12(d) are interchanged by the rotation cycle $H_a \rightarrow H_b \rightarrow H_c \rightarrow H_a$ around the N—CH₃ bond for sites 1, 2, and 3. Rotation of all three methyl groups is assumed. Only one methyl group is labelled, the other two are identical to the first and have the same ^1H hf couplings. The best-fit m_I and temperature dependent linewidths employed are: $\Delta B_{pp} = 0.12$ mT ($m_I = \pm 1$) and 0.08 mT ($m_I = 0$) for the 300 K spectrum; $\Delta B_{pp} = 0.20$ mT ($m_I = \pm 1$) and 0.10 mT ($m_I = 0$) for the spectra above 200 K; $\Delta B_{pp} = 0.20$ mT ($m_I = \pm 1$) and 0.20 mT ($m_I = 0$) below the 200 K spectrum. The other ESR parameters used are the same as those of [(CH₃)₃NCH₂]⁺ in the SAPO-42 system. The evaluated exchange rates are in the order of SAPO-37 (with sodalite cages) < Al-offretite (with gmelinite cages or main channels) < SAPO-42 (with α-cages) in the temperature range 110–300 K.

6.5   Titanium Dioxide (TiO₂) Semiconductor Photocatalysis

Semiconductor-based heterogeneous photocatalysts have been interested by a large number of scientists. Titanium dioxide (TiO₂), which is inexpensive, nontoxic, resistant to photo-corrosion, and has high oxidative power, is the most widely used

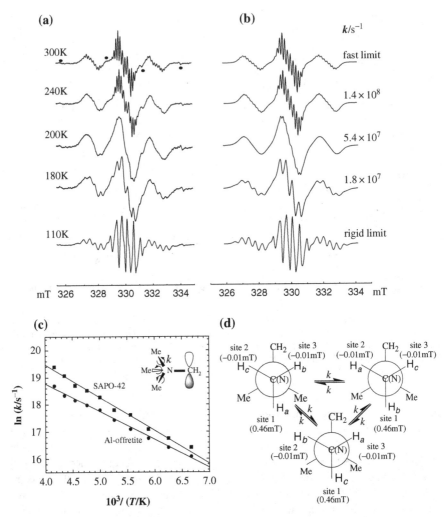

Fig. 6.12 (a) Temperature dependent X-band ESR spectra of $[(CH_3)_3NCH_2]^+$ generated and stabilized in γ-irradiated Al-offretite. The lines marked as a *closed circle* (•) are attributed to $[(CH_3)_3N]^+$. (b) ESR spectra simulated using the three-site exchange model for the hydrogens of the rotating methylgroup of $[(CH_3)_3NCH_2]^+$. (c) Arrhenius plots of the exchange rates, $\ln k(s^{-1})$ vs. $T^{-1}(K^{-1})$ for $[(CH_3)_3NCH_2]^+$ in SAPO-42 (■) and Al-offretite (•). (d) Three-site exchange model for ESR lineshape simulation of $[(CH_3)_3NCH_2]^+$. The figure is adapted from [61] with permission from the Royal Society of Chemistry

material for heterogeneous photocatalysis. The properties of TiO_2 have made it a target for industrial uses including chemical synthesis, solar energy conversion and storage, environmental remediation, odor control, sensors, and protective coatings [15, 73–77].

The advantage of TiO_2 as a semiconductor photocatalyst comes from its ability of converting photon energy into chemical energy. The absorption of photons with band-gap energy excites an electron (e^-) from the semiconductor valence band (VB) to the conduction band (CB), leaving a positively charged hole (h^+) in the valence band. Many of the charged pairs recombine with each other, accounting in part for low photo-efficiency and differences in the photo-activity of various catalysts [78]. A fraction of the electrons and holes moves to the surface and reacts with an adsorbed compound or migrates into trapping sites (e_t; h_t) prior to surface reaction and/or recombination (Fig. 6.13).

A number of TiO$_2$ samples have been developed as photocatalysts and shown to exhibit various photoefficiencies and chemical selectivities [79]. The anatase phase of TiO$_2$ has been in general considered to have the higher activity as an oxidative photocatalyst in comparison with the rutile phase [80, 81], see Appendix A6.2. Furthermore, in addition to the single phase TiO$_2$, a mixed phase TiO$_2$ is also commercially available. For example, Degussa P25 is produced by the addition of rutile to anatase with the ratio of *ca.* 1:4 and shows an unusually high activity [73, 74, 82]. Factors contributing to the increased activity include high surface areas, high adsorption affinity for organic compounds and lower recombination rates. Each of these surface and interface dependent factors increases the capability of chemical reaction by the photogenerated holes and electrons.

An important drawback of TiO$_2$ for photocatalysis is that its band-gap is rather large, 3.0–3.2 eV (λ: 380–410 nm), and only a small fraction of the solar spectrum ($\lambda < 380$ nm, corresponding to the UV region) is absorbed. Sunlight can be more efficiently used in photocatalysis under visible light, rather than UV light. The potential technological impact of this system is huge and, to lower the threshold energy for photoexcitation, a large number of studies have focused on TiO$_2$ doped with both transition metal and non-metal impurities [16, 83–96].

Due to its high sensitivity and its capability of an accurate characterization of paramagnetic species, ESR spectroscopy is particularly useful for the investigation of paramagnetic centers in the solid state. Thus, the ESR method has been extensively applied to the photo-catalytic TiO$_2$ system and the identification and characterization of paramagnetic species such as electrons, holes and their reaction

Fig. 6.13 Conversion of photon energy into chemical energy by semiconductors. Absorbed photons ($h\nu$) lead to charged ion pairs (e^-; h^+) that can be separated and transferred to electron acceptors (A) and electron donors (D) at the surface. The figure is adapted from [15] with permission from Elsevier

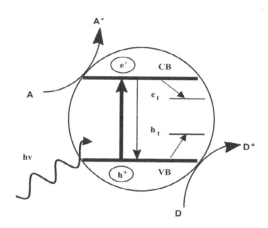

products generated by photocatalytic reactions [15, 16, 96–102]. We start with ESR applications to the nitrogen-doped TiO_2 system.

6.5.1 Nitrogen-Doped TiO₂

One of the most promising and widely investigated systems is nitrogen-doped titanium dioxide, $N-TiO_2$, which shows a significant catalytic activity in various reactions performed under visible light irradiation [83–95]. A highly efficient dye-sensitized solar cell (DSC) has been recently fabricated using a nanocrystalline nitrogen-doped titania electrode [103].

Various preparation methods have been employed to dope nitrogen into TiO_2 either based on chemical reactivity (sol-gel synthesis, chemical treatments of the bare oxide, oxidation of titanium nitride, etc.) or on physical methods (ion implantation, magnetron sputtering) [16]. These different procedures lead, at least in some cases, to materials with somewhat different chemical and physical properties. In addition to the preparation method, many studies have addressed the electronic states associated to the N-impurities including the questions of localized or delocalized states.

Giamello et al. have extensively carried out ESR studies on paramagnetic species in N-doped anatase TiO_2 powders obtained by sol-gel synthesis [16, 96, 102, 104]. Based on a combination of theoretical DFT calculations the paramagnetic N-impurity dopant has been successfully connected to the absorption of visible light and to the photoinduced electron transfer from the bulk to a surface-adsorbed electron scavenger such as molecular oxygen (O_2).

6.5.1.1 ESR Spectra of Nitrogen Centered Radical (N_b˙)

The UV-visible diffuse reflectance spectrum of N-doped TiO_2 (anatase), which was prepared by the sol-gel method, is compared with that of bare TiO_2 (anatase) in Fig. 6.14. The two spectra essentially differ for the broad absorption in the

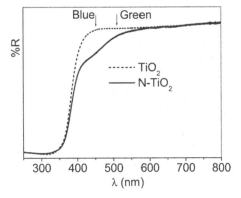

Fig. 6.14 UV-visible diffuse reflectance spectra of bare and N-doped TiO_2 (anatase). The diagram is adapted from [96] with permission from the American Chemical Society

visible region centered at about 450 nm (blue), which characterizes the doped material [96].

The ESR spectra of the N-doped TiO$_2$ sample are shown in Fig. 6.15. The X-band spectra (ν : 9.5 GHz) were recorded for the sample containing either ^{14}N($I = 1$) or ^{15}N($I = 1/2$) isotope and the Q-band spectrum (ν : 35 GHz) for the ^{14}N isotope species.

The ESR spectrum is characterized by the rhombic $\boldsymbol{g}$ tensor ($g_1 = 2.0054$, $g_2 = 2.0036$, and $g_3 = 2.0030$) and the hyperfine $\boldsymbol{A}$ tensor with a large splitting constant of $A_3(^{14}$N$) = (\pm)32.3$ G in the direction of the g_3 component. The other two smaller hf splittings, $A_1(^{14}$N$)$ *and* $A_2(^{14}$N$)$, are given in Table 6.4. Both X- and Q-band spectra are well simulated by using the same set of $\boldsymbol{g}$ and $\boldsymbol{A}$ tensors as shown in Fig. 6.15.

Based on a close relationship with the theoretical ones by the DFT computations the observed hyperfine $\boldsymbol{A}$ tensor components are attributable to nitrogen centered radical (N$_b$˙) ("b" stands for "bulk", see below) located at either substitutional or interstitial site of anatase TiO$_2$ lattice (Fig. 6.16) [16, 104]. Furthermore, the DFT computations suggest a localized electronic state of these species, whose energy levels are located somewhat above the top of the valence band for N$_s$ and N$_i$, respectively [105]. The ESR spectra of N$_b$˙ in Fig. 6.15 show the presence of two slightly different species in the system, one of which is 3.5–4 times more intense than the

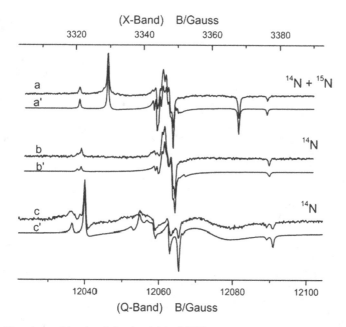

Fig. 6.15 Experimental (**a–c**) and simulated (**a′–c′**) ESR spectra of the N-containing paramagnetic species in N-TiO$_2$. experimental and simulation. (**a, a′**) X-band spectrum of the species containing ^{15}N and ^{14}N isotopes (70 and 30%, respectively). (**b, b′**) X-band spectrum of the species containing ^{14}N (99.63% in natural abundance). (**c, c′**) Q-band spectrum of the species containing ^{14}N. The spectra are adapted from [104] with permission from the American Chemical Society

Table 6.4 ESR parameters of various paramagnetic species identified in N-doped TiO$_2$ system[a]

Radical	g-values			A-values/G[b]				B-values/G[b]			ρ_{2p}
	g_1	g_2	g_3	A_1	A_2	A_3	a_{iso}	B_1	B_2	B_3	
N$_b$˙	2.005	2.004	2.003	2.3	4.4	32.2	13.0	−10.7	−8.6	19.3	0.54
N$_{sub}$˙ (cal)				2.5	2.8	38.2	14.5	−12.0	−11.7	23.7	0.87
N$_{int}$˙ (cal)				0.2	1.8	33.4	11.8	−11.6	−10.0	21.6	0.67
O$_2$˙⁻ (1)[c]	2.023	2.009	2.004								
O$_2$˙⁻ (2)[c]	2.025	2.009	2.003								
NO˙[d]	2.001	1.998	1.927	<1	32.2	9.6					
NO$_2$˙[d]	2.004	2.001	1.990	2.3	4.4	32.3					

[a] Data taken from refs. [16, 104].
[b] 1 G (gauss) = 0.1 mT (T: tesla).
[c] The two signals are due to superoxide radicals stabilized on two distinct surface Ti^{4+} ions which slightly differ in terms of coordinative environment.
[d] By-products of the nitrogen incorporation in the solid and do not directly influence the electronic structure of the system.

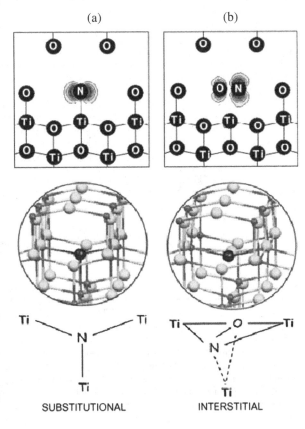

Fig. 6.16 Partial geometry and schematic sketch (*middle*) and electron spin density (SD) maps (*top*) of the unpaired electron of the model for (**a**) substitutional and (**b**) interstitial N-doping in an anatase TiO$_2$ crystal lattice. In the middle the N atom is represented by a black sphere, O atoms are represented by yellow spheres, and Ti atoms are represented by small grey spheres. The SD plot is in the plane perpendicular to that containing the three Ti atoms bound to the N species. The figure is adapted from [104] with permission from the American Institute of Physics

other. The case of two or more species, which have the same chemical nature, but slightly different in the coordinative environment, is rather common in solid state chemistry.

The electron spin density (SD) in the $2p$ orbitals, ρ_{2p}, is evaluated to be $\rho_{2p} = 0.505$ in one $2p$ nitrogen orbital and $\rho_{2p'} = 0.035$ in a second one ($2p'$) by comparison of the experimental anisotropic (or dipolar) hf values (B_3) with the corresponding atomic value (B_0): $\rho_{2p} = B_3/B_0$, where B_0 is $= 39.62$ G [34]. Furthermore, a small spin density of $\rho_{2s} = 0.02$ in the $2s$ nitrogen orbital is evaluated from the isotropic hf splitting of $a = 13$ G with the corresponding atomic value ($a_0 = 646.2$ G) [34]. The total spin density on the nitrogen atom of the observed species amounts therefore to 0.56, with the larger contribution due to a single p orbital of the nitrogen atom in the center. The experimentally obtained hf value does not account for the whole unpaired electron spin density in the radical system. The unaccounted spin density is possibly delocalized on other atoms of the species having zero nuclear spin. It should be noted that the ESR parameters of the present $N_b\dot{}$ radical are quite different from those of an isolated nitrogen atom with a quartet electronic ground state (with $S = 3/2$) [106].

No dipolar broadening was observed for the ESR spectra of the $N_b\dot{}$ species when paramagnetic O_2 molecules are adsorbed at low temperature on the surface [16]. Furthermore, the $N_b\dot{}$ radical is stable upon washing and calcination in air up to 773 K. These experimental results suggest a deep interaction of the $N_b\dot{}$ species with the TiO$_2$ matrix; for this reason the present nitrogen centered radical is written as "$N_b\dot{}$" with a subscript "b" referring to "bulk". The intensity of the ESR signal from the $N_b\dot{}$ species dramatically decreases when the N-doped TiO$_2$ is outgassed at 773 K, the treatment being known to cause oxygen depletion and thus reduction of TiO$_2$, and reversibly reappears after re-oxidation in O_2 atmosphere at the same temperature. These results indicate that the energy levels of the N-species are part of the electronic structure of the solid and that their population is affected by the structural or electronic modifications induced by the reduction of TiO$_2$.

ESR spectra of nitric oxide (NO) and nitrogen dioxide (NO$_2$) radicals trapped in micro-voids of the solid were observed when the N-TiO$_2$ samples were prepared and treated in different ways; the ESR parameters are listed in Table 6.4. The NO radical was found to be a product of the complex oxidation process of ammonium salts occurring upon calcinations of the solid. NO$_2$ was formed only when nitrates or nitric acid were used as nitrogen source and could be thus considered to derive from their decomposition. Due to their nature of the trapped species, it is concluded that both NO and NO$_2$ do not directly influence the electronic structure of the system.

6.5.1.2 Paramagnetic $N_b\dot{}$ Species and Visible Light Illumination

To clarify the role of $N_b\dot{}$ species in the photo-activity of N-doped TiO$_2$, ESR spectra were measured under illumination in vacuum and in oxygen atmosphere, using visible light with two different wavelengths, one in the blue with $\lambda = 437$ nm (corresponding to the maximum absorption of the sample), and the other in the green with

Fig. 6.17 ESR signal
intensity of N_b and $O_2{}^{\cdot-}$ in
various conditions. $I/I_0 = 1$
corresponds to the $N_b{}^{\cdot}$
intensity in the
non-illuminated sample. The
diagram is adapted from [96]
with permission from the
American Institute of Physics

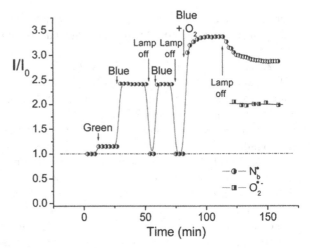

$\lambda = 500$ nm (corresponding to the tail of the absorption band), see Fig. 6.14 [96]. Changes in the ESR signal intensity of $N_b{}^{\cdot}$ in various conditions are summarized together with those of the superoxide ion radical ($O_2{}^{\cdot-}$ or $O_2{}^-$) in Fig. 6.17.

Illumination with green light causes only a small increase in the ESR signal of $N_b{}^{\cdot}$. With blue light illumination, however, the signal intensity grows by a factor of about 2. This higher intensity remains constant until the lamp is turned off, when the signal returns to the initial value. This behavior is reversible, and the signal grows again as the lamp is turned on. When the sample is illuminated with blue light in oxygen atmosphere, the increase in the $N_b{}^{\cdot}$ signal is accompanied by the simultaneous appearance of new ESR signals due to $O_2{}^{\cdot-}$ ion radical (see Section 6.6). The $O_2{}^{\cdot-}$ ion radical was observed with the two different g_{zz} values of 2.025 and 2.023 as shown in Fig. 6.18 (b and b′). This suggests that the $O_2{}^{\cdot-}$ radicals have been stabilized at two distinct surface Ti^{4+} ions, which slightly differ from each other in terms of coordination environment [16, 104].

Fig. 6.18 Experimental
(**a–c**) and simulated (**a′–c′**
and **c″**) ESR spectra of $N_b{}^{\cdot}$
and surface adsorbed $O_2{}^{\cdot-}$
radicals. (**a, a′**) $N_b{}^{\cdot}$ radical.
(**b, b′**) $N_b{}^{\cdot}$ and $O_2{}^{\cdot-}$ radicals.
(**c, c′, c″**) Deconvolution of
spectrum (**b**) (see ESR
parameters of $N_b{}^{\cdot}$, $O_2{}^{\cdot-}$(1),
and $O_2{}^{\cdot-}$(2) in Table 6.4).
The figure is adapted from
[96] with permission from the
American Institute of Physics

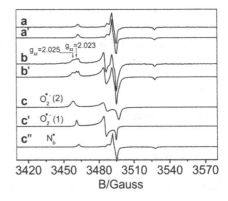

Fig. 6.19 Sketch of the proposed mechanism for the processes induced by visible light illumination of the N-doped TiO$_2$ sample in O$_2$ atmosphere. The diagram is adapted from [96] with permission from the American Institute of Physics

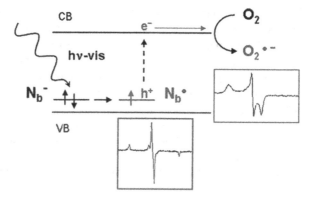

6.5.1.3 A Mechanistic View

Based on the above experimental and theoretical results the following schematic picture can be drawn for the mechanism connected to the interaction of N-TiO$_2$ with visible light (Fig. 6.19) [16, 96].

Upon illumination in vacuum or in O$_2$ atmosphere the ESR signal intensity of N$_b$ increased. This can be explained in terms of electron transfer from diamagnetic N$_b$$^-$ species, *i.e.*, the illumination with visible light at 437 nm (2.84 eV) selectively promotes electrons from the N$_b$$^-$ states to the conduction band according to the following process:

$$N_b^- \rightarrow N_b^\bullet + e^- \tag{6.6}$$

$$N_b^\bullet \rightarrow N_b^+ + e^- \tag{6.7}$$

The visible light has not sufficient energy to excite electrons from the valence band, as the anatase TiO$_2$ has a band gap of 3.2 eV. However, as noted above, the DFT calculations show that the N-induced defect states lie a few tenths of an eV above the valence band edge for both paramagnetic N$_b$ and diamagnetic charged N$_b$$^-$ species. The N$_b$$^-$ species are expected to be definitely more abundant because they are energetically favored and are thus preferentially excited (Eq. 6.11) so that under illumination the equilibrium attained involves an increase in the paramagnetic N$_b$ species (Fig. 6.17). The equilibrium conditions in the dark are recovered instantaneously when the light illumination is stopped.

The presence of oxygen molecules modifies the reactions as a fraction of photoexcited electrons is scavenged by O$_2$, generating O$_2$$^{\bullet-}$ radicals adsorbed on the surface:

$$N_b^- + O_2(\text{gas}) \rightarrow N_b^\bullet + O_2^{\bullet-}(\text{surf}) \tag{6.8}$$

$$N_b^\bullet + O_2(\text{gas}) \rightarrow N_b^+ + O_2^{\bullet-}(\text{surf}) \tag{6.9}$$

This shifts the equilibrium with formation of a larger amount of paramagnetic $N_b^{\cdot}$ centers with respect to the illumination in vacuum. At this stage of the experiment the number of paramagnetic states observed in the system reaches its maximum value as the increase in $N_b^{\cdot}$ concentration is accompanied by $O_2^{\cdot -}$ formation. In other words, a photo induced charge separation has occurred. Stopping illumination, the electrons scavenged by O_2 remain in the adsorbed-layer so that the initial concentration of $N_b^{\cdot}$ centers is not recovered.

The picture proposed here is based on the experimental results for the systems prepared by sol-gel reactions and, very likely, for other chemically prepared N-TiO$_2$ systems. It cannot be excluded, however, that for systems prepared by radically different techniques other types of nitrogen centers are formed, and other mechanisms of photo-activation may apply.

6.5.2 Reversible Photoinduced Electron Transfer in TiO$_2$ (Rutile)

It was shown in the foregoing section that the presence of oxygen molecules can modify the reactions as a fraction of photoexcited electrons is scavenged by O_2 molecules to generate adsorbed superoxide (O_2^- or $O_2^{\cdot -}$) ion radicals. The O_2^- ion radicals have been further suggested to play an important role in the photocatalytic oxidative and reductive degradations of organic pollutants under UV irradiation [74, 107, 108] and also in the cleavage of the conjugated structures of dyes on TiO$_2$ illuminated with visible light [109, 110]. Komaguchi et al. [111] have recently studied photoinduced electron transfer reaction of O_2 species formed on the H$_2$-reduced surface of TiO$_2$ (rutile) by monitoring the ESR spectra of the O_2^- species and Ti^{3+} ions. The ESR study demonstrates that the photoreaction occurs under sub-band gap illumination by the visible light, i.e. < 2.5 eV (> 500 nm), and the reverse process takes place after the illumination.

Figure 6.20(a) shows the 77 K CW-ESR spectrum of the rutile TiO$_2$ sample, which has been thermally reduced under H$_2$ atmosphere. The spectrum consists of a broad anisotropic singlet with a line-width of 3 mT at $g \approx 1.96$, attributable to Ti^{3+} ion (with d^1 electronic configuration) formed on the TiO$_2$ surface [99, 108, 109, 112]. By exposing the sample to O$_2$ at room temperature the spectral intensity of Ti^{3+} decreased to ca. 1/20 of the original, and signals of surface O_2^- radical with orthorhombic g-tensor ($g_x = 2.003$, $g_y = 2.010$, and $g_z = 2.020$ and 2.023) appeared, as seen in Fig. 6.20(b). There are two g_z peaks at 2.023 and 2.020 for the O_2^- ion, suggesting the presence of two kinds of O_2^- species, i.e. O_2^-(A) and O_2^-(B).

Upon visible-light illumination at 77 K, the ESR spectra of the TiO$_2$ sample markedly changed (Fig. 6.20(c)): both O_2^-(B) and Ti^{3+} signals increased in intensity whereas the O_2^-(A) signal remained almost constant. Figure 6.20(d) depicts a correlation between the numbers of the O_2^-(B) and Ti^{3+} ions generated by illumination. A concomitant and equimolar formation of O_2^-(B) and Ti^{3+} ions on the TiO$_2$ surface was suggested by a plot showing a straight line with a slope of unity. Once the illuminated sample was kept at room temperature for several minutes, the ESR

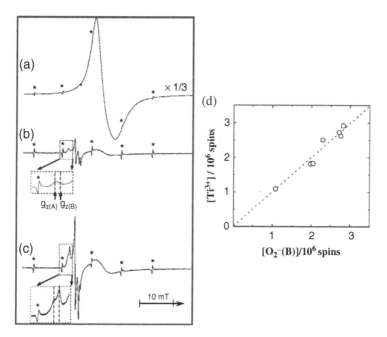

Fig. 6.20 ESR spectra observed at 77 K for rutile TiO$_2$ nanoparticles treated in the following sequence: (**a**) preheated at 573 K in air followed by thermal treatment at 773 K under H$_2$ atmosphere (27 kPa), (**b**) exposed to O$_2$ (4.0 kPa) for 5 min at room temperature, and (**c**) illuminated with visible light for 10 min. The O$_2^-$ line shapes around the g_z component in spectra (**b**) and (**c**) are expanded for clarification. The peaks at $g_{z(A)}$ and $g_{z(B)}$ are designated by O$_2^-$(A) and O$_2^-$(B), respectively. Asterisks indicate signals due to a standard marker, Mn^{2+}/MgO. (**d**) Relationship between the numbers of O$_2^-$(B) radicals and of Ti^{3+} ions increased by visible-light illumination ($\lambda > 500$ nm) at 77 K. The figure is adapted from [111a] with permission from the American Chemical Society

spectra were perfectly restored to the original one, showing that this photoinduced (ESR) spectral change is reversible.

Two different Ti^{3+} sites were assumed to be formed in the thermally treated rutile TiO$_2$ sample under H$_2$ atmosphere; (a) five-coordinate Ti^{3+} site and (b) oxygen vacancy Ti^{3+} site consisting of two Ti^{3+} ions adjacent to each other, see Scheme 6.3. Then, the following reaction scheme was proposed for the O$_2$ adsorption and photoinduced reactions by taking the experimental results into accounts.

An O$_2$ molecule attached to the oxygen vacancy site (b) forms a peroxide O$_2^{2-}$ ion by coordination with the two Ti^{3+} ions, in accord with theoretical studies [113]. When one electron is transferred from O$_2^{2-}$ to one of the two oxidized Ti^{4+} ions by visible-light illumination, the corresponding O—Ti coordination is broken, and a pair of O$_2^-$(B) and Ti^{3+} species are concomitantly formed as observed. On the other hand, adsorption of O$_2$ at the five-coordinate site (a) may generate O$_2^-$(A) by one-electron transfer from Ti^{3+} to the adsorbed O$_2$ molecule. The paramagnetic O$_2^-$(A) species appears to be inactive under visible-light illumination.

(a)

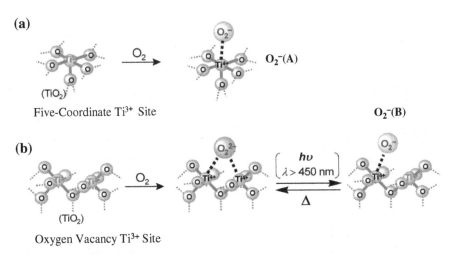

Five-Coordinate Ti³⁺ Site

(b)

Oxygen Vacancy Ti³⁺ Site

Scheme 6.3 Schematic representation of (**a**) five-coordinate Ti^{3+} sites and (**b**) oxygen vacancy sites. O_2^-(A) and O_2^-(B) species are proposed to be formed by visible-light illumination to the O_2 molecules adsorption on (**a**) and (**b**) sites, respectively. The diagram is adapted from [111a] with permission from the American Chemical Society

The ESR spectral response for the generation rate of O_2^-(B) is shown in Fig. 6.21. The generation rate was found to increase with the photon energy and had a maximum at around 480 nm. On the other hand, the diffuse-reflectance vis-NIR(near infrared) spectra show a sharp absorption band around 460 nm for the TiO$_2$ sample treated at 773 K under vacuum; the 460 nm band having been attributed to the oxygen vacancies (*F*-type color centers) in subsurface layers [114–118]. Based on the close correspondence between the 480 nm wave length giving the maximum ESR spectral response and the 460 nm absorption, it was concluded that *F*-type color centers generated in subsurface layers of TiO$_2$ absorb the visible light

Fig. 6.21 ESR intensity of O_2^-(B) plotted against the wavelength of light in samples illuminated at 77 K. The wavelength was controlled by passing the source light through a series of interference filters combined with appropriate cutoff filters. The ESR intensities were measured after 50 s illumination at each wavelength. The diagram was obtained from Dr. K. Komaguchi [111b]

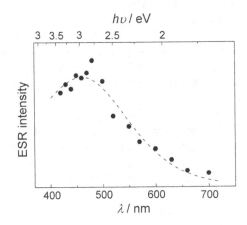

to induce indirectly the electron transfer reaction from O_2^{2-} to Ti^{4+} at the surface oxygen vacancy site.

6.5.3 Electron Transfer in Mixed Phase of Anatase and Rutile

Degussa P-25 consists of a mixed phase of anatase and rutile TiO$_2$ nanoparticles in the ratio of *ca.* 4:1, and exhibits a higher photocatalytic activity than that of each of the pure phases [119, 120]. A number of studies have been carried out to elucidate the mechanism of the synergetic effect in the mixed TiO$_2$ particles. Although it has been generally accepted that the enhanced activity is caused by an efficient charge separation, the detailed mechanism, however, has never been clarified. Recently Komaguchi et al. [121] have reported an "in situ" ESR study on the photo-effects of Ti^{3+} formed in partially reduced TiO$_2$ nanoparticles. A preferential electron transfer from the Ti^{3+} ($3d^1$) ion in anatase to the Ti^{4+} ($3d^0$) ion in rutile phases in the TiO$_2$ (P-25) particles took place upon light illumination with an energy lower than the band gap. An advantage of using partially reduced TiO$_2$ is that the electron transfer is possible to observe by ESR without any serious interference due to the charge recombination with positive holes as the electron is released only from the paramagnetic Ti^{3+} ions by visible light illumination.

6.5.3.1 ESR Spectra of Ti^{3+} in Partially Reduced TiO$_2$

The ESR spectra shown in Fig. 6.22 were observed at 77 K for anatase, rutile, and P-25 TiO$_2$ particles, which had been partially reduced in H$_2$ atmosphere at 773 K. Both pure anatase and rutile TiO$_2$ samples show a singlet ESR spectrum with a small *g*-anisotropy whose lineshapes are almost independent of the observation temperature in the range of 3.6–77 K. Another important spectral characteristic is that the linewidth is considerably broader for the anatase sample than for the rutile, which makes it possible to evaluate relative amount (concentration) of ESR signals from the anatase and rutile particles in the mixed phase TiO$_2$ sample (for example, Degussa P-25) by the ESR spectral simulation method. The following ESR *g*-tensors and the peak-to-peak linewidth (ΔB_{pp}) were evaluated by ESR spectral simulations: $g_\perp = 1.953$, $g_{||} = 1.886$, and $\Delta B_{pp} = 8.0$ mT for the anatase, and $g_\perp = 1.963$ and $g_{||} = 1.931$, and $\Delta B_{pp} = 3.1$ mT for the rutile. The observed *g*-values are attributable to the paramagnetic Ti^{3+} ions located on the surface of the anatase and rutile particles [119, 122]. The assignment to surface Ti^{3+} ions is supported by the experimental result showing that an electron is irreversibly moved from the Ti^{3+} to O$_2$ to generate a superoxide (O$_2^-$) anion radical when the O$_2$ molecules are introduced to the sample.

The ESR spectrum of the P-25 sample was well simulated by the superposition of two distinct Ti^{3+} signals from the anatase and rutile phases with a 1:1 concentration ratio, see Fig. 6.22(c). The P-25 sample originally consists of a phase composition ratio of 4:1 for the anatase and rutile. The increased ratio of the rutile Ti^{3+} ions in comparison to the original phase composition indicates that in the mixed phase the

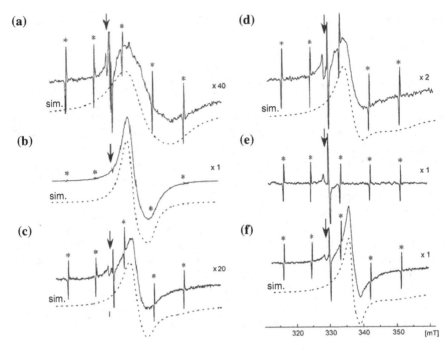

Fig. 6.22 ESR spectra observed at 77 K for partially reduced TiO_2 nanoparticles, which were prepared by heating the TiO_2 sample at 773 K in vacuum and then in H_2 gas atmosphere for 1 h. (a) Anatase, (b) Rutile, (c) Degussa P-25, (d) Degussa P-25 after exposure to air (i.e., a trace amount of air was admitted for several seconds and evacuated at room temperature prior to illumination). (e) Upon visible light ($\lambda = 500$–800 nm) illumination of (d). (f) In the dark after illumination of (e). The simulated ESR line shapes are shown by dotted lines. The ESR parameters used for the simulation are given in the text. The peaks due to the Mn^{2+}/MgO standard marker are denoted by asterisks. The signals around the field denoted by an arrow are due to defects in the glass of the liquid N_2 Dewar. The spectra are adapted from [121] with permission from Elsevier

Ti^{4+} ions in the rutile phase are more easily reduced to the Ti^{3+} ions than in the anatase phase. Note that the relative concentration of the Ti^{3+} ions was changed to a 3:1 ratio from the 1:1 ratio when the H_2-reduced P-25 sample was contacted to a trace amount of air prior to illumination, see Fig. 6.22(d).

6.5.3.2 Synergetic Effects on Visible Light Illumination

The ESR spectral intensity and the line-shape of the Ti^{3+} ions in the partially reduced pure phase rutile and anatase samples changed upon visible light illumination in the ESR cavity at 77 K [121]. The signal intensity decreased and disappeared entirely after several minutes of illumination with white light. When the light was turned off, the signals started to reappear and increased in the dark at 77 K. Finally, the signals were restored completely to the initial intensity within 50 min. The overall photoresponse of the Ti^{3+} ESR signal intensity is shown in Fig. 6.23.

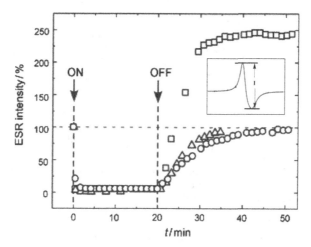

Fig. 6.23 Photoresponse of the ESR spectral intensity (I: the peak-to-peak height in the first derivative spectrum) of Ti^{3+} observed for partially reduced TiO_2 nanoparticles at 77 K. (Δ) anatase (AMT-100), (Δ) rutile (STR-60C), (□) Degussa P-25 after air treatment. The samples were illuminated with white light from $t = 0$ min to $t = 20$ min. The diagram is adapted from [121] with permission from Elsevier

The Ti^{3+} signal restoration of the air contacted P-25 sample shows a striking contrast to those of the pure phases as shown in Fig. 6.23. The enhanced spectral intensity was attributed to the increase in the rutile Ti^{3+} ions with the narrow linewidth. In fact the relative concentration of rutile Ti^{3+} ions increased two times after the restoration, whereas the total amount of Ti^{3+} ions remained unchanged before and after the illumination. Taking the TiO_2 band gap of 3.0–3.2 eV (λ: 380–410 nm) into account, the observed photo-response can be caused by the excitation of electrons from the surface Ti^{3+}, but not from the valence band. This is in accord with the suggested energy level of the Ti^{3+} ions being 0.3–0.8 eV below the conduction band edge [123]. The electrons in the conduction band, which are generated from the Ti^{3+} ions by the photo excitation, are ESR silent until they relax back so as to be re-trapped by the Ti^{4+} ions as Ti^{3+} ions. Thus it was concluded that the electrons of Ti^{3+} are preferentially transferred from the anatase to the rutile by photoexcitation and retrapped at the rutile surface in dark as illustrated in Fig. 6.24 [121].

6.6 Superoxide ($O_2{}^-$) Ion Radical

The superoxide ($O_2{}^-$) ion radical is one of the most important oxide radical intermediates in catalysis and has been extensively studied by means of ESR spectroscopy. Känzig and Cohen [124] have reported the first ESR spectrum of $O_2{}^-$ trapped in a single crystal of alkali halides five decades ago. The reported g-value analysis has been well accepted and adapted by Lunsford [3], Kasai [4] and other scientists (for

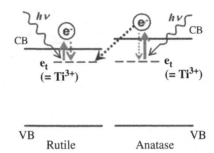

Fig. 6.24 Sketch of the proposed mechanism for the processes induced by visible light illumination of partially reduced a mixed phase TiO_2 sample (P-25). The electrons are excited from the surface Ti^{3+} ion to the conduction band (CB) upon illumination of light energy lower than the band gap and are re-trapped as Ti^{3+} ions in the dark. For the partially reduced P-25 sample, a certain amount of the excited electrons in the anatase phase can transfer through the interfacial boundaries to be re-trapped by the surface Ti^{4+} ions in the rutile phase. The diagram is adapted from [121] with permission from Elsevier

example, Shiotani [125]) in studies of the ion radicals adsorbed on various metal oxides and supported catalysts. Here we summarize some characteristics of the g-values and the hyperfine spectrum due to ^{17}O ($I = 7/2$) labeling obtained from an ESR study on O_2^- adsorbed on Ti^{4+} ions on oxide supports [125, 126].

6.6.1 g-Values of O_2^-

We start by comparing the electronic structure of the homonuclear O_2^- ion radical with that of the heteronuclear NO radical discussed in Section 6.2.1 because of their close similarity in electronic structure. The O_2^- radical is a 17 electron system having two additional electrons compared with the NO radical (15 electrons system). The degenerate anti-bonding $2p\pi_g^*(x,y)$ orbitals are occupied by three electrons in a "free" (not adsorbed) O_2^- radical, whereas the $2p\pi^*(x,y)$ orbitals contain one electron in the NO radical. Thus, for O_2^- a hole resides in a $2p\pi_g^*(x)$ orbital after the orbital degeneracy is lifted by the crystal field (Δ) due to the Ti^{4+} ion on which O_2^- is adsorbed (or bonded), see Fig. 6.25. For NO the unpaired electron resides in a $2p\pi^*(y)$ orbital after the orbital degeneracy being removed. This leads to a change in the sign of spin-orbit coupling (λ), i.e., minus sign for O_2^- instead of plus sign for NO. Thus the g-component along the inter-nuclear axis, g_z, are shifted in opposite directions from the g_e value, i.e., $g_{zz} = g_e + \dfrac{2l|\lambda|}{\sqrt{\lambda^2+\Delta^2}}$ for O_2^- and

$g_{zz} = g_e - \dfrac{2l|\lambda|}{\sqrt{\lambda^2+\Delta^2}}$ for NO, refer to Eq. (6.3).

The previously reported g_z values of O_2^- range from 2.03 to 2.02. Consistent with this, for the present $O_2^-/Ti^{4+}/oxide$ support system, the following g-values were derived from the ESR spectrum recorded at 4.2 K: $g_x = 2.0027$, $g_y = 2.0092$ and $g_z = 2.0268$. The small shift of g_x and g_y from g_e and the order of $g_x < g_y$ are also supported by the theory [3c, 4d].

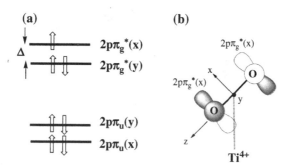

Fig. 6.25 The O_2^- ion radical adsorbed at Ti^{4+} ion on oxide supports. (**a**) Illustration of the unpaired electron $2p\pi_g^*(x)$ orbital in the ground electronic state in which the molecular coordinates and the principal axes of the g-tensor coincide. (**b**) Energy levels of O_2^- interacting with the Ti^{4+} ion. The term Δ stands for the energy splitting between $2p\pi_g^*(x)$ and $2p\pi_g^*(y)$ orbitals. The reader is referred to Fig. 6.2. for a comparison with the NO radical adsorbed at a metal ion

6.6.2 ^{17}O Labeling Study

The ^{17}O labeling is a powerful method for identifying oxygen species and their structure in the adsorbed phase on oxides and other catalysts by ESR. An observation of ^{17}O ($I = 7/2$) hyperfine structure can give an unambiguous assignment of the ESR spectrum, e.g. if it is due to a superoxide ion O_2^-, and not other oxide radicals such as O^- and O_3^-. Furthermore, it can provide important experimental evidence if the O_2^- consists of either equivalent or non-equivalent oxygen nuclei.

The ESR spectrum of ^{17}O enriched O_2^- adsorbed on Ti^{4+} ions on oxide supports is shown in Fig. 6.26(b). The ESR spectrum is well simulated by a superposition of three different spectra of $(^{17}O-^{18}O)^-$, $(^{17}O-^{17}O)^-$, and $(^{18}O-^{18}O)^-$; the g-values and ^{17}O hf splittings employed are given in Table 6.5. An expanded spectrum of the central part is shown together with the simulated ones of $(^{17}O-^{18}O)^-$ and $(^{17}O-^{17}O)^-$ species in Fig. 6.27.

Comparing the experimental ESR spectra with the simulated ones the following conclusions were suggested. (1) The principal values of the ^{17}O hf tensor are almost axially symmetric with the perpendicular component $A(\approx A_2 \approx A_3)$ being less than the linewidth of ca. 3 G. (2) Two different parallel components of ^{17}O hf splittings were observed for both $(^{17}O-^{18}O)^-$ and $(^{17}O-^{17}O)^-$ radical ions with $A_1 (= A_{//}) = 74.8$ G and $A_1' (= A_{//}) = 80.3$ G. (3) The observation of the

Table 6.5 ESR parameters of O_2^- adsorbed on Ti^{4+}/oxide supports[a]

	g_x, g_y, g_z	^{17}O A_1, A_2, A_3/G[b]	Abundance/%
(1) $(^{17}O-^{18}O)^- -Ti^{3+}$	2.0027, 2.0092, 2.0268	74.9, 0, 0	50
(2) $(^{18}O-^{17}O)^- -Ti^{3+}$	2.0027, 2.0092, 2.0268	80.3, 0, 0	50

[a] Data taken from ref. [125].
[b] 1 G (gauss) = 0.1 mT (T: tesla).

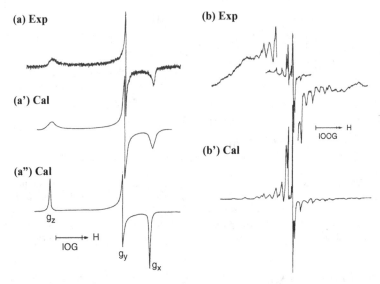

Fig. 6.26 (a) X-band ESR spectrum of O_2^- adsorbed on Ti^{4+} ions on oxide supports recorded at 4.2 K. (a') The best fit simulated spectrum using one set of rhombic g-values listed in Table 6.5 and an orientation dependent Lorentzian linewidth 0.95, 0.18, and 1.35 G for the x, y, and z component, respectively. (a") The spectrum calculated using the same ESR parameters in (a'), but an orientation independent linewidth of 0.2 G. (b) X-band ESR spectrum of ^{17}O enriched O_2^-; the experiments were carried out using ^{17}O enriched O_2 with 70% of ^{17}O with $I = 7/2$ and 30% of ^{18}O with $I = 0$). (b') Spectrum simulated by superposing three different spectra of $(^{17}O—^{18}O)^-$, $(^{17}O—^{17}O)^-$, and $(^{18}O—^{18}O)^-$ with a relative intensity of 0.42: 0.49: 0.09. See Fig. 6.27 for the expanded spectra of the central part in (b). The spectra are adapted from [125] with permission from the American Institute of Physics

nonequivalent ^{17}O *hf* splittings suggests that the inter-nuclei axis (z-axis) of O_2^- is tilted slightly from the surface and/or one oxygen is closer to the Ti^{4+} ion. A molecular orbital study resulted in a small tilting of the O_2 axis from the parallel conformation (*i.e.*, less than 10°) [126]. (4) The observed ^{17}O *hf* splitting of A_1 and A_1' correspond to the minimum g-tensor component, $g_x = 2.0027$.

The ESR spectra of O_2^- adsorbed on supports sometimes show strong temperature dependency. Such ESR spectral changes are generally accompanied by shifting and/or broadening of certain features due to g-tensor anisotropy and give very rich information about the motional dynamics of the O_2^- on oxide surface [125].

6.7 Summary

Paramagnetic chemical species with unpaired electron(s) are involved as reaction intermediates and/or active sites in many catalytic reactions. Thus ESR spectroscopy has played an important role to obtain valuable experimental information on catalytic and/or surface reactions with high selectivity and high sensitivity,

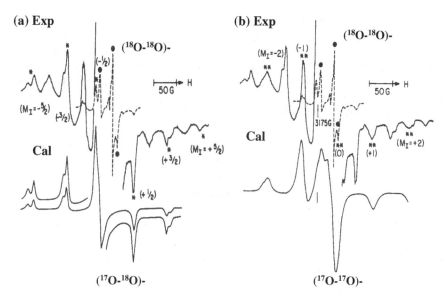

Fig. 6.27 *Upper spectra*: Expanded ESR spectra of the central part in Fig. 6.26(b). The peaks marked as (*), (**) and (•) correspond to resonances due to $(^{17}O-^{18}O)^-$, $(^{17}O-^{17}O)^-$ and $(^{18}O-^{18}O)^-$ ion radicals, respectively. *Lower spectra*: The best fit simulation spectra calculated for the rigid limit spectra (including second-order corrections) with the two different sets of ^{17}O *hf* splittings listed in Table 6.5. The spectra are adapted from [125] with permission from the American Institute of Physics

which has not been achieved by any other methods. With recent advancement in the measurement techniques and the data analysis methods the ESR method is an increasingly important tool in the studies on heterogeneous catalysis and solid surfaces. This chapter consisted of five topics relevant to ESR application to catalysis and environmental science.

The interaction of nitric oxide (NO) with metal ions in zeolites has been one of the major subjects in catalysis and environmental science and the first topic was concerned with NO adsorbed on zeolites. NO is an odd-electron molecule with one unpaired electron and can be used here as a paramagnetic probe to characterize the catalytic activity. In the first topic focus was on a mono NO-Na$^+$ complex formed in a Na$^+$-LTA type zeolite. The experimental ESR spectrum was characterized by a large g-tensor anisotropy. By means of multi-frequency ESR spectroscopies the g tensor components could be well resolved. The ^{14}N and ^{23}Na hyperfine tensor components were accurately evaluated by ENDOR spectroscopy. Based on these experimentally obtained ESR parameters the electronic and geometrical structures of the NO-Na$^+$ complex were discussed. In addition to the mono NO-Na$^+$ complex the triplet state $(NO)_2$ bi-radical is formed in the zeolite and dominates the ESR spectrum at higher NO concentration. The structure of the bi-radical was discussed based on the ESR parameters derived from the X- and Q-band spectra. Furthermore the dynamical ESR studies on nitrogen dioxides (NO_2) on various zeolites were briefly presented.

The second topic is an extension of the first one and was concerned with ESR studies of the Cu(I)-NO complexes. Copper ion exchanged high siliceous zeolites such as ZSM-5 and MCM-22 have been considered as a promising environmental catalyst for the NO decomposition. The Cu(I)-NO complex has attracted special interest because of its important intermediate in the catalytic NO decomposition. Pöppl and other scientists have extensively applied multi frequency ESR, pulsed ENDOR and HYSCORE methods to clarify the local structure of Cu(I)-NO adsorption complexes.

The third topic was concerned with ESR studies on the structure and the dynamics of organic radicals in zeolites. ESR methods were applied to investigate microenvironment effects on molecular dynamics and thermal stability of relatively large organic molecules such as triethylamine $((CH_3CH_2)_3N^{+\bullet})$ and tripropylamine $((CH_3CH_2CH_2)_3N^{+\bullet})$ cation radicals used as spin probes. The cation radicals were generated by γ-ray irradiation of amines and related ammonium ions in various zeolites and subjected to X-band CW-ESR studies. The experimentally observed temperature dependent ESR spectral lineshapes were successfully analyzed by assuming a two-jump exchange process of the methylene hydrogens next to the nitrogen. The exchange rates and the barriers heights evaluated were discussed in terms of interaction with the surrounding zeolite wall by referring to theoretical DFT calculations. Furthermore the ESR studies using amine cation radicals as a spin probe were extended to investigate "cage effects" on stability and molecular dynamics of organic molecules in zeolites.

The fourth topic was concerned with titanium dioxide (TiO_2) semiconductor photocatalysis. ESR spectroscopy has been extensively applied to the TiO_2 systems and played an important role in the identification and characterization of paramagnetic species such as electrons, holes and their reaction products generated by photocatalytic reactions. The band-gap of pure TiO_2 is 3.0–3.2 eV and only a small fraction of the solar spectrum ($\lambda < 380$ nm, corresponding to the UV region) is absorbed. This is an important drawback of TiO_2 for photocatalysis using sunlight. A large number of modified TiO_2 systems have been prepared to reduce the threshold energy for photoexcitation so as to use sunlight more efficiently in photocatalysis under visible light. One of the most promising and widely investigated systems is nitrogen-doped titanium dioxide (N-TiO_2). The N-TiO_2 samples were illuminated by visible light and subjected to X- and Q-band ESR studies. Based on the ESR studies it was revealed that substitutionally or interstitially doped diamagnetic nitrogen ion (N_b^-) absorbs visible light at 437 nm (2.84 eV) so as to promote an electron to the conduction band, generating N-centered neutral radical ($N_b^{\bullet}$). Furthermore, the ESR studies clarified that the presence of O_2 molecules modifies the reactions as a fraction of photoexcited electrons is scavenged by O_2, generating superoxide (O_2^-) ion radicals. In addition to the N-TiO_2 system ESR spectroscopy was applied to the photoinduced electron transfer reaction of O_2 species formed on partially reduced TiO_2 (rutile). By monitoring ESR spectra of paramagnetic O_2^- species and Ti^{3+} ions it was revealed that the electron transfer could occur under sub-band gap illumination by visible light, and the reverse process takes place after the illumination. Furthermore, synergetic effects on the electron transfer upon illumination with visible light in a mixed phase of anatase and rutile were also presented by monitoring the ESR intensity of Ti^{3+} ions in partially reduced TiO_2 nanoparticles.

The last topic was concerned with the ESR spectra of the superoxide ion radical (O_2^-) which is one of the most important oxide radical intermediates in catalysis and has been extensively studied by means of ESR spectroscopy. Some important characteristics of the g-values and the hyperfine structure due to ^{17}O ($I = 7/2$) labeling were presented by exemplifying an ESR study on O_2^- adsorbed on Ti^{4+} ions on oxide supports.

Appendices

A6.1 Isotropic Hyperfine Splittings of β-Hydrogens

The isotropic *hf* splitting of β-hydrogens, $a_{\beta\text{-}H}$, arises, in principle, due to a hyperconjugative mechanism [56b]. To illustrate the general ideas here we consider triethylamine radical cation, $(CH_3CH_2)_3N^{\cdot+}$, described in Section 6.4. The hydrogens of $(CH_3CH_2)_3N^{\cdot+}$ are conventionally labeled with *Greek* letters: β for one bonded to carbon adjacent to the π-radical center (α-position; N atom) and γ for a hydrogen one carbon further out as $C_\gamma H_3\text{-}C_\beta H_2\text{-}N_\alpha{}^{\cdot+}\text{-}(CH_3CH_2)_2$. It is found experimentally that the β-hydrogen *hf* splitting is formulated as follows [127]:

$$a_{\beta-H} = (A + B\cos^2\theta)\rho \tag{6.10}$$

where θ is the angle between the unpaired electron $2p_z$ orbital at the π-radical center and the $C_{(\beta)}$-H bond and ρ is the spin density on the nitrogen $2p_z$ orbital, see Fig. 6.8(c). Coefficient B reflects the spin density arising from hyperconjugation and should be positive; the positive sign has been confirmed by NMR experiments. On the other hand coefficient A accounts for that arising from orientation-independent mechanisms such as spin polarization. If there is free rotation about the C_β—N_α bond then we observe an orientationally averaged *hf* splitting:

$$\left\langle a_{\beta-H}\right\rangle_{av} = \left(A + B\left\langle\cos^2\theta\right\rangle_{av}\right)\rho = (A + (1/2)B)\rho \tag{6.11}$$

Studies of many systems suggest that the value of A is much smaller than B, i.e., less than *ca.* 0.3 mT and Eq. (6.10) can be approximated as:

$$a_{\beta-H} = B\rho\cos^2\theta \tag{6.12}$$

From the ESR spectral analysis of $(CH_3CH_2)_3N^{\cdot+}$ in AlPO$_4$-5 the following isotropic *hf* splittings have been identified: $a_{\beta\text{-}H} = 2.0$ mT for six equivalent β-hydrogens at 300 K corresponding to an averaged structure and $a_{\beta\text{-}H} = 3.6$ mT for three equivalent hydrogens, one from each β-methylene group, at a low temperature of 77 K corresponding to a rigid limit structure, see Fig. 6.8(a) and (b). Combining Eq. (6.12) with the experimental β-hydrogen *hf* splittings at 77 and 300 K, we have the following four equations:

$$a_{\beta-H}(1) = B\rho \cos^2 \theta(1) = 3.6\text{mT (experimental value at 77 K)} \qquad (6.13)$$

$$a_{\beta-H}(2) = B\rho \cos^2 \theta(2) \qquad (6.14)$$

$$\left(a_{\beta-H}(1) + a_{\beta-H}(2)\right)/2 = 2.0\text{mT (experimental value at 300 K)} \qquad (6.15)$$

$$\theta(1) + \theta(2) = 120° \text{ (assumption)} \qquad (6.16)$$

By solving the equations we obtain: $B\rho = 6.9$ mT, $a_\beta{}^H(2) = 0.4$ mT, $\theta(1) = 44°$ and $\theta(2) = 76°$, see Fig. 6.8(c). Consistent with the experimental result the value evaluated for $a_{\beta H}(2)$, 0.4 mT, is much smaller than the experimental line-width of 1.2 mT at 77 K and too small to be resolved in the spectrum.

We close this section by repeating that Eq. (6.10) for the angular dependence on β-hydrogens is very useful in discussing geometrical structures (conformations) of not only the amine radical cations exemplified here but also for other many alkyl-type radicals [56b] including those stabilized in organic solid polymers as described in Chapter 7.

A6.2 Anatase and Rutile TiO₂

Titanium dioxide has several polymorphs. Among them, rutile ($P4_2/mnm$ space group, $D_{4\,h}$) and anatase ($I4_1/amd$ space group, $D_{4\,h}$) are well known from viewpoint of crystal structure [128] and have been intensively studied for photocatalysis. Experimentally, anatase has a slightly larger band gaps than rutile, 3.2 *vs.* 3.0 eV [129, 130]. Ab-initio calculations show that the primary structural difference between the anatase and rutile phases is that the former is 9% less dense than the latter, and has larger Ti-Ti distances, a more pronounced localization of the Ti $3d$ states and a narrower $3d$ band [131]. This can be a reason why the carrier (electron) generated in anatase by UV excitation is less mobile than in rutile. Also the O $2p$-Ti $3d$ hybridization is different in the two structures (more covalent mixing in rutile), with anatase exhibiting a valence and a conduction band with more pronounced O $2p$ and Ti $3d$ characters, respectively [131]. In rutile the greater Pauli repulsion among the oxygen $2p$ electrons results in a larger O $2p$ bandwidth. Experimentally, the bandwidth in rutile is 6 eV while it is 4.7 eV in anatase [132, 133]. The calculated values are 5.3 and 4.5 eV, respectively; the values nicely reflect the important difference in the electronic structures [105].

References

1. E. Angelescu, C. Nicolau, Z. Simon: J. Am. Chem. Soc. **88**, 3910 (1962). (b) E.H. Adema: J. Polym. Sci. C **16**, 3643 (1968).
2. D.E. O'Reilly, D.S. MacIver: J. Phys. Chem. **66**, 277 (1962).
3. J.C. Lunsford: Adv. Catal. **22**, 265 (1972). (b) J.C. Lunsford: J. Phys. Chem. **72**, 4163 (1968). (c) J.C. Lunsford: Catal. Rev. Sci. Eng. **12**, 137 (1975).
4. (a) P.H. Kasai, R.J. Bishop Jr.: In ' *Zeolite Chemistry and Catalysis*', ACS Monograph 171, ed. by J.A Rabo, American Chemical Society, Washington, DC (1976), p. 350. (b) P.H. Kasai, R.J. Bishop Jr.: J. Am. Chem. Soc. **94**, 5560 (1972). (c) P.H. Kasai, R.M. Gaura: J. Phys. Chem. **86**, 4257 (1982). (d) P.H. Kasai: J. Chem. Phys. **43**, 3322 (1965).
5. M. Che, E. Giamello: In *'Spectroscopic Characterization of Heterogeneous Catalysis'*, Vol. 57, ed. by J.L.G. Fierro, Elsevier, Amsterdam (1993), p. 265.
6. K. Dyrek, M. Che: Chem. Rev. **97**, 306 (1997).
7. A. Lund, C. Rhodes (eds.): *'Radicals on Surfaces'*, Kluwer Academic Publisher, Dordrecht (1994).
8. (a) D. Biglino, H. Li, R. Erickson, A. Lund, H. Yahiro, M. Shiotani: Phys. Chem. Chem. Phys. **1**, 2887 (1999). (b) M. Shiotani, H. Yahiro, A. Lund: Zeolites 104 (1998).
9. M. Hartmann, L. Kevan: Chem. Rev. **99**, 635 (1999).
10. T. Rudolf, A. Pöppl, W. Brunner, D. Michel: Magn. Reson. Chem. **37**, 93 (1999).
11. Z. Sojka, M. Che: Appl. Magn. Reson. **20**, 433 (2001).
12. W. Liu, A. Lund, M. Shiotani, J. Michalik, D. Biglino, M. Bonora: Appl. Magn. Reson. **24**, 285 (2003).
13. H. Yahiro, A. Lund, M. Shiotani: Spectroschim. Acta. A **60**, 1267 (2004).
14. P. Decyk: Catal. Today **114**, 142 (2006).
15. D.C. Huruma, A.G. Agrios, S.E. Crist, K.A. Gray, T. Rajh, M.C. Thurnauer: J. Electron Spectros. Relat. Phenomena **150**, 155 (2006).
16. C.D. Valentin, E. Finazzi, G. Pacchioni, A. Selloni, S. Livraghi, M.C. Paganini, E. Giamello: Chem. Phys. **339**, 44 (2007).
17. T. Nagano, T. Yoshimura: Chem. Rev. **102**, 1235 (2002).
18. Y. Li, W.K. Hall: J. Phys. Chem. **94**, 6145 (1990).
19. (a) M. Iwamoto, H. Yahiro, N. Mizuno: Nippon Kagaku Kaishi 574 (1991). (b) M. Iwamoto, H. Hamda: Catal. Today **10**, 57 (1991). (c) S. Sato, Y. Yu-u, H. Yahiro, N. Mizuno, M. Iwamoto: Appl. Catal. **70**, 11 (1991). (d) H. Yahiro and M. Iwamoto: Appl. Catal. A **222**, 103 (2001).
20. Y. Traa, B. Burger, J. Weitkamp: Microporous Mesoporous Mater. **30**, 3 (1993).
21. M. Shelef: Chem. Rev. **95**, 209 (1995).
22. (a) G. Centi, S. Perathoner, F. Vazzana, M. Marella, M. Tomaselli, M. Mantegazza: Adv. Envir. Res. **4**, 325 (2000). (b) G. Centi, P. Ciambelli, S. Perathoner, P. Russo: Catal. Today, **75**, 3 (2002).
23. M.M. Hoffman, N.J. Nelson: J. Chem. Phys. **50**, 2598 (1969).
24. D. Biglino, H. Li, R. Erickson, A. Lund, H. Yahiro, M. Shiotani: Phys. Chem. Chem. Phys. **1**, 2887 (1999).
25. A. Pöppl, T. Rudolf, D. Michel: J. Am. Chem. Soc. **120**, 4879 (1998).
26. A. Pöppl, T. Rudolf, P. Manikandan, D. Goldfarb: J. Am. Chem. Soc. **122**, 10194 (2000).
27. T. Rudolf, A. Pöppl, W. Brunner, D. Michel: Magn. Reson. Chem. **37**, S93 (1999).
28. T. Rudolf, A. Pöppl, W. Hofbauer, D. Michel: Phys. Chem. Chem. Phys. **3**, 2167 (2001).
29. E. Giamello, D. Murphy, G. Magnacca, C. Morterra, Y. Shioya, T. Nomura, M. Anpo: J. Catal. **136**, 510 (1992).
30. T.I. Barry, L.A. Lay: J. Phys. Chem. Solids **29**, 1395 (1968).
31. V. Umamaheswari, M. Hartmann, A. Pöppl: J. Phys. Chem. B **109**, 1537 (2005).
32. Y.-J. Liu, A. Lund, P. Persson, S. Lunell: J. Phys. Chem. B **109**, 7948 (2005).

33. (a) B.M. Hoffman, J. Martinsen, R.A. Venters: J. Magn. Reson. **59**, 110 (1984). (b) G.C. Hurst, T.A. Henderson, R.W. Kreilick: Am. Chem. Soc. **107**, 7294 (1985). (c) A. Kreiter, J. Hüttermann: J. Magn. Reson. **93**, 12 (1991).

34. J.R. Morton, K.F. Preston: J. Magn. Reson. **30**, 577 (1978).

35. C.A. Hutchison Jr., D.B. McKay: J. Chem. Phys. **66**, 3311 (1977).

36. H. Yahiro, A. Lund, R. Aasa, N.P. Benetis, M. Shiotani: J. Phys. Chem. A **104**, 7950 (2000).

37. H. Yahiro, K. Kurohagi, G. Okada, Y. Itagaki, M. Shiotani, A. Lund: Phys. Chem. Chem. Phys. **4**, 4255 (2002).

38. A. Volodin, D. Biglino, Y. Itagaki, M. Shiotani, A. Lund: Chem. Phys. Lett. **327**, 165 (2000).

39. H. Zeldes, R. Livingston: J. Chem. Phys. **35**, 563 (1961).

40. M. Shiotani, J.H. Freed, J. Phys. Chem. **85**, 3873 (1981).

41. For example, see J.H. Freed: In '*Spin Labelling: Theory and Applications*', Vol. 1, ed. by L. Berliner, Academic Press, New York, NY (1976), Chapter 3.

42. (a) H. Yahiro, M. Shiotani, J.H. Freed, M. Lindgren, A. Lund: Stud. Surf. Sci. Catal. **94**, 673 (1995). (b) M. Nagata, H. Yahiro, M. Shiotani, M. Lindgren, A. Lund: Chem. Phys. Lett. **256**, 27 (1996).

43. H. Li, H. Yahiro, K. Komaguchi, M. Shiotani, E. Sagstuen, A. Lund: Microporous Mesoporous Mater. **30**, 275 (1999).

44. H. Li, A. Lund, M. Lindgren, E. Sagstuen, H. Yahiro: Chem. Phys. Lett. **271**, 84 (1997).

45. H. Yahiro, M. Nagata, M. Shiotani, M. Lindgren, H. Li, A. Lund: Nukleonika **42**, 557 (1997).

46. H. Li, H. Yahiro, M. Shiotani, A. Lund: J. Phys. Chem. B **102**, 5641 (1998).

47. B.M. Hoffman, T.B. Eames: J. Am. Chem. Soc. **91**, 5186 (1969).

48. G.P. Lozos, B.M. Hoffman: J. Phys. Chem. **78**, 2110 (1974).

49. E.V. Lunina, G.L. Markaryan, O.O. Parenago, A.V. Fionov: Colloids Surf. A **72**, 333 (1993).

50. M. Gutjahr, A. Pöppl, W. Böhlmann, R. Böttcher: Colloids Surf. A **189**, 93 (2001).

51. A. Pöppl, M. Hartmann, L. Kevan: J. Phys. Chem. **99**, 17251 (1995).

52. W. Bohlmann, A. Pöppl, D. Michel: Colloids Surf. A **158**, 235 (1999)

53. V. Umamaheswari, M. Hartmann, A. Pöppl: J. Phys. Chem. B **109**, 10842 (2005).

54. V. Umamaheswari, M. Hartmann, A. Pöpp: J. Phys. Chem. B **109**, 19723 (2005).

55. V. Umamaheswari, M. Hartmann, A. Pöppl: Magn. Reson. Chem. **43**. S205 (2005).

56. (a) A. Carrington, A.D. Mclachlan: '*Introduction to Magnetic Resonance*', A Harper Intel. Edition, Harper and Row, London (1967), p. 147. (b) *ibid.* p. 83 and p. 109.

57. R. Ramprasad, K.C. Hass, E.F. Schneider, J.B. Adams: J. Phys. Chem. B **101**, 6903 (1997).

58. C. Freysoldt, A. Pöppl, J. Reinhold: J. Phys. Chem. A **108**, 1582 (2004).

59. (a) P. Pietrzyk, W. Piskorz, Z. Sojka, E. Broclawik: J. Phys. Chem. B **107**, 6105 (2003). (b) J. Dědeček, D. Kaucký, B. Wichterlová, O. Gonsiorová: Phys. Chem. Chem. Phys. **4**, 5406 (2002).

60. W. Liu, P. Wang, K. Komaguchi, M. Shiotani, J. Michalik, A. Lund: Phys. Chem. Chem. Phys. **2**, 2515 (2000).

61. W. Liu, S. Yamanaka, M. Shiotani, J. Michalik, A. Lund: Phys. Chem. Chem. Phys. **3**, 1611 (2001).

62. W. Liu, M. Shiotani, J. Michalik, A. Lund: Phys. Chem. Chem. Phys. **3**, 3532 (2001).

63. '*Handbook of Chemical and Physics*', 83rd Edition, CRC Press, Boca Raton, FL (2002–2003).

64. (a) D.H. Aue, H.M. Webb, M.T. Bowers: J. Am. Chem. Soc. **98**, 311 (1976). (b) Y.L. Chow, W.C. Danen, S.F. Nelsen, D.H. Rosenblatt: Chem. Rev. **78**, 243 (1978).

65. (a) R.M. Krishna, V. Kurshev, L. Kevan: Phys. Chem. Chem. Phys. **1**, 2833 (1999). (b) B. Xiang, L. Kevan: Langmuir **10**, 2688 (1994). (c) B. Xiang, L. Kevan: J. Phys. Chem. **98**, 5120 (1994).

66. (a) R.J.H. Clark, R.E. Heste: '*Advances in Spectroscopy; Spectroscopy of New Materials*', Vol. 22, Wiley, New York, NY (1993). (b) S. Oliver, A. Kuperman, G.A. Ozin: Angew. Chem. Int. Ed. **37**, 46 (1998).

67. M. Lindgren, N.P. Benetis, M. Mastumoto, M. Shiotani: Appl. Magn. Reson. **9**, 45 (1995).

68. (a) J. Heinzer: Mol. Phys. **22**, 167 (1971). (b) L. Sjöqvist, N.P. Benetis, A. Lund, J. Maruani: Chem. Phys. **156**, 457 (1991).

69. R. Szostak: In '*Handbook of Molecular Sieves*', Van Nostrand Reinhold, New York, NY (1992).

70. L.S. de Saldarriaga, C. Saldarriaga, M.E. Davis: J. Am. Chem. Soc. **109**, 2686 (1987).

71. J. Michalik: Appl. Magn. Reson, **10**, 507 (1996).

72. J.S. Yu, J.W. Ryoo, C.W. Lee, S.J. Kim, S.B. Hong, L. Kevan: J. Chem. Soc. Faraday Trans. **93**, 1225 (1997).

73. A. Fujishima, K. Honda: Nature **238**, 37 (1972).

74. A.L. Linsebigler, G. Lu, J.T. Yates: Chem. Rev. **95**, 735 (1995).

75. (a) A.G. Agrios, K.A. Gray, E. Weitz: Langmuir **19**, 1402 (2003). (b) ibid. **19**, 5178 (2003).

76. M.A. Fox, M.T. Dulay: Chem. Rev. **93**, 341 (1993).

77. A. Hagfeldt, M. Grätzel: Chem. Rev. **95**, 49 (1995).

78. A. Mills, S. Le Hunte: J. Photochem. Photobiol. A **108**, 1 (1997).

79. R.R. Bacsa, J. Kiwi: Appl. Catal. B **16**, 19 (1998).

80. (a) T. Watanabe, A. Nakajima, R. Wang, M. Minabe, S. Koizumi, A. Fujishima, K. Hashimoto: Thin Solid Films **351**, 260 (1999). (b) T. Sumita, T. Yamaki, S. Yamamoto, A. Miyashita: Appl. Surf. Sci. **200**, 21 (2002).

81. (a) U. Stafford, K.A. Gray, P.V. Kamat, A. Varma: Chem. Phys. Lett. **205**, 55 (1993). (b) G. Riegel, J.R. Bolton: J. Phys. Chem. **99**, 4215 (1995).

82. M.R. Hoffmann, S.T. Martin, W. Choi, D.W. Bahnemann: Chem. Rev. **95**, 69 (1995).

83. S. Sato: Chem. Phys. Lett. **123**, 126 (1986).

84. R. Asahi, T. Morikawa, T. Ohwaki, K. Aoki, Y. Taga: Science **293**, 269 (2001).

85. S. Sakthivel, M. Janczarek, H. Kisch: J. Phys. Chem. B **108**, 19384 (2004).

86. H. Irie, Y. Watanabe, K. Hashimoto: J. Phys. Chem. B **107**, 5483 (2003).

87. O. Diwald, T.L. Thompson, T. Zubkov, E.G. Goralski, S.D. Walck, J.T. Yates Jr.: J. Phys. Chem. B **108**, 6004 (2004).

88. M. Miyauchi, A. Ikezawa, H. Tobimatsu, H. Irie, K. Hashimoto: Phys. Chem. Chem. Phys. **6**, 865 (2004).

89. J.L. Gole, J.D. Stout, C. Burda, Y. Lou, X. Chen: J. Phys. Chem. B **108**, 1230 (2004).

90. Z. Lin, A. Orlov, R.M. Lambert, M.C. Payne: J. Phys. Chem. B **109**, 20948 (2005).

91. S. Sato, R. Nakamura, S. Abe: Appl. Catal. B **284**, 131 (2005).

92. M. Sathish, B. Viswanathan, R.P. Viswanath, C.S. Gopinath: Chem. Mater. **17**, 6349 (2005).

93. M. Alvaro, E. Carbonell, V. Fornés, H. Garcia: Chem. Phys. Chem. **7**, 200 (2006).

94. Y. Nosaka, M. Matsushita, J. Nasino, A.Y. Nosaka: Sci. Technol. Adv. Mater. **6**, 143 (2005).

95. Y. Irokawa, T. Morikawa, K. Aoki, S. Kosaka, T. Ohwaki, Y. Taga: Phys. Chem. Chem. Phys. **8**, 1116 (2008).

96. S. Livraghi, M.C. Paganini, E. Giamello, A. Selloni, C. Di Valentin, G. Pacchioni: J. Am. Chem. Soc. **128**, 15666 (2006).

97. P.F. Cornaz, J.H.C. Van Hooff, F.J. Pluijm, G.C.A. Schuit: Discuss. Faraday Soc. **41**, 290 (1966).

98. R.D. Iyenger, M. Codell: Adv. Colloid Interface Sci. **3**, 365 (1972).

99. R.F. Howe, M. Grätzel: J. Phys. Chem. **89**, 4495 (1985).

100. Y. Nakaoka, Y. Nosaka: J. Photochem. Photobiol. A **110**, 299 (1997).

101. S.W. Ahn, L. Kevan: J. Chem. Soc. Faraday Trans. **94**, 3147 (1998).

102. S. Livraghi, A. Votta, M.C. Paganini, E. Giamello: Chem. Commun. **4**, 498 (2005).

103. T.L. Ma, M. Akiyama, E. Abe, I. Imai: Nano Lett. **5**, 2543 (2005).

104. C. Di Valentin, G. Pacchioni, A. Selloni, S. Livraghi, E. Giamello: J. Phys. Chem. B **109**, 11414 (2005).

105. C. Di Valentin, G. Pacchioni, A. Selloni: Phys. Rev. B **70**, 085116 (2004).

106. Y. Itagaki, K. Nomura, M. Shiotani, A. Lund: Phys. Chem. Chem. Phys. **3**, 4444 (2001).

107. A. Heller: Acc. Chem. Res. **28**, 503 (1995).

108. D.-R. Park, J. Zhang, K. Ikeue, H. Yamashita, M. Anpo: J. Catal. **185**, 114 (1999).

109. J. Yang, C. Chen, H. Ji, W. Ma, J. Zhao: J. Phys. Chem. B **109**, 21900 (2005).
110. M. Stylidi, D. Kondarides, X.E. Verykios: Appl. Catal. B **47**, 189 (2004).
111. (a) K. Komaguchi, T. Maruoka, H. Nakano, I. Imae, Y. Ooyama, Y. Harima: J. Phys. Chem. C **113**, 1160 (2009). (b) *ibid.* C **114**, 1240 (2010).
112. E. Carter, A.F. Carley, D.M. Murphy: J. Phys. Chem. C **111**, 10630 (2007).
113. M.P. De Lara-Castells, J.L. Krause: Chem. Phys. Lett. **354**, 483 (2002).
114. D.N. Mirlin, I.I. Reshina, L.S. Sochava: Sov. Phys. Solid State **11**, 1995 (1970).
115. T. Sekiya, K. Ichimura, M. Igarashi, S. Kurita: J. Phys. Chem. Solids **61**, 1237 (2000).
116. V.N. Kuznetsov, T.K. Krutitskaya: Kinet. Catal. **37**, 446 (1996).
117. J. Chen, L.-B. Lin, F.-Q. Wing: J. Phys. Chem. Solids **62**, 1257 (2001).
118. (a) A.A. Lisachenko, V.N. Kuznetsov, M.N. Zakharov, R.V. Mikhailov: Kinet. Catal. **45**, 189 (2004). (b) A.A. Lisachenko, R.V. Mikhailov: Tech. Phys. Lett. **1**, 21 (2005).
119. S.W. Ahn, L. Kevan: J. Chem. Soc. Faraday Trans. **94**, 3147 (1998).
120. J. Jia, T. Ohno, M. Matsumura: Chem. Lett. 908 (2000).
121. K. Komaguchi, H. Nakano, A. Arakia, Y. Harima: Chem. Phys. Lett. **428**, 338 (2006).
122. Y. Nakaoka, Y. Nosaka, J. Photochem. Photobiol. A **110**, 299 (1997).
123. S. Leytner, J.T. Hupp: Chem. Phys. Lett. **330**, 231 (2000).
124. W. Känzig, M.H. Cohen: Phys. Rev. Lett. **3**, 509 (1959).
125. M. Shiotani, G. Moro, J.H. Freed: J. Chem. Phys. **74**, 2616 (1981).
126. K. Tatsumi, M. Shiotani, J.H. Freed: J. Phys. Chem. **87**, 3425 (1983).
127. C. Heller, H.M. McConnell: J. Chem. Phys. **32**, 1535 (1960).
128. M. Mikami, S. Nakamura, O. Kitao, H. Arakawa, X. Gonze: Jpn. J. Appl. Phys. **39**, 847 (2000).
129. H. Tang, H. Berger, P.E. Schmid, F. Lévy, G. Burri: Solid State Commun. **23**, 161 (1977).
130. V.E. Henrich, R.L. Kurtz: Phys. Rev. B **23**, 6280 (1981).
131. R. Asahi, Y. Taga, W. Mannstadt, A.J. Freeman: Phys. Rev. B **61**, 7459 (2000).
132. J.C. Woicik, E.J. Nelson, L. Kronik, M. Jain, J.R. Chelikowsky, D. Heskett, L.E. Berman, G.S. Herman: Phys. Rev. Lett. **89**, 077401 (2002).
133. R. Sanjinés, H. Tang, H. Berger, F. Gozzo, G. Margaritondo, F. Lévy: J. Appl. Phys. **75**, 2945 (1994).

Chapter 7
Applications to Polymer Science

Abstract ESR spectroscopic applications to polymer science are presented. ESR parameters used for the molecular and material characterization of polymer materials are reviewed. It is emphasized that ESR studies of the polymer science are particularly effective in three areas. (1) Intermediate species such as free radicals produced in chemical reactions of polymer materials can be directly detected. (2) The temperature dependent ESR spectra of free radicals trapped in the polymer matrices are very effective for the evaluation of molecular mobility (molecular motion) of polymer chains. (3) The mobility of electron, the structure of solitons, and the doping behavior in conduction polymers can be observed in detail in order to clarify the mechanism of conduction.

7.1 Introduction

Electron spin resonance (ESR) spectroscopy has been used in polymer science for half a century. Two major areas have been investigated. One is the study of mechanisms of chemical reactions in polymerization and the effects of radiation. Intermediate species such as neutral and ionic radicals produced by exposure to ionizing radiation and ultraviolet light, mechanical fracture, deterioration of polymers and polymerization of monomers have been identified. Many kinds of reactions such as decay and conversion of the free radicals to different species, have also been observed.

The other area has been the elucidation of relaxation phenomena of polymer chains by observing temperature and pressure dependent ESR spectra of radical species trapped in solid and liquid polymers.

Polymer science has also been developing over this half century. Great progress has occurred in the field of functional polymers, biopolymers with concurrent advances in molecular design and molecular characterization. It can be expected that the ESR techniques will give valuable insight in the area of molecular characterization. It is well known that solid polymers have many kinds of structural heterogeneities which lead to phenomena like the distribution of relaxation times

A. Lund et al., *Principles and Applications of ESR Spectroscopy*,
DOI 10.1007/978-1-4020-5344-3_7, © Springer Science+Business Media B.V. 2011

correlated with the molecular disorder of a polymer chain in crystalline and non-crystalline regions and the distribution of the stereo-tacticity of the chain.

The ESR method is effective for the characterization of the inhomogeneous structure of polymers. Physical and chemical information about three particular features can be studied using the ESR technique. First, ESR parameters such as the hyperfine splitting, the fine or zero-field splitting and the g-values for the spectrum, make it possible to approach the detailed molecular structures at a molecular level giving accessibility to both intra- and inter-chain interactions. Secondly, the line shape of the spectrum is sensitive to molecular motion and molecular orientation being influenced by the structure of the environment surrounding the radical species. Thirdly, the total intensity and the line shape of the spectra reflect directly the chemical reactions occurring in solid polymers. A particularly attractive feature of the ESR technique is thus that three kinds of information, structure, molecular motion and chemical reaction, are obtained simultaneously.

These characteristic features originate from the interaction between the electron spin and the nuclear spins and with other surrounding electron spins. It is well known that paramagnetic relaxation kinetics of radical species is closely related to relaxation phenomena of polymers. The electron spin echo (ESE) method can for example directly observe the relaxation behavior of electron spins.

In this chapter, we present typical applications of ESR spectroscopy to polymer science, which include structures and molecular motions of polymer chains and chemical reactions in the polymer material. The first part deals with ESR parameters derived from spectra and the molecular information of polymer chains. In the second part, examples of applications to polymer science are introduced. In the following section, the close relations between structure, molecular motion and chemical reactions are discussed. The narrative will illustrate how these studies have made a considerable contribution to polymer physics and chemistry especially describing mechanisms of deteriorations, polymerization and relaxation phenomena of solid polymers.

7.2 What Can We Obtain from the ESR Parameters and Their Changes in Polymer Materials?

Several parameters can be derived from the ESR spectrum and be used for molecular characterization of polymer materials. The temperature and time dependent parameters reflect the dynamical behaviors of chemical reaction and physical properties in the polymer matrices. In the classification system used here the ESR parameters are divided into a group related to molecular characterization of polymer materials (A) and examples of applications to polymer science (B).

(1) Hyperfine splitting (*hfs*):

    (A-1) Chemical structure of free radical is assigned from the magnitude of the *hfs* which is related to the strength of the interaction between the electron and nuclear spins and from the number of the splittings.

Fig. 7.1 Steric conformation of end alkyl radicals (plane ABCD is perpendicular to direction of $(C_\alpha\cdot)$—C_β bond) Broken lines and primes indicate the projection of the corresponding items on the plane ABCD

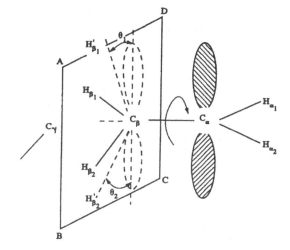

(A-2) Stereo-structure of free radical can be obtained from McConnell's relation [1]. This is the dependence of the isotropic *hfs* interaction, A_β on dihedral angle, θ, shown in Fig. 7.1 as an example in the case of polyethylene end radical.

$$A_\beta = A_0 + B\cos^2\theta. \tag{7.1}$$

In this equation, A_0 is a small constant. B is also a constant. We can determine the twisted structure of the radical by evaluating the angle θ from the experimental value of A_β. McConnell's relation can be explained by hyper-conjugation, which allows some spin density to be located in the β-hydrogen 1s orbital. The β-hydrogen is bonded to C_β which is bonded to $(C_\alpha\cdot)$ having an unpaired electron.

(B-1) Assignment of unstable free radicals produced by irradiation of polymers with ionizing radiation and ultraviolet light, mechanical fracture, and deterioration of polymers and polymerization.

(B-2) Stereo-structure of polymeric free radicals reflected on the polymer chain conformation in the polymer matrices.

(2) Fine structure (*fs*) or zero-field splitting (*zfs*):

(A-1) Structure of excited triplet, quintet, and other multiplets and energy levels between the ground and excited states.

(A-2) The distance between free radicals (unpaired electrons)

(B-1) Structures of conducting organic polymers

(B-2) Excited structures of polymer chains upon photo-irradiation

(B-3) Aggregated structures of metal ions (Cu^{2+}, Mn^{2+}) in ionomer polymer materials

(B-4) Free radical pairs produced by irradiation of polymers with ionizing radiation and the distance between the paired radicals.

(B-5)   Local concentration of free radicals produced by irradiation of polymers with ionizing radiation

(3) g-value:

(A)   Electronic structures of ground and excited states
(B-1)   Electronic structures of polymer complexes containing transition metal ions
(B-2)   Electronic structures of neutral, anion and cation free radicals and of trapped electrons in polymer matrices.

(4) Anisotropic features of the ESR parameters

(A-1)   Orientation of free radicals
(A-2)   Anisotropic molecular motion
(B-1)   Degree of orientation of polymer chains in crystalline and non-crystalline regions
(B-2)   Molecular motion of polymer chains in crystalline and non-crystalline regions and in polymer alloys and polymer blends.

(5) Line width and line shape

(A-1)   Molecular mobility
(A-2)   Distribution of structures of free radicals
(B-1)   Rotation and vibration rate of polymer chains
(B-2)   Distribution of a polymer chain structures with regard to intra-chain conformation and inter-chain distances

(6) Longitudinal (T_1) and transverse (T_2) relaxation time

(A-1)   Molecular mobility
(A-2)   Average distance between free radicals (unpaired electrons)
(B-1)   Average and local concentration of free radicals produced by irradiation of polymers with ionizing radiation and ultraviolet light, mechanical fracture, and deterioration of polymers and polymerization.

(7) Relative and absolute concentration of free radicals

(A-1)   The average distance between free radicals (unpaired electrons)
(A-2)   Stability of free radicals
(B-1)   The average and local concentration of free radicals produced by irradiation of polymers with ionizing radiation and ultraviolet light, mechanical fracture, and deterioration of polymers and polymerization.
(B-2)   Mechanism of deterioration of polymers by auto-oxidization
(B-3)   Mechanism of free radical type polymerization by grafting initiators
(B-4)   Diffusion controlled reaction of free radical decay in solid polymers

7.3 Polymerization Mechanism

7.3.1 Radical Polymerization in the Liquid State

One of the fascinating features of the ESR technique is that it allows direct observation of growing radicals in radical polymerization. It is usually difficult to observe them under stationary conditions as the stationary concentrations are very low. We can, however, detect the short lived radicals in the liquid state by a rapid mixing of the initiator and monomer solutions in the ESR cavity. Figure 7.2 shows a well-resolved spectrum obtained during the polymerization of vinyl acetate (VAc) with the Ti^{3+}/H_2O_2, [2, 3] redox initiator as an example of what can be observed using the rapid mixing continuous method. The spectrum is a double triplet of a narrow quartet having hyperfine splitting of 2.10, 1.25, and 0.14 mT for the α-proton, two β-methylene protons, and methyl protons of the ester group, respectively. The splitting confirms that the spectrum can be assigned to the VAc monomer radical, $((OH)C_\beta H_{\beta 2}-(C_\alpha \cdot)H (OCOCH_3)OH)$. The isotropic spectrum with narrow line widths is caused by the rapid motion of the radicals in the liquid solution. The highly reactive VAc radical is preferentially deactivated by chain termination with OH before the monomer radical propagates to form dimer radicals. When a small amount of a second monomer is added to VAc, the ESR spectra for the copolymer radical $(OH)-VAc-M_2$ and for the co-monomer radical $(OH)-M_2$ overlap with that of the monomer radical.

Figure 7.3. shows the ESR spectrum obtained for the polymerization system vinyl acetate-acrylnitrile (VAc-AN), the spectrum is due to three kinds of free radicals, the VAc monomer radical, $(OH)CH_2-(C \cdot)$—CH—$(OCOCH_3)$. VAc-AN copolymer radical, $(OH)CH_2$—CH $(OCOCH_3)$ CH_2—$(C \cdot)$ H(CN), and the AN

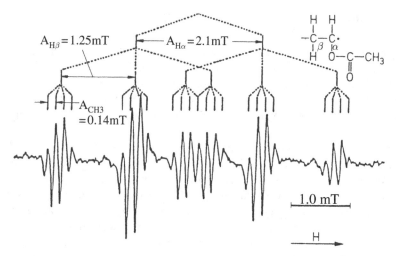

Fig. 7.2 ESR spectrum of VAc in aqueous redox sysem. The figure is adapted from [2] with permission from John Wiley & Sons Inc.

Fig. 7.3 ESR spectrum for VAc copolymerized with AN in aqueous solution. The dominant spectral component is assigned to radical (OH)CH$_2$—CH (OCOCH$_3$) CH$_2$—(C ·) —CH(CN). The figure is adapted from [3] with permission from John Wiley & Sons Inc.

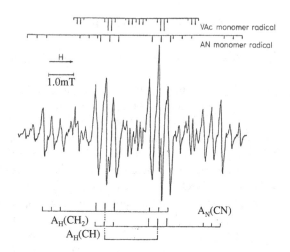

monomer radicals, (OH)CH$_2$—(C ·) H (CN). The stick spectrum pictured under the experimental spectrum represents the copolymer radical, where the hyperfine splitting is lower than that for the AN monomer radical because of the low spin density on the α-carbon due to the delocalization of the unpaired electron. The relative concentrations of the three radical species are a function of the mole fraction of AN monomer. The concentration of VAc monomer radical decreases and those of copolymer radical and AN monomer radical increase with an increase of the mole fraction of AN. The relative rates of conversion of VAc to VAc-M$_2$ can be estimated for the various copolymer systems using the ESR techniques.

7.3.2 Radical Polymerization in the Solid State

Many authors have studied radical polymerization in the solid state by using the ESR technique. An advantage of studying polymerizations in the solid state is the long life time of free radicals leading to much higher concentrations than that in the liquid state. As an example, the benzoyl peroxide (BPO)-initiated polymerization of methyl methacrylate (MMA) in frozen aromatic solvents with irradiation of light is shown in Fig. 7.4 [4].

The nine lines, designated as the (5 + 4) lines are assigned to the propagating radical, —C$_\beta$H$_{\beta2}$(C$_\alpha$ ·)R(CH$_3$). Details about the radical were obtained from the shape of the spectrum, especially the relative intensity of the lines which has been interpreted in terms of various factors such as that there are more than two conformations, there is site exchange motion around the (C$_\alpha$ ·)—C$_\beta$ bond, and that there is a distribution of dihedral angles for one conformation. Figure 7.4 shows the temperature and solvent dependency of the intensity ratio of the 9 lines.

Kamachi et al. [5] calculated the conformation energy of the propagating radical of MMA and found that two stable conformations with different dihedral angles between the (C$_\alpha$ ·)—H$_\beta$ bond and the p-orbital were possible. There are two possible

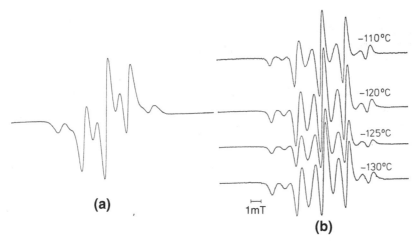

Fig. 7.4 Temperature dependent ESR spectra of PMMA propagating radical. (**a**) In anisole at – 120°C. (**b**) In benzonitrile at temperatures shown in the figure. The figure is adapted from [4] with permission from the Society of Polymer Science, Japan

ways of monomer addition to the propagating end, resulting in isotactic and syndio-tactic configurations. Because the molecular motion of the propagating radicals are in the frozen state, the solvent dependent spectra cannot be explained by the rotation about the $(C_\alpha \cdot)$—C_β bond. The different intensity ratios of the spectra between two solvents, anisole and benzonitrile, were therefore attributed to the different concentration ratios of the two stable radicals whose propagating rate constants are different, for instance, in the two solvent because of their different conformations. The temperature dependent spectra can be related to the change of the concentration ratio of the two radicals because of their different conformation energies. The ESR spectra of the methacrylate propagating radical depend also on the size of any bulky side group and the segmental mobility. Iwasaki et al. [6] have given other interpretations for the MMA propagating radical. They obtained simulated spectra by a model of site exchange between two stable conformations having the dihedral angles of 65° and 55°. The simulated spectra shown in Fig. 7.5 depend on the life time (τ) of the two conformations (the inverse of the exchange rate) and are very similar to the temperature dependent spectra of the radicals produced by the γ-irradiation of MMA monomer.

7.3.3 Radical Polymerization of Macro-Monomers

The number of studies about syntheses and applications of macro-monomers has been increasing since macro-monomers are very useful in the preparation of various kinds of functional graft copolymers having well-defined structures. One can clarify the mechanism of polymerization of macro-monomers in comparison with small monomers by the ESR method, which is also a good example of the application to

Fig. 7.5 The dependency of the ESR spectra of PMMA propagating radicals on the life time τ(s) (the inverse of the exchange rate of the two conformations). The figure is adapted from [6] with permission from John Wiley & Sons Inc.

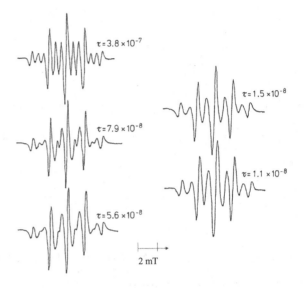

the kinetics of the polymerization. The propagating rate constant, k_p can be estimated by evaluating the absolute concentration of the propagating radical, $[M^*]$. The rate k_p, the termination rate constant, k_t, and the radical life time, τ, can for instance be directly determined from the time dependence of $[M^*]$ and the degree of polymerization, D_p by the following relation.

$$-\frac{d[M]}{dt} = k_p[M^*](M) \tag{7.2}$$

$$k_t = -\frac{d[M]}{dt}\left(\frac{1}{D_p[M^*]}\right) \tag{7.3}$$

$$\tau = \frac{D_p}{k_p[M]} \tag{7.4}$$

Here $[M]$ is the concentration of the macro-monomer. Tsukahara et al. [7] have observed ESR spectra of propagating radicals during the polymerization of a styrene macro-monomer having a metacyloyl end group.

Figure 7.6 shows typical examples of the ESR spectra for a macro-monomer MMA. The five and four (5+4) lines measured at 60°C and −196°C are due to the MMA propagating radical described in the previous section. From the ESR signal, the concentration of the propagating radical of the macro-monomer $[M^*]$ can be evaluated. The value of k_p was determined to be 37 (L/mol·s). Rate constants for the polymerization of the MMA monomer at different conditions are shown in Table 7.1. It is seen from this table that k_p is considerably decreased in the macro-monomer system compared to polymerization of the small monomers. This is clearly due to the high viscosity of polymerization media and the specific multi-branched structure of the propagating radicals.

Fig. 7.6 Change of ESR
spectra of MMA
macro-monomer propagating
radicals by quenching and
aging: (**a**) at 60°C, (**b**) at
−196°C after (**a**), and (**c**) at
23°C after (**b**). The spectra
were measured at the
temperature of the respective
heat treatment. The figure is
adapted from [7] with
permission from the
American Chemical Society

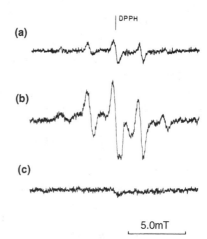

Table 7.1 Propagating rate constant, k_p for MMA polymerization. The table is adapted from [7] with permission from the American Chemical Society

k_p, L/mole s	Method for evaluation of k_p	Polymerization condition Initiator, temperature, matrices
37	ESR	AIBN, 60°C benzene solution, macro-monomer
720	ESR	Dibenzoil peroxide, 60°C, bulk
187	ESR	UV, 30°C, benzene solution
790	ESR	50°C, emulsion
260	Rotating sector	25°C, benzene solution

7.4 Radiation Effects: Radiation Physics and Chemistry of Polymer Materials

We present ESR applications to radiation effects of polymer materials in this section. The ESR spectral analyses of the polymer radicals produced are described in relation to the geometrical structure of the free radicals and the polymer structure of the crystalline and non-crystalline regions. The decay reaction, radical migration and pair-wise trappings of the polymer radicals are discussed. Furthermore, inhomogeneous "spur-like" trapping of polymer radicals by irradiation are presented. For simplicity, the study of the free radicals in solid polyethylene is the primary example given.

7.4.1 Free Radicals Produced by Irradiation of Polymers with Ionizing Radiation

Cross-linking and degradation are well known radiation effects in polymers. The molecular weight increases and gels are formed for the cross-linked type polymers with increasing radiation dosage. For the degradation type polymer the length of the polymer chain and the molecular weight decrease with increasing radiation.

(1) Polyethylene (PE) (cross-linked type polymer)

ESR spectra of irradiated polyethylene have been observed and analysis and identification of the corresponding free radicals have been carried out over an extended time. In early studies, the broad and not well resolved sextet was assigned to alkyl type radicals inside the chain, $-C_\beta H_{\beta 2}-(C_\alpha \cdot)H_\alpha-C_\beta H_{\beta 2}-$. The sextet is due to the hyperfine interaction between the unpaired electron and one α- and four β-hydrogen nuclei. In later studies ESR spectra of the alkyl radicals were related to the structures of polyethylene chains thereby incorporating molecular conformation and molecular orientation in the analysis. Figure 7.7 shows ESR spectra of free radicals in solid polyethylene as an example of the cross-linked type polymer. The spectra are composed of a sextet having various kinds of finer structures.

(a) *Anisotropic hyperfine splitting due to the α-hydrogen nuclei*: The small splitting of the wing peak of the sextet spectrum (a) shown in Fig. 7.7 [8] is assumed to reflect the anisotropy of coupling constants due to α-hydrogen and the pattern is referred to as being amorphous. Figure 7.7(a) was observed in a urea-polyethylene complex (UPEC) in which a PE chain was surrounded by urea molecules in a complex which will be described below. Figure 7.7(b) was recorded for solution grown crystals of polyethylene. Both samples are powdered. Rapid molecular motion of PE in UPEC narrows the spectrum.

In order to confirm the anisotropy due to the α-hydrogen, Salovey et al. [9] and Shimada et al. [10] studied the patterns of the ESR spectra from irradiated solution grown crystals of polyethylene. The crystal c-axis was oriented perpendicular to the plane of the sample while the a- and b-axes were randomly oriented in the plane as shown in Fig. 7.8. Six- and ten-line spectra were observed when the c-axis of the crystal was set to be parallel (Fig. 7.9(a)) and perpendicular (Fig. 7.9(b)) to the direction of the applied magnetic field, respectively. From these results, the anisotropic hyperfine splitting due to the α-hydrogen $A_Y = 0.75$ mT, $A_X = 1.72$ mT and $A_Z = 3.70$ mT were determined and were related to the molecular orientation of the crystal. The x-, y- and z-axes coincide with the directions of the p-orbital, the $(C_\alpha \cdot)-H_\alpha$ bond, and the main chain axis, respectively, as shown in Fig. 7.10.

(b) *Different hyperfine splitting (hfs) due to the β-hydrogen nuclei*: The precise structure of the alkyl radicals was determined by extensive computer simulation. In order to elucidate the structure, it was necessary to determine the exact values of

Fig. 7.7 Comparison of the ESR spectrum of PE alkyl radicals in UPEC (**a**) with that of solution grown crystals (**b**) observed at 320 K. The figure is adapted from [8] with permission from the American Chemical Society

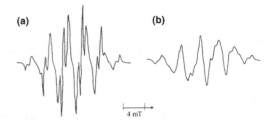

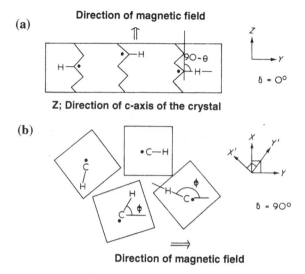

Fig. 7.8 Schematic representations of the angles (φ) between the $(C_\alpha \cdot)$—H_α bonds direction and the applied magnetic field in a PE single crystal. (**a**) Projection on the (y-z) plane containing the polymer chains (the crystal c-axis) which are parallel to the applied magnetic field (z), i.e. $\delta = 0°$. (**b**) Projection on the (x-y) plane containing the $(C_\alpha \cdot)$—H_α bonds which are perpendicular to both the polymer chain axis (z-axis) and the magnetic field (y-axis), i.e. $\delta = 90°$. See Fig. 7.10 for the x, y, and z coordinate system and the angle θ. Here, a uniform distribution of the angle, φ is assumed. The x′ and y′ axes indicate an example of another orientation of the $(C_\alpha \cdot)$—H_α bond and π-orbital. The figure is adapted from [10] with permission from Elsevier

the line widths and coupling constants due to the α- and β- hydrogen nuclei. It was also assumed that the *hfs* due to the two β-hydrogen atoms, $A_{\beta 1}$ and $A_{\beta 2}$ need not be equal to each other. Figure 7.11 shows a comparison of the calculated spectrum with the experimental spectrum of alkyl radicals trapped in the UPEC at 107 K. The calculated spectrum gives a good fit with respect to the overall line shape and the peak positions. It was found that the *hfs* interaction of $A_{\beta 1}$ (3.08 mT) was not equal to that

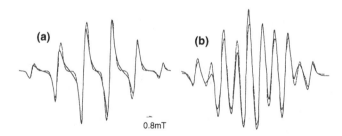

Fig. 7.9 Comparison of the experimental ESR spectra ($\delta = 0°$ (**a**), $\delta = 90°$ (**b**) of PE alkyl radicals observed at 254 K (*solid lines*) for solution grown crystals and calculated spectra (*dashed lines*) ($A_{\beta 1} = 3.0$ mT, $A_{\beta 2} = 3.4$ mT)). The angle between the crystal c-axis and the direction of the applied magnetic field is designated δ. The figure is adapted from [10] with permission from Elsevier

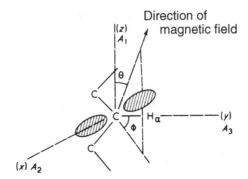

Fig. 7.10 Molecular axes of the PE crystal. The x-, y-, and z-axes are parallel to the axis of the π-orbital, the C—H bond, and the chain axis, respectively and θ is the angle between the chain axis and the applied magnetic field. The figure is adapted from [10] with permission from Elsevier

of $A_{\beta 2}$ (3.67 mT) at low temperatures [11]. Therefore by applying equation (1) the alkyl radicals were considered to have a structure deformed from the symmetric β–H position at low temperatures. In contrast, $A_{\beta 1}$ and $A_{\beta 2}$ at 321 K were both found to have the same value, 3.0 mT from the simulation The temperature dependence of the interaction A_β, was thus concluded to be the result of rapid oscillation around the $(C_\alpha \cdot)$—C_β bond, which released the structure deformation. A decrease of the line width and an averaging out of the anisotropy of the *hfs* due to the α-hydrogen were also observed at higher temperature. These phenomena are a reflection of the rapid molecular motion around the chain axis.

(2) Polymethylmethacrylate (PMMA) (degradation type polymer)

One usually observes chain scission type radicals, —$C_\beta H_{\beta 2}$ $(C_\alpha \cdot)$ R(CH$_3$) in γ-irradiated PMMA. Is the chain end radical the primary species of γ-irradiation? Ichikawa et al. [12] studied the mechanism of radiation-induced degradation of PMMA by ESR and electron spin echo (ESE) methods. Figure 7.12 shows ESR spectra of γ-irradiated PMMA observed at 77 K. They assigned spectrum (a) to three kinds of radicals, $(C \cdot)HO$, $(C \cdot)H_3$, and —$(C \cdot)(O^-)(OCH_3)$, which were a doublet

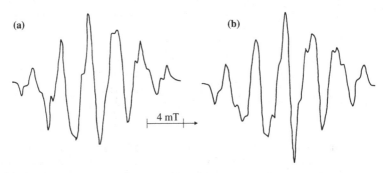

Fig. 7.11 Comparison of the experimental ESR spectrum of PE alkyl radicals in UPEC observed at 107 K (**a**) with the calculated spectrum, using $A_{\beta 1} = 3.0 8mT$, $A_{\beta 2} = 3.67$ mT (**b**). The figure is adapted from [8] with permission from Elsevier

Fig. 7.12 ESR spectra of
γ-irradiated PMMA observed
at 77 K before (**a**) and after
(**b**) photo-bleaching with
visible light at 77 K. The
figure is adapted from [12]
with permission from John
Wiley & Sons Inc.

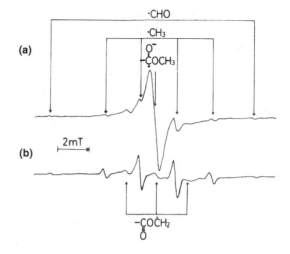

with a *hfs* of 13 mT, a quartet with the *hfs* of 2.3 mT, and a singlet, respectively. The radical —C(O)O(C ·)H$_2$ (triplet with a *hfs* of 2 mT) was also detected. Spectrum (b) was recorded at 77 K after photo-bleaching with visible light. It was found that (C ·)(O$^-$)(OCH$_3$) disappeared and the amount of (C ·)H$_3$ increased. A part of (C ·)(O$^-$)(OCH$_3$) was converted to (C ·)(O$^-$)(O) and (C ·)H$_3$. Figure 7.13 shows the ESE-detected ESR spectra of a γ-irradiated PMMA sample. The ESE-detected ESR spectra were measured with the 90°-τ-180° two-pulse sequence at various times, t_1 on the echo decay curve. The (C ·)H$_3$ radical could not be detected because of a short longitudinal relaxation time. However, other radicals with longer longitudinal relaxation were observed in the ESE measurements. One was attributed to a radical having a doublet spectrum with a *hfs* of 2 mT due to a —CH— group. The broad singlet at the center of the spectrum in Fig. 7.13(a) is due to the PMMA anion radical which disappears after photo-bleaching. The triplet spectrum is assigned to be —C(O)O(C ·)H$_2$ radical. In conclusion, γ-irradiation of PMMA does not produce the scission type radical, —C$_\beta$H$_{\beta2}$(C$_\alpha$ ·)R(CH$_3$) but five primary radical species.

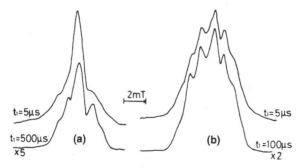

Fig. 7.13 ESE detected ESR spectra observed at 77 K of γ-irradiated PMMA recorded at different times t_1 of echo decay before (**a**) and after (**b**) photo-bleaching with visible light at 77 K. The figure is adapted from [12] with permission from John Wiley & Sons Inc.

The primary radical species —C(O)O(C ·)H$_2$ converts to the secondary product —C$_\beta$H$_{\beta2}$(C$_\alpha$ ·)R(CH$_3$) upon warming the sample above 180 K. Various degradation mechanisms starting with the initial formation of primary species such as the cation and anion of PMMA were suggested.

(3) Polytetrafluoroethylene (PTFE) (degradation type polymer)

ESR spectra of irradiated PTFE have been observed and analysis and identification of the corresponding free radicals have been carried out over an extended period. The broad and not well resolved spectrum observed after the γ-irradiation suggests the production of many kinds of radical species including the scission type radicals. The unstable radicals decay during the heat treatments. Figure 7.14 shows ESR spectra of free radicals in solid PTFE after the sample was warmed up to 450 K [13]. The ESR spectrum observed at 300 K was composed of a double (8.7 mT) quintet (3.2 mT), which was assigned to fluoro-alkyl type radicals inside the chain, —C$_\beta$F$_{\beta2}$(C$_\alpha$ ·)F$_\alpha$C$_\beta$F$_{\beta2}$—. The *hfs* splitting of the α-fluorine in perfluoro alkyl radicals is larger than that of the α-hydrogen in alkyl radicals. The double-quintet is due to the hyperfine interaction between the unpaired electron and one α- and four β-fluorine nuclei. The temperature dependence of the spectrum is remarkable as shown in Fig. 7.14. With decreasing temperature, the widths of the component lines become at first broader and then the splitting of the hyperfine structure begins to vary. At 77 K, where the PTFE is in the rigid state of the overall splitting is estimated to be 37.5 mT, which is greater than 21.5 mT at room temperature. These changes are completely reversible. The transitional decrease around 295 K is ascribed to the onset of rotational motion of chains in crystalline regions. The decrease of the overall splitting is caused by the motional averaging of the angular dependences of the β-fluorines which is a function of the dihedral angle, θ in analogy with that shown in Fig. 7.1. The scission type radical, —C$_\beta$F$_{\beta2}$(C$_\alpha$ ·)F$_{\alpha2}$ is obtained by UV-illumination of the peroxy-radical, —C$_\beta$F$_{\beta2}$C$_\alpha$F$_\alpha$(OO ·)C$_\beta$F$_{\beta2}$— and by mechanical fracture of PTFE as described in Section 7.5.

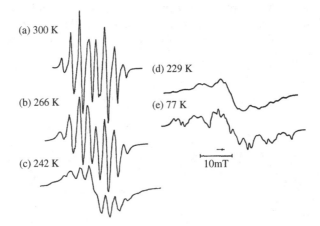

Fig. 7.14 Temperature dependent ESR spectra of γ-irradiated PTFE in a temperature range from 77 to 300 K. The figure is adapted from [13] with permission from the American Institute of Physics

7.4.2 Decay of Free Radicals Produced by γ-Irradiation of Polymers

Radiation processing has been carried out in order to modify various polymers. For example, radiation cross-linked polyethylene and polytetrafluoroethylene having high performance have been developed. Graft copolymerization by the initiation of radicals is applied to the modification of the surface of the polymer film. For the design of polymer processing, it is very important to know the mechanism of chemical reaction of the free radicals. Other applications are the deterioration and mechanical fracture of the polymers and polymerization in the solid state. It can be considered that chemical reactions in solid polymers have the following features. First, the intermediate radical species in solid polymers have a long lifetime in comparison with that of low molecular weight compounds in solution because the polymer matrix trapping the radical species is relatively rigid and the molecular mobility is low. This fact suggests that the chemical reactions should be strongly influenced by the structure and mobility of the solid polymers, which have many kinds of heterogeneities. Second, we can generate various radical species in particular regions of the polymers. The ESR spectra of the radicals can be easily observed because of the long lifetime. The variation of the intensity and the spectrum shape during heat treatment can be readily obtained. In this section, we discuss the kinetics of the reactions in solid polymers and how the time constants and activation energies are related to the structure and molecular motion of the polymers.

(1) Decay of the alkyl free radicals in polyethylene crystals
In general, chemical reactions of the type $P + Q \rightarrow R + T$ consist of three stages:

(a) Approach of the molecules P and Q
(b) Generation of the molecules R and T
(c) Separation of the molecules R and T

The decay reaction of the alkyl radical, $R + R \rightarrow R\text{--}R$ is very simple because the activation energy of the reaction in stage (b) is nearly equal to zero and the probability of the back reaction $R - R \rightarrow R + R$ is also nearly equal to zero. In the case of polyethylene, the products, $R - R$ contain inter chain cross-links and double bonds. Therefore, we can analyze the decay reaction with a scheme involving a diffusion-controlled bimolecular reaction.

The fundamental diffusion equation [14,15] used for the decay reaction of the polymer radicals is

$$\frac{C_0}{C} = 1 + 8\pi r_0 D C_0 \left[1 + \frac{\sqrt{2} r_0}{(\pi D t)^{1/2}} \right] t = 1 + A t^{1/2} + B t \qquad (7.5)$$

$$A = 8\sqrt{2} r_0^2 C_0 \sqrt{\pi D} \qquad (7.6)$$

$$B = 8\pi r_0 D C_0 \qquad (7.7)$$

$$D = \frac{2}{\pi} \left(\frac{B}{A} \right)^2 r_0^2 \qquad (7.8)$$

where C indicates the concentration of the radicals studied at a certain time t, C_0 is the initial value of C, D is the diffusion constant, and A and B are constants satisfying Eqs. (7.6) and (7.7). The constant r_0 is the capture radius which is defined as the distance at which two radicals can combine with minor activation energy. In the decay of the alkyl radicals, —$C_\beta H_{\beta 2}(C_\alpha \cdot)H_\alpha C_\beta H_{\beta 2}$—, "diffusion" does not necessarily mean mass translation in three dimension space, but the migration of the radical site. Migration of the radical site requires the abstraction of hydrogen by an unpaired electron. This in turn requires a distortion of the backbone chain of the trapping matrix. Therefore, a "distorted region" must move with the process of radical migration. Figure 7.15 [15] shows the decay curve of the alkyl radicals trapped in solution grown polyethylene crystals. Zero-time is the time at which the ESR measurement started.

Equation (7.5) can actually be applied to estimate A and B. An approximate value of B can be obtained from the slope of C_0/C against time in the long time region. The exact values of A and B are determined by the method of least squares using computer modeling. The solid line in Fig. 7.15 shows the calculated decay curve obtained for A = 1.2×10^{-2} ($s^{-1/2}$) and B = 4.5×10^{-5} (s^{-1}). The simulated curve is in good agreement with data based on diffusion-controlled reaction theory and improves the interpretation of the decay reaction for free radicals trapped in a solid polymer. The values of D and the time constant of diffusion can also be determined from the data of the decay reaction by using Eq. (7.5). The relaxation time is also considered to be a time constant of the molecular motion causing a slight distortion associated with the "diffusion" of the free radical. Thus, the relaxation time of molecular motion associated with the decay reaction can be estimated. In order to validate this procedure, the diffusion constant was estimated from the known relaxation time obtained in dynamic mechanical studies of polyethylene within the temperature region of the so-called α-relaxation process in a crystalline phase. The value was compared with the diffusion constant obtained from the decay reaction of

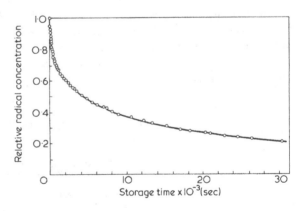

Fig. 7.15 Decay curves of alkyl radicals trapped in γ-irradiated PE single crystals: *Circles*, experimental; *solid curve*, calculated. The figure is adapted from [15] with permission from Elsevier

the radicals in the same temperature region. The values obtained were in good agreement with each other. The relaxation times from the decay reaction of the radicals trapped in the crystalline region are in good agreement with the data from mechanical studies of molecular motion in the same region, called α-relaxation [16]. The activation energies for the diffusion of the radicals and the molecular motion agree with each other. These facts suggest that the decay reaction of the radicals trapped in single crystals is closely related to the molecular motion of polymer chains in the crystalline phase. Three dimensional migration of the free radicals can also be concluded because the decay temperature region is extremely lower than the melting temperature of the PE crystals. As mentioned earlier the diffusion of free radicals does not necessarily mean mass translation in three dimensional space, but rather the migration of the radical site.

(2) Is the migration of alkyl free radicals in polyethylene crystals intra-chain or inter-chain?

Three dimensional migration of the free radicals trapped in the crystalline region was considered in the previous section and diffusion-controlled reaction was shown to be a good description of the decay reaction of the free radicals. The following equations can be applied depending on whether the migration of the alkyl free radicals in polyethylene crystals is due to their abstraction of hydrogen from the intra-chain or inter-chain segments? [15]. Migration along the chains is described by Eq. (7.9) and inter-chain segment migration by Eq. (7.10):

$$-C_\beta H_{\beta 2} \overset{\bullet}{C}_\alpha H_\alpha C_\beta H_{\beta 2} CH_{\gamma 2}-. \quad \Longrightarrow \quad -C_\beta H_{\beta 2} C_\alpha H_{\alpha 2} \overset{\bullet}{C}_\beta H_\beta CH_{\gamma 2}-. \quad (7.9)$$

$$\begin{aligned} -C_\beta H_{\beta 2} \overset{\bullet}{C}_\alpha H_\alpha C_\beta H_{\beta 2} CH_{\gamma 2}-. &\qquad -C_\beta H_{\beta 2} C_\alpha H_{\alpha 2} C_\beta H_{\beta 2} CH_{\gamma 2}-. \\ &\Longrightarrow \\ -C_\beta H_{\beta 2} C_\alpha H_{\alpha 2} C_\beta H_{\beta 2} CH_{\gamma 2}-. &\qquad -C_\beta H_{\beta 2} \overset{\bullet}{C}_\alpha H_\alpha C_\beta H_{\beta 2} CH_{\gamma 2}-. \end{aligned}$$

$$(7.10)$$

If only migration along the chain occurs, three dimensional considerations cannot be applied, at least for the decay process in the crystalline regions. On the other hand, even if migration takes place only through inter-chain hydrogen abstraction Eq. (7.10) three dimensional migration is possible.

An answer to the question posed above can be found in the following study of free radicals trapped in a urea-polyethylene complex (UPEC). Figure 7.7 in Section 7.4.1 showed ESR spectra observed at 320 K which demonstrated the difference between the spectra of the alkyl radicals trapped in UPEC and those in solution grown crystals. The conclusion was that the radical sites in the UPEC were more mobile than in the crystals. The difference between the decay behavior of the alkyl radicals trapped in the complex and in the solution grown crystals is shown in Fig. 7.16. It can be said that the free radicals in the complex have a very long life time at 318 K and the decay rate in the complex at 411 K is of the same order as that in

Fig. 7.16 The decay curves of alkyl radicals trapped in γ-irradiated UPEC (*open circles*) and PE single crystal (*closed circles*). *A*, 308 K; *B*, 411 K; *C*, 327 K. The figure is adapted from [15] with permission from Elsevier

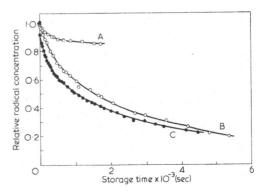

the solution grown crystals at 327 K. Thus the decay reaction of the radicals in the UPEC was found to be very slow. Figure 7.17 illustrates a scheme of the probable orientation in the solution grown crystals and that in the UPEC. In the UPEC the molecules of polyethylene are wholly surrounded by urea molecules which inhibit inter-chain migration. If radical migration along the chain is a main process of the decay reaction of the alkyl radicals, a significant decay of the radicals in the complex must occur because of the high mobility of the polyethylene in the complex. The experimental facts show, however, that this not the case. This means that free radical migration along the chain hardly occurs at room temperature in the UPEC. Therefore, it can be concluded that the rate of radical migration along the chain is very small in the solution grown crystals of polyethylene where the molecular mobility is even lower than that in the UPEC complex and the decay in the crystal should originate from inter-chain free radical migration.

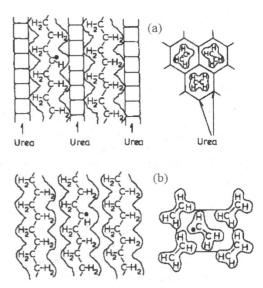

Fig. 7.17 Schematic models of alkyl radicals in UPEC structure (**a**) and PE single crystals (**b**). The figure is adapted from [15] with permission from Elsevier

7.4.3 Free Radical Pairs Produced by Irradiation of Polymers with Ionizing Radiation

The pair-wise formation of radicals in polymer materials is important as an elementary process for radical formation by irradiation. Iwasaki et al. [17] found direct evidence for pair-wise formation in various polymers.

Figure 7.18 depicts the typical spectra of the $\Delta m_s = 2$ transition observed at $g = 4$ for irradiated polyethylene and polyoxymethylene. Some other polymers also yield similar, poorly resolved spectra similar to those in the figure. By the $\Delta m_s = 2$ transition two unpaired electrons of the radical pair are prompted simultaneously from the β-spin state in low energy level to the α-spin state in high energy level. The spectrum for polyethylene can be assigned to paired radicals of main chain alkyl radicals. The $\Delta m_s = 2$ signals disappeared when the polymers were warmed up to room temperature, indicating the recombination and/or the separation of the paired radicals. According to the theory of radical pairs, the hyperfine coupling is half that of the single radicals. In fact, the *hfs* of the $\Delta m_s = 2$ signal of polyethylene is 1.4 mT, which is approximately half the hydrogen coupling constant usually observed in similar compounds at $g = 2$. Evidence of *hfs* in the $\Delta m_s = 1$ spectrum of the radical pair was also obtained. By the $\Delta m_s = 1$ transition only one of the two unpaired electrons in the radical pair is prompted from the β state to the α state. Figure 7.19 shows ESR spectra of the $\Delta m_s = 1$ transition for oriented polyethylene in a drawn film [18]. When the magnetic field was parallel to the draw direction or the chain axis of polyethylene, well resolved eleven lines were clearly observed on both sides of the main sextet assigned to the single radicals. The separation between the two sets of eleven lines is 37.1 mT. The *hfs* of the eleven lines is 1.6 mT similar to the value in the $\Delta m_s = 2$ spectrum. When the magnetic field was perpendicular to the draw direction, the shape and position of the outer features greatly changed and the *hfs* disappeared.

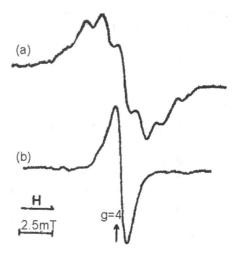

Fig. 7.18 ESR spectra due to $\Delta m_s = 2$ transitions: 77 K γ-irradiated PE (**a**) and γ-irradiated polyoxymethylene (**b**). The ESR measurements were made at 77 K with a power of 15 mW. The figure is adapted from [17] with permission from the American Institute of Physics

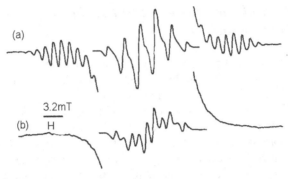

Fig. 7.19 ESR spectra due to $\Delta m_s = 1$ transitions: Oriented PE γ-irradiated at 77 K. ESR measurements were made at 77 K. The magnetic field was parallel (**a**) and perpendicular (**b**) to the chain axis. Outer parts of the spectra were recorded with the gain increased one hundred times. The figure is adapted from [18] with permission from John Wiley & Sons Inc.

These results indicate that the outer parts of the $\Delta m_s = 1$ spectrum undoubtedly comes from the radical pairs. When the magnetic field (H_0) was parallel to the chain axis, the fine splitting, d, was easily determined to be 37.1 mT. The splitting d (mT) changes with the orientation according to the equation:

$$d = \frac{2780(3cos^2\theta - 1)}{r^3} \tag{7.11}$$

where θ is the angle between the axis of symmetry of the radical pairs and the direction of the magnetic field, and r is the distance (Å) between the two interacting electron spins.

Using the crystallographic data for polyethylene, the values of r and θ can be calculated for various sets of radical pairs and the corresponding values of d are compared with the observed value, 37.1 mT. The observed radical pair is of the intra chain type which is produced four bonds apart in the same chain: —CH_2—(C·)H—CH_2—CH_2—CH_2—(C·)H—CH_2—. The calculated distance between the pair is 0.510 nm, which is close to the observed value, 0.531 nm (5.31 Å).

7.4.4 Inhomogeneous Spur-Like Trapping of Free Radicals by Irradiation of Polymers with Ionizing Radiation

In general, the intensity of an ESR spectrum increases with an increase in the microwave power P. When the applied power level is sufficiently low, thermal relaxation processes can, to a good approximation, maintain the Boltzmann equilibrium between spin levels. When the power level exceeds that amount, the ESR spectrum broadens and its intensity begins to decrease and eventually disappears. This phenomenon is called the power saturation or saturation broadening effect and depends on the magnetic relaxation times as well as on the microwave power level. For long spin-lattice relaxation time (T_1) and spin-spin relaxation time (T_2) the Boltzmann equilibrium cannot be established at high microwave power. Because polymer free radicals have long relaxation times, low microwave power levels are preferred. As an example, ESR spectra were recorded with various microwave power levels and the intensities of the spectra (integrated intensity) were plotted against the square-root of the power applied in order to estimate the factors affecting the power saturation. In general T_1 and T_2 are good measures of the saturation effect. Figure 7.20 shows the saturation behavior of alkyl radicals in UPEC produced by irradiation of polymers with ionizing radiation as an example [19].

Similar saturation effects were also detected for the solution grown crystal of PE. The straight line in Fig. 7.20 indicates the case of no saturation. Power saturation occurred more easily after the heat treatments. Since the total concentration of radicals does not change after the heat treatments at these temperatures, an average increase in the T_2 values is less probable if the radicals were trapped uniformly. On the other hand, power saturation characteristics are determined by the values of both T_1 and T_2 for a certain value of the microwave power. Therefore, easy saturation after the heat treatments reflects either (a) an increase in T_1 due to the change of physical state around the radical site by heat treatment, (b) an increase in T_2, due to local diffusion of the radicals out of the regions (spurs) of high concentration without change of the total radical concentration, or (c) effects of both (a) and (b) together. Case (b) seems to be most plausible, making a discussion of the local concentration of free radicals necessary. According to Portis' analysis [20], Eq. (7.12) can be applied.

$$T_1 T_2 \propto H_{1/2}^{-2} \propto P_{1/2}^{-1} \tag{7.12}$$

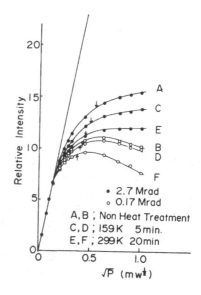

Fig. 7.20 Effects of heat treatment and irradiation dose on the saturation behavior of the ESR spectrum of alkyl radicals in UPEC. Temperatures and periods indicated in the figure designate the heating temperature and durations. The points indicated by arrows designate the powers at which the spectrum intensity was one half of the value compared to the case of no saturation. The figure is adapted from [19] with permission from Elsevier

Table 7.2 Relative values of total concentration [R] and local concentration [M] of the radicals trapped in UPEC. 1 Mrad $=10$ kGy. The table is adapted from [15] with permission from Elsevier

Condition of heat treatment	D [Mrad]	[R]	[M] $\propto$ P$_{1/2}$	(P$_{1/2}$)$^{1/2}$
No annealing	2.7	1.00	1.00	0.56
	0.17	0.08	0.77	0.43
159 K, 5 min	2.7	1.00	1.00	0.50
	0.17	0.08	0.82	0.41
299 K, 20 min	2.7	1.00	1.00	0.45
	0.17	0.08	0.84	0.38

In the equation, $H_{1/2}$ designates the strength of the oscillating microwave magnetic field at which the intensity of the saturated spectrum becomes half of the intensity of the unsaturated spectrum, and $P_{1/2}$ is the corresponding value of the microwave power. If T_1 is assumed to be constant in a certain system, the local concentration of radicals, [M], is proportional to the inverse of T_2. Therefore, Eq. (7.13) can be utilized.

$$[M] \propto P_{1/2} \tag{7.13}$$

The assumption that T_1 is constant was made for comparison of [M] in identical materials irradiated with different doses. Based on the data in Fig. 7.20, local concentrations of radicals can be compared and are tabulated in Table 7.2. Comparison of the values for [M] in the materials irradiated with low and high doses are mainly made. As shown in Table 7.2, the relative values of the local concentrations in the materials irradiated with low doses are of the same order, mostly 70–80%, compared with the values for the materials with high doses. This is conflict with the fact that the total radical concentration, [R], is much less, of the order 1/10 of that at high doses. This would seem to indicate that the free radicals in the materials are trapped in the tracks of the radiation forming spurs even after thermal treatment at relatively high temperatures.

7.5 Mechanical Destruction of Solid Polymers

Since destruction of polymer materials is very important for practical purposes, a large number of investigations on fracture phenomena in polymers have been carried out from both the experimental and theoretical points of view. Several reports provide indirect evidence for main chain scissions, for example decreases in molecular weight or initiations of the graft or block copolymerization after mastication. Direct evidence for chemical bond scission can be obtained from ESR measurements on fractured polymer materials [21]. The high reactivity and high mobility of free radicals produced by mechanical fracture (mechano-radicals) can also be followed. The ESR application to mechanical destruction of polymer materials is presented below. Temperature-dependent ESR spectra of polymer radicals produced by the destruction are analyzed in terms of molecular motion. Furthermore, ESR studies on their chemical reactivity are presented.

7.5.1 Free Radicals Produced by Mechanical Destruction of Solid Polymers: Mechano-Radicals

(1) Polyethylene (PE)

The ESR spectra obtained from PE sawed while immersed in liquid nitrogen are shown in Fig. 7.21 [22]. All spectra were observed at 77 K. The pattern (a) is the spectrum without any heat treatment, and the spectra (b), (c) and (d) are those observed after heat treatment for 5 min at the temperatures indicated in the figure. Spectrum (a) is assigned to be scission type free radicals, mainly $—C_\beta H_{\beta 2}$ $(C_\alpha \cdot)$ $H_{\alpha 2}$.

The sextet is due to the hyperfine interaction between the unpaired electron and two α- and two β-hydrogen nuclei. The shape of the ESR spectra of the scission radicals depends on the conformational structure of the chain end site and on the molecular motion as described in detail in a later section. The spectrum obviously changes with increased temperature and this fact indicates that the majority of the scission radicals are converted to peroxy radicals by heat treatments when the initial mechanical treatment is carried out in liquid nitrogen containing a small amount of

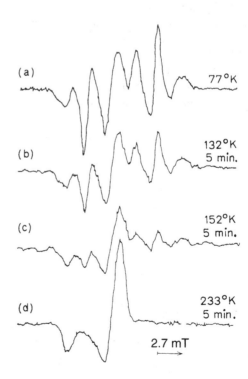

Fig. 7.21 ESR spectra observed at 77 K from polyethylene sawed in liquid nitrogen: (**a**) without heat treatment; (**b**), (**c**), and (**d**) after heat treatments at the temperatures indicated in the figure. The figure is adapted from [22] with permission from the Society of Polymer Science, Japan

oxygen. The spectrum (d) is a well known spectrum of the peroxy-radical of PE and it can be seen making a minor contribution to the spectrum (a). The above identification of the radicals leads us to the conclusion that the macroscopic destruction of PE primarily produces the scission of the polymer chain and that unpaired electrons at the chain ends react with oxygen molecules.

(2) Polymethylmethacrylate (PMMA)

PMMA is one of the heterogeneous polymers which are suitable for the detection of pair formation of mechano-radical. An ESR spectrum observed at 77 K from the PMMA milled for 24 h is shown as (a) in Fig. 7.22 [23]. It is well known that the PMMA radical produced by the main chain scission shows the characteristic (4+5 lines) spectrum similar to that when PMMA is irradiated by γ-rays, mentioned in Section 7.3.

It is reasonable to assume that one partner of the pair formation of the PMMA mechano-radical is the species, $—C_\beta H_{\beta 2} (C_\alpha \cdot) R(CH_3)$, yielding the characteristic spectrum shown in the figure. Based on this assumption, one can simulate the spectrum by a (4+5 lines) spectrum superposed with a doublet, as shown (b) in Fig. 7.22. The relative intensity of the two components which gave the best simulated spectrum was found to be equal within experimental error. This means that the two radicals corresponding to the two components of the observed spectrum were produced in equal amounts and in pairs. One species of the pair was naturally identified as $—C_\beta H_{\beta 2}(C_\alpha \cdot) R(CH_3)$, which is a primary product of the main chain scission, and consequently the other radical might reasonably be presumed to be the complementary product of the scission, $—C_\beta R(CH_3) (C_\alpha \cdot)H_{\alpha 2}$. However, the ESR spectrum anticipated for the latter radical is not a doublet but a triplet. A radical produced by a hydrogen shift to the scission site of the latter partner is the species $—C_\beta R(CH_3) (C_\alpha \cdot) H_\alpha C_\beta R(CH_3)—CH_3$ and this secondary radical would yield a doublet. INDO calculation for a model compound indicated that the secondary radical was more stable than the primary one [24]. Furthermore, the theoretical value of the doublet separation obtained from the INDO calculation on the model compound

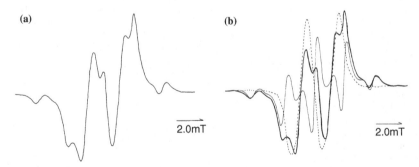

Fig. 7.22 (a) ESR spectra observed at 77 K from PMMA milled in vacuum at 77 K. (b) Superposition of the quintet-quartet spectrum (*thin line*) and the doublet (*dotted line*) at equal intensities. The *bold line* is the superposed spectrum. The figure is adapted from [23] with permission from John Wiley & Sons Inc.

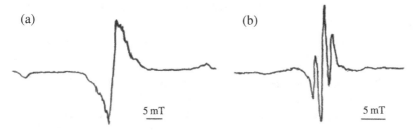

Fig. 7.23 (**a**) ESR spectrum observed at 77 K from PTFE milled in vacuum at 77 K. (**b**) The spectrum observed at 243 K. The figure is adapted from [23] with permission from John Wiley & Sons Inc.

was 2.06 mT, which is close to the experimental value 2.4 mT. From this analysis of the spectrum, it was concluded that mechanical fracture of PMMA produced a pair of radicals by main chain scission.

(3) Polytetrafluoroethylene (PTFE)

Figure 7.23(a) shows an ESR spectrum of PTFE fractured by ball-milling [23]. A weak but clear doublet was apparent with a separation of about 45 mT, although the strong central band was not well resolved. An ESR spectrum observed at 243 K is also shown in Fig. 7.23(b). The central part was a well resolved triplet with intensity ratio 1:2:1. This temperature variation of the spectrum was found to be reversible. It was established that a radical having two α-fluorine atoms was the source of the strong central band and the weak doublet with a wide separation of about 45.0 mT at low temperature where the fluoro radicals were immobilized in an amorphous matrix. When molecular motion of the polymer was stimulated at higher temperature, the separation of the doublet was reduced due to averaging of the anisotropic part of the fluorine hyperfine splitting. It was reported that the molecular motion in solid PTFE became active at room temperature. The triplet structure in the central band of the spectrum was attributed to the coupling with two β-fluorine atoms showing less anisotropy of the hyperfine coupling. The argument presented above leads to the conclusion that the radical species, $—C_\beta F_{\beta 2}$ $(C_\alpha \cdot)$ $F_{\alpha 2}$ is produced and this identification demonstrates that scissions of C—C bonds occur by mechanical fracture of solid PTFE at 77 K.

7.5.2 High Chemical Reactivity of Mechano-Radicals

High chemical reactivity of mechano-radicals is detected in various reactions by using the ESR technique. The high reactivity of mechano-radicals is caused by the conditions at the trapping region with new fresh surface formed by the mechanical destruction and the high molecular mobility of the radicals. We present the following examples of chemical reaction in order to show the effectiveness of the ESR method. Here, we can distinguish between homogeneous and inhomogeneous scissions of the polymer main chains, which generate neutral free radicals and ion species, respectively.

(1) Conversion of mechano-radical

    (A) The direct product from the homogeneous scission of the PE main chain, $—C_\beta H_{\beta 2} (C_\alpha \cdot) H_{\alpha 2}$ converts to $—C_\beta H_{\beta 2} (C_\alpha \cdot) H_\alpha (CH_3)$ by mild heat treatment. This is an intra chain migration of an unpaired electron (a hydrogen atom) along the chain, which is never observed for the $—C_\beta H_{\beta 2} (C_\alpha \cdot) H_\alpha —C_\beta H_{\beta 2}$, produced by the γ-irradiation of PE.

    (B) The radical produced directly by the homogeneous scission of the PMMA main chain, $—C_\beta R(CH_3)(C_\alpha \cdot) H_{\alpha 2}$. converts to the secondary radical, $—C_\beta R(CH_3) (C_\alpha \cdot) H_\alpha C_\beta R(CH_3)—CH_3$ during the milling.

(2) Reactivity of mechano-radicals with oxygen

    The PE mechano-radicals react easily with a small amount of oxygen molecules, during the sawing in liquid nitrogen mentioned in Section 7.5.1. In contrast to mechano-radicals, PE radicals formed by ionizing radiation do not react readily with oxygen but rather decay before formation of peroxy radicals as shown in warming experiments in the presence of oxygen. High reactivity of mechano-radicals of polypropylene and PTFE with oxygen has also been found.

(3) Reactivity of mechano-ions

    The anions and cations produced by an inhomogeneous scission of the polymer main chain during milling convert to neutral free radicals. The reactions with the highest reactivity occur through release and capture of an electron by mild heat treatment and photo bleaching. The phenomena are detected by the increasing intensity of the ESR spectrum during the treatments because the ion species show no ESR spectra. It has been found that the ion products produced by the inhomogeneous scission also initiate ionic polymerization.

(4) Initiation of copolymerization by mechano-radicals

    Mechano-radicals of high chemical reactivity can readily initiate various copolymerization processes. Figure 7.24 [25] shows the ESR spectrum observed at 77 K from fractured PTFE with ethylene monomer in vacuum at 77 K. In the experiment, the monomer in gas phase had been brought into the

Fig. 7.24 ESR spectrum from fractured PTFE with ethylene monomer in vacuum at 77 K (*solid line*) and the simulated spectrum of the chain end radicals of PE molecules based on free rotation around the C—C bond axis (*dashed line*). The figure is adapted from [25] with permission from American Chemical Society

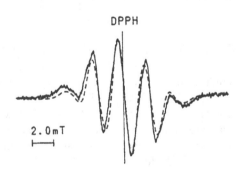

DPPH

2.0 mT

ball-milling chamber of the evacuated ampoule and frozen together with PTFE flakes before the fracture by milling. Thus, all flakes of the PTFE sample were covered with a frozen layer of the ethylene monomer, and the milling was carried out for the monomer-covered flakes at 77 K. The fracture of PTFE without the ethylene monomer generated the scission type radicals, $-C_\beta F_{\beta 2}$ $(C_\alpha \cdot)$ $F_{\alpha 2}$ as mentioned in Section 7.5.1. The ESR spectrum of the radicals is shown in Fig. 7.24.

It was confirmed that the fracture of the ethylene monomers at 77 K produced no free radicals. The quintet ESR spectrum shown in Fig. 7.24 can be undoubtedly attributed to the propagating radical of polyethylene, $-C_\beta H_{\beta 2}$ $(C_\alpha \cdot)$ $H_{\alpha 2}$, when both the polymers and the monomers are simultaneously fractured. The quintet is due to hyperfine splitting of two α- and two β-hydrogen nuclei. No trace of the PTFE radical was detected in the observed spectrum. Accordingly the polymerization of ethylene, which was proved by ESR, had been initiated not by the ethylene radicals but by the PTFE mechano radicals at as low a temperature as 77 K. This extremely high reactivity of the radicals is rather surprising because both PTFE and ethylene react in the solid state at 77 K. The mechano radicals newly created by the chain scission are surrounded by the monomer molecules because the radicals are trapped in the fresh surface formed by the mechanical destruction.

7.5.3 High Molecular Mobility of Mechano-Radicals

The ESR spectrum of PE fractured by ball-milling in vacuum consists of more than two components. The main product is the bond scission type radical, $-C_\beta H_{\beta 2}$ $(C_\alpha \cdot)$ $H_{\alpha 2}$, which decays at lower temperature than the other free radicals. Figure 7.25(a) [25] shows a sextet spectrum assigned to $-C_\beta H_{\beta 2}$ $(C_\alpha \cdot)$ $H_{\alpha 2}$ observed at 77 K. The spectral line shape was changed from the sextet (a) to a quintet (b) upon warming the sample from 77 to 104 K. The temperature dependence of the spectra can be interpreted in terms of the site exchange model as described in detail later. It was found that the scission type radical had a structure that was twisted by 15° from a symmetrical position, as was described in Appendix A7.1 and revealed by the spectral simulation of Fig. 7.25(a).

The quintet spectrum observed at 104 K (Fig. 7.25(b)) was successfully simulated by considering a more rapid exchange rate (4.0×10^8 s^{-1}) between the two equilibrium and twisted conformational states than that (7.1×10^7 s^{-1}) at 77 K. The quintet spectrum of the propagating ethylene radical on the PTFE surface was caused by the high rate of the exchange even at 77 K as shown in Fig. 7.24. The chemical structure of the propagating radical is just the same as the bond scission type radical. Figure 7.26 shows the observed and simulated spectra of the tethered PE chain end to a PTFE surface. The quintet-like spectrum observed at 2.8 K shows a relative intensity which is considerably different from that of the sextet in Fig. 7.25(a). This indicates that the propagating radical did not freeze even at a low temperature of 2.8 K. A computer program based on a site exchange between two

Fig. 7.25 ESR spectra observed at 77 K (**a**) and 104 K (**b**) from PE milled in vacuum at 77 K and the simulated spectra considering a site exchange of two β-hydrogen atoms between two equilibrium states at a rate of 7.1×10^7 s^{-1} (**a**) and 4.0×10^8 s^{-1} (**b**) (*dashed lines*). The figure is adapted from [25] with permission from the American Chemical Society

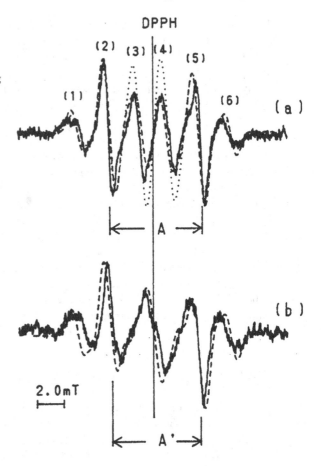

or more states employing the Liouville density matrix theory has been developed by Heinzer. Sakaguchi et al. [26] modified the computer program in order to simulate the ESR spectra due to the powder pattern of the chain end radicals that have two H$_\alpha$ atoms with anisotropic hyperfine splitting (*hfs*) constants and two H$_\beta$ atoms with isotropic *hfs* constants. The two sites and the coordinate system are described in Appendix A7.1.

Figure 7.26 shows the temperature dependence of ESR spectra of PE chain end type of propagating radical tethered to a PTFE surface [27]. The radicals were produced by ball-milling of PTFE with a large amount of ethylene monomer. The dotted spectra at 15 and 30 K were calculated using the site exchange rates of 10 and 56 MHz, respectively. The spectrum observed at 95 K could be simulated by the rotation of the chain end about the chain axis along with the site exchange motion. The oscillation amplitude was also estimated from the values of SPLIT A and B, shown in Fig. 7.26. The ESR spectra of the end type of scission radicals produced by the mechanical fracture of PE were also simulated by the site exchange model between the sites 1 and 2 as described in Fig. 7.50 (in Appendix A7.1). The reason

Fig. 7.26 Observed (*solid line*) and simulated (*dashed*) spectra of tethered PE chain ends. The measurements were carried out at the temperatures indicated in the figure. The simulated spectra were calculated by considering a site exchange of two β-hydrogen atoms between two equilibrium states at rates of 10–1,100 MHz. The figure is adapted from [26] with permission from the American Chemical Society

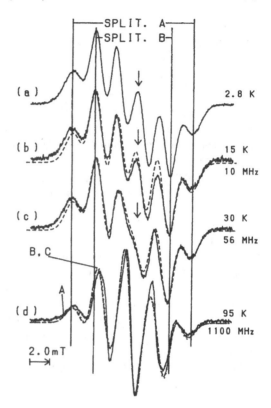

why high exchange rates were observed for the PE end type of the scission radical tethered to the PTFE surface can originate from a large void around the chain. Large voids are possibly formed due to the poor chain aggregation because the concentration of the chain ends is low and/or due to the poor miscibility between PE and PTFE. When the radicals were produced by ball-milling of PTFE with an extremely small amount of ethylene monomer, a quintet spectrum was observed even at 2.6 K. The spectrum was simulated by using a model of a mutual exchange motion between the two α hydrogen nuclei in a tunneling process. The phenomenon reflects that isolated single PE chains tethered on the PTFE surface display a free rotation in three dimension space similar to the molecular motion in the gas phase.

7.6 Auto-Oxidation Mechanisms of Polymer Systems

7.6.1 ESR Spectrum of Peroxy Radicals in Polymer Matrices

In general, alkyl radicals trapped in polymer materials convert to peroxy radicals when oxygen molecules are introduced. The disappearance of the double quintet due to the chain type fluoro alkyl radicals in polytetrafluoroethylene (PTFE) described in Section 7.4.1 and the growth of the asymmetric pattern due to the peroxy radicals are for example observed after the admission of oxygen:

$$- CF_2 - (C\cdot)F - CF_2- + O_2 \rightarrow -CF_2 - CF(OO\cdot) - CF_2- \qquad (7.14)$$

Iwasaki et al. [28] observed the ESR spectra of peroxy radicals in irradiated powders and oriented sample of PTFE using a K-band (24 GHz) ESR spectrometer and determined the principal values and directions of the g-tensor, both at room temperature and at 77 K.

Measurements at a higher frequency made it possible to obtain good resolution of the spectrum with respect to the g-anisotropy. Figure 7.27 shows the variation with temperature of the ESR powder spectra of peroxy radicals trapped in PTFE. The following principal values of the g-tensor were obtained at 77 K from the line shape of these spectra: $g_3 = 2.038$, $g_2 = 2.007$ and $g_1 = 2.002$; g_1 was considered to correspond approximately to $g_{//}$, while g_2 and g_3 represent $g_\perp$. On the other hand, the spectrum measured at room temperature (Fig. 7.27(a)) was split into two components. The symmetric component (thin line) was attributed to the peroxy radical which had enough motional freedom to average out the entire anisotropy in g-tensor.

The asymmetric component (dashed line) has a characteristic line shape arising from the axially symmetric g-tensor. From the line shape one can obtain the principal values: $g_{//}^r = g_1 \fallingdotseq 2.006$ and $g_\perp^r = \frac{g_1 + g_2}{2} \approx 2.022$, where $g_{//}^r$ and $g_\perp^r$ represent the apparent principal values at room temperature. The averages of the principal

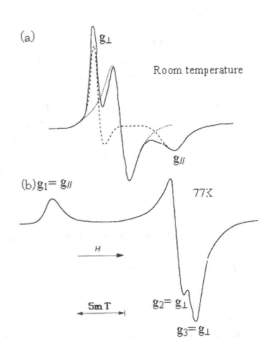

Fig. 7.27 ESR spectra of peroxy radicals in γ-irradiated PTFE powders observed at room temperature (**a**) and 77 K (**b**). Measurements were made at 24 GHz. The figure is adapted from [28] with permission from John Wiley & Sons Inc.

values of the g-tensors of the asymmetric component are $2.017 \approx \frac{g_{//}^r + g_{\perp}^r \cdot 2}{3} = \frac{2.006 + 2.0022 \times 2}{3}$ at room temperature and $2.016 \approx \frac{g_3 + g_2 + g_1}{3} = \frac{2.002 + 2.007 + 2.038}{3}$ at 77 K. The observed value of the symmetric component at room temperature was also 2.016. These facts suggest that the apparent spectral changes at room temperature and 77 K were due to complete or partial averaging of the tensor elements by molecular motion.

The ESR spectra for stretched PTFE films were also measured at room temperature and 77 K. The room temperature spectra obtained at various orientations of the stretch axis to the magnetic field are shown in Fig. 7.28. In addition to a strong signal with large anisotropy, a broad weak signal can be seen at $g = 2.016$. Since this peak is nearly isotropic and its g value is very close to the g value of the symmetric component of the powder samples, we can again attribute this peak to the peroxy radicals of PTFE. The value of 2.005, which was found with the field parallel to the stretch axis, is very close to $g_{//}^r = 2.006$ obtained for the powder spectrum, while 2.021 measured for the perpendicular direction, is close to $g_{\perp}^r = 2.022$, This means that the symmetry axis of the g tensor is parallel to the molecular chain axis at room temperature.

An angular dependence of the spectra was also observed at 77 K. As a result, the parallel spectrum has a narrow symmetric peak at $g = 2.003$, while the perpendicular spectrum still extends from $g = 2.005$–2.038. Therefore, the direction of the maximum principal value is perpendicular to the stretch axis. This means that the direction of the O—O bond is perpendicular to the molecular chain axis. It is concluded that the COO group lies in the plane perpendicular to the molecular chain axis, as shown in Fig. 7.29. If rapid motion around the chain axis takes place, the g tensor should be axially symmetric about the molecular chain axis, which is in

Fig. 7.28 Angular dependence of the ESR spectra of peroxy radicals in γ-irradiated oriented films of PTFE, observed at room temperature. The angles between the magnetic field and the stretching direction are indicated. The symbols "i" and "a" indicate the isotropic and anisotropic components, respectively. Measurements were made at 24 GHz. The figure is adapted from [28] with permission from John Wiley & Sons Inc.

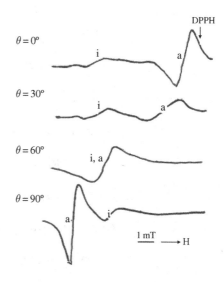

Fig. 7.29 Structure of peroxy
radicals in γ-irradiated PTFE
The figure is adapted from
[28] with permission from
John Wiley & Sons Inc.

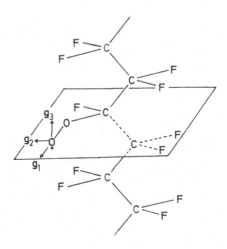

agreement with the experimental results, $g'_\perp \approx \frac{2.038(g_1)+2.007(g_2)}{2} \approx 2.022$. A rotation around the chain axis, rather than rotation around the O—O bond, is concluded.
Similar structures and molecular motions of peroxy radicals trapped in polyethylene
and polypropylene solids have been observed as discussed in Section 7.6.3.

7.6.2 Oxidation Processes in Irradiated Polymers and Chemical Reactions of Peroxy Radicals

The oxidation of polymers has been studied by many authors. The mechanisms of
oxidation were presented in the studies, and the following reaction equations are
generally accepted for all cases of oxidation in polyethylene.

$$(A)\ RH \rightarrow activation \rightarrow (R\cdot) + H \tag{7.15}$$

$$(B)\ Diffusion\ of\ O_2 \rightarrow (R\cdot) + O_2 \rightarrow ROO\cdot \tag{7.16}$$

$$(C)\ ROO\cdot + R'H \rightarrow ROOH + (R'\cdot) \tag{7.17}$$

$$(D)\ (R\cdot) + (R\cdot) \rightarrow Vanishing\ of\ radicals \tag{7.18}$$

$$(E)\ (R\cdot) + ROO\cdot \rightarrow Vanishing\ of\ radicals \tag{7.19}$$

$$(F)\ ROO\cdot + ROO\cdot \rightarrow Vanishing\ of\ radicals \tag{7.20}$$

$$(G)\ ROOH \rightarrow RO\cdot + OH \tag{7.21}$$

$$(H)\ ROOH \rightarrow ROH, RCHO, RCO, etc. \tag{7.22}$$

The equations described above constitute the schematic oxidation process for polyethylene but observation and detailed study of each reaction are very important in order to elucidate the oxidation mechanism. Many investigations of polyethylene oxidation have been based on analysis of the final products described in Eq. (7.22). In this section, direct ESR observations of the radical species in the reaction equations are introduced. The direct observation is very important to clarify the mechanism of oxidation and to design methods to prevent deterioration. Reaction Eqs. (7.15)–(7.21) can be examined either qualitatively or quantitatively based on ESR results.

(1) Reaction (A): The structures of the free radicals in Eq. (7.15) are clarified by the assignment of ESR spectra produced by ionizing irradiation, UV-irradiation and mechanical fracture as indicated in Sections 7.4.1 and 7.5.1.

(2) Reaction (D): The kinetics of decay of the polymer radicals is studied by observing the time dependent concentration of the radicals (the intensity of ESR spectra). Equation (7.18) can be modeled by using a diffusion-controlled bimolecular reaction. Equation (7.19) and (7.20) can be addressed in the same way using a diffusion-controlled model with the reactivity of $ROO \cdot$ being higher than $R \cdot$.

(3) Reaction (B): Hori et al. [29] studied Eq. (7.16) in detail. They made quantitative measurements of the reaction of oxygen with allyl-type radicals trapped in the amorphous regions of polyethylene. The kinetic data were analyzed based on the diffusion-controlled process theory. Diffusion of the oxygen into the amorphous regions of polyethylene was discussed and diffusion constants at low temperatures were estimated. As an example, the diffusion constant at 201 K has been found to be 6.3×10^{-10} cm^2/s. Quantities such as the activation energy of the diffusion process and the solubility constant were also estimated. The oxidization process for alkyl radicals was also studied for the solution grown single crystals of polyethylene by Hori et al. [30] and Seguchi et al. [31]. The order of magnitude of diffusion constants of oxygen in the crystalline region was found to be 10^{-16} cm^2/s at 320 K, which was extremely low in comparison with those in the amorphous region.

(4) Reaction (C): The reaction is important in a chain mechanism during the auto oxidization process. Direct observation of the reaction was successfully carried out by Hori et al. [32] for a urea-polyethylene complex (UPEC). Figure 7.30 shows the variations of ESR spectra for peroxy radicals in the UPEC due to heat treatment in vacuum at 361 K. Spectrum (a) was obtained from peroxy radicals and spectrum (d) was that obtained from alkyl radicals as shown in Section 7.4.1. Figure 7.31 shows the variation of the concentrations of various radicals with the time of heat treatment at 361 K. It was found that 20% of the radicals decayed, but ca. 80% of the decaying peroxy radicals were converted into alkyl radicals. A few authors reported that peroxy radicals trapped in polytetrafluoroethylene converted into alkyl fluoro radicals by heat treatment in a similar manner. Some authors claimed that the back reaction of Eq. (7.16) occurred in that case:

$$ROO \cdot \rightarrow R \cdot + O_2 \tag{7.23}$$

Fig. 7.30 Change of the ESR
spectra of free radicals
trapped in UPEC by the heat
treatment in vacuum at
361 K. The figure is adapted
from [32] with permission
from Elsevier

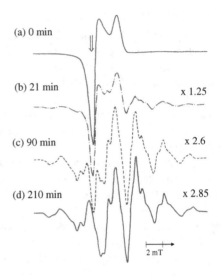

Fig. 7.31 Change of the
concentrations of various free
radicals in UPEC with the
time of heat treatments at
361 K in vacuum; *squares*,
total radicals; *solid circles*,
peroxy radicals; *triangles*,
alkyl radicals. The figure is
adapted from [32] with
permission from Elsevier

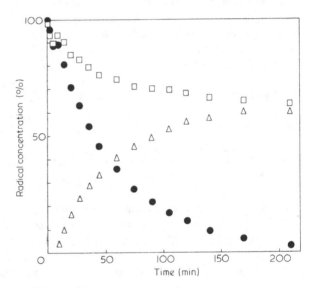

7.6.3 Structure and Molecular Motion of Peroxy Radicals in Polymer Matrices

7.6.3.1 Calculation of the ESR Spectrum of the Peroxy Radical

ESR spectra of peroxy radicals show g-anisotroy when the molecular motion is
frozen at low temperatures. The spectra can then be calculated on the assumption
that the radicals have completely random orientation in three-dimensional space.

In order to simulate the ESR spectrum of a peroxy radical, one must calculate the resonance magnetic field, $H_r(\Omega)$ and the intensity of the ESR transition, $I(H_r)$ which is proportional to the number of radicals for a particular arbitrary orientation with respect to the external field, Ω represents the set of direction cosines of the external field, $(\cos\gamma_1, \cos\gamma_2, \cos\gamma_3)$ expressed in the principal axis system of the g-tensor. $H_r(\Omega)$ is represented by the following Eq. (7.24):

$$H_r(\Omega) = \frac{h\nu}{g(\Omega)\mu_\beta} \qquad (7.24)$$

where $h\nu$ and μ_β are the microwave energy and the Bohr magneton, respectively.
The direction cosines, $\cos\gamma_1$, $\cos\gamma_2$, and $\cos\gamma_3$ are given by:

$$\cos\gamma_1 = \sin\theta\cos\varphi \quad \cos\gamma_2 = \sin\theta\sin\varphi \quad \cos\gamma_3 = \cos\theta \qquad (7.25)$$

here, θ and φ are the polar and azimuthal angles of the directions of the external field, in the principal axis system, respectively. The g-value, $g(\Omega)$ for each orientation of the radical can be computed from the following equation:

$$g^2(\Omega) = g_1^2\cos^2\gamma_1 + g_2^2\cos^2\gamma_2 + g_3^2\cos^2\gamma_3 \qquad (7.26)$$

The ESR spectrum, $I(H)$ can be calculated from the line shape function G as follows,

$$I(H) = \int_0^{\pi/2} d\varphi \int_0^{\pi/2} G((H - H_r(\Omega), D_h)\sin\theta\, d\theta \qquad (7.27)$$

where D_h is the line width. In the employed computer program, g-values can actually be computed for 30,000 sets of (θ,φ) in a solid angle of $\pi/2$ and the spectra are obtained by summation of the line shape function $G(H-H_r(\Omega), D_h)$ for all orientations.

7.6.3.2 Structure and Molecular Motion of Peroxy Radicals of Polyethylene

In Section 7.6.1, principal directions of the g-tensor of PTFE peroxy radicals were determined and rotation of the radicals around the PTFE chain axis, rather than around the O–O bond, was demonstrated. In this section, we describe a detailed study of the ESR spectra and the molecular motion of peroxy radicals when they are trapped in various regions with many kinds of heterogeneities in polymer matrices. Temperature dependent ESR spectra of peroxy radicals in the powdered polyethylene having a low crystallinity are shown in Fig. 7.32 [33].
The spectra consist of two components, as can be seen from the figure, in particular for spectra (b) and (c). One component, which we call the A radical, is similar to the amorphous patterns observed for the rigid peroxy radicals. The other component, due to the so-called B-radical, has a broad singlet-like pattern. The A radical was

Fig. 7.32 Temperature dependent ESR spectra of peroxy radicals in PE in the range from 118 to 297 K. The figure is adapted from [33] with permission from Elsevier

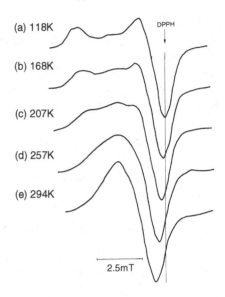

readily assigned to the peroxy radical, while the B-radical was also assigned to the same radical, because of the reversibility of the spectral change with temperature. It should be noted that both kinds of radicals exist in the same amorphous region of PE under these experimental conditions. A computer simulation of the spectra was carried out and an example of a simulated spectrum is shown in Fig. 7.33.

The good agreement obtained between the simulated and experimental spectra confirmed the coexistence of the two kinds of radicals. The g-anisotropy for both radicals averaged out as shown in Fig. 7.34 and the fraction of the mobile component (B-radical) increased with an increase in temperature. These facts show that the

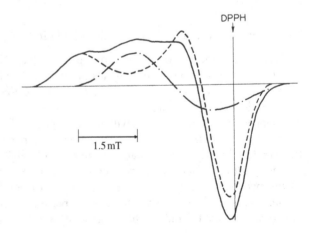

Fig. 7.33 An example of the simulated spectrum of peroxy radicals with superposition of two kinds of spectra. The figure is adapted from [33] with permission from Elsevier

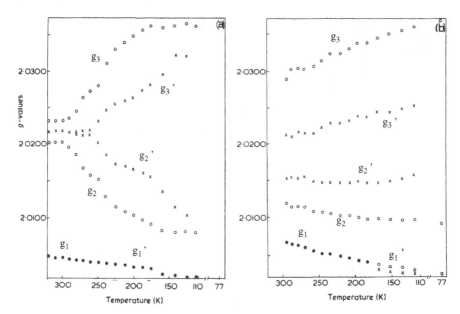

Fig. 7.34 Change in the principal values of the g-tensor with temperature. (**a**) UPEC (**b**) normal PE *Open circles* and *crosses* indicate the changes of the A-(g_1, g_2, g_3) and B-($g_1{}'$, $g_2{}'$, $g_3{}'$) radicals, respectively. The figure is adapted from [33] with permission from Elsevier

structure and mobility of the peroxy radicals in the amorphous region had heterogeneity. The B-radicals begin to move at lower temperatures while the A-radicals (rigid radical) convert to the B-radical (mobile radical). The peroxy radicals have actually a distribution of mobility, which is described here with two representative components. Clearer temperature dependent ESR spectra of the peroxy radicals have been observed in the urea–PE complex (UPEC) as shown in Fig. 7.35 [34]. The spectra could be simulated with two components due to A- and B-radicals as described above. This suggests that the radicals had a distribution of mobility, even in the UPEC crystals. For the peroxy radicals produced from the alkyl radicals in the UPEC, the principal values of the g-tensor at 77 K were $g_1 = 2.0022$, $g_2 = 2.0081$, and $g_3 = 2.0366$.

The motional averaging of the g-values for the A- and B-radicals occurred at lower temperatures than those in the normal PE as shown in Fig. 7.34. The value of g_1 changed slightly, whiles g_2 increased and g_3 decreased to reach the same value with an increase in temperature. Thus, in the polymer chain two main motions of the peroxy radicals have to be taken into consideration.

The first is a rotation of the OO group around the C—O bond axis, and the second is a rotation or vibration of the entire COO group around the chain axis. If rotation of the OO group around the C—O bond would occur, all of the g-values, g_1, g_2, and g_3, are averaged with the same relaxation time, because the g_3-axis is along the O—O bond direction. This conflicts with the experimental results. Therefore, the rotation of the COO group around the chain axis has been concluded to be faster and the

Fig. 7.35 Temperature
dependences of the observed
(*dashed*) and calculated
(*solid*) ESR spectra of peroxy
radicals in UPEC. The figure
is adapted from [34] with
permission from Elsevier

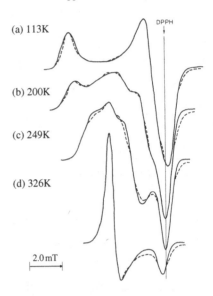

direction of the g_1-axis should be along the chain axis as for the PTFE peroxy radicals. The g_1 value for normal PE shows much larger temperature dependence than the same quantity for the UPEC, as shown in Fig. 7.34. This difference must be a reflection of a different situation, where three-dimensional motion of the peroxy radicals is marginally allowed in normal PE, while in the UPEC only two dimensional motion is possible. Rotation around the g_1-axis occurs more rapidly in the UPEC than in normal PE at the same temperature. This rotation of the complex completely averages out the anisotropy of the g_2- and g_3-values at room temperature and gives an axial symmetric spectrum, as shown in Fig. 7.35 (d), $g_{//}{}^r = g_1 = 2.0022 \cong 2.0048$ and $g_{\perp}{}^r = (g_2 + g_3)/2 = (2.0081 + 2.0366)/2 = 2.0223 \cong 2.0213$. This result indicates that rotation around the chain axis in the complex is faster than in normal PE. Schlick and Kevan [35] concluded from computer simulation that 180° rotational jumps of the OO fragment of the peroxy radical take place around the C—O bond. However, the results obtained for a single crystal of a urea-n-tetracosane complex indicated conclusively that rotation around the chain axis is much more rapid than that of the rotation of the OO fragment around the CO bond.

7.6.3.3 Structure of Peroxy Radicals of Polypropylene

In order to determine a more precise structure for the peroxy radicals of isotactic polypropylene (i-PP), angular dependent ESR spectra of peroxy radicals trapped in an elongated film of i-PP were investigated [36, 37]. The tertiary alkyl radicals, $—C_{\beta}H_{\beta2}—(C_{\alpha}\cdot)(CH_3)—C_{\beta}H_{\beta2}—$ formed by γ-irradiation were stably trapped in the crystalline region of elongated PP. The alkyl radicals converted to the peroxy radicals when oxygen molecules were introduced. A very marked angular dependence was observed for the spectra measured at 77 K. A computer simulation of

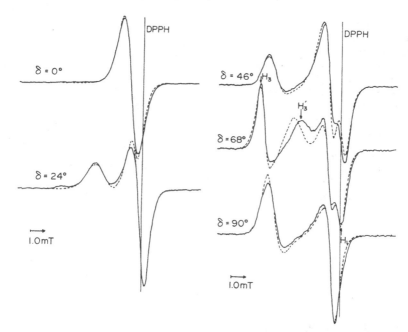

Fig. 7.36 Angular dependence of the observed (*solid*) and calculated (*dotted*) ESR spectra of the peroxy radicals trapped in isotactic polypropylene upon rotation angle δ between the direction of elongation and that of the magnetic field. The ESR spectra were observed at 77 K. The figure is adapted from [36] with permission from the American Chemical Society

the angular dependent spectra was also carried out by using the method described in Appendix A7.2. The profiles of the experimental spectra showed good agreement with those of the calculated spectra, as shown in Fig. 7.36, where the best fit simulation spectra were calculated employing the degree of orientation, $f_\alpha = 0.968$, and the angles $\lambda_1 = 39°$ and $\lambda_3 = 68°$. Here λ_1 and λ_3 stand for the angles between the principal directions of the g tensor components, g_1 and g_3 and the polymer chain axis. The angles are of relevance to discuss the structure of the peroxy radical, as schematically illustrated in Fig. 7.37. The (C–O–O) bond angle was also obtained and its value was supported by ab initio calculations, with the latter giving a value of 111.0°. The dihedral angle for internal rotation angle, φ was determined to be 55.5°. This value is close to 60°, corresponding to the gauche conformation. The deviation of 4.5° from the gauche conformation may be caused by the fact that the 3^1-helical structure deviates very much from a planar zig-zag conformation. This work has been extended to determine the structure of the mobile peroxy radical in the inner crystalline region of i-PP. The angular dependent ESR spectra gave $f_\alpha = 0.536$, $\lambda_1 = 24°$ and $\lambda_3 = 73°$. These values of λ_1 and λ_3 correspond to $\varphi = -103°$ and a bond angle C–O–O of 111°. The value of φ is fairly close to $-120°$, corresponding to a skew conformation, which indicates the *cis*-form. The small value of $f_\alpha = 0.536$ suggests that the mobile radicals reside in molecularly disordered, crystalline defect regions.

Fig. 7.37 Schematic
illustrations of the
conformation and mobility of
peroxy radicals of isotactic
polypropylene. λ_1, λ_2, and λ_3
are the angles between the
principal directions of the g
tensor and the polymer chain
axis: g_1 is perpendicular to
the COO plane; g_3 is along
the O—O bond direction. The
figure is adapted from [36]
with permission from the
American Chemical Society

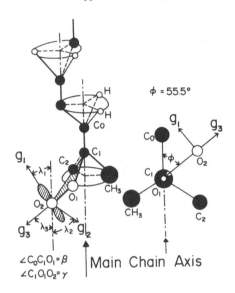

7.7 Conducting Polymers

Conducting polymers are very important and interesting materials having at the
same time flexible mechanical properties. ESR has been applied to the characteristics and understanding of the mechanisms of conduction. Various applications to
the conduction polymers are presented in this section. For simplicity, the ESR study
of solitons (electrons) in solid polyacetylene (PA) is mainly given as an example.
ESR line shapes of the Dysonian type and ENDOR spectra for conduction electrons
and solitons are presented in detail. It is also very interesting that ESR power saturation [38] and line width [39] change remarkably with the doping process of PA.
Those matters are, however, outside the scope of this chapter.

7.7.1 Diffusive Motion of the Soliton in Pristine Polyacetylene
Detected by the ESR Line Width

The unpaired electron spin in trans type polyacetylene (PA) plays an important role
as a soliton for the conduction. The PA developed by Shirakawa [40] is a semiconductor, but changes to a conducting polymer by adding dopants. The distribution of
the spin along the chain is symmetrical like a wave centering around the midpoint
of the soliton. The neutral soliton of the excited state is not a conduction carrier.
However, when dopants like As P_5, I_2 and similar agents are introduced and abstract
electrons from the solitons, the formed carbanium ion solitons are converted to conduction carriers. When dopants like Li, Na and similar substances add electrons to
the solitons, carboanion solitons also change to conduction carriers.

Fig. 7.38 The frequency (f) dependence of the ESR line width (ΔH_{pp}) for trans-PA with observation temperature. Frequency, f ($= \omega/2\pi$) is used in place of ω in the text. 1 Oe (Gauss) = 0.1 mT. The figure is adapted from [41] with permission from the American Physical Society

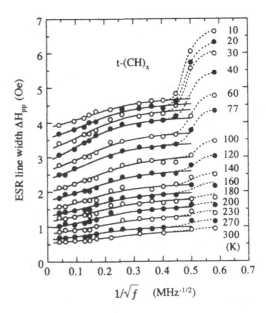

The characterization of solitons by the ESR method is described in this section. Mizoguchi et al. [41] obtained interesting results concerning the dynamics of electron spins in the pristine PA from a study of the ESR line width of the soliton. Figure 7.38 shows the frequency dependence of the ESR line width for trans PA with the observation temperature as parameter. This result suggests three kinds of interesting and important characteristics of the soliton.

(1) One dimensional (1D) diffusion of the electron spin along the chain

The spectral density $\varphi_{Q1D}(\omega)$ as a function of frequency, ω, can be obtained by a solution of the 1D diffusion equation:

$$\varphi_{Q1D}(\omega) = \frac{1}{\sqrt{4D_{//}/\tau_\perp}} \left[\frac{1 + \sqrt{1 + (\omega\tau_\perp/2)^2}}{1 + (\omega\tau_\perp/2)^2} \right]^{1/2} \tag{7.28}$$

where, $D_{//}$ is the diffusion rate along the chain and $\tau_\perp$ is the mean life time of 1D correlation. $D_{//}$ can for instance be the rate of motion along the chain and $1/\tau_\perp$ the rate of the inter-chain hopping of the spins.

The following expressions are valid in different frequency regions.

$$\varphi_{Q1D}(\omega) = \frac{1}{\sqrt{2D_{//}\omega}} \quad \text{for } 1/\tau_\perp \ll \omega \ll D_{//} \tag{7.29}$$

$$\varphi_{Q1D}(\omega) = \frac{1}{\sqrt{2D_{//}/\tau_\perp}} = \text{const. for } \omega \ll 1/\tau_\perp \tag{7.30}$$

The line width, T_2^{-1} is given as follows:

$$T_2^{-1} \propto (0.3\varphi(0) + 0.5\varphi(\omega_0) + 0.2\varphi(2\omega_0)) \tag{7.31}$$

The linear relationship between T_2^{-1} and $\omega^{-1/2}$ in the higher ω regime and the independence of T_2^{-1} on ω in the lower frequency range in Fig. 7.38 indicate that the spins diffuse on a one dimensional chain with a diffusion coefficient, $D_{//}$. We can estimate $D_{//}$ from the slope of the line, T_2^{-1} vs. $\omega^{1/2}$ in the high frequency regime of $(0 < f^{-1/2} < 0.2)$. On the other hand, $\tau_{\perp}$ is calculated by using the constant value of T_2^{-1} in the lower frequency regime $(0.4 < f^{-1/2} < 0.45)$ of Fig. 7.38.

Figure 7.39 shows the temperature dependence of the diffusion rate $D_{//}$ along the polyacetylene chain, together with that derived from NMR. $D_{//}$ is of the order of magnitude 10^{13} (rad/s) whereas $1/\tau_{\perp}$ is estimated to be of the order 10^7 (rad/s). These results suggest that the diffusion rate of the spins along the chain is much higher than the rate of the inter-chain hopping of the spins.

(2) Diffuse-trap model in the diffusion of the spin

Nechtschein et al. [42, 43] have proposed a diffusion-trap model for the mechanism of the line broadening at X-band ESR. The diffuse-trap model assumes that the neutral soliton will visit some trapping sites and stay for a long time compared with the time spent at the normal sites. The resulting line width is the sum of the broadening caused by two different mechanisms. One is the 1D diffusive motion of the neutral soliton that works predominantly when the soliton diffuses in the normal sites. It broadens the ESR line. Another mechanism works when the soliton is in the trapped state where the static hyperfine coupling broadens the ESR line width by $c_{tr}\Delta H_{hyp}$ and the static electron dipolar coupling between the trapped solitons does it by $c_{tr}^2\Delta H_{dip}$, while the diffuse-trap model predicts an increase in the line width, ΔH_{trap} (T) in the trapped state,

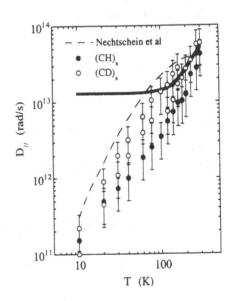

Fig. 7.39 The temperature (T) dependence of the diffusion rate along the chain ($D_{//}$) for the trans-PA together with that obtained by ^{1}H-NMR. The thick *solid curve* indicates the variation of $D_{//}$ for the trans-PA without the diffuse-trap model correction. The figure is adapted from [41] with permission from the American Physical Society

$$\Delta H_{trap}(T) = c_{tr}\Delta H_{hyp} + c_{tr}^2 \Delta H_{dip} \qquad (7.32)$$

The probability for the soliton being in the trapped state, c_{tr} is 1 at T = 0 K, decreases with increase in T and reaches 0 at infinite T. ΔH_{hyp} and ΔH_{dip} are the line widths due to the hyperfine coupling and the interaction of the electron dipoles, respectively when the electron movement is frozen. Nechtschein et al. [43] obtained a theoretical temperature dependence of the line width by the diffuse-trap model and compared it with the experimental one of the pristine PA. The theoretical dependence is in good agreement with the experimental one when the distribution of the trapping energy is considered. The solid curve in Fig. 7.39 represents the variation of $D_{//}$ without the trapping correction. This result demonstrates the applicability of the diffuse-trap model and that the effect due to trapping is negligibly small above 100 K. The effect of the neutral soliton trapping also appeared in the anomalous broadening at low frequency observations, mentioned in the following section.

(3) Structure of soliton detected by anomalous broadening in trans-PA

An anomalous broadening of the line width below 6 MHz has been found at extremely low temperature in trans PA as shown in Fig. 7.38. The broadening is strongly dependent on the observation temperature and the orientation of the main chain. The maximum and minimum line widths at 10 K, 0.98 and 0.66 mT are detected when the magnetic field is applied parallel and perpendicular to the stretch direction of the PA film, respectively. These results suggest that the broadening originates from $\Delta H_{trap}(T) \approx c_{tr}(T)\Delta H_{hyp}$ while $c_{tr}(T)\Delta H_{hyp} \gg c_{tr}^2(T)\Delta H_{dip}$.

The anisotropic hyperfine splittings due to an α-hydrogen of the alkyl type free radical, parallel and perpendicular to the chain axis are 3.70 mT and 1.25 mT, respectively, as described in Section 7.4.1. The anisotropy is reflected on the anisotropic broadening of the line width. As a conclusion the origin of the anomalous broadening in trans-PA could be assigned to a crossover from "unlike spin" to "like spin" as the ESR resonance frequency was decreased and the value of $f^{-1/2}$ was increased. An effective coupling constant with $a_{eff} = A_0 \cdot \rho$, should be considered. Here, A_0 is the isotropic hyperfine constant of a π-electron radical with spin density equal to one and ρ is the spin density on the central carbon site in the soliton. The a_{eff} value determines the hyperfine splitting of the ESR spectrum. In Fig. 7.38 the anomalous broadening is found below 6 MHz ($f^{-1/2}$~0.4). From the threshold frequency of 6 MHz, one can estimate the maximum spin density for the neutral soliton as $\rho_{max} \geq 0.1$, based on the isotropic hyperfine splitting due to an α-hydrogen ($\rho = 1$), 2.3 mT corresponding to 64 MHz. This result provides strong evidence for the SSH-type [44] localized wave function for the neutral soliton, where the electron moves rapidly along the chain.

7.7.2 Spin Density Distribution of the Soliton in Pristine Polyacetylene Detected by ENDOR

The detailed structure of the soliton in PA can be clarified by using the ENDOR technique as shown by Kuroda et al. [45, 46].

Fig. 7.40 Anisotropic
ENDOR line shape of a 190%
stretched cis-rich PA The
external magnetic field is
applied parallel (//) and
perpendicular (⊥) to the
stretch direction of the film.
The figure is adapted from
[45] with permission from the
Physical Society of Japan

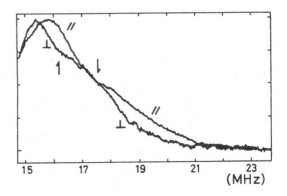

(1) Determination of the spin density in PA

Figure 7.40 shows the high frequency side of the ENDOR signals of a 190% stretched cis-rich PA at 4 K. The ENDOR spectra show a nearly symmetrical doublet around the free proton frequency, 14 MHz.

The anisotropy of the line width is clearly manifested. When the external magnetic field is applied parallel to the stretching direction of the film, the signal intensity has a higher intensity in most frequency regions than that in the case of the field being perpendicular to the direction. The spectra seem to be composed of more than two components. The ENDOR features at low temperatures can be interpreted as the direct evidence of the "soliton like" spin density by the simulation of the anisotropic spectrum. The maximum frequency of the ENDOR spectrum is related to the spin density, $\rho(0)$ at the central carbon of the soliton as indicated in Fig. 7.41.

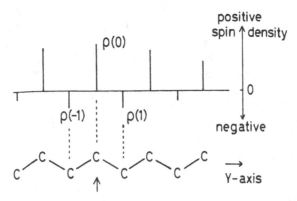

Fig. 7.41 Schematic view of the spin density, ρ of the soliton, which is symmetrical centering around the position 0 along the polymer chain axis (y-axis). The densities at even positions, $\rho(0)$, $\rho(\pm 2)$ etc. are positive whereas those at odd positions $\rho(\pm 1)$, $\rho(\pm 3)$ etc., are negative. The figure is adapted from [45] with permission from the Physical Society of Japan

The anisotropy of the ENDOR spectra is attributed to the corresponding anisotropies of the C—H protons of the soliton. The directions of the principal tensor axes are parallel to the C—H axis (x), the chain axis (y), and the axis of the π-orbital (z), respectively in accordance with the π-electron radical structure of the soliton. The ENDOR frequency due to the proton of the n-th C—H unit with the magnetic field along a principal direction is given by a simple equation. For the magnetic field along the y-axis one obtains for instance Eq. (7.33):

$$\nu_\pm = \nu_0 \pm 0.5 \cdot \rho(n) \cdot A_{yy} \qquad (7.33)$$

A_{yy} is the principal value for the C—H proton in a π-electron radical with spin density $\rho = 1$, and ν_0 is the free proton frequency. When the external field is applied parallel to the stretching direction (y) the maximum ENDOR frequency ν_+ is estimated to be c.a. 21 MHz at which point the signal changes its slope. By substituting the experimental values of A_{yy} and $(\nu_\pm - \nu_0)$ in Eq. (7.33), $\rho(0)$ is evaluated. Due to the uncertainty of the exact value for A_{yy} the value of $\rho(0)$ ranges between 0.11 and 0.19. The spin density, $\rho(1)$ depicted in Fig. 7.41 is negative as discussed below and estimated to be –0.05 by using the observed frequency of 17.5 MHz at the turning point of the inner part of the ENDOR spectrum indicated by an arrow in Fig. 7.40.

(2) Simulation of anisotropic ENDOR

Kuroda et al. [45] also calculated the anisotropic ENDOR line shape in partially oriented fibril systems. Samples of PA are known to consist of a fibril structure of polymers. The fibrils are partially oriented along the stretching direction of the film. The summation of the spectrum for each orientation of the polymer chain taking into account the preferential orientation of the fibrils gives the ENDOR spectrum of the partially oriented system. This simulation method is similar to the calculation of the ESR spectrum of partially oriented peroxy radicals described in Appendix A7.2. Figure 7.42 shows examples of the calculated spectra of the parallel (thick line) and perpendicular (thin line) directions for frequencies higher than ν_0 with appropriate parameter values. The anisotropy and the occurrence of turning points are well reproduced. The arrows in the figure show the ENDOR frequencies of the tensor components due to $\rho(0)$ and $\rho(1)$. The stick diagram indicates the ENDOR frequencies along the polymer axis (y-axis) for the positive and negative spin densities Fig. 7.41. The positive spin sites are at the central and even numbered positions, i.e. at, 0, ±2 ±4, etc., with positive spin densities on the carbon atoms, whereas the negative spin sites are the neighbor carbon atoms, ±1, ±3, ±5, etc., spin densities of which are negative.

The origin of the turning point is understood as the frequency due to $\rho(1)$ at which the contributions from negative spin sites start to overlap with those from positive spin sites. While the signs of $\rho(0)$ and $\rho(1)$ are not determined by the present experiment, their sign should be opposite as discussed in the next section. Concerning the details of the spin density, the observation of the clear turning points is another important result of the ENDOR study and in fact the turning points of the parallel direction were found to be directly correlated to the negative spin density by the simulation study which also explained the small structures of the spectrum. A similar

Fig. 7.42 Calculated line shapes of the parallel and perpendicular directions of preferentially oriented polymer fibril. *Arrows* show frequencies of corresponding hyperfine tensor components. The figure is adapted from [45] with permission from the Physical Society of Japan

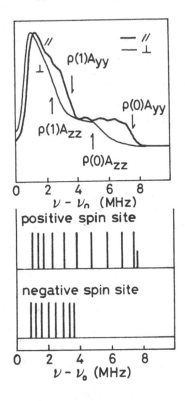

observation of π-electron defect states was also made by well-resolved ENDOR in undoped poly (paraphenylene vinylene) [46].

(3) Sign of spin density and triple resonance

It is well known that the presence of spin densities with opposite signs can be studied by an electron-nuclear-nuclear triple resonance (TRIPLE) experiment in which an additional radio-frequency (RF) source is applied in comparison with the normal ENDOR experiments. A detailed comparison of TRIPLE and ENDOR spectra of the ν_+ band at 4 K is shown in Fig. 7.43 [47]. The pumping frequency was first applied at 18.3 MHz (Fig. 7.43(a)) and 19.1 MHz (Fig. 7.43(b)), of the high frequency region, $(\nu > \nu_{1+})$. While the TRIPLE intensity is lower than the ENDOR intensity at most parts of the high frequency region, it is higher in a certain range $\nu_p < \nu < \nu_{1+}$. ν_{0+} and ν_{1+} indicate the frequencies corresponding to the peak values of the spin density sites, $\rho(0)$ and $\rho(1)$. This situation cannot happen unless the spin densities are of opposite sign.

When, the pumping frequency was next applied in the range $\nu_p < \nu < \nu_{1+}$, at 16.2 MHz (Fig. 7.43(c)), the TRIPLE intensity did not change. In this range the signals of both positive and negative densities are contained as indicated in Fig. 7.41. In that situation, the TRIPLE effect may be nearly canceled due to the competing effect of the pumping of both regions of the spin density and less amount of change would be observed. On the other hand, the TRIPLE intensity is slightly higher for the region, $\nu_{1+} < \nu$. The effect may be understood as a slightly larger efficiency of the

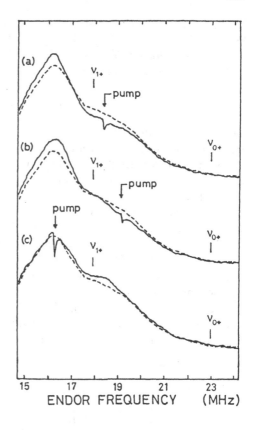

Fig. 7.43 Detailed comparison of TRIPLE (*solid*) and ENDOR (*dashed*) spectra of ν_+ branch at 4 K. The pumping frequencies are (**a**) 18.3 MHz, (**b**) 19.1 MHz (**c**) 16.2 MHz. The figure is adapted from [47] with permission from the Physical Society of Japan

pumping of negative spin site because the spectral density of negative sites is high for $\nu_p < \nu < \nu_{1+}$ as seen in Fig. 7.41. These observations confirm the opposite spin density for even and odd sites of carbon in Fig. 7.41.

7.7.3 Dyson's Theory of ESR Line Shapes in Metals

The theory of paramagnetic resonance absorption by the conduction electrons in a metal was developed at an early stage by Dyson [48] taking into account the diffusion of the electron in and out of the thin skin into which the radio-frequency (RF) field penetrates. The classical skin depth, δ is given by

$$\delta = (c^2/2\pi\sigma\omega)^{1/2} \tag{7.34}$$

where c is the speed of light, σ is the conductivity of the metal and ω is the fixed frequency of the radio micro-wave field. Under the influence of the radio frequency magnetic field, $H_1(r)$ a certain macroscopic magnetization, $M(r)$ will be created in the metal as a result of the turning of the magnetic moments of the conduction electrons. The field $H_1(r)$ in turn depends on M because the penetration of the radio-frequency field into the skin of the metal is affected by the magnetization. As a result, the electron diffusion should shift the resonance frequency and change the ESR line shape. Dyson calculated both H_1 and M as functions of an external field, H and derived a theoretical line shape of the conduction electrons. An average electron will diffuse across the skin depth in a time.

$$T_D = (3\delta^2/2v\Lambda) \tag{7.35}$$

where T_D is the diffusion time, v is the velocity of the electron and Λ is the mean free path of the electron. When the electron relaxation time, T_2 (for metals $T_1 = T_2$) is much shorter than the time of the electron diffusion, T_D, the ESR absorption line is entirely independent of the diffusion and has a symmetrical Lorlentzian shape with half-width T_2^{-1}.

On the other hand, if T_D is shorter than T_2, the diffusion tends to increase the width of the resonance line from T_2^{-1} to T_D^{-1}. When the conductivity, σ increases, the skin depth, δ decreases and then the diffusion time, T_D also decreases according to Eqs. (7.34) and (7.35). As a result, the condition with $T_D < T_2$, is realized. The line shape theory by Dyson is then applicable. The exact line shape can be derived as functions of T_D, T_2, and T_T where T_T is the time it takes for the electron to pass through the sample. Feher and Kip [49] calculated the line shapes for different ratios of the diffusion time T_D to the relaxation time T_2, which are shown in Fig. 7.44 in the case of $T_T \gg T_D$, $T_T \gg T_2$. A typical Dysonian line shape is that depicted in the

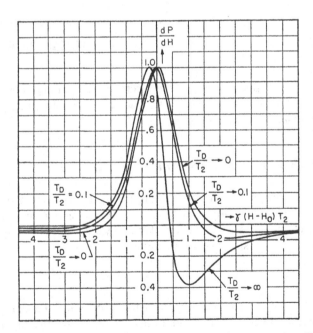

Fig. 7.44 Derivative of the power observation due to electron spin resonance in a thick metal plate for different diffusion time T_D to relaxation time T_2 ratios. The figure is adapted from [49] with permission from the American Physical Society

case of $T_D/T_2 \to \infty$. In those conditions with $T_T \gg T_D$, $T_T \gg T_2$ and $T_D/T_2 \gg 1$, the following absorption line shape, P is obtained.

$$P = \left[\frac{\omega H_1^2}{4}(\delta A)\omega_0\chi_0 T_2\right]\left\{(1/2)\frac{1 - T_2(\omega - \omega_0)}{1 + T_2^2(\omega - \omega_0)^2}\right\} \qquad (7.36)$$

where H_1 = amplitude of the linearly polarized magnetic rf field, χ_0 = paramagnetic part of the static susceptibility, A = area of surface, and ω_0 = the resonance frequency satisfying $h\omega_0/2\pi = g\beta H_0$; where, h = Planck constant, β = the Bohr magneton, H_0 = the resonance magnetic field, g = $(h/2\pi\beta)\gamma$, γ = the gyro magnetic ratio. This is the thick-plate case with slowly diffusing magnetic dipoles. The absorption line as a function of $(1-T_2(\omega-\omega_0))$ is no longer symmetric.

Figure 7.45 shows the changes in the asymmetry of the line, A/B when the ratio T_D/T_2 is varied. The absorption is drawn as a function of $\gamma(H-H_0)T_2 = T_2(\omega-\omega_0)$.

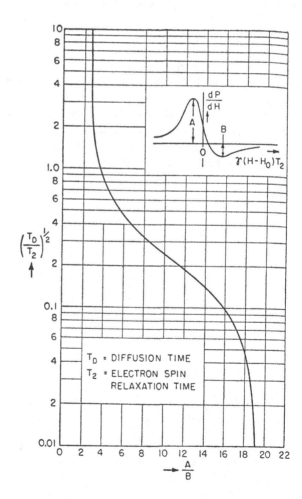

Fig. 7.45 A/B vs. $(T_D/T_2)^{1/2}$ for the derivative of the power absorption due to electron spin resonance in thick metal plate. The figure is adapted from [49] with permission from the American Physical Society

We can evaluate T_2 from the line width and T_D from the values of T_2 and T_D/T_2 obtained by the asymmetry of A/B in Fig 7.45.

7.7.4 ESR Spectra of Pristine and AsF₅ Doped Polyacetylene (PA)

Polyacetylene (PA), is in principle the simplest of the linear conjugated polymers. Interestingly, through chemical doping of PA films the electrical conductivity, σ can be varied at room temperature over twelve orders of magnitude with properties ranging from that of an insulator ($\sigma \sim 10^{-9}$ S/cm) to a metal ($\sigma \sim 10^3$ S/cm). The ESR spectrum should accordingly change with the degree of the doping. The line shape has a strong dependency on the conductivity, σ originated from the skin effect as discussed in the previous section. The ESR spectrum can therefore be varied from Lorentzian type (low conductivity) to Dysonian type (thin skin depth and high conductivity). The doping process can be monitored by an in situ doping technique. Goldenberg et al. [50] observed the AsF₅ doping process of trans polyacetylene by the ESR technique. Figure 7.46 shows ESR spectra of trans PA before doping (pristine PA). The excellent agreement between the experimental and calculated spectra indicates that the line shape is Lorentzian. It was found from quantitative measurements that there is one unpaired electron for every 3,200 carbon atoms. The dependency of ESR spectra on the weight uptake of AsF₅ is shown in Fig. 7.47. The line shape changes significantly after doping with AsF₅. The line width decreases, the line amplitude increases, and the line shape becomes unsymmetrical.

As shown in Fig. 7.47 the asymmetry increases with the amount of AsF₅ uptake. The doping should result in free carriers with an associated increase in the spin

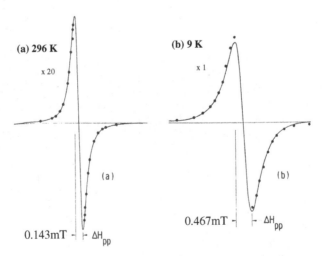

Fig. 7.46 ESR spectra of PA films observed at 296 K (**a**) and 9 K (**b**). The *circles* show the theoretical line shape (Lorentz type). The figure is adapted from [50] with permission from the American Institute of Physics

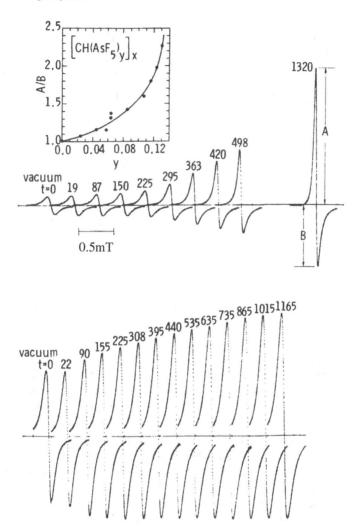

Fig. 7.47 Time dependence of the ESR spectrum upon exposure to AsF$_5$ at the pressures of 120 Torr (*above*) and 60 Torr (*below*). The figure is adapted from [50] with permission from the American Institute of Physics

susceptibility while the asymmetric line shape reflects the change of the conducting properties during doping, as discussed previously.

An example of an ESR spectrum of AsF$_5$ doped PA is also shown in Fig. 7.48. The excellent agreement between the experimental and calculated spectra indicates that the line shape is Dysonian. The highest doping levels result in a skin depth less than the sample thickness. For $\sigma \sim 10^2$ S/cm, the skin depth was calculated

Fig. 7.48 ESR spectrum
observed after doping with
AsF$_5$ at 90 Torr for 45 min.
The circles show the
theoretical line shape
(Dysonian type). The figure is
adapted from [50] with
permission from the
American Institute of Physics

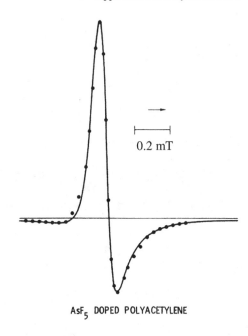

0.2 mT

AsF$_5$ DOPED POLYACETYLENE

to be approximately 5×10^{-3} cm, which is smaller than the thickness of the sample 10^{-2} cm. This condition is satisfied with sufficiently thick sample which give Dysonian line shape. The value of T_2, 0.52 μs is evaluated from the line width for the ratio of A/B = 2.7. According to Feher and Kip this corresponds to the point at which $(T_D/T_2)^{1/2} = 1.6$. At the point of greatest mobility of conduction electrons, $T_D = 1.3$ μs. Since this film is highly conductive, the minimum value of T_D suggests that the transport mechanism occurs through individual chains. The time of electron diffusion through the skin depth, $T_D = 1.3$ μs corresponds to a velocity of electron movement, 3.8×10^{10} nm/s, which is two or three orders of magnitude lower than that of the movement along the chain, $\sim 10^{13}$ nm/s. This fact is consistent with the observation of partial anisotropy of the film conductivity with the greatest conductivity in the direction of the alignment.

7.8 Summary

We can relate structure and molecular motion of polymer chain to chemical reaction in solid polymers by the ESR method. In this chapter, we present ESR studies that are particularly effective in three areas of polymer science.

(1) Direct detection of intermediate species in chemical reactions of polymer materials. (a) The identification and the structure of the initiating and propagating radicals and the kinetics of the formation are important for clarifying polymerization mechanisms. (b) The free radicals formed by irradiation of polymers with ionizing

radiation and photo-irradiation are intermediate species of cross-linking, degradation, grafting, and auto-oxidation deterioration. (c) The mechano-radicals produced by mechanical destruction are trapped on the newly exposed damaged surface and are extremely reactive.

(2) The evaluation of molecular mobility (molecular motion) of polymer chains. (a) The decay of the radicals trapped in polyethylene can be interpreted in terms of a diffusion controlled reaction. The decay reaction is closely related to the molecular motion of polymer chains in the crystalline region, attributed to so-called α-relaxation because the time constants of the molecular motion agree with those of the diffusion. (b) The high molecular mobility of isolated polyethylene chains tethered on polytetrafluoroethylene surfaces has been identified.

(3) The mobility of electron, the structure of solitons, and the doping behavior in conduction polymers can be observed in detail in order to clarify the mechanism of conduction.

Finally, we present an example of the relation between structure, molecular motion and chemical reaction in solid polymers studied by the ESR method as one of the conclusions in this chapter.

The conclusion reached is that chemical reactions are closely related to molecular motion and structures, where molecular conformation and molecular orientation are important parameters. We have discussed whether the hydrogen abstraction reaction of peroxy radicals in the inner crystalline region of polypropylene is due to their abstraction of the hydrogen from the intra-chain segment or inter-chain one. (1) It was confirmed that the unstable and mobile radicals participated in a rotation or a rotatory vibration of the COO group around the main chain axis. (2) The conformational structure of the mobile peroxy radicals was close to a skew conformation. (3) It could be safely concluded that the mobile peroxy radicals were trapped in disordered sites in the inner crystalline region.

In Fig. 7.49, a model is given schematically for hydrogen abstraction by peroxy radicals in isotactic polypropylene crystals [36]. In this figure the distance between the orbital containing the unpaired electron and typical hydrogen atoms are shown in the case of a mobile peroxy radical. For the calculation, the conformation structure of the radicals was considered. The distance between the unpaired electron orbital and a hydrogen in a methyl group was applied to be a minimum because of the free rotation of the methyl group. Surprisingly, the distances from the unpaired electron orbital to the hydrogen atoms, H_{im}, H_2 and H_o of the next chain were of the same order of magnitude as that from the orbital to the hydrogen atom H_{3m} of the same chain and they are not widely different from the O—H bond distance, 0.96 Å. Therefore, the peroxy radicals should abstract inter-molecularly the hydrogen atoms, H_{im}, $H_{2'}$ and $H_{o'}$. The hydrogen atoms might be able to easily reach the $p\pi$ orbital of the peroxy radical by rotation or rotatory vibration but the hydrogen atom, H_{3m} might not be able to reach it because of the small amplitude of the inter molecular vibration. It can be concluded that the mobile peroxy radicals should abstract hydrogen atoms on adjacent chains.

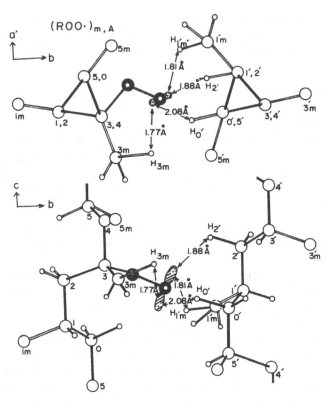

Fig. 7.49 Model of peroxy radical of isotactic polypropylene and its surroundings. Representative distances between the pπ orbital on the oxygen atom and hydrogen atoms are also shown. The *large open circles* and the *small open circles* indicate carbon and hydrogen atoms, respectively. The circles with a darkened exterior indicate oxygen atoms. The figure is adapted from [36] with permission from the American Physical Society

The distribution of molecular disorder and relaxation times of polymer chains should also affect chemical reactions in solid polymers. The kinetics of reactions in heterogeneous structures of polymer blends is an especially important research area in polymer physics and chemistry.

Appendices

A7.1 Two Site Model and Coordinate System of Polyethylene (PE) Chain End Radical, —$C_\beta H_{\beta 2} (C_\alpha \cdot)H_{\alpha 2}$: (Bond Scission Type Radical and Propagating Radical)

Figure 7.50 shows the two sites at the ends of isolated single PE chains and the employed coordinate system. The molecular coordinate system (x, y, z) fixed to the

Fig. 7.50 Two sites and the coordinate system at the ends of an isolated single PE chain. The figure is adapted from [26] with permission from the American Chemical Society

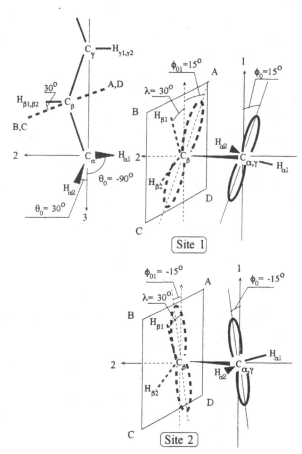

principal axes (A_x, A_y, A_z) of the *hfs* tensor of H_α is usually defined as follows: the x-axis is parallel to the direction of the p_z orbital occupied by the unpaired electron at the carbon, $C_{\alpha'}$. The z-axis is along the direction of the ($C_{\alpha'}$)—H_α bond. The y-axis lies in the $H_{\alpha 1}$—($C_{\alpha'}$)—$H_{\alpha 2}$ plane and is perpendicular to both x- and y-axes. A new coordinate system (1, 2, 3) indicated in Fig. 7.50 was employed to treat the *hfs* terms due to the two H_α atoms with different principal axes The *hfs* can be represented using spherical polar angles (θ and φ) for the applied magnetic field in the coordinate system (1, 2, 3) as follows:

$$A(\theta, \varphi) = A_x \sin^2(\theta + \theta_0) + A_y \sin^2(\theta + \theta_0)\sin^2(\varphi + \varphi_0) + A_z \cos^2(\theta + \theta_0) \quad (7.37)$$

where θ_0 and φ_0 are the spherical polar angles for the principal axes, A_z and A_x with respect to axes 3 and 1.

Axis 2 lies in the plane of ($C_{\alpha'}$)—C_β—C_γ and bisects the dihedral angle of $H_{\beta 1}$—C_β—$H_{\beta 2}$ (120°). Axis 1 is perpendicular to both the plane of ($C_{\alpha'}$)—C_β—C_γ and axis 2. Axis 3 is perpendicular to both axis 1 and axis 2 and is parallel to the chain axis of PE. Two sites, called site 1 and site 2, are assumed for the end of the tethered PE chain. Site 1:Each x-axis with respect to $H_{\alpha 1}$ or $H_{\alpha 2}$ is parallel to the direction of p_z and tilts from axis 1 with an angle of $\varphi_0 = 15°$. Each z-axis lies in the $H_{\alpha 1}$—($C_{\alpha}\cdot$)—$H_{\alpha 2}$ plane and makes an angle of $\theta_0 = -90°$ relative to axis 3 for $H_{\alpha 1}$ and $\theta_0 = 30°$ for $H_{\alpha 2}$. Site 2: y- and z-axes are the same as those in site 1, except that the x-axis tilts $\varphi_0 = -15°$ from axis 1. The isotropic *hfs* constant (A_β) can be estimated by the Mc'Connell equation; $A_\beta = B\cos^2(\lambda + \varphi_{01})$ where B is an empirical constant. Here, λ is the angle between the axis 1 and the C_β–$H_{\beta 1}$ or C_β–$H_{\beta 2}$ bonds projected on the ABCD plane. The plane is perpendicular to the $(C_{\alpha}\cdot)$–C_β–C_γ plane and the $(C_{\alpha}\cdot)$–C_β bond as indicated in Fig. 7.50. The value of λ is 30° in the case of a trans zigzag chain of PE. φ_{01} is the projection angle between axis 1 and the p_z orbital axis to the ABCD plane. The computer simulations were carried out by considering the inter exchange motion between the two α-hydrogen nuclei, the site exchange motion between sites 1 and 2 as shown in figure A-1 and free rotational motion for calculating the anisotropic and isotropic *hfs* couplings due to α- and β-hydrogen nuclei, respectively.

A7.2 The Calculated ESR Spectrum of Partially Oriented Peroxy Radicals

In the case of partially and axially symmetrically oriented free radicals, the distribution function of the radicals in two dimensional space must be considered [36]. Figure 7.51 represents the geometrical relationship for a main chain axis (OC) in a laboratory system. The Z axis is the direction of the external field (OZ) and the stretching direction (ON) is in the YZ plane. The angle γ_0 between OZ and OC can be obtained from Eq. (7.38)

$$\cos \gamma_0 = \cos \delta \cos \alpha + \sin \delta \sin \alpha \cos \varphi \qquad (7.38)$$

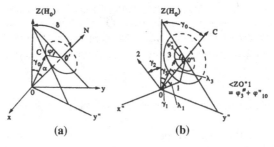

(a)   (b)

Fig. 7.51 Geometrical relationship for the main chain axis (OC) in a laboratory axis system (OX, OY, OZ) (**a**) and in the principal axis system (O1, O2, O3) of the g tensor (**b**). The external field (OZ) and the stretching direction are located in the plane, YZ. The figure is adapted from [36] with permission from the American Chemical Society

where δ is the experimental rotation angle between ON and OC. The uniaxial orientation of cylindrical symmetry around the stretching direction means that the distribution of chain axes has an equal probability around the direction ON. The observed spectrum can be simulated by integrating over all possible orientations of the g and/or *hfs* magnetic tensors in the X, Y, Z system, as represented in Fig. 7.51 with the g tensor as an example.

In order to calculate the value of $H_r(\Omega)$, the angle Ω can be obtained by the following equations; where λ_1, λ_2, and λ_3 are defined as the angles between the principal directions of the g tensor and the polymer chain axis

$$\cos \gamma_3 = \cos \gamma_0 \cos \lambda_3 + \sin \gamma_0 \sin \lambda_3 \cos \varphi'' \tag{7.39}$$

$$\cos \gamma_1 = \cos \gamma_0 \cos \lambda_1 + \sin \gamma_0 \sin \lambda_1 \cos(\phi_3'' + \phi_{10}'') \tag{7.40}$$

$$\cos \varphi_{10}'' = \frac{-\cos \lambda_3 \cos \lambda_1}{\sin \lambda_3 \sin \lambda_1} \tag{7.41}$$

$$\cos \gamma_2 = \pm\sqrt{1 - \cos \gamma_1^2 - \cos \gamma_3^2} \tag{7.42}$$

The intensity, $I(H_r)$ is proportional to the number of radicals contributing to the spectrum at the field H_r. This number was given by a distribution function $P(\alpha)$, for instance assumed to be a Gaussian distribution, with the distribution function, $\exp(a\sin^2\alpha)$. A computer simulation was carried out for the orientation function, f_α, and the angles λ_1, λ_2, and λ_3. The ESR spectrum, $I(H)$ can be calculated from the convolution of $P(\alpha)$ and the line shape function G as follows.

$$I(H) = \int_0^{2\pi} \int_0^{\pi} \int_0^{\pi/2} P(\alpha)G(H - H_r(\Omega)), D_h)d\alpha\alpha d \, d\varphi_3'' \tag{7.43}$$

The line width, D_h and the principal values, g_1, g_2 and g_3 were determined from the simulation of the powdered sample. From the determined value of a in $\exp(a \sin^2\alpha)$, the orientation function, f_α was calculated as follows.

$$f_\alpha = \frac{\int_0^{\pi/2} ((3 \cos^2 \alpha - 1)/2) \sin \alpha \, \exp(a \sin^2 \alpha)d\alpha}{\int_0^{\pi/2} \sin \alpha \, \exp(a \sin^2 \alpha)d\alpha} \tag{7.44}$$

References

1. C.H. Heller, H.M. McConnell: J. Chem. Phys. **32**, 1535 (1960).
2. M. Kamachi: Adv. Polym. Sci. **82**, 207 (1987).
3. H. Yoshida, B. Ranby: J. Polym. Sci. C **16**, 1333 (1967).

4. M. Kamachi, M. Kohno, D.J.L. Liaw, Y. Katsuki: Polymer J. **10**, 69 (1978).
5. M. Kamachi, Y. Kuwae, S. Nozakura: Polymer J. **13**, 919 (1981).
6. Y. Sakai, M. Iwasaki: J. Polym. Sci. Part A-1 **7**, 1749 (1969).
7. Y. Tsukahara, K. Tutsumi, Y. Yamashita, S. Shimada: Macromolecules **22**, 2869 (1989).
8. S. Shimada: Prog. Polym. Sci. **17**, 1045 (1992).
9. R. Salovey, Y. Yager: J. Polym. Sci. A **2**, 219 (1964).
10. S. Shimada, H. Kashiwabara: Polymer **22**, 1385 (1981).
11. S. Shimada, Y. Hori, H. Kashiwabara: Polymer **19**, 763 (1978).
12. T. Ichikawa, H. Yoshida: J. Polym. Sci. Part A, Polym. Chem. **28**, 1185 (1990).
13. N. Tamura: J. Chem. Phys. **37**, 479 (1962).
14. T.R. Waite: J. Chem. Phys. **32**, 21 (1962).
15. S. Shimada, Y. Hori, H. Kashiwabara: Polymer **18**, 25 (1977).
16. S. Shimada, Y. Hori, H. Kashiwabara: Polymer **22**, 1377 (1978).
17. M. Iwasaki, T. Ichikawa: J. Chem. Phys. **46**, 2851 (1967).
18. T. Fujimura, N. Tamura: Polymer Lett. **10**, 469 (1972).
19. S. Shimada, Y. Hori, H. Kashiwabara: Radiat. Phys. Chem. **19**, 33 (1982).
20. A.M. Portis: Phys. Rev. **115**, 1506 (1959).
21. J. Sohma, M. Sakaguchi: Adv. Polym. Sci. **20**, 109 (1976).
22. T. Kawashima, S. Shimada, H. Kashiwabara, J. Sohma: Polym. J. **5**, 135 (1973).
23. M. Sakaguchi, J. Sohma: J. Polym. Sci. B **13**, 1233 (1975).
24. M. Sakaguchi, S. Kodama, O. Edlund, J. Sohma: J. Polym. Sci. B **12**, 609 (1974).
25. M. Sakaguchi, T. Yamaguchi, S. Shimada, Y. Hori: Macromolecules **26**, 2612 (1993).
26. M. Sakaguchi, S. Shimada, K. Yamamoto, M. Sakai: Macromolecules **30**, 8521 (1997).
27. M. Sakaguchi, S. Shimada, K. Yamamoto, M. Sakai: Macromolecules **30**, 3620 (1997).
28. M. Iwasaki, Y. Sakai, J. Polym. Sci. Part A-2 **6**, 265 (1968).
29. Y. Hori, S. Shimada, H. Kashiwabara: Polymer **18**, 151 (1977).
30. Y. Hori, S. Shimada, Z. Fukunaga, H. Kashiwabara: Polymer **20**, 181 (1979).
31. T. Seguchi, N. Tamura: J. Phys. Chem. **77**, 40 (1977).
32. Y. Hori, S. Shimada, H. Kashiwabara: Polymer **20**, 406 (1979).
33. Y. Hori, S. Shimada, H. Kashiwabara: Polymer **18**, 567 (1977).
34. Y. Hori, S. Shimada, H. Kashiwabara: Polymer **18**, 1143 (1977).
35. S. Schlick, L. Kevan: J. Am. Chem. Soc. **102**, 4622 (1980).
36. S. Shimada, Y. Hori, H. Kashiwabara: Macromolecules **18**, 170 (1985).
37. S. Shimada, Y. Hori, H. Kashiwabara: Macromolecules **17**, 1104 (1984).
38. J.C.W. Cien, G.E. Wnek, F.E. Karasz, J.M. Waraknsky, L.C. Dickison, A.J. Heeger, A.G. MacDiarmid: Macromlecules **15**, 614(1982).
39. K. Tanaka, T. Koike, T. Yamabe, J. Yamauchi, Y. Deguchi, S. Yata: Phys. Rev. B **35** 8368 (1987).
40. C.K. Chiang, C.R. Fincher Jr., Y.W. Park, A.J. Heeger, H. Shirakawa, E.J. Louis, S.C. Gau, A.G. MacDiarmid: Phys. Rev. Lett. **39**, 1098 (1977).
41. K. Mizoguchi, S. Masubuchi, K. Kume, K. Akagi, H. Shirakawa: Phys. Rev. B **51**, 8864 (1995).
42. M. Nechtsheim, F. Devreux, R.L. Greene, T.C. Cllarke, G.B. Street: Phys. Rev. Lett. **44**, 356 (1980).
43. M. Nechtsheim, F. Devreux, T.C. Cllarke, M. Guglielmi: Phys. Rev. B. **27**, 61 (1983).
44. P. Su, J.R. Schreffer, A.J. Heeger: Phys. Rev. B **22**, 2099 (1980).
45. S. Kuroda, H. Bando, H. Shirakawa: J. Physical Soc. Japan **54**, 3956 (1985).
46. S. Kuroda, T. Noguchi, T. Ohnishi: Phys. Rev. Lett. **72**, 286 (1994).
47. S. Kuroda, H. Shirakawa: J. Physical Soc. Japan **61**, 2930 (1992).
48. F.J. Dyson: Phys. Rev. **98**, 349 (1955).
49. G. Feher, A.F. Kif: Phys. Rev. **98**, 337 (1955).
50. I.B. Goldberg, H.R. Croue, P.R. Newman: J. Chem. Phys. **70**, 1132 (1979).

Chapter 8
Spin Labeling and Molecular Dynamics

Abstract Measurements of temperature dependent ESR spectra of spin labels trapped in the polymer matrices is very effective for the evaluation of molecular mobility (molecular motion) of polymer chains. We can characterize the molecular motion of the particular sites in different region by the ESR method. It is possible to detect the mobility of polymer material at the segment or atomic level. ESR studies help to relate the features of the nanometer scale to macroscopic properties of the polymer materials. We present examples of spin labeling studies in the polymer science to help readers new to the field understand how and for what areas the method is effective. In the first part, we give a simple review of the spin labeling method and present applications in the polymer physics related to relaxation phenomena in various systems. In the second part applications to biopolymer system are introduced to help the clarification of various mechanisms of bio-membranes.

8.1 Introduction

In this chapter, we deal with applications of spin labeling to molecular dynamics in polymer science. The radical species in these studies are stable so-called spin labels and spin probes. The success of these studies has made the ESR method increasingly popular for polymer investigations. We can characterize the molecular motion and the structure of the polymer chain in a wide temperature range including that above the melting temperature. The ESR method is effective for the characterization of the heterogeneity in mobility and structure of polymer materials as mentioned in Chapter 7. The heterogeneity is one of the most important features in polymer materials including biopolymers and related to the broad distribution of relaxation time of physical properties such as stress and dielectric relaxations, creep etc. Selective spin labeling is very effective for the detection of the molecular motion in the various sites and regions of the polymer systems. We can for example selectively observe the molecular motion of the particular sites in a chain and particular chains in the copolymer and blend systems.

In this chapter we present typical applications of the spin labeling method to polymer physics and biopolymer systems.

A. Lund et al., *Principles and Applications of ESR Spectroscopy*,
DOI 10.1007/978-1-4020-5344-3_8, © Springer Science+Business Media B.V. 2011

8.2 Molecular Motion in Solid Polymers

8.2.1 Characteristics of ESR for Studying Molecular Motion in Solid Polymers

ESR techniques have been applied over an extended period to evaluate motion in solid polymer at a molecular or an atomic level. More than forty yeas ago, in Kivelson's famous paper [1] the dependency of line width on the magnetic nuclear spin quantum number was proposed. Stone et al. [2] applied Kivelson's theory and reported an important relationship between the triplet line shape of the hyperfine splitting due to nitrogen nuclei and the correlation time of the rotational diffusion motion of macromolecules containing a nitroxide radical as a spin probe, as will be mentioned later. A few years later, stable radicals containing nitroxide radicals were used as a spin label reagent and the molecular motion of spin labeled molecules with nitroxide radicals were widely studied. An earlier application of the spin label technique investigated bio-related materials. In the late 1960 the method was applied to molecular motion studies of synthetic high polymers to clarify physical properties obtained by other techniques such as mechanical and dielectric measurements. Recently, the spin-labeled technique was applied to characterizations of structure and molecular mobility in complicated polymer systems such as polymer blends and micro-phase separated copolymer structures, having concentration fluctuation. It is very important to be able to study molecular motion of a particular chain in a particular region of the complicated system. Techniques have been developed to allow selective spin labeling with stable nitroxide radicals at particular sites on polymer chains, for example inside segments and end sites. The nitroxide labels can also be selectively bonded at particular regions such as crystalline and amorphous domains in semi-crystalline polymers and in different phases in polymer blends. This gives the ability to detect molecular motion at particular sites in the different regions and to relate the molecular mobility with the structure at the chain sites in the regions.

8.2.2 Evaluation of Correlation Time for Molecular Motion from ESR Spectra

Freed et al. [3] evaluated the relaxation rate of polymer chains by the analysis of the anisotropic spectra of nitroxide spin labels. The main triplet spectrum was due to hyperfine coupling caused by the nitrogen nucleus. It narrowed with an increase in mobility of the radicals because of motional averaging of the anisotropic hyperfine interaction. The rotational correlation time of nitroxide spin labels can be estimated using the procedure of Freed et al. by taking into account the anisotropic rotational motion. The ESR line width, ΔH_{msl} of the spectrum can be expressed as follows:

$$\Delta H_{msl} = A + Bm_I + Cm_I^2 (m_I = -1, 0, 1) \tag{8.1}$$

$$A = \frac{2\pi}{15\sqrt{3}} \frac{g_0\mu_\beta}{h} B_0^2 \left\{ (F_g^0)^2 \tau_0 + 2(F_g^2)^2 \tau_2 \right\} +$$
$$m_I(m_I + 1) \frac{\pi}{5\sqrt{3}} \cdot \frac{g_0\mu_\beta}{h} B_0^2 \left\{ (F_A^0)^2 \tau_0 + 2(F_A^2)^2 \tau_2 \right\} \tag{8.2}$$

$$B = \frac{8\pi}{15\sqrt{3}} \frac{g_0\mu_\beta}{h} B_0 \left\{ (F_g^0 F_A^0) \tau_0 + 2(F_g^2 F_A^2) \tau_2 \right\} \tag{8.3}$$

$$C = \frac{2\pi}{6\sqrt{3}} \frac{g_0\mu_\beta}{h} \left\{ (F_A^0)^2 \tau_0 + 2(F_A^2)^2 \tau_2 \right\} \tag{8.4}$$

$$F_A^0 = (2/3)^{1/2}(A_z - 1/2(A_x + A_y)) \tag{8.5}$$

$$F_g^0 = (2/3)^{1/2}(g_z - 1/2(g_x + g_y)) \tag{8.6}$$

$$F_A^2 = (1/2)(A_x - A_y) \tag{8.7}$$

$$F_g^2 = (1/2)(g_x - g_y) \tag{8.8}$$

The nomenclature is as follows. B_0 is the resonance magnetic field, g_0 is the g value at B_0, μ_β is the Bohr magneton, and h is the Planck constant. A_i and g_i (i = x, y, and z) are the principal values of the hyperfine- and g-tensor, respectively. The (x, y, z) coordinate system is fixed to the axes frame of an axially symmetric rotor. The values of A, B, and C in Eq. (8.1) are obtained from Eq. (8.9):

$$(\Delta H_{msl}(m_I))^2 I_{m_I} = const.$$

$$B = 1/2 \left(\sqrt{\frac{I_0}{I_1}} - \sqrt{\frac{I_0}{I_{-1}}} \right) \Delta H_{msl}(0)$$
$$C = 1/2 \left(\sqrt{\frac{I_0}{I_1}} + \sqrt{\frac{I_0}{I_{-1}}} - 2 \right) \Delta H_{msl}(0) \tag{8.9}$$

Here, $I_{m_I}(m_I = -1, 0, 1)$ are the peak height as indicated in Fig. 8.11, below, while τ_0 and τ_2 are expressed as the following equations:

$$\frac{1}{\tau_m} = 6D_\perp + m^2(D_{//} - D_\perp)(m = 0, 2)$$

$$\tau_{//} = \frac{1}{6D_{//}}, \tau_\perp = \frac{1}{6D_\perp}, \tau_c = \sqrt{\tau_{//}\tau_\perp} \tag{8.10}$$

here, $D_{//}$ and $D_\perp$ are rotational diffusion coefficients about the major and minor axes of an axially symmetric rotor, respectively, and τ_c is the average rotational correlation time of the anisotropic rotational motion. For the evaluation of isotropic molecular mobility, a short correlation time of 10^{-9}–10^{-12} (s) is sometimes calculated from the following simple equation, given by Stone et al. [2]:

$$\tau_c = \left(\frac{-15\pi^3}{8b\Delta\gamma B_0} \right) \left(\sqrt{\frac{I_0}{I_1}} - \sqrt{\frac{I_0}{I_{-1}}} \right) \Delta H_{msl}(0) \tag{8.11}$$

$$\tau_c = \left(\frac{4\pi^3}{b^2} \right) \left(\sqrt{\frac{I_0}{I_1}} + \sqrt{\frac{I_0}{I_{-1}}} - 2 \right) \Delta H_{msl}(0) \tag{8.12}$$

where b and $\Delta\gamma$ are quantities depending on the anisotropies of the g- and A-tensors.

Freed et al. [4, 5] proposed also an equation for the calculation of the correlation time based on the simulation of the spectra. This equation is widely applied even in the case of slow rate of molecular motion ($10^{-7} > \tau_c(s) > 10^{-9}$):

$$\tau_c = a(1 - S)^b \quad (S = A_z/A_{z0}) \tag{8.13}$$

where S is the ratio of the principal value A_z of the hyperfine coupling-tensor at a certain temperature, to that of the rigid state, A_{z0}. The constants "a" and "b" depend on the motional model. For example, $a = 1.10 \times 10^{-9}$ and $b = -1.01$ for "moderate jump" diffusion. The extreme separation width as shown in Fig. 8.2, below is a good measure of molecular mobility of the polymer chain. The extreme separation width decreases with temperature because the anisotropic values average out gradually.

8.2.3 Glass-Rubber Transition Detected by the Spin Label Method for Polystyrene (PS): Molecular Weight Dependence

The phenomenon of the glass-rubber transition is one of the most important properties in polymer material characterization. Many authors have studied its origin theoretically. Recently, calorimetric and relaxation equipments were developed and polymer morphology and molecular motion of polymer chains could be detected directly. As a result, the phenomenon of glass-rubber transition of a particular site in a particular region was clarified and discussed. As an example, Kajiyama et al. [6] reported that polymer surface enriched with segregated chain ends exhibited higher molecular mobility, in other words, lower glass-rubber transition temperature (T_g) as evidenced by atomic force measurement. Miwa et al. [7] spin-labeled selectively at particular sites and detected the molecular mobility of each site. Figure 8.1 shows chemical structures of spin-labeled polystyrene (PS) at the inside and the end of chain segments. ESR spectra of the spin-labeled PS at the chain end (number average molecular weight, $M_n = 21$ kDa) and the inside sites ($M_n = 22$ kDa) in the temperature range 77–471 K are shown in Fig. 8.2. In general, the splitting between

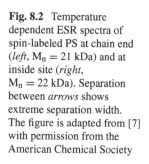

Fig. 8.1 Chemical structure of spin-labeled PS at inside (*left*) and end (*right*) of chain. The figure is adapted from [7] with permission from the American Chemical Society

the outer-most lines of the main triplet spectrum due to the nitrogen hyperfine coupling becomes narrower with an increase in the mobility of the radicals because of the motional averaging of the anisotropy.

The extreme separation width shown with arrows in Fig. 8.2 gradually narrows and steeply drops with an increase in temperature. The temperature dependence of the extreme separation width for each specimen is shown in Fig. 8.3. The steep drop is caused by a micro–Brownian type molecular motion reflecting the glass-rubber transition. We can estimate a transition temperature of the molecular motion, $T_{5.0mT}$ where the extreme separation width is equal to 5.0 mT. The $T_{5.0mT}$ values of the spin-labeled PS at the chain end ($T_{5.0mT,e}$) and the inside sites ($T_{5.0mT,i}$) are 423 and 435 K, respectively. The $T_{5.0mT}$ of the spin-labeled PS at the chain end is thus lower

Fig. 8.2 Temperature dependent ESR spectra of spin-labeled PS at chain end (*left*, $M_n = 21$ kDa) and at inside site (*right*, $M_n = 22$ kDa). Separation between *arrows* shows extreme separation width. The figure is adapted from [7] with permission from the American Chemical Society

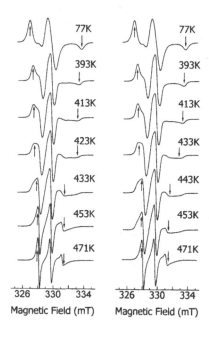

Fig. 8.3 Temperature
dependence of extreme
separation width for
spin-labeled PS at chain end
(*solid*) and at inside site
(*open*). The figure is adapted
from [7] with permission
from the American Chemical
Society

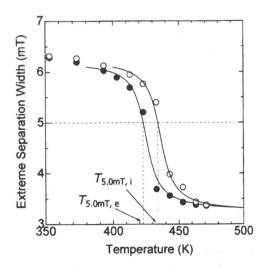

than that of the spin-labeled PS at the inside site. This is caused by the large free volume at the chain end in comparison with that at the inside of the chain.

The transition temperature, $T_{5.0mT}$ was observed to be higher than the glass-rubber transition temperature (T_g) for the PS sample. This can occur if the rate of motion at T_g is too slow for the averaging of the anisotropy of the hyperfine splitting. In other words the spin labeling method (ESR) detects the molecular motion at a higher frequency than the thermal analysis.

The T_g of the PS estimated from thermal analysis, T_{gDSC}, has been obtained from Modulation Differential Scanning Calorimeter (MDSC) curves at a rate of 2.0 K/min. The transition temperatures, T_{gDSC}, $T_{5.0mT,e}$, and $T_{5.0mT,i}$, are plotted against M_n in Fig. 8.4. The most successful model for the description of the

Fig. 8.4 Plots of $T_{5.0mT,e}$
(*solid circle*) $T_{5.0mT,i}$ (*open
circle*), and T_{gDSC} (*triangle*)
vs M_n. The figure is adapted
from [7] with permission
from the American Chemical
Society

relationship between the temperature dependence for viscous flow, and the dielectric dispersion of polymers and super-cooled liquids is the Williams-Landel-Ferry (WLF) equation [8]:

$$\log \frac{\tau(T)}{\tau(T_0)} = -\left(\frac{C_1(T - T_0)}{(C_2 + T - T_0)}\right) \qquad (8.14)$$

Here, τ is the relaxation time, T_0 is a chosen reference temperature, and C_1 and C_2 are universal constants. The quantities T_0 and T, were set equal to the values of T_{gDSC} and $T_{5.0mT}$, respectively.

When T_0 equals T_g, the constants C_1 and C_2 have the universal values of 17.44, and 51.6, respectively. The temperature difference ΔT ($= T_{5.0mT} - T_{gDSC}$) was calculated to be 72 K from Eq. (8.14) when the relaxation times of 7×10^{-9} s and 100 s were substituted in $\tau(T)$ and $\tau(T_0)$, respectively. The ΔT obtained from the experiments, 66 K, agreed with the calculated value, 72 K. The experimental temperature, $T_{5.0mT\infty,i}$, is adopted as an asymptotic value of $T_{5.0mT}$ for the infinite molecular weight in Fig. 8.4. The values of $T_{5.0mT,e}$ and $T_{5.0mT,i}$ were identified with the transition temperatures, T_{ge} and $T_{g,i}$, by using the WLF equation [8].

It is well known that the glass transition temperature, T_g of a given polymer is a function of its molecular weight. Unberreiter and Kanig [9] suggested the dependence of T_g on the number average molecular weight (M_n) for polystyrene;

$$\frac{1}{T_g} = \frac{1}{T_{g\infty}} + \frac{A}{M_n} \qquad (8.15)$$

This relationship indicates that a reduction of the molecular weight increases the number of chain ends per volume and the free volume fraction. When the inverses of T_{gDSC}, T_{ge} and $T_{g,i}$ were plotted against the reciprocal of the molecular weight, M_n as shown in Fig. 8.5 straight lines were obtained. From the linear relationship of Eq. (8.15), the $T_{g\infty,DSC}$ and A_{DSC} were determined to be 375 K and 0.404, respectively.

$T_{g\infty,i}$ and A_i were also estimated to be 366 K and 0.518, respectively. Similar relationships of the transition temperature against the molecular weight demonstrate that the $T_{5.0mT}$ for the inside label correlates with the temperature T_g where the micro Brownian type molecular motion starts. The motion is attributed to so-called α-relaxation of PS. The molecular weight dependence of T_{ge} was also in agreement with Eq. (8.15). The parameters $T_{g\infty,e}$ and A_e were 361 K and 0.920, respectively. $T_{g\infty,e}$ was 5 K lower than $T_{g\infty}$, while the value of A_e is roughly 2 times larger than A_i. As mentioned previously, the chain end has higher molecular mobility than the inside segment due to the larger free volume around the chain end than that around the inside segment. The strong molecular weight dependence of $T_{g,e}$ demonstrates that $T_{5.0mT,e}$ also reflects the α-relaxation of the PS because the local β- and γ-relaxation modes of the PS have no molecular weight dependence. These results indicate that the molecular mobility of the chain end can also be interpreted in terms of the free volume fraction which is affected by the surrounding segments. This

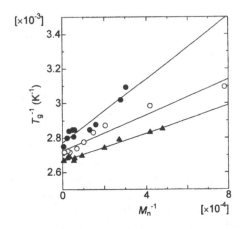

Fig. 8.5 Molecular weight dependence of $T_{5.0mT,e}$ (*solid circle*) $T_{5.0mT,i}$ (*open circle*), and T_{gDSC} (*triangle*) for mono-disperse PS. Solid lines represent Eq. (8.15) for all data with $T_{g\infty,e} = 361$ K, $A_e = 0.920$, $T_{g\infty,i} = 366$ K, $A_i = 0.518$, $T_{g\infty,DSC} = 375$ K, and $A_{DSC} = 0.404$. The figure is adapted from [7] with permission from the American Chemical Society

behavior is caused by the cooperative motion of the chain end with the surrounding segments at the glass transition. The lower T_{ge} and the stronger molecular weight dependence of T_{ge} can be interpreted in term of the larger free volume generated by two chain ends, which is estimated by the calculation of the probabilities of each end and inside segment. It has also been observed that approximately nine monomeric unit size segments undergo a segmental motion at T_g.

8.2.4 Glass-Rubber Transition Detected by the Spin Label Method for Polyethylene (PE): Crystallinity Dependence

Temperature dependent ESR spectra of high density PE (HDPE) and low density PE (LDPE) labeled at the inside segments of chains with different crystallinity (X_c) are shown in Fig. 8.6 [10]. The temperature dependence of the line shape of the ESR spectra is due to the change in the motional correlation time, τ_c of the nitroxides. The extreme separation width between arrows in Fig. 8.6 of the main triplet narrows with an increase in mobility of the radicals. Complete averaging of the anisotropic hyperfine interaction gave rise to isotropic narrowed spectra at increased temperature, such as those at 383 K in Fig. 8.6. Two spectral components, a "fast" and a "slow" component, were observed in a certain temperature range. The slow component with the larger extreme separation width and the fast component with small extreme separation width and narrow line width can be attributed to radicals trapped in less mobile and mobile regions, respectively. This is a reflection of a broad distribution of rotational correlation times, τ_c, arising from a heterogeneous structure in the polyethylene. It is, generally hard to extract two rotational correlation times from a spectrum without a spectrum simulation, however.

The narrowing of the extreme splitting has been used to analyze the transition temperature, $T_{5.0mT}$ as defined in the previous section. Figure 8.7 shows the dependence of crystallinity, X_c, on the $T_{5.0mT}$ values obtained by the ESR method.

Fig. 8.6 Temperature
dependent ESR spectra of
spin-labeled PE of different
crystallinity. Separation
between *arrows* shows
extreme separation width.
The figure is adapted from
[10] with permission from the
American Chemical Society

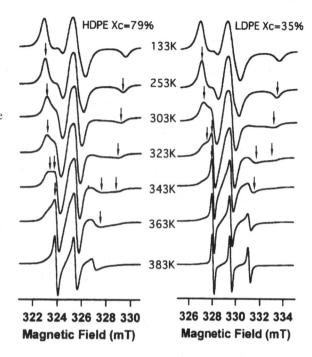

Steling and Maderkern [11] measured the thermal expansion of linear polyethylene
with different X_c. Boyer re-plotted the amorphous relaxation (glass transition) data
as a function of X_c. The temperatures, T_{gdil} estimated by Boyer [12] have been
inserted in the figure for comparison with the ESR data. It was concluded that the
$T_g(U)$ and $T_g(L)$ data were the result of a glass transition of the chain trapped in
the different amorphous regions, indicating a heterogeneous structure for the PE.

Fig. 8.7 Crystallinity(X_c)
dependence of transition
temperature. Open circles are
$T_{5.0mT}$ obtained from ESR.
Solid circles and *open
squares* are T_g values
measured by a thermal
expansion method. The figure
is adapted from [10] with
permission from the
American Chemical Society

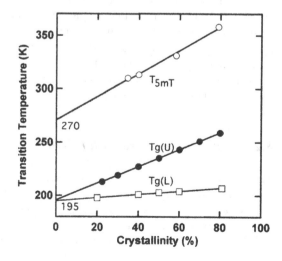

Two ESR spectral components (fast and slow components) may be related to two phases in the amorphous region, as observed in the thermal expansion experiment. The regions trapping less mobile and mobile spin labels may for example be at the crystalline surface and at locations apart from the crystalline phase. The molecular motion of the chain in the latter region is hardly influenced by the crystalline phase. Interestingly, the crystallinity dependence of $T_{5.0mT}$ was very close to that of $T_g(U)$, except for the transition temperature. The difference of these transition temperatures was ascribed to a frequency difference between the two observations. The extrapolated temperature at $X_c = 0$ is interpreted as the transition temperature of "a perfect amorphous polyethylene", corresponding to a "region apart from the crystalline" mentioned above. The extrapolated temperatures estimated from $T_{5.0mT}$ and $T_g(U)$ are 270 K and 195 K, respectively. The relaxation time, τ_{ESR} at $T_{5.0mT}$ was evaluated to be 4.3×10^{-9} s by using the observation frequency of X-band ESR. Then, the τ_{ESR} value and the relaxation time, τ_g of 100 s at T_{gdil}, measured by a dilatometric method were substituted for $\tau(T)$ and $\tau(T_0)$ in Eq. (8.14), respectively in order to estimate $\Delta T (= T_{5.0mT} - T_{gdil})$.

The values of $\Delta T (= T_{5.0mT} - T_{gdil})$ obtained from Eq. (8.14) was estimated to be 76 K in agreement with the value $\Delta T = 75$ K ($X_c = 0$) obtained from the experiments, as described in Fig. 8.7. In addition, the value of T_g from ESR is expected to be 194 K which is close to the extrapolated T_g (195 K) at $X_c = 0$. Therefore, the molecular motion detected by the ESR spin-labeled method here is associated with the glass-rubber transition (β-relaxation, $T_g(U)$) proposed by Boyer et al. The spin-labeled sites are likely located at the amorphous region because labeled segments can reasonably be excluded from the crystalline region. Thus it can be said that the labeled segments reflect the motion of amorphous segments. The crystallinty dependence of $T_{5.0mT}$ is obtained as

$$T_{5.0mT}(K) = 1.1X_c + 270 \qquad (8.16)$$

The effect of crystallinity on the mobility was the result of various effects originating from the crystalline size and the fraction of amorphous region which might affect the mobility of the amorphous chains. The difference between the β-relaxations around the inside and end segments was studied by the selective spin-labeling method and the analysis of the molecular weight dependence of the glass-rubber transition. The characteristic length, ξ of glass transition corresponds to the size cooperative rearranging region. The value of ξ was determined to be 1.6 nm.

8.2.5 Glass-Rubber Transition Detected by the Spin Probe Method

The spin-probe method detects molecular motions of solid polymers from ESR spectra of paramagnetic probes which are usually stable nitroxide radicals embedded in the polymer matrices. The spin-probe does not bond to the polymer chain and thus does not necessarily move with the polymer chain segment. Smaller probes only weakly interact with the chain and thus move independently with the chain

Fig. 8.8 Schematic model of jumping of probe and segment: A, probe (volume V_p); B, segment (volume V_f); C, hole (volume $V'_f > V^*_m$): D, hole (volume $V'_f > V^*_p$); E, packet (volume v_f). The figure is adapted from [14] with permission from Elsevier

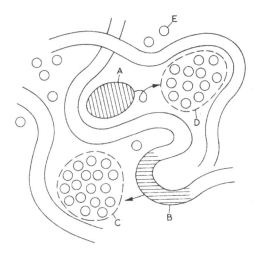

segment. However, when the probe has a size above a limiting value the motion of the probe is dominated by that of the polymer matrices. In terms of the hole theory, each segment of the polymer chains begins to jump into adjacent holes formed by the displacement of other segments and by the thermal expansion of the polymer matrix at temperatures around T_g. At the same time, if the spin probe has a comparable size to that of a segment, it will follow into an adjacent hole along with the polymer segment. According to the theory of Buche, [13] we can consider a polymer system which consists of polymer molecules with n segments, probes, and the free volume, V_f as shown in Fig. 8.8.

Kusumoto et al. [14] calculated the probability, $p(q)$ that the q packets (volume v_f) are gathered around a certain segment to make a hole of the size, V_f. A packet was defined as a vacancy surrounded by polymer atoms when the molecular motion was frozen. The segment or the probe can jump into the hole when the hole volume, V_f exceeds the activation volume for jumping, V^*_m and V^*_p which are for the segment and the probe, respectively. Some packets are gathered to a hole by thermal energy. We can calculate the sum of the probabilities that a hole has a size larger than that of the spin probe at a certain temperature. Kusumoto et al. obtained the following Eq. (8.17) which relates the volume ratio of V_p/V_m to the difference between the transition temperature, T_n (corresponding to $T_{5.0mT}$ in Section 8.2.3) for the spin probe and T_g:

$$T_n - T_g = 52(2.9f(ln\frac{1}{f} + 1) - 1) \tag{8.17}$$

where $f = V^*_p/V^*_m = V_p/V_m$.

Of course, when the size of the spin probe is smaller than the segment size at T_g, and when the jumping mode occurs by the local motion of the segment, the ESR spectrum of the spin probe sometimes narrows at a temperature lower than T_g. When T_n and T_g are determined experimentally and the molecular size of the

Fig. 8.9 Chemical structures of the spin probes and the spin-labeled PMMA. The figure is adapted from [15] with permission from John Wiley & Sons

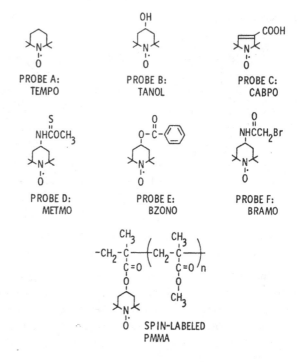

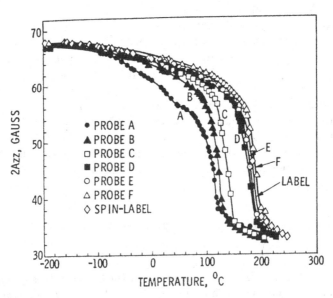

Fig. 8.10 Temperature dependence of ESR extreme separation width for the spin-probed and spin-labeled PMMA films. The figure is adapted from [15] with permission from John Wiley & Sons

probe is known, the segmental size of the polymer can be estimated using Eq. (8.17). Tsay et al. [15] found a strong dependence of $T_{5.0mT}$ on the molecular size of the spin probe. Figure 8.9 shows chemical structures of the spin probes and the spin labeled polymethylmethacrylate (PMMA). Figure 8.10 shows also the temperature dependence of the extreme separation width for the spin-probes and for the spin-labeled PMMA. $T_{5.0mT}$ increased with an increase in the molecular size of the spin label and approached that of the spin-labeled PMMA. These experimental facts are in agreement with Kusumoto's theory. Care must be taken to prevent the large probe from destroying the micro structure of the polymer substance.

8.2.6 Molecular Mobility of an Amorphous Chain in the Crystallization Process

Many authors have developed various devices to perform in situ and time-resolved measurements of the crystallization process. For example, long- and short-range orders were detected simultaneously by the Small Angle Scattering (SAXS) and Wide Angle Scattering (WAXS) time-resolved measurements. In the process, the mobility of the polymer chains in the amorphous region as well as in the crystalline region may change with an increase in crystallinity. A few authors also studied the molecular mobility during the structural change in the crystallization. Hara et al. [16] evaluated the molecular mobility of spin-labeled poly(ε-caprolactone) (PCL) by the ESR method during the crystallization at 47 and 50°C. The structural change in the crystalline process was also studied by the time dependence of the intensity of SAXS and WAXS at the same temperatures. Figure 8.11 shows the time dependence of ESR spectra of the spin-labels bonded to the PCL chain end during the crystallization process at 47°C. A slight change of the triplet spectrum with the storage time was detected. The relative intensity ratios of the three lines change with time. The rotational correlation time, τ_c was calculated by using Eqs. (8.9) and (8.10) and its time dependence is shown in Fig. 8.12. To clarify the correlation between the molecular mobility and structural changes in the crystallization process the time evolution of the normalized SAXS intensity is also depicted in Fig. 8.12. It was found that the crystalline structures and the higher-order structures were formed simultaneously because the intensities of SAXS and WAXS increased in the same manner. The correlation time τ_c also increased with an increase in the crystallization time, indicating that the mobility of the polymer chain decreased during the crystallization process.

In addition, the remarkable changes of the increase of τ_c agree with the normalized SAXS intensity. Therefore, these changes of τ_c in the crystallization process originate from the structural changes of PCL. However, at early stages of less than ca. 700 s at 50°C, no change of the normalized SAXS intensity from the alternative lamella structure was observed, even through the correlation time τ_c increased gradually with increasing crystallization time, At the beginning of the crystallization process, the mobility of the polymer chains decreased even before the crystalline structure had begun to form. When the normalized SAXS intensity

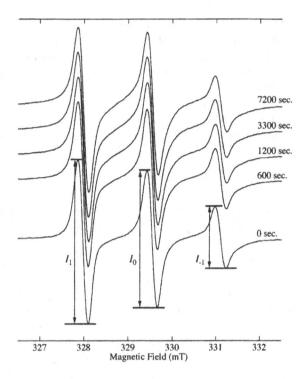

Fig. 8.11 Time dependence of ESR spectra of the spin-labels bonded to the PCL chain end, during the crystallization process at 47°C. The figure is adapted from [16] with permission from the American Chemical Society

started to be detected as the crystalline structure appeared, τ_c increased considerably. Finally, no further increase of the SAXS intensity was detected, but τ_c continued to increase gradually. At the early stage in the crystalline process, structural changes were generated, for example, spinodal decomposition, conformational change, and nucleation. Therefore, the increase in τ_c at the early stage originates from structural changes such as the self-aggregation, which indicate the existence of the precursory aggregation of the polymer chains in the crystalline process. The remarkable decrease in the mobility (the increase in τ_c) with the change of the

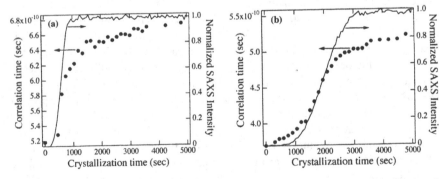

Fig. 8.12 Normalized SAXS intensity (*solid line*) and the correlation time (*closed circle*) plotted against the crystallization time at 47°C (**a**) and 50°C (**b**). The figure is adapted from [16] with permission from the American Chemical Society

SAXS intensity was caused by the growth of the crystalline regions. The mobility of the polymer segments in the amorphous layer between the crystalline layers should be suppressed by the immobile segments in the crystalline regions. A structure where the crystalline and amorphous layers lined up in a queue alternately started to develop. The amorphous chains were then considerably restricted and the correlation time increases with an increase in crystallinity. At the final stage in the crystallization process, all spherulites collided and the crystallization rate was considerably decreased, as confirmed by the absence of structural change being clearly evidenced by X ray scattering measurements. The mobility of he polymer chains in the amorphous layer was also strongly influenced by the crystal growth at the late stages. Consequently, the τ_c detected by the ESR measurements continue to increase with the crystallization time.

8.3 Applications of Spin Labeling Method to Biopolymer Systems

The cell membrane is one of the most significant constituents for the life. In general, the cell can absorb required substances and exclude wrong ones through the membrane. In order to understand the mechanism of the peptide transport for example, we have to clarify the structure of the membrane, the fluidity of the lipid chain and the interaction between the chain and the peptide in the membrane. The spin label method is very effective for the characterization of the cell membrane because we can allow for selective spin labeling with nitroxide radicals at particular sites in the lipid chain. The labels can also be selectively bonded at particular molecules such as lipid chains and at peptides. The fluidity of membrane by the spin-labeling method is reviewed by Toyoshima [17].

8.3.1 Structure of Biological Cell Membrane

Many models for the membrane of a biological cell have been proposed. A variety of physical techniques confirm the lipid bilayers structure, where the hydrocarbon chains with a hydrophilic head of polar groups line up and face each other

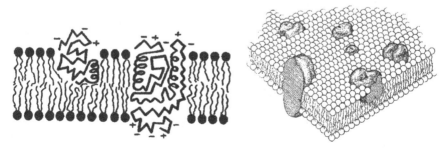

Fig. 8.13 Mosaic model of a lipid bilayer for a cell membrane. The figure is adapted from [18] with permission from the American Association for the Advancement of Science

as shown in Fig. 8.13. It is also proposed from biochemistry and physical chemistry points of view that the peptides are incorporated in the membrane because of their hydrophobic character. Singer and Nicholson [18] proposed the lipid bilayer mosaic model, where the protein molecules are floating in the bilayer membrane as shown in Fig. 8.13.

8.3.2 Fluidity of Biological Membranes and the Order Parameter of a Lipid Chain

A spin label such as I(m,n), as indicated in Fig. 8.14 is used for the determination of the degree of orientation. The so called order parameter can be estimated by the anisotropic ESR spectra of spin label I(13,2) in an oriented sample of smectic liquid crystal, as shown in Fig. 8.15 [19]. There is a dramatic difference in the hyperfine splitting and position of the spectrum, when the axis of the long chains is oriented parallel and perpendicular to the direction of the applied magnetic field.

Figure 8.16 shows the geometrical relationship for the molecular axis system (x, y, z) and the optical axis (ζ) parallel to the orientation axis.

The oxazolidine ring of I(m,n) attaches the NO group rigidly to the hydrocarbon chain, and the nitrogen $2p\pi$ orbital will be oriented to the molecular chain axis. The molecular axis system is then coincident with the principal axis system of the hyperfine tensor. The anisotropic hyperfine splitting is described by an axial symmetric tensor having the principal values of T_{xx}, T_{xx}, and T_{zz}. Molecular motion around the ζ axis leads to an average tensor having the principal values of $T_\perp$, $T_\perp$ and $T_\parallel$:

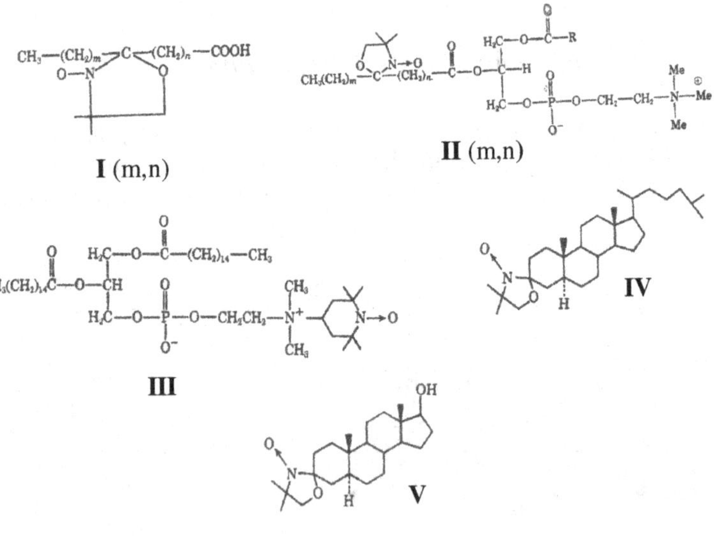

Fig. 8.14 Chemical structures of spin-labels

Fig. 8.15 ESR spectra of spin label I (13, 2) in an oriented sample of smectic liquid crystal: The magnetic field is applied parallel (**a**) and perpendicular (**b**) to the oriented axis of the molecular chain. The figure is adapted from [19] with permission from the American Chemical Society

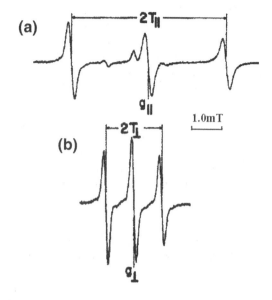

Fig. 8.16 Geometrical relationship for the molecular axis system (x, y, z) and the optical axis (ζ)

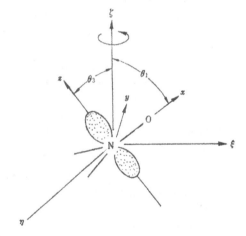

$$T_\perp = \frac{1}{2}(1+ <\cos^2 \theta_3 >)T_{xx} + \frac{1}{2}(1- <\cos^2 \theta_3 >)T_{zz} \qquad (8.18)$$

$$T_{//} = (1- <\cos^2 \theta_3 >)T_{xx}+ <\cos^2 \theta_3 > T_{zz} \qquad (8.19)$$

θ_3 is the angle between the z axis and ζ axes. $<\cos^2\theta_3>$ therefore determines the mean square deviation of the nitrogen 2pπ orbital from the optical axis. The degree of order of the z axis with respect to the ζ direction is defined by:

$$S_3 = \frac{1}{2}(3 < \cos^2 \theta_3 > -1) \qquad (8.20)$$

Combination of Eqs. (8.18) and (8.19) results in a very convenient expression for the experimental determination of S_3:

$$S_3 = \frac{T_{//} - T_{\perp}}{T_{zz} - T_{xx}} \qquad (8.21)$$

Therefore, the order parameter of the lipid chain can be estimated by Eq. (8.21). The order parameter is a good measure of the fluctuation of the chain orientation which is strongly correlated with the fluidity of the lipid.

8.3.3 Dependence of Molecular Mobility of Lipid Bilayers on Position of the Methylene Group

When the labels, I (m,n), are incorporated into aqueous dispersions of the lipid bilayers such as egg lecithin-cholesterol, carboxy and other polar groups of the spin labels should be trapped near the polar head group of the lipid bilayers. Therefore, the N-oxyl-4′,4′dimethyl oxazolidine (NODO) group is positioned at sites farther away from the polar head group of the bilayers with an increase of the n number. We can then detect the dependence of molecular mobility of the lipid bilayers on the number n of methylene groups in the lipid chain. A phospholipid spin label (m,n) of the type indicated in Fig. 8.14 is also used for the estimation of molecular motion [20]. Table 8.1 shows resonance data for I (m,n) in egg lecitin-cholesterol bilayers. The isotropic hyperfine splitting, a' decreases with an increase of n. This fact suggests that the value of a′ is affected by the polarity of the local environment. It is well known that there is a small dependence of the isotropic hyperfine interaction of nitroxide radicals on solvent polarity. The cation of the polar group can for example attract the electrons at the oxygens in the NO group. As a result, the spin density on the nitrogen nucleus increases and then, the hyperfine splitting also increases. The splitting is 1.41 mT for the NODO ring in spin label IV in hydrophobic hexane, whereas the value observed in the resonance of V in distilled water is 1.52 mT. Consequently, it is found that the NO group in the spin label II (10,3) should be

Table 8.1 Resonance data for phospholipid spin labels II (m, n) in Egg-lecitin-cholesterol

Spin label	$T_{//}'$ (mT)	$T_{\perp}'$ (mT)	a′ (mT)	S
II (10,3) 0.70	2.78		0.90	1.52
II (7,6) 0.62	2.60		0.95	1.50
II (5,10) 0.46	2.18		1.03	1.41

The table is adapted from [19] with permission from the American Chemical Society.

trapped near the polar head group of the bilayers whereas the NO group in the spin label II (5,10) is positioned at the hydrophobic methylene group of the lipid chain.

It is then necessary to modify the expression for the order parameter S in Eq. (8.21) to obtain Eq. (8.22) where the denominator is the quantity $(T_{zz}-T_{xx})$ in a hydrophobic environment

$$S_3 = \frac{T'_{//} - T'_{\perp}}{T_{zz} - T_{xx}}\left(\frac{a}{a'}\right) \qquad (8.22)$$

$T_{//}'$ and $T_{\perp}'$ are the principal values of the anisotropic hyperfine splitting and a' is the isotropic value for the spin label at the position of the lipid chain. The isotropic coupling a in a hydrophobic environment equals 1.41 mT. T_{zz} and T_{xx} are obtained from the ESR spectra of spin label II (m,n) in cholesterol chloride to be 3.08 and 0.58 mT, respectively. Figure 8.17 gives a semilog plot of the order parameter S_n vs. n, where n is the number of methylene groups in II (m,n) and I (m,n) that separate the paramagnetic NODO ring from the carbonyl group. Two of the plots in Fig. 8.17 indicate that the order parameter S_n depends exponentially on n. Even though this simple exponential dependence does not hold for the important phospholipid spin labels II (m,n), the exponential dependence of S_n on n is nevertheless of considerable interest from a theoretical point of view, and serves as a useful starting point for a more elaborate treatment of molecular motion in bilayers. The anisotropic molecular motion determined from the resonance spectra can be accounted for by a realistic model of rapidly inter-converting isomeric states of the polymethylene chain.

The analysis of the resonance data in terms of this model makes it possible to estimate the probabilities of gauche and trans conformations about carbon-carbon single bonds at various positions in fatty acid chains in phospholipid bilayers. The probability of the gauche conformation increases for instance with increasing n by a rotation about the single bond when one end of the chain, the carboxy group, is

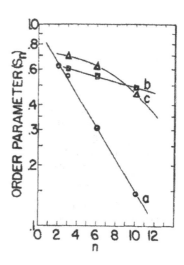

Fig. 8.17 The order parameter S_n for fatty acid spin label I(m,n) in smectic liquid crystals (*a*), fatty acid spin label I (m,n) (*b*), and phospholipid spin label II (m,n) (*c*) in egg-lecitin-cholesterol. The figure is adapted from [20] with permission from the American Chemical Society

"anchored" or rigidly fixed at the bilayer-water interface. McConnell et al. carried out a theoretical analysis of the degree of order, S_n as a function of n:

$$\log S_n = n \log P_t + C \qquad (8.23)$$

P_t is the probability that any single carbon-carbon bond is a trans conformation, which is a decreasing function of temperature. The probability of a gauche conformation, P_g is an increasing function of temperature:

$$P_t + P_g = 1, \quad P_t > P_g \qquad (8.24)$$

The value of C in Eq. (8.23) is also dependent on n and P_t and is small in comparison with the first $n\log P_t$ term. Hence, the exponential dependence of S_n can be interpreted and the values of P_t and P_g are obtained by an iteration procedure. It can be concluded that the fluctuation and molecular mobility of the lipid chains in the bilayers increase with an increase of distance from the hydrophilic head group.

The temperature dependence of the order parameter for the label II (m,n) in dipalmitoyllecithin-H_2O showed an abrupt decrease at 39°C, suggesting a phase transition. The transition was attributed to the mesomorphic phase transition in phospholipids. Figure 8.18 shows semilog plots of S_n vs. n for two temperatures, one just above the transition temperature (40°C), another just below (35°C). It is found that the order parameter near the polar head group changes slightly whereas this parameter changes drastically in the inside of the membrane. This fact suggests that the hydrophilic head group is rigidly fixed at the bilayer interface and the polymethylene in the central part of the membrane has a structure that is much disturbed by a rapid molecular motion.

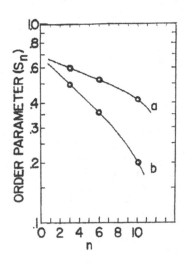

Fig. 8.18 The order parameter S_n as a function of n phospholipid spin label II (m,n) in aqueous dispersions of dipalmitoyllecitin (*a*) 4°C below and (*b*) 1°C above the transition temperature at 39°C. The figure is adapted from [20] with permission from the American Chemical Society

8.3.4 Lateral Phase Separation in Phospholipid Membranes Caused by Lateral Diffusion of Lipid Chains

The fluidity of the membrane of a biological cell is a very important factor in many essential processes of biological significance. Molecular motion of membrane components is for example required for the active transport of ions and molecules through the membrane. There is much interest in the question of the rate of lateral diffusion of molecules in membranes. This question arises in connection with the possible lateral diffusion of "messenger" molecules in membranes. In this section, we study the rate of the lateral diffusion of a spin labeled lipid in a phospholipid bilayer system. It is well known that the spin-spin interaction broadens the ESR spectrum at higher concentration of free radicals. As the line width of a spectrum with hyperfine splitting increases due to an increase of the radical concentration, the splitting eventually disappears and a broad singlet spectrum is finally observed. An example of this so-called exchange broadening is shown in Fig. 8.19 [21]. Spin label III and dihydrosterculoyl phosphatidylcholine (DPC) of different molar ratios were then mixed homogeneously on flat quartz plates. The magnetic field was applied

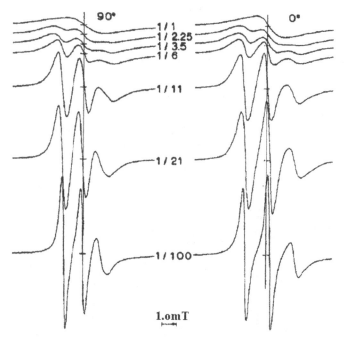

Fig. 8.19 Normalized reference spectra of spin label III in bilayers of the host lipid DPC corresponding to the homogeneous mixtures having different ratios. The molar ratio is represented as a/b (spin label III/DPC) The magnetic field was applied perpendicular (90°) to the planes of the bilayers in one set of spectra (*left*) and was parallel (0°) in the other spectra (*right*). The figure is adapted from [21] with permission from the American Chemical Society

perpendicular (90°) and parallel (0°) to the planes of the phospholipid bilayers. A remarkable exchange broadening was observed. In order to detect the diffusion of the lipid chain, a sample with an inhomogeneous concentration of the spin label was prepared. The unlabeled lecithin was added carefully surrounding the central patch of the lecithin with high concentration of spin labels. The membrane was pressured between two quartz plates.

Figure 8.20 shows a typical set of spectra for two different orientations of the magnetic field. The initial exchange broadened spectrum changes to one with well resolved hyperfine splitting. The local concentration of spin labels decreases with an increase of the storage time. These facts suggest that the spin-labeled lecithin diffuses laterally in the membrane. The time dependent spectrum is analyzed by assuming a mixture of spin labels having a distribution of concentration. For instance, the sum of seven spectra with the different molar ratios of spin labeled to non-labeled lecithine described in Fig. 8.19 is calculated considering the relative intensities of the spectra and compared with the time dependent spectrum. Consequently, the concentration of the spin labeled lecithin and the diffusion constant, D are obtained by the diffusion equation:

$$D\Delta^2 P(r,t) = \frac{\Delta P(r,t)}{\Delta t} \tag{8.25}$$

Here P(r,t) gives the concentration of spin labeled molecules at a point r away from the center of the circle at time t. The results lead to a lateral diffusion coefficient:

$$D = 1.8 \times 10^{-8} cm^2/sec \tag{8.26}$$

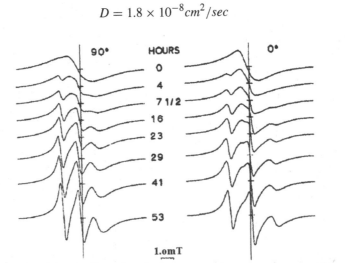

Fig. 8.20 Dependences of ESR spectra of spin label III in bilayers of the host lipid DPC on storage time. The molar ratio is represented as a/b (spin label III/DPC). The magnetic field was applied perpendicular (90°) to the planes of the bilayers in one set of spectra (*left*) and was parallel (0°) in the other spectra (*right*). The figure is adapted from [21] with permission from the American Chemical Society

How many times per second does the lipid chain jump and how long distance does the chain move per second? For the sake of discussion the lattice is assumed to be hexagonal and the diffusion in a two-dimensional lattice of phosphatidylcholine phospholipids is considered. For two-dimensional diffusion:

$$(\bar{x})^2 + (\bar{y})^2 = 4Dt \tag{8.27}$$

where $(\bar{x})^2$ and $(\bar{y})^2$ are the mean square displacements in the x and y directions. If τ^{-1} is the probability per unit time that a lipid molecule jumps from one site to any one of the six neighboring sites, then:

$$\tau = \frac{a^2}{4D} \tag{8.28}$$

where a is the lattice constant of the hexagonal lattice. Thus the jumping rate τ^{-1} is calculated to be of the order of magnitude of 10^7 s^{-1}. The lipid chain moves 2×10^{-4} cm/s from Eq. (8.27). The viscosity, η of the lipid bilayers is calculated to be 3.3 poise at 40°C from the Einstein-Stokes equation:

$$D = \frac{kT}{6\pi\eta l} \tag{8.29}$$

where k, l and T are the Boltzmann constant, the distance of movement and the temperature, respectively. It is found that the bilayers are very fluid comparable to liquid glycerin. The lateral diffusion rate, 10^7/s is also much higher than the exchange rate of the lipid chain between the layers of the order, 2×10^{-4}/s. Figure 8.21

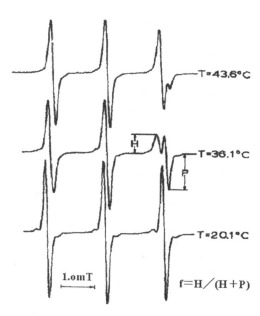

Fig. 8.21 ESR spectra of TEMPO dissolved in an aqueous dispersion of DPPC above the transition temperature. The figure is adapted from [22] with permission from the American Chemical Society

shows ESR spectra of TEMPO in an aqueous dispersion of O-(1,2-Dipalmitoyl-X-glycero-3-phosphoryl)choline (DPPC) at three temperatures [22]. Each spectrum is a superposition of two components: one is due to TEMPO dissolved in the fluid, hydrophobic region of the lipid; the other is due to TEMPO in the aqueous phase. Small differences in the isotropic hyperfine coupling constant and g factors for the spin label in each environment result in a partial resolution of the high field hyperfine line, whereas the low and central lines are not resolved. It is well known that the isotropic hyperfine interaction increases and the g value decreases with an increase of solvent polarity.

The two amplitudes, H and P of the high-field hyperfine signals, indicated in Fig. 8.21 are proportional to the amounts of the spin label in the membrane bilayer and in the aqueous phase, respectively. Figure 8.22(b) shows a spectral parameter f, equal to H/(H+P), which is approximately the fraction of the TEMPO spin label dissolved in the membrane bilayer as a function of 1/T for binary mixtures of phospholipids in the O-(1,2-Dimyristoyl-sn-glycero-3-phosphoryl)choline/DPPC (DMPC/DPPC) system. The value of f increases abruptly at a specific temperature upon warming. The observed transition is caused by the change of the 'gel' phase designated S, where he hydrocarbon chains are packed into a hexagonal subcell to the 'fluid' phase where the fatty acid chains are more flexible. The transition temperature depends strongly on the composition of DPPC. We can make a phase diagram as shown in Fig. 8.22(a) by a quantitative analysis. The two components observed in the temperature range for the transition correspond to the mixed phase, F + S in the phase diagram. It can be considered that the rapid lateral diffusion of the lipid chain leads to phase separation of the lipid bilayers. For thermodynamic reasons, a biological membrane whose lipids are partially in the F state and partially in the

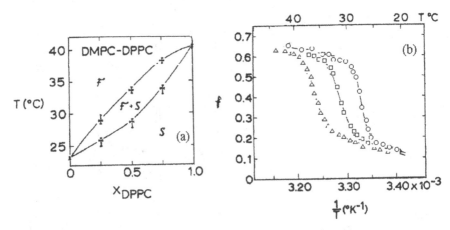

Fig. 8.22 (a) The fluid-solid equilibrium phase diagram for aqueous dispersions of the DMPC-DPPC binary system: temperature vs. mole fraction of DPPC. The fatty acid chains are stiffer (less fluid) in the F region and more flexible (fluid) in the S region. (b) Experimental TEMPO spectral parameter, f, as a function of 1/T for :76 ($\triangle$), 51 ($\bigcirc$) and 26 ($\square$) mol% of DPPC. The figure is adapted from [22] with permission from the American Chemical Society

S state should have a high lateral compressibility and extensibility. A lateral compressibility would facilitate the insertion of newly synthesized protein, lipid or new membrane into an old membrane without a corresponding expansion in the area of the membrane. Ohnishi et al. also studied calcium induced phase separation in phosphatidylserine-phosphatidylchloline membrane by the spin label method [23].

8.3.5 Lipid-Protein Interactions and Rotational Diffusion

How molecular motions of the protein and lipid chains are correlated with each other is a very important issue. The interaction between the protein and the lipid is strongly related to the mechanism of transportation of the protein molecules which are floated in the bilayer membrane. Figure 8.23 shows the conventional ESR spectra of spin labeled stearic acid, 14-SASL, in complexes of a 26-residue peptide (K26)) with DMPC of different lipid/peptide ratios [24]. The spectra consist of two components: one corresponding to the fluid bilayer regions of the lipid complexes and the other, with the larger hyperfine anisotropy (visible in the outer wings of the spectra), corresponding to the motionally restricted lipid environment at the intra-membranous surface of the incorporated peptide.

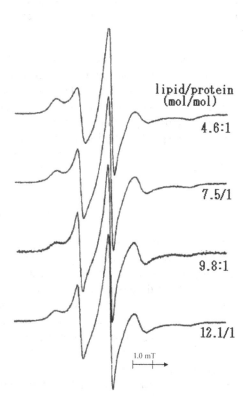

lipid/protein
(mol/mol)

4.6:1

7.5/1

9.8:1

12.1/1

1.0 mT

Fig. 8.23 ESR spectra of the 14-stearic acid spin label (SASL) in complexes of the K26 peptide with DMPC at lipid/peptide ratios of 4.6, 7.5, 9.8, and 12.1 mol/mol. The figure is adapted from [24] with permission from the American Chemical Society

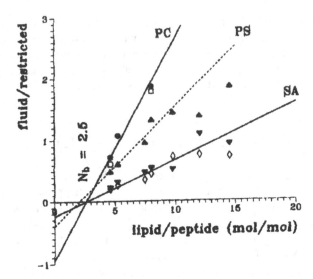

Fig. 8.24 Lipid/peptide titration of the K26 peptide complexes with DMPC obtained from ESR difference spectra of various spin labels: 14-PASL phosphatidic acid (◇), 14-SASL Stearic acid (▼), 14-PSSL phosphatidylserine (▲), 14-PGSL, phosphatidylglycerol (□), and 14-PCSL phosphatidylcholine (●) spin labels. The ratio of nf*/nb* (fluid/restricted) is calculated from the double integrated intensity of the fluid and motionally restricted components in the ESR spectra. The figure is adapted [24] by permission of the American Chemical Society

The amount of the restricted component increases with an increase of the peptide. It is found that the peptide molecules slow down the rate of rotational motion of the lipid chains when the interaction is strong. The dependence of the ratio of the intensity of the fluid component to that of the motionally restricted component (n_f^*/n_b^*) is given as a function of the lipid/peptide ratio of the complexes (n_t) for the different spin-labeled lipid as shown in Fig. 8.24.

The data are displayed in terms of the equation for equilibrium lipid-peptide association:

$$n_f^*/n_b^* = \frac{n_f/N_b - 1}{K_r} \qquad (8.30)$$

where N_b is the number of lipid association sites per peptide, and K_r is the association constant of the spin-labeled lipid with the peptide, relative to that for the unlabeled host lipid (DMPC). The linear dependences cross the x-axis at a common value, $N_b \cong 2.5$ lipid/peptide monomer, as would be expected on structural grounds. For instance, approximately two lipids per peptide monomer are associated with the hydrophobic surface of the peptide in the bilayer. The values of K_r differ for the different spin-labeled lipids, reflecting the lipid selectivities for interaction with the peptide. The result suggests that the K26 peptide must be positioned in the lipid in such a way that one or both of the positively charged lysine and arginine residues at the N- and C-termini, respectively, of the peptide must be located in a region close

Fig. 8.25 Second harmonic, 90° out of phase, absorption saturation transfer spectra of 5-InVSL spin label covalently bound to the K26 peptide. The sharp signals marked as * were quenched by addition of 5 mM Ni²⁺ in the aqueous phase. The figure is adapted from [24] with permission from the American Chemical Society

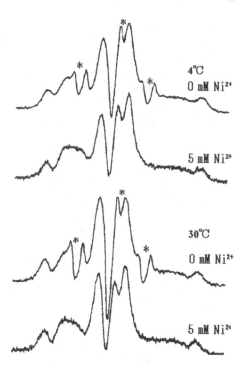

to the negatively charged lipid head groups. How fast is the molecular motion of the peptide in the lipid bilayer system? Saturation transfer ESR (STESR) spectra of spin-labeled K26 peptide floated in the lipid bi-layers has been employed to investigate the issue. The STESR spectrum shape is very sensitive for slow motions having a correlation time of the order of magnitude of μs. Figure 8.25 shows the second harmonic, 90° out-of-phase absorption STESR spectra of K26 peptide labeled with 2-[(-oxy-2,2,5,5-tetramethyl-3-pyrrolin-3yl)]Indian-1,3-dione (5-InVSL) in a complex with DMPC. The effective rotational correlation times were estimated to be approximately of the order 100–150 μs. It can be considered that the slow motion of the peptide affects the mobility of the lipid chains. The spectra also contain a sharp component, most probably arising from spin labels that released slowly from the peptide, which can be quenched by addition of 5 mM Ni²⁺ in the aqueous phase.

8.4 Summary

The ESR method is effective for the characterization of the heterogeneity in mobility and structure of polymers materials. For instance, selective spin labeling is very effective for the detection of the molecular motion in the various sites and regions of the polymer systems. The experimental facts obtained by the spin labeling with high selectivity and high sensitivity have never been achieved by other methods. In the present chapter, the following examples are introduced.

(1) The molecular motion of a segment at a chain end is more rapid than that in an inner chain site. The fact is clarified by the selective spin labeling in an amorphous polymer. The molecular weight dependencies of glass-rubber transitions at two kinds of sites, the chain end and inner chain sites give a free volume size of micro Brownian motion.

(2) The molecular motion of a segment in the non-crystalline region is strongly affected by the rigidity of a crystalline region. The fact is clarified by the selective labeling of the non-crystalline region. (a) The effect of the micro-Brownian molecular motion on the rigidity in the crystalline region is evaluated from the crystallinity dependence of the transition temperature. Then the free volume size of micro Brownian motion can be estimated in the crystalline polymer. (b) The molecular mobility of the segment in non-crystalline region decreases with increasing crystallization time. The decrease in molecular mobility before crystallization has been shown to originate from structural changes such as the self-aggregation, which indicates the existence of the precursory aggregation of polymer chains in the crystallization process. Multi stages of the crystalline process are detected.

(3) The dependence of the molecular mobility of lipid bilayers on the positions of the methylene group is detected by the selective labeling of the positions away from the surface of the bilayers. It was found that hydrophilic head groups are rigidly fixed at the bilayer interface and the polymethylene in the central part of the membrane has a much disturbed structure caused by rapid molecular motion.

(4) Lateral phase separation in phospholipid membranes caused by lateral diffusion of lipid chain is detected by the time dependence of the local concentration which is estimated from the spin-spin exchange interaction.

(5) Lipid-protein interactions can be estimated from the molecular mobility of spin labeled lipid chains. The mobility is lowered by attractive interactions. It is found by the selective labeling method that the lipid chain has extremely high mobility in comparison with that on the peptide in the membrane.

References

1. D. Kivelson: J. Chem. Phys. **33**, 1094 (1960).
2. T.J. Stone, T. Buckman, P.L. Nordio, H.M. McConnell: Proc. Natl. Acad. Sci. USA **54**, 1010 (1965).
3. J.H. Freed: J. Chem. Phys. **41**, 2077 (1964).
4. J.H. Freed, G.V. Bruno, C.F. Polanaszek: J. Phys. Chem. **75**, 3385 (1971).
5. S.A. Goldman, G.V. Bruno, J.H. Freed: J. Phys. Chem. **76**, 1858 (1972).
6. T. Kajiyama, K. Tanaka, A. Takahara: Macromolecules **30**, 2809 (1997).
7. Y. Miwa, T. Tanase, K. Yamamoto, M. Sakaguchi, M. Sakai, S. Shimada: Macromolecules **36**, 3235 (2003).
8. M.L Williams, R.F. Landel. J.D. Ferry: J. Am. Chem. Soc. **77**, 3701 (1955).
9. K. Unberreiter, G.J. Kanig: J. Colloid Sci. **7**, 569 (1952).
10. K. Yamamoto, K. Kato, Y. Sugino, S. Hara, Y. Miwa, M. Sakaguchi, S. Shimada: Macromolecules **38**, 4737 (2005).

11. F.C. Stehling. L. Maderkern: Macromolecules **3**, 242 (1970).
12. R.F. Boyer: Macromolecules **6**, 288 (1973).
13. F. Buche: *'Physical Properties of Polymers'*, Wiley, New York, NY (1962).
14. N. Kusumoto, S. Sano, N. Zaitsu, Y. Motozato: Polymer **17**, 448 (1976).
15. F.-D. Tsay, G. Amitava: J. Polym. Sci. Part B, Polym. Phys. **25**, 855 (1987).
16. S. Hara, K. Yamamoto, S. Okamoto, S. Shimada, M. Sakaguchi: Macromolecules **37**, 5323 (2004).
17. Y. Toyoshioma: In *'Function of Membrane'* ed. by M. Senoo, Kyouritu Shuppann, Tokyo (1977), Chapter 3. (In Japanese)
18. S.J. Singer, G.L. Nicolson: Science **175**, 720 (1972).
19. J. Seelig: J. Am. Chem. Soc. **92**, 3881 (1970).
20. W.L. Hubbell, H.M. McConnell: J. Am. Chem. Soc. **93**, 314 (1971).
21. P. Devaux, H.M. McConnell: J. Am. Chem. Soc. **94**, 4475 (1972).
22. E.J. Shimshick, H.M. McConnell: Biochemistry **12**, 2351 (1973).
23. S. Ohnishi, T. Ito: Biochemistry **13**, 881 (1974).
24. LI. Horvath, T. Heimburg, P. Kovacher, J.B.C. Findlay, K. Hideg, D. Marsh: Biochemistry **34**, 3893 (1995).

Chapter 9
Applications of Quantitative ESR

Abstract ESR applications concerned with the measurement of the amount of paramagnetic species are the common theme of the chapter. Procedures to obtain absolute concentrations are outlined. Error sources are discussed and procedures to reduce the uncertainties are reviewed. The principles of ESR-dosimetry are presented. Strategies to increase the dose response by using materials other than L-alanine, by isotopic substitution, metal ion doping, and instrumental developments are briefly described. The measurement of the spatial distribution of radiation dose by the ESR imaging (ESRI) method is discussed. Geological dating by ESR using the additive dose method is applicable for periods up to two million years. Procedures to estimate doses by ESR in contaminated areas and after radiological accidents are described as well as ESR analyses for test of irradiated food and of medical equipment. Methods to analyze the ESR line-shapes are considered in the context of obtaining the integrated spectral intensity.

9.1 Introduction

ESR provides a direct method of measuring the presence and concentrations of free radicals and other paramagnetic species in materials. Applications in free radical research and in biophysics were published at an early stage [1–3]. Recent progress has been reviewed [4]. Compared to other methods the following advantages have been emphasized.

- *Selectivity*: No background signal occurs from the diamagnetic host material containing the paramagnetic species.
- *Sensitivity*: At a typical width of 1–10 mT of a free radical spectrum, milli-molar concentrations are detectable with a regular X-band CW spectrometer.
- *Non-destructivity*: The sample is not destroyed by recording the spectrum, and can be re-measured many times if necessary.

Quantitative ESR was used in early studies to determine the amount of transition metal ions in frozen solutions of metallo-proteins [3], to obtain the yields of free radicals in solids [5] and for the quantitative analysis of metals in aqueous solutions [6].

A. Lund et al., *Principles and Applications of ESR Spectroscopy*,
DOI 10.1007/978-1-4020-5344-3_9, © Springer Science+Business Media B.V. 2011

In these cases the absolute concentrations usually have to be determined. Relative concentrations can be measured more accurately than the absolute values. Internationally recognized applications are concerned with the effects induced in the solid state by high-energy radiation. One example is ESR-dosimetry, which gives the possibility to estimate the absorbed dose of high-energy radiation by the alanine dosimeter [7]. Methods to increase the accuracy over that obtained in earlier studies by improved calibration procedures have been developed in ESR dosimetry in standard applications and in emergency situations after an accident [4]. Other applications provide procedures for dating of ancient objects [8] and for identification of irradiated food by ESR.

9.2 Absolute Concentration Measurements

The concentration, expressed e.g. as the number of spins/kg in a solid sample or mol/l in a liquid has in nearly all cases been obtained by comparison with a standard sample of known concentration, see, however, [9] for a recent development that does not require a reference.

The intensities of the reference (R), and sample (X) are determined by double integration of the first derivative spectra over their spectral ranges. The procedure is usually computer controlled using commercial or home-made software for the integration. Scan width and amplification spectrometer settings that had to be taken into account in hand calculation, see Appendix F in [10], are usually automatically compensated for by the software in modern instruments.

The sample concentration C_X can be estimated from the corresponding C_R as:

$$C_X = C_R \cdot \frac{I_X}{I_R} \cdot \frac{S_R(S_R+1)}{S_X(S_X+1)} \cdot \frac{\Gamma_X}{\Gamma_R} \tag{9.1}$$

The first ratio is between the experimentally determined intensities, the second compensates for the difference in spin quantum numbers. The third ratio contains a correction due to differences in g-factors. For isotropic systems it has been customary to put $\frac{\Gamma_X}{\Gamma_R} = \left(\frac{g_R}{g_X}\right)^2$, but as explained by Aasa and Vänngård [3] this is not correct for CW (magnetic field swept) ESR spectra. A factor $\frac{\Gamma_X}{\Gamma_R} = \frac{g_R}{g_X}$ is appropriate for isotropic g-factors. The importance of this observation was illustrated for the first time by work on the iron-transporting protein transferrin. The protein contains two ferric ions per molecule. The ESR spectrum is an isotropic line with $g = 4$. To decide if this line was due to one or both Fe^{3+} ions, the spin concentration was determined with a standard having $g = 2$. Applying the correction factor due to the difference in g-factors derived in [3] it was concluded that the signal corresponded to all Fe in the protein. The assumption that $\frac{\Gamma_X}{\Gamma_R} = \left(\frac{g_R}{g_X}\right)^2$ would in this case lead to the wrong conclusion that only one Fe^{3+} contributed to the signal.

For solids a correction for the g-anisotropy may also be needed. This correction factor is particularly important in case of the significant anisotropy often occurring in transition metal ions. The correction factor depends on the symmetry of the g-tensor. The value for a powder sample differs from that of a single crystal in which case the factor is orientation dependent [11]. We refer to the literature for values of the correction factor due to g-anisotropy in powder samples [3, 12].

Calibration can be employed to compensate for differences in signal strength between different sample tubes containing identical samples of liquids or frozen solutions [13]. The variation of spectrometer sensitivity may be monitored by a sample mounted permanently in the ESR cavity. Alternatively a dual cavity can be employed. A common material for this purpose is Mn^{2+} diluted in MgO. It can be used both to correct for variations in sensitivity between different samples and for magnetic field calibration, using the known g-factor (g = 2.0014) and the hyperfine coupling ($a = 8.67$ mT, $I(^{55}Mn) = 5/2$) of this substance.

For samples of polycrystalline or disordered solids the concentration of spins per unit weight is usually the quantity of interest. Errors due to different densities of the sample and the reference and of different samples can then occur. A procedure to normalize the signal intensity by the mass to length ratio of the sample has therefore been recommended for quantitative measurements on powder samples [14]. The procedure may be motivated for a sample contained in a normal ESR sample tube as follows:

$$I \propto C^{(m)} \times A \times \ell_{eff} \times \frac{m}{\ell \times A} = C^{(m)} \times l_{eff} \times \frac{m}{\ell} \qquad (9.2)$$

$C^{(m)}$ is the concentration by weight of paramagnetic species, the next two factors give the effective volume exposed to the microwaves in the tube while the last ratio is the density of the sample. The cross section A of the tube, assumed to be constant, cancels out. The effective length of the sensitive region of the cavity, ℓ_{eff}, can be considered as a constant for all samples, provided that the sample tube is filled to a length ℓ that is longer than ℓ_{eff}, of the order 10 mm at X-band. The intensity I is either the peak-peak amplitude of the derivative spectrum as in alanine dosimetry or the integrated signal when the absolute concentration is determined. In the latter case an additional measurement with a standard is made and the unknown concentration is determined using a modification of Eq. (9.1):

$$C_X^{(m)} = C_R^{(m)} \cdot \frac{I_X}{I_R} \cdot \frac{S_R(S_R + 1)}{S_X(S_X + 1)} \cdot \frac{\Gamma_X}{\Gamma_R} \cdot \frac{\ell_X}{m_X} \cdot \frac{m_R}{\ell_R} \qquad (9.3)$$

The measured weight (m) to length (ℓ) ratios of the sample and reference are introduced in the formula. The superscript (m) indicates that concentrations by weight are employed.

As mentioned previously the amount of paramagnetic species in a sample is of interest in biophysical applications [3]. The radical yields determined by ESR have also been measured to help elucidate the radiation chemistry of the solid state and in several other applications referred to in Table 9.1. The concentration of radicals at

Table 9.1 Concentration standards used in quantitative ESR

Instrument	Standard sample	Comment	Application
X-band	TEMPOL	Reduction by ascorbate	Determination of ascorbate in leaf[a].
X-band	Tempone		ESR Dosimetry[b]
	Nitroxide in polar solvents	Standardised by optical absorption	Solutions[c]
X-band	Vanadyl sulfate in K_2SO_4	Vanadium content by polarography	Preparation and evaluation of standards[d]
	DPPH in benzene; carbamoyl-2,2,5,5-tetramethylpyrrolidin-1-oxyl in water.	Error estimate	Method development[e]
	potassium perchromate	Easy to prepare in pure form	Free radicals in fossil fuels[f]
X- and Q-band	Picein 80 cement	Small space, adjustable concentration	Single crystals[g]
X-band	Cu(II) salt in H_2O	Frozen solution	Transition metal enzymes[h]
X-band	Cu(II)dithiocarbamate	Standardised by optical absorption	Yield of radicals by irradiation[i]
X-band transmission	DCCN, DPPH	Error estimate[j]	
X-band	4-hydroxy-TEMPO and Reckitt's ultramarine		Radicals in humic acids[k]

[a] Y. Lin et al.: Anal. Sci. **15**, 973 (1999).
[b] M. Brustolon et al.: J. Magn. Reson. **137**, 389 (1999).
[c] R.G. Kooser et al.: Concepts in Magnetic Resonance **4**, 145 (1992).
[d] A. Madej et al.: Fresenius' J. Anal. Chem. **341**, 707 (1991).
[e] A. Redhardt, W. Daseler: J. Biochem. Biophys. Methods **15**, 71 (1987).
[f] N.S. Dalal et al.: Anal. Chem. **53**, 938 (1981).
[g] B. Schmitz et al.: Z. Naturforsch **34A**, 906 (1979).
[h] R. Aasa, T. Vänngård: J. Magn. Reson. **19**, 308 (1975).
[i] P.-O. Kinell et al.: i1) Adv. Chem. Ser. **82**(II), 311 (1972). i2) Arkiv Kemi **23**, 193 (1964).
[j] H.J.M. Slangen: J. Phys. E **3**, 775 (1970).
[k] A. Jezierski et al.: Spectrochim. Acta A **56**, 379 (2000).

a fixed radiation dose is also of importance for the development of ESR dosimetry discussed later in this chapter. The yield (G-value) given as the number of radicals formed per 100 eV absorbed radiation energy is low for aromatic substances but can exceed unity for saturated organic compounds. G-values obtained in early studies are summarized in [5], see Table 9.2 for additional data. In modern applications the yields are usually reported in SI units as μmol/J.

A commercial system to determine the absolute number of spins in a sample has recently become available. It does not require a reference sample, see [9] for details.

Table 9.2 Yields and relative sensitivity of materials tested as ESR-dosimeters

Sample	G radicals/100 eV	Relative sensitivity
L-α-alanine[a–e]	2–7.7	1.0
MgSO$_4$[f]		≥ 1
Lactates[g]	0.46–0.78	10
Dithionates[h]		> 10
Li-acetate·2H$_2$O[i]	0.4	
Li$_3$PO$_4$[i]	1.02	
Ammonium tartrate[j]		2–3
Dimethylalanine[k]		2–3
Formates[l, m]		4–8

[a] T. Henriksen et al.: Radiat. Res. **18**, 147 (1963).
[b] D.F. Regulla, U. Deffner: Appl. Radiat. Isot. **33**, 1101 (1982).
[c] P.P. Panta et al.: Appl. Radiat. Isot. **40**, 971 (1989).
[d] K. Nakagawa et al.: Appl. Radiat. Isot. **44**, 73 (1993).
[e] Z. Stuglik, J. Sadlo: Appl. Radiat. Isot. **47**, 1219 (1996).
[f] J.R. Morton, F.J. Ahlers: Radiat. Prot. Dosimetry **47**, 263 (1993).
[g] M. Ikeya et al.: Appl. Radiat. Isot. **52**, 1209 (2000).
[h] S.E. Bogushevich, I.I. Ugolev: Appl. Radiat. Isot. **52**, 1217 (2000).
[i] G.M. Hassan, M. Ikeya: Appl. Radiat. Isot. **52**, 1247 (2000).
[j] S.K. Olsson et al.: Appl. Radiat. Isot. **50**, 955 (1999), ibid. **52**, 1235 (2000).
[k] S. Olsson et al.: Radiat. Res. **157**, 113 (2002).
[l] T.A. Vestad: On the development of a solid-state, low dose EPR dosimeter for radiotherapy, Dissertation, University of Oslo 2005.
[m] A. Lund et al.: Spectrochim. Acta A **58**, 1301 (2002).

9.2.1 Reference Samples

Reference samples of stable free radicals have been mostly employed to determine the concentrations in samples containing free radicals. Diphenyl-1-picrylhydrazyl, DPPH, was popular in early applications. In this case solutions with known concentrations were prepared by dissolving the solid material in an organic solvent, alternatively solid DPPH diluted with an inert material was employed. Solid samples of pitch intended for sensitivity calibrations of commercial instruments have also been used to determine spin concentrations. More recently spin label compounds have been employed. The concentration of transition metal ions in a sample might be most accurately determined by using a reference containing the same ion; Cu(II) salts in frozen water solution has for example been employed to determine the amount of Cu(II) in enzymes [3]. The sample and the reference are then held at low temperature, usually at 77 K or below, otherwise dielectric losses occur. A non-exhaustive list of concentration standards is given in Table 9.1. Several other concentration standards have been employed depending on the application [13].

9.2.2 Accuracy

The accuracy that can be achieved in quantitative ESR depends on the instrument used to record the spectra, and on the numerical processing to extract the numbers from the experimental data. To reduce the measurement errors several precautions have been recommended, see Appendix F in [10]:

- Same sample shape for reference and examined sample
- Microwave power sufficiently low to avoid saturation
- The same temperature for reference and sample
- Identical positioning of reference and sample in microwave cavity

Additional recommendations of value for the non-specialist to improve accuracy can be found in an early review [13]. The largest source of uncertainty was attributed to differences in the microwave field strength even with identical microwave power input for different samples. The uncertainty can be reduced by using similar materials for the reference as for the sample, both contained in sample tubes of the same diameter and wall thickness, to minimize differences in the cavity Q factor between the measurements. A modern guide to quantitative ESR by the same authors is in preparation [15].

Instrumental sources of errors must also be taken into account [14].

- *Poorly calibrated receiver*, particularly in older instruments. The receiver gains may be calibrated by recording the signal of a strong sample held fixed in the ESR resonator at different receiver settings.
- *Instability of spectrometer sensitivity over time*. The instability may be corrected for by a reference sample permanently placed in the ESR resonator.
- *Nonuniform microwave field*. Pellet samples must be located at the centre, powder samples should fill the whole length of the resonator.
- *Numerical processing*. The accuracy of the numerical integration of the 1st derivative spectrum to obtain the spin concentration depends on the ESR line-shape. The subject has been treated in early reports that may not be easily available [2, 16, 17 (Appendix 5)], and more recently, see Appendix F in [10]. The dependency on the line-shape is apparent in the numeric calculation using the formula $I = \Delta B^2 \cdot \sum_{n=-N}^{N} y_n' \cdot n$, with y_n' = amplitude of first derivative spectrum at $2 \cdot N + 1$ equidistant field points separated by ΔB that was employed for integration by hand of symmetric spectra before computers were available [17 (Appendix 5)]. This 1st moment method shows that the outer portions of a signal with long tails can contribute significantly to the integrated intensity because of the large n values away from the centre. An uncertainty occurs if the outer tails are left out because of low signal/noise ratio. The error is of the order 10 and 20% for Gauss and Lorentz lines with a signal/noise ratio of 10 [16]. The error can be compensated for if the line-shape is known. Procedures to determine line-shapes are summarised in Section 9.2.3.

On one hand measurements on a series of samples using a fixed procedure in a single laboratory can give reproducible results with relatively small scatter in the determined concentrations. On the other hand large deviations have been obtained of the absolute concentrations determined at different laboratories for a specific sample. Thus, the yield of free radicals in irradiated L-α-alanine determined by different authors, vary from 2 to 7.7 radicals per 100 eV absorbed radiation energy [4]. The difference was partly ascribed to the sample preparation method, partly to the fact that several of the measurements were made at a time when the quantitative aspects of ESR spectrometry were not well established. Uncertainties of the order 25% were estimated for the concentration of free radicals in solid systems [5] and for the pitch sample employed in sensitivity tests of commercial ESR instruments. Even if accuracy to within a few % has been indicated more recently [13], it seems that consensus about the errors in absolute concentrations by the ESR method has not yet been reached. The largest source of uncertainty has been attributed to differences in the cavity Q factor at measurements of sample and reference. An increased accuracy would therefore result with a recently advertised method that eliminates the need of a reference [9]. A much needed practitioner's guide in quantitative ESR that is now in preparation [15] should also serve to reduce measurement errors. The uncertainty in the absolute concentration is fortunately not a major problem for the applications of ESR in dosimetry and dating where relative measurements are always employed.

9.2.3 Line Shapes

ESR lines in solution can almost always be approximated by a Lorentz function. In the solid state the line-shape can in general be reproduced by a Gauss curve. In some instances a so-called Voigt profile can give a better approximation to the experimental line-shape. A Voigt line is a convolution of a Lorentz and a Gauss line. The shape is determined by the ratio $\Delta B_L/\Delta B_G$ of the respective line-widths. The shapes of the 1st derivative lines of these types are given in Fig. 9.1.

The Gauss curve decays faster to zero away from the centre than the Lorentz line. The tails of the Voigt line are between those of the Gauss and Lorentz lines. The tails of the latter give a significant contribution to the integrated intensity, which is 3.51 times larger than for a Gaussian with the same peak-peak amplitude of the derivative line [13]. The variation of the integrated intensity of an inhomogeneously broadened line with a constant amplitude as shown in Fig. 9.1(b) is typical for a convolution of Gauss and Lorentz lines to a Voigt shape.

Numerical integration of the 1st derivative spectrum to obtain the spin concentration of a spectrum with a Lorentz line-shape is more difficult and often less accurate than with a Gauss shape. The error in the integrated area of a Lorentz line is, for example, of the order 5% even by recording over a range of ca 50 times the peak-peak line-width of the 1st derivative spectrum [2]. Knowledge of the experimental line-shape can help to estimate and eventually reduce the error [16].

(a)   **(b)**

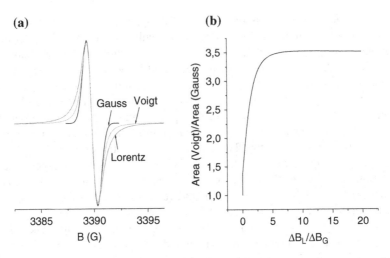

Fig. 9.1 (**a**) Gauss, Lorentz and Voigt ($\Delta B_L/\Delta B_G = 1.0$) lines as 1st derivatives adjusted to the same peak–peak amplitudes. (**b**) Integrated area of derivative lines with the same amplitude as function of the $\Delta B_L/\Delta B_G$ ratio of a Voigt line. For $\Delta B_L/\Delta B_G \gg 1$ the line approaches a Lorentz shape with an area which is 3.51 times larger than that of a Gaussian with the same amplitude [13]

The line-shape of an experimental spectrum can in principle be determined by the procedure illustrated in Fig. 9.2. The 2nd derivative of the resonance line is then recorded. For a Gauss line the ratio h_1/h_2 between the minimum and maximum amplitudes of the 2nd derivative (Fig. 9.2(a)) equals 2.24 [18], while for a Lorentz shape it approaches the value 4. The h_1/h_2 ratio for a Voigt profile varies

(a)   **(b)**

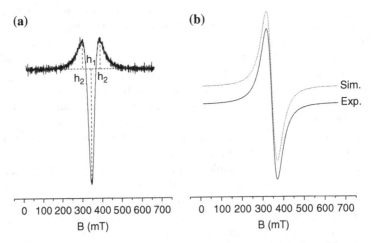

Fig. 9.2 ESR spectra of a GdCl$_3$ water solution. (**a**) 2nd derivative with $h_1/h_2 = 3.7$ corresponding to $\Delta B_L/\Delta B_G = 3.9$. (**b**) 1st derivative experimental and simulated spectra with a Lorentz line-shape. The spectra were obtained from Dr. H. Gustafsson

with the ratio $\Delta B_L/\Delta B_G$ between the Lorentz and Gauss line-widths. This method to determine the line-shape has been successfully applied [19] but requires a good signal/noise ratio and no interference with overlapping lines. These conditions cannot always be realized. The method of simulation of the 1st derivative spectrum is more generally applicable. Simulation with a Lorentz line in Fig. 9.2(b) gave a nearly perfect fit as expected for the ESR line shape of a liquid sample.

9.3 ESR Dosimetry

The method employed in ESR dosimetry is based on the fact that the free radicals formed by irradiation of organic and inorganic solid substances can be stable up to several years as for instance in the L-α-alanine amino acid. To be suitable for ESR dosimetry several additional properties are desirable [4]:

(1) Signal intensity should increase linearly with the radiation dose.
(2) Signal intensity should remain constant after irradiation.
(3) Signal intensity should not show saturation at available microwave powers.
(4) ESR spectrum should be simple, ideally a single line due to one kind of radical.
(5) If the dosimeter is to be applied in medicine one has further to consider that the material has similar atomic composition as tissue or be "tissue-equivalent".

Several factors affect the sensitivity of an ESR-dosimeter.
Microwave saturation: The amplitude of an unsaturated ESR line increases with the square root of the applied microwave power. For high sensitivity one therefore attempts to use as high power as possible without saturating the sample. High microwave power can be applied to samples with short spin-lattice relaxation; for alanine a value of up to about 20 mW has been employed with standard X-band ESR equipment.
Line-width: The amplitude of an ESR 1st derivative line is inversely proportional to the line-width squared at a fixed concentration of free radicals. A narrow line-width is therefore desirable. The anisotropy of the g-factor and/or the hyperfine coupling causes line broadenings in several of the commonly used dosimeter materials, e.g. alanine and Li-formate. The line-width also tends to increase at microwave saturation, which is an additional reason for not increasing the power excessively. In practice the signal of an ESR-dosimeter may be distributed over several hyperfine lines, causing loss of sensitivity. For alanine five lines are present (Fig. 9.3).
The line-width can have a strong influence on the sensitivity as shown in Fig. 9.3 for the ESR spectra of L-α-alanine and ammonium dithionate irradiated to the same dose. Due to its sulfur content the latter material is not tissue-equivalent, however, which makes it unsuitable for medical applications.
Attempts to decrease the line-width artificially by deuteration and other isotopic substitutions of tissue-equivalent materials have resulted in an increase of the sensitivity with up to a factor 2.

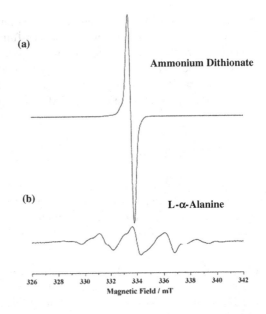

Fig. 9.3 ESR spectra of ammonium dithionate (*above*) and L-α-alanine (*below*) irradiated and recorded under the same conditions. The figure is reproduced from [M. Danilzcuk et al.: Spectrochim. Acta A **69**, 18 (2008)] with permission from Elsevier

Free radical yield: The sensitivity depends on the yield (*G*) of radicals for a certain dose. The yield (*G*-value) given as the number of radicals formed per 100 eV absorbed radiation energy is low for aromatic substances but can exceed unity for saturated organic compounds. Values up to $G \approx 5$ have been reported. Early results are summarized in [5]. In modern applications the yields are usually reported in SI units as μmol/J. Some values obtained for different materials are collected in Table 9.2.

The free radical yields of the potential ESR-dosimeter materials in Table 9.2 are in several cases not known. In other cases *G*-value data obtained in independent studies of a specific substance disagree, indicating that the techniques of quantitative ESR spectrometry were not generally established at the time the measurements were done [4].

The sensitivity *S* of potential ESR dosimeter materials can be estimated by an equation showing the dependence on the microwave power (P), the line-width (w) and the radiation yield (G) [20]:

$$S \propto \frac{\sqrt{P}}{\sqrt{1 + \dfrac{P}{P_0}}} \cdot \frac{1}{w^2} \cdot G \tag{9.4}$$

The power dependence in the first factor applies for the saturation of a so-called inhomogeneously broadened ESR line (see Section 9.5.1) where P_0 is a measure of the microwave saturation. At an applied microwave power $P = P_0$ the ESR signal is $1/\sqrt{2} = 0.707$ of that of an unsaturated signal, while the signal strength becomes independent of the microwave power for $P \gg P_0$, see Fig. 9.4. The signal may also decrease with increasing power for $P > P_0$. A procedure to analyse this behavior is presented in Appendix A9.1 where P_0 is referred to as the microwave power at saturation. It is an advantage if this parameter is large, which means that the magnetic relaxation should be short. Doping with transition metal salts can in some cases improve the relaxation, see Section 9.3.3.2. In practice an optimum value of the microwave power is usually determined experimentally by recording spectra at different microwave powers. The magnetic field modulation amplitude can also be varied to obtain the optimal spectrometer settings for highest sensitivity. The procedure is illustrated in Fig. 9.4 for lithium formate, one of the new materials that have been suggested as a solid-state, low dose ESR dosimeter for radiation therapy [21].

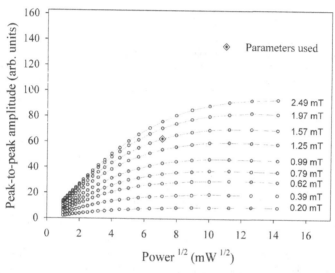

Fig. 9.4 Dependence of the peak-to-peak amplitudes of the ESR spectra of an irradiated lithium formate ESR dosimeter on microwave power and modulation amplitude. The settings for the experiments are marked in the figure. The diagram is reproduced from [21c] with permission from Dr. T.A. Vestad

9.3.1 The Alanine ESR Dosimeter

L-α-alanine, $H_2NCH(CH_3)COOH$, is the most commonly used material for ESR dosimetry. Alanine has excellent characteristics for dosimetric purposes in several respects [4]: (1) high free radical yield (G-value), (2) short relaxation times so that high microwave power can be applied, (3) linear ESR response with radiation dose up to $5 \cdot 10^4$ Gy, (4) high stability of the radiation induced free radicals so that the dosimeter can be kept as a document of the radiation dose, (5) tissue equivalence for medical and biological applications, (6) non-destructive read-out of dosimeter so that the dose accumulation at repeated exposures can be monitored.

9.3.1.1 Radical Structure

X-irradiated polycrystalline L-α-alanine yields the ESR spectrum at room temperature shown in Fig. 9.5.

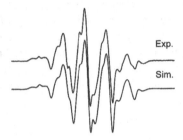

Fig. 9.5 X-band ESR spectra of L-α-alanine powder at room temperature. *Exp.*: experimental powder spectrum X-irradiated to 60 kGy, *Sim.*: reconstructed powder spectrum due to three radicals, R1, R2, R3, with parameters determined by single crystal ENDOR measurements and with relative intensities 0.589: 0.335: 0.076, see text for assignment. Spectra are adapted from [M.Z. Heydari et al.: J. Phys. Chem. A **106**, 8971 (2002)] with permission from the American Chemical Society

It consists of five lines with additional substructure and extends over 11 mT. The spectrum was originally assigned to the radical $\dot{C}H(CH_3)COOH$, abbreviated R1 [22]. Recent single crystal ENDOR studies [23] have, however, shown that there are at least two major (R1 and R2) and one minor (R3) paramagnetic species in the irradiated alanine sample. R2 has been assigned to the radical $^+H_3N\dot{C}(CH_3)COO^-$, R3 to $H_2N\dot{C}(CH_3)COOH$. The experimental ESR spectrum can be reconstructed with R1: 0.589; R2: 0.335; R3: 0.076, Fig. 9.5. Those radicals are probably formed from unstable radicals referred to as oxidation and reduction products as discussed in a recent review [24]. The R1 radical that is stable at room temperature has thus been assumed to be formed by deamination of the protonated radical anion (a reduction product) according to the reaction (9.5). The reaction occurs in the μs range at room temperature [25].

$$H_3NCH(CH_3)\dot{C}OOH \rightarrow NH_3 + CH_3\dot{C}HCOOH \qquad (9.5)$$

We refer to the literature for a discussion of the mechanism of formation for the R2 and R3 radicals [23, 24].

9.3.1.2 Measurements of High Doses

L-α-alanine has been employed as an ESR dosimeter since 1962 [26]. Regulla et al. [7, 27] have reported detailed studies on its dosimetric properties. It is the most commonly used material in the field of ESR dosimetry being accepted by the International Atomic Energy Agency (IAEA) [28] and other agencies as a secondary reference and transfer dosimeter for high (industrial) dose irradiation.

Alanine dosimeters are commercially produced in the form of pellets, rods, films and cables. For calibration a set of standards irradiated in advance with known doses and prepared from the same alanine dosimeter batches as those under study are employed. The radiation dose of the unknown sample is obtained by comparison of the heights of the central line of the sample and the standards. Usually a reference ESR standard is recorded simultaneously with the alanine spectra to correct for variations in spectrometer sensitivity.

A procedure using so called "self-calibrated" dosimeters has been recently proposed [4, 29]. Each dosimeter incorporates an internal standard permanently present in the pellet. The signal is recorded together with the signal of the radiation sensitive alanine material. The ratio between the heights of the central line of alanine to that of the internal standard depends on the radiation dose. The dose is read out using calibration with known doses made by the user, or alternatively supplied by the manufacturer. The material commonly used as the internal standard in the self-calibrated dosimeter is Mn^{2+} diluted in MgO. Figure 9.6 shows the full ESR spectrum of a self-calibrated alanine dosimeter. As seen from the figure it contains the six ESR lines of Mn^{2+}. The alanine lines appear in the central part of the Mn^{2+} spectrum. The outer Mn^{2+} lines are used for the calibration.

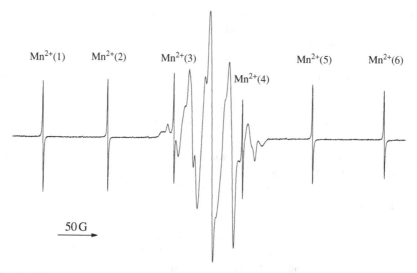

Fig. 9.6 ESR spectrum of irradiated self-calibrated alanine dosimeter. The diagram is reproduced from [4] with permission from Springer. An editing error in [4] was corrected

Radiation processing by electron beam or γ-irradiation is a commonly employed method for the sterilization of medical devices. The method has on one hand the advantage that sterilization can be carried out with the items in their original packages. On the other hand, dosimetry is required to ensure that the radiation treatment is at a tolerable level to avoid toxicological hazard as emphasized in the standards on radiation sterilization drafted by international standards organizations. Dosimetry plays key roles in characterization of the facility, in qualification of the process and in routine process control, see e.g. [4, 30].

ESR-dosimetry with alanine has been one of the standard dosimetry methods for several years. Dosimetry intercomparisons involving 38 radiation sterilization facilities using dosimeters supplied by accredited calibration laboratories indicated an accuracy of the order of 25% that was considered satisfactory agreement [31]. The effect of irradiation temperatures between ambient and 80°C on the response of alanine dosimeters is of concern for high dose irradiations, particularly with electron beams [32]. This may be a disadvantage in comparison with radiochromic dosimeters that colours upon exposure to ionizing radiation for use with optical read-out in radiation processing applications [33].

9.3.1.3 Measurements of Low Doses

Sophisticated measurement protocols have made it possible to measure low doses with an impressive precision using the alanine ESR dosimeter. Nagy et al. [34] have reported measurements within 1.5–4% for doses between 1 and 5 Gy, and Hayes et al. [35] have demonstrated measurement of doses in the range 0.01–1,000 Gy with an uncertainty of 1% for high doses and 10 mGy for low doses. An uncertainty of less than 0.5% was achieved for doses between 5 and 25 Gy [36].

Data processing sources of errors are of concern when high accuracy is needed [14] as in medical applications of dosimetry.

(1) Non-linearity between ESR signal and dose. Statistical tests for the linearity of the dose curve can be applied.
(2) Calibration errors. Errors can be reduced by calibration design of the dose curve, in certain applications by calibrating at the lower and upper endpoints of the dose.
(3) Errors in calibration doses. A modified least squares method, taking into account that calibration doses also contain uncertainties, can be applied [37].
(4) Analytical statistical treatments are not always possible. Monte Carlo simulations may then be applied [14].

The applied measuring protocols required measuring times up to several hours with the common alanine dosimeter. New materials have therefore been suggested as more practical dosimeters for clinical applications.

9.3.2 New Materials for ESR Dosimetry

The alanine ESR dosimeter is less convenient for low dose dosimetry due to the long measurement times. New materials for the dose determinations in the interval 0.2–10 Gy may therefore be needed if the ESR method shall become a realistic alternative to existing dosimetry systems, e.g. ionisation chambers, Si-diodes, thermoluminiscence or chemical dosimetry like Fricke solution. Strategies for finding new materials for ESR dosimeters were first discussed by Ikeya et al. [38]. The proposed criteria were high yield of radicals, sharp spectral lines and thermal stability of the radicals at room temperature. Similar criteria have been applied in recent work [20].

The disadvantage of alanine as low-dose dosimeter is mainly attributed to the unfavorable shape of the ESR spectrum extending over ca. 11 mT with the intensity distributed over five hyperfine lines (Fig. 9.5). Using only the central line for dosimetric purposes only a fraction of its radiation induced ESR response is employed. The hyperfine lines are broad due to hyperfine coupling anisotropy. At equal ESR spectrometer settings the increased sensitivity of the recently studied materials is thus mainly due to more suitable ESR spectra with narrow lines and/or no hyperfine splitting. Several new materials have been investigated for example sugar particularly as a personnel monitor for radiation emergencies [39], lactates and phosphates [38, 40, 41], tartrates [42–44], sulfates [38, 45], dithionates [46–50], carbonates [51] and formates [20, 21, 52]. Some of them are several times more sensitive than alanine, Table 9.2.

9.3.2.1 The Lithium Formate Dosimeter

Lithium formate is the most extensively studied new material for ESR-dosimetry [21]. It has also been tested for use in clinical applications [53, 54]. Thus, doses in the range 0.2–3.5 Gy due to 6 MeV photons were determined in blind tests shown in Fig. 9.7. A deviation of less than 1.2% from those by ionization chamber measurements was obtained, which is well within the uncertainty of the measurements. It was concluded that no trend could be seen in the ESR dosimeter response, regarding either the dose rate or the beam quality. Each measurement took 15 min, while several hours would be required with the common alanine dosimeter.

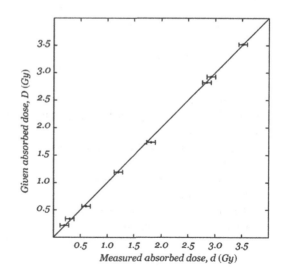

Fig. 9.7 Comparison between absorbed dose determined by lithium formate ESR dosimeters, d, and given absorbed dose, D. Error bars mark the uncertainty in the measured dose. The diagram is reproduced from [H. Gustafsson et al.: Phys. Med. Biol. **53**, 4667 (2008)] with permission from the Institute of Physics publishing

9.3.2.2 Recent Developments

Measurements at frequencies lower than X-band, isotopic substitution in existing materials, and addition of sensitizing additives have recently been suggested as means to increase the dose response.

Low frequency: Most measurements for dosimetric purposes have so far been made by standard ESR spectrometers at X-band. The diameter of the dosimeter pellets used in these spectrometers is limited to ca 5 mm. Film dosimeters have a maximum width of 5–6 mm and a thickness less than 0.5 mm. Due to these size limitations at X-band, ESR spectrometers working at lower frequency might be more useful. Such spectrometers allow larger sample volumes, are smaller and therefore more portable, and presumably less expensive. The loss of sensitivity, compared to the X-band, may be compensated for by the increased sample dimensions, and for samples with significant g-anisotropy also by a narrower line-width as schematically shown in Fig. 9.8 for a g-anisotropy typical for the spectrum observed in lithium formate.

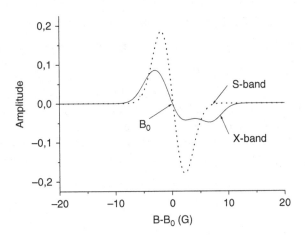

Fig. 9.8 Signal amplitudes at X- and S-bands at equal integrated intensities for a g-anisotropy typical of the CO_2^- radical anion observed in lithium formate

Experimental studies are scarce, but an analysis of the frequency dependence on the sensitivity in fact suggests that frequencies lower than X-band might give higher sensitivity in quantitative ESR [55]. Measurements at low frequency are not an alternative in clinical applications, however, where even the dosimeters for X-band (5·5 mm) in many cases are too large.

Isotopic substitution: The sensitivity of ESR-dosimeters in the low-dose range can in some cases be increased by isotope exchange of the materials. Samples enriched in certain isotopes like ^2H or ^6Li often have narrower line-widths, and accordingly stronger amplitudes than those of the same material with natural isotope abundance. An approach to increase the sensitivity of ammonium formate for X- and γ-irradiation by recrystallisation in D_2O has been described [52].

Fig. 9.9 ESR spectra of lithium formate irradiated with γ-rays and measured at room temperature, (**a**) ^{7}LiOOCH·H$_2$O, (**b**) ^{6}LiOOCH·H$_2$O. Errors caused by spectrometer instability were corrected by a standard reference of Mn^{2+}/MgO measured together with the sample. The figure is adapted from [56] with permission from Elsevier

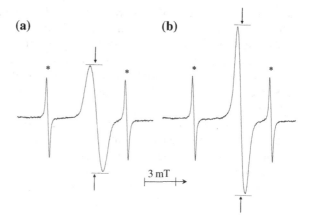

The dose-response for γ-irradiation was enhanced by approximately a factor of 2 by using 95% ^{6}Li enriched lithium formate in place of the natural composition with 7.5% ^{6}Li and 92.5% ^{7}Li as seen in Fig. 9.9. This method to increase the dose-response is expensive, which may limit its applications in routine measurements.

Sensitizing agents: The sensitizing effect of metal ion dopants observed in initial studies by Ikeya et al. of certain inorganic substances does not seem to have been further explored. Attempts to increase the dose response of the alanine dosimeter by similar treatments have also not been entirely successful. An increase in the sensitivity of some of the new materials by transition metal ion dopants has, however, been reported [57]. An increase in the free radical yield by up to a factor two and a shortening of the relaxation times contributed to the increase in sensitivity. Those studies are still on an experimental level.

ESR dosimetry in neutron beams has been reported recently using alanine, ammonium tartate, and lithium formate ESR dosimeters [58–60]. A sensitivity improvement by gadolinium addition was observed experimentally and examined theoretically by Monte Carlo simulations. Those materials have also been considered for dosimetry in a mixed radiation field of photons and neutrons.

ESR dosimetry of ion beams has been explored in the context of ESR imaging.

9.3.3 ESR Imaging of Radiation Dose

The spatial distribution of radiation dose can in principle be obtained by measurements on pieces cut from an irradiated sample. The same type of information can also be obtained in a non-destructive way and at higher resolution by ESR imaging (ESRI). The method was proposed already in 1973, but has developed slower than the NMR imaging (MRI) method. The distribution of electron spins in the samples is determined using a magnetic field gradient. Alanine was used in the first application to measure the dose distribution from a β radiation source [61]. We refer to the literature for reviews of the development of the technique and of applications in polymer science [62, 63]. An application of ESRI to measure the dose distribution after irradiation with ion beams is described below [64].

The dose distribution in potassium dithionate tablets along the beam of C^{6+} and N^{7+} ions accelerated to 25 MeV was examined using the ESRI methods developed by Schlick et al. [63, 65]. Spectra were collected using an X-band spectrometer equipped with two Lewis coils powered by regulated DC power supplies to produce a variable field gradient up to 20 mT/cm applied along the direction of the tablets. Details of the experimental procedure can be found in an article by Gustafsson et al. [64]. Information about the analysis is described in [63, 65].

The intensity profiles in potassium dithionate tablets (Fig. 9.10) clearly show Bragg peaks occurring at the end of the range of the particles. Note that no paramagnetic centres were detected beyond the penetration depth (approximately 2.3 mm).

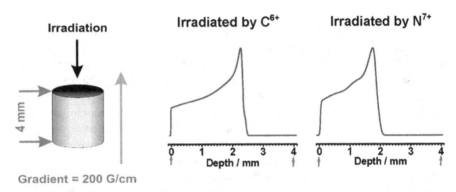

Fig. 9.10 Distribution of paramagnetic centers in potassium dithionate tablets after irradiation with C^{6+} and N^{7+}-particle beams. The diagram is reproduced from [64] with permission from the American Chemical Society

The 2D spectral-spatial ESRI diagram in Fig. 9.11 was obtained from a potassium dithionate sample that had been irradiated to a dose of ca 400 Gy with a C^{6+} particle beam. The ions had an estimated range of 3.8 mm at a linear energy transfer (LET) of 60 keV/μm in water. The ESRI measurements were made with a field gradient of 165 G/cm along the direction (y) of irradiation of the sample. The two peaks were attributed to SO_3^- ion radicals trapped in different environments. A closer analysis indicated a change in the peak ratio of the two radicals as a function of increasing depth (y) along the sample. We refer to the literature for a study of this effect in the context of determining the dose distribution and LET by ESR imaging in a potassium dithionate dosimeter irradiated with C^{6+} and N^{7+} ions [64].

ESRI spectrometers have been constructed for different purposes, at X-band for small samples, e.g. in ESR dosimetry, at longer wave-lengths for biological applications [62, 63]. Spectrometers at different wave-length bands are also available, commercially. Those spectrometers are of the CW-type.

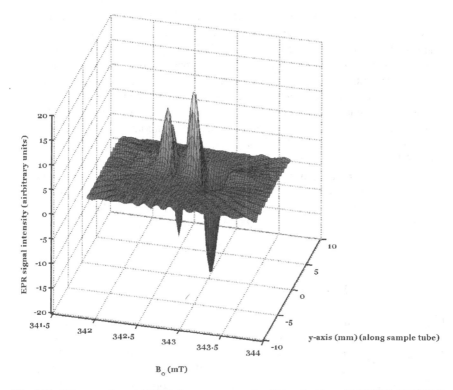

Fig. 9.11 2D spectral-spatial ESRI diagram measured at X-band with an ELEXSYS E500/E540 X-EPRI spectrometer of potassium dithionate tablets irradiated with a C^{6+}-particle beam. H. Gustafsson and E. Lund are acknowledged for permission to reproduce spectra prior to publication

9.4 ESR Dating

Geological dating by ESR is based on the measurement of the radiation-induced stable paramagnetic defects formed by natural radioactivity. The age of the specimen is obtained from measurements of the past dose, estimated by the additional dose method described in Section 9.4.1, and of the dose-rate at the site. Dating of bones and teeth is made with radicals derived from carbonate present in small amounts in the hydroxyapatite structure [66]. The ESR signal in tooth enamel is extremely long-lived, up to 10 million years. The signal can be observed down to a dose of 0.1 Gy. ESR dating of tooth enamel has been used in the time range beyond that of the ^{14}C method and up to at least two million years. Dating of materials other than tooth enamel has generally been restricted to ages less than one million years. The intrinsic and impurity ESR signals of silicates and quartz might be suitable for dating of older materials.

The paramagnetic species of interest are those created as a result of radiation damage, which implies that ESR has many similarities as a dating technique to thermoluminescence (TL). It has some advantages over TL, mainly because the signal is not destroyed during measurement, and so it is easier to study a sample under a variety of experimental conditions. Geological dating with ESR using hydroxyapatite and other materials has therefore been suggested as a viable alternative to other methods [66, 67]. ESR is, however, at least in some cases less sensitive that TL. Thus, the sensitivity of detection puts a lower limit on the time frame for which the method is applicable. The signal must be inducible by radiation, and must grow monotonically with the applied radiation dose. The material will eventually reach saturation, when an increase in applied radiation no longer increases the signal, limiting the time range of ESR dating for that type of sample.

The practical application of ESR as a dating method began with the work of Ikeya in 1975 [67]. Since then, there have been substantial contributions based on carbonate materials, bones, and quartz. The method of ESR dating has been outlined in a monograph by the pioneer of the field [8]. We refer to this work for details about principles and applications.

9.4.1 Additive Dose Method

The additive dose method is used to estimate the dose of samples previously irradiated to an unknown dose. The method has been applied e.g. in retrospective dosimetry using tooth enamel and in geological and archaeological dating. The sample is then further irradiated with a series of well-known additional doses. The ESR signal is measured before the first additional irradiation and after each of the following irradiations. The resulting dose response curve is extrapolated to zero ESR signal to obtain the unknown dose. The method is illustrated in Fig. 9.12.

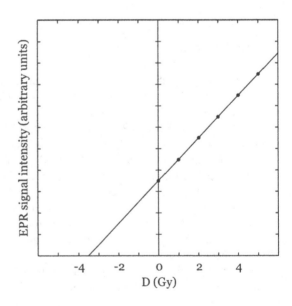

Fig. 9.12 The additive dose method. The magnitude of the intercept on the horizontal axis gives an estimate of the unknown dose. The figure is reproduced from [H. Gustafsson, "Development of sensitive EPR dosimetry methods", Linköping University medical dissertations, No. 1044 (2008)] with permission from Dr. H. Gustafsson

The additive dose method can be performed with high accuracy since only a single sample is used. Materials like sugar compounds, shell buttons, tooth enamel, quartz grains and finger nails have been measured with this method [4]. It was applied to measure the radiation doses caused by the atomic bombs in Hiroshima and Nagasaki [8]. The radiation doses of citizens living in the contaminated areas of Belarus after the Chernobyl accident were also obtained using ESR measurements of tooth enamel. Dose estimation with this material must take into account that the ESR signal of an irradiated tooth consists of two main overlapping components, that of radicals induced by radiation in enamel and another due to organic radicals in dentine. The enamel signal intensity that gives information about the radiation dose is therefore difficult to measure directly. Procedures to obtain the true radiation induced signal involving computer subtraction, microwave saturation of the background signal, chemical and light treatments have been applied. Those methods, based on visual inspection, are somewhat subjective techniques. An automatic deconvolution of the ESR spectra in three components, due to native-, radiation- and mechanically induced radicals was therefore employed with software designed for the reconstruction of individual absorbed doses by the additive dose method [68].

The additive dose method is also of interest in readiness for possible accidents with ionising radiation. The dose is then usually above the limits of the personal dosimeters used in radiological work. ESR dosimetry may then be applied for the reconstruction of the absorbed dose using suitable samples. Materials like shell buttons, sugar compounds and finger nails have been employed. Another method to measure the radiation dose was applied in an early report concerned with a radiological accident [69]. Identical tablets as those found with the victim were irradiated to obtain a calibration curve of signal intensity against radiation dose. The signal was due to radicals from sugar compounds, which were the main constituents of the tablets. The received dose obtained in the accident was then obtained by fitting the measured intensity from the accident into the calibration diagram.

9.5   Irradiated Food and Radiation Processed Materials

Irradiation of food is approved in many countries [70], while import and export of irradiated food products are forbidden in other. Methods to determine if a food sample has been treated by irradiation are therefore required. The analysis is qualitative to show if a piece of food has been irradiated or not. In samples containing hard tissues ESR analysis gives an unambiguous proof of radiation because the signals are stable with time. In other foodstuffs the radiation generated ESR signals can be recorded only during a limited period of time.

The inorganic part of bones and teeth, hydroxyapatite, $Ca_{10}(PO_4)_6(OH)_2$, contains some amount of carbonate. After irradiation an anisotropic ESR signal with $g_{||} = 1.996$ and $g_{\perp} = 2.002$ appears assigned to the $\dot{C}O_2^-$ anion radical [4, 71]. The signal is stable even when the bone is boiled. The ESR signals appearing after irradiation of fish bones disappear shortly after the radiation processing, however.

Shells contain $CaCO_3$ as a mineral part with some $CO_3{}^{2-}$ exchanged with $SO_3{}^{2-}$. The radiation induced ESR signal is attributed to sulfur-centered radicals $SO_2{}^-$ and $SO_3{}^-$ with long lifetimes and provides unambiguous evidence for previous radiation treatment. No special preparation procedure is required and a measurement takes only ca. 30 min. The alternative gas chromatography method takes much longer.

The applicability of ESR to analyze the radiation history of foodstuffs that only contain soft tissues is limited by the lifetime of the radiation-induced free radicals. In the case of fruits having a shell or containing stones the radiation induced ESR signal remains unchanged for more than one year [72]. In cellulose-containing food the radiation induced signals disappear within a few weeks or months depending on storage conditions. The work by Yordanov and Gancheva may be consulted for additional details and references [4].

The study of irradiated wheat flour, with main components of starch and protein provides examples of difficulties in the ESR analysis, since the spectra in some cases depended on the orientation of the sample in the spectrometer. The wheat grains have some degree of crystallinity, and the ESR spectrum of each crystallite shows orientation dependence because of g and/or hyperfine coupling anisotropy. The problem was overcome by passing the flour through a sufficiently fine mesh, until no orientation effect remained among the spectra for different rotation angles of the sample tube with respect to the magnetic field [73–75].

The separation of signals from different species by means of different microwave saturation has been applied in work on irradiated spices [76]. A typical spectrum may contain contributions from transition metal ions, particularly, Mn^{2+}, and a radiation organic radical, sometimes overlapped with another naturally occurring free radical spectrum. It was recommended that the microwave saturation properties should be considered in future protocols to survey the radiation history of soft foodstuffs.

9.5.1 Microwave Saturation Properties

CW-microwave saturation studies were performed at an early stage to obtain the spin-lattice and spin-spin relaxation times of paramagnetic centres in inorganic solids [77, 78]. Investigations involving CW-microwave saturation measurements of paramagnetic products trapped in solids and frozen glasses were also made at an early stage. The interest in this old method has been revived in several applications of quantitative ESR. As discussed in the previous sections the microwave saturation properties are of interest for the performance of potential dosimeter materials. Those properties have also been used to characterize the radicals found in food products that have been treated by irradiation (Section 9.5) and in similar applications.

The theory to obtain relaxation times T_1 and T_2 from CW saturation data was developed many years ago [77–79]. The amplitude can reach a maximum at a particular microwave power and approach zero at saturation for a homogeneous line

Fig. 9.13 Schematic shapes of saturation curves for homogeneous, inhomogeneous and Voigt ESR-lines. A homogeneous line has the Lorentz shape usually occurring in liquids. An inhomogeneous line is an envelope of narrow homogeneous lines, with the envelope usually approximated by a Gaussian, while the Voigt line is an envelope of homogeneous lines with an appreciable line-width

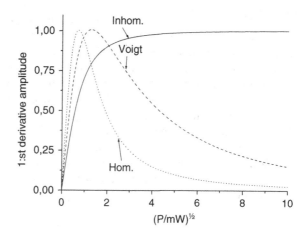

or come to a constant level in the inhomogeneous case, Fig. 9.13. An inhomogeneous line is assumed to be an envelope of homogeneous lines. The envelope is approximated by a Gaussian function while a homogeneous line has the Lorentz shape usually occurring in liquids as shown in Fig. 9.14. This line-shape was first explained by a treatment of the corresponding case for NMR [80].

The line-width ΔB_L measured at half height of the absorption amplitude is related to the spin-spin relaxation time by $T_2 = 1/(\gamma \cdot \Delta B_L)$, where the gyromagnetic ratio $\gamma = g\mu_B/\hbar$ equals $176 \cdot 10^6$ $(s \cdot mT)^{-1}$ for a species with $g = 2.00232$. The line is said to be homogeneously broadened. The opposite case with an inhomogeneous

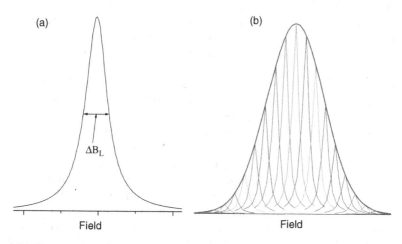

Fig. 9.14 Homogeneous (**a**) and inhomogeneous (**b**) ESR lines. The spin-spin relaxation time $T_2 = 1/(\gamma \cdot \Delta B_L)$ is obtained from the line-width ΔB_L measured at half height of the absorption amplitude of the Lorentz line (**a**). The inhomogeneous line (**b**) is an envelope of overlapping homogeneous line. The shape of the envelope approaches a Gauss curve

broadened line occurs generally in solids. The line-shape is an envelope of so-called spin-packet lines centred at slightly different fields. It is usual to assume that the spin-packets are Lorentz lines. The envelope tends to a Gauss curve when the spin-packet line-width, ΔB_L, is much narrower than that of the envelope. The line-shape is often approximated with a Voigt line for appreciable values of ΔB_L. The spectral lines observed in samples of interest in quantitative ESR are often assumed to be of this type. An automatic procedure to analyse such lines based on the theory in ref. [77, 78, 81] is summarized in Appendix A9.1. The relaxation times T_1 and T_2 are obtained by a least-squares fit using all experimental data of the saturation curves, alternatively T_2 and the microwave power P_0 at saturation are specified, see further Appendix A9.1 for an application to obtain the relaxation parameters of a proposed ESR dosimeter.

9.6 Summary

ESR applications concerned with the measurement of the absolute or relative amounts of paramagnetic species are the common theme of the chapter. Applications in biophysics were published in early works, emphasizing that the g-factor dependency on the measured concentrations is correctly taken into account by a 1/g-factor in place of $1/g^2$ traditionally employed. Applications in free radical research are briefly summarized. Procedures described in the literature to obtain absolute concentrations by calibration with reference samples are outlined. Error sources are discussed and procedures to reduce the uncertainties are reviewed. Methods to obtain the ESR line-shapes are described in the context of obtaining the integrated spectral intensity. The principles of ESR-dosimetry using the ESR-signal due to long-lived radicals induced by ionizing radiation in solid materials are presented in considerable detail. Suitable materials with sharp ESR lines, high radical yield and fast magnetic relaxation to avoid saturation by the applied microwave are considered. With the most studied material, L-α-alanine, $H_2NCH(CH_3)COOH$, dose measurements down to 1 Gy are feasible using commercially available or home-made ESR-dosimeters. Several alternative materials with narrower ESR spectra show a stronger dose response suitable for the low dose range employed in medicine. With the most extensively studied new material, lithium formate, doses in the range 0.2–3.5 Gy can be determined with a deviation of less than 1.2%. Isotopic substitution, metal ion doping, and instrumental developments to further increase the sensitivity are briefly described. The spatial distribution of radiation dose can be obtained by ESR imaging (ESRI). The method proposed already in 1973, has only recently become commercially available. The method is illustrated by an application to measure the dose distribution after irradiation with ion beams. Geological dating by ESR using the additive dose method is feasible for periods up to two million years. Procedures to estimate doses by ESR in contaminated areas and occasionally after radiological accidents are described as well as ESR analyses for test of irradiation treatment of food products and of medical equipment.

Appendix

A9.1 Relaxation Times by CW Microwave Saturation Measurements

The method to obtain relaxation times from CW microwave saturation measurements for an inhomogenously broadened line is based on assumptions given in the literature [77, 78]. The ESR line shape is then expressed as a convolution of a Gauss and a Lorentz function [81]:

$$g(r) \propto \frac{B_0 \beta}{\Delta B_G} \int_{-\infty}^{\infty} \frac{e^{-(ar')^2}}{t^2 + (r - r')^2} dr' \tag{9.6}$$

Here β is the transition probability of the line $g(r)$ centered at field B_0. The variables r and r' are defined in terms of the corresponding magnetic fields B and B' as:

$$r = \frac{B - B_0}{\Delta B_L} \qquad r' = \frac{B' - B_0}{\Delta B_L} \tag{9.7}$$

ΔB_L and ΔB_G are the widths of the unsaturated Lorentzian, L(B), and Gaussian, G(B), line shapes:

$$L(B) = \frac{1}{\pi \Delta B_L} \frac{1}{1 + \left(1 + \dfrac{B - B_0}{\Delta B_L}\right)^2}$$

$$G(B) = \frac{1}{\sqrt{\pi} \Delta B_G} e^{-\left(\frac{B - B_0}{\Delta B_G}\right)^2} \tag{9.8}$$

The parameters t^2 and a affecting the shape of the saturation curve are given by:

$$t^2 = 1 + \gamma^2 B_1^2 \beta T_1 T_2 = 1 + s^2 \tag{9.9}$$

$$a = \frac{\Delta B_L}{\Delta B_G} \tag{9.10}$$

The saturation factor "s" contains the gyromagnetic ratio γ, the amplitude of the microwave magnetic field B_1, and the spin-lattice and spin-spin relaxation times T_1 and T_2. The amplitude is related to the input microwave power by an expression of the type:

$$B_1 = k\sqrt{Q_L P} = K\sqrt{P} \tag{9.11}$$

The constant K depends on the type of resonator and its quality factor Q_L with the sample inserted. It may often be difficult to estimate its value precisely, therefore the quantity P_0 is introduced:

$$P_0 = \frac{1}{K^2 \gamma^2 \beta T_1 T_2} \qquad (9.12)$$

Using the microwave power P as variable and P_0 as a relaxation dependent parameter one obtains:

$$t = \sqrt{1 + P/P_0} \qquad (9.13)$$

The conventional saturation parameter s equals 1 for $P = P_0$. P_0 was therefore referred to as the microwave power at saturation in [82].

The line-shape given by (9.6) is a Voigt function that can be evaluated numerically by a standard procedure [81]. For a single inhomogenously broadened line, the transition probability β is set to 1 as in a simple two-level system. The relaxation times are given by:

$$T_2 = \frac{1}{\gamma \Delta B_L} \qquad (9.14)$$

$$T_1 = \frac{\Delta B_L}{K^2 \gamma P_0} \qquad (9.15)$$

The estimate of the spin-lattice relaxation time T_1 depends on the factor K to calculate the B_1 field from the corresponding microwave power according to equation (9.11).

The saturation curves must be recorded under slow passage conditions, that is, conditions such that the time between successive field modulation cycles is sufficiently long for each spin packet to relax between cycles. The spin system is then continually in thermal equilibrium and the true line shape is observed. A convenient formulation of the slow passage condition is given by the expression (9.16) [79]:

$$\omega_m B_m << \Delta B_L / T_1 \qquad (9.16)$$

Here ω_m is the modulation circular frequency and B_m is the modulation amplitude. The equation illustrates intuitively that the modulation rate must be much slower than the relaxation rate $(1/T_1)$. Typical values may be $\Delta B_L = 0.01$ mT, $T_1 = 10$ μs and $B_m = 0.1$ mT, requiring the modulation frequency $\nu_m = \omega_m/2\pi$ being less than 1.5 kHz, a condition which often is difficult to achieve with many modern commercial spectrometers.

The formulae were incorporated in a computer program [82]. The input is an experimental array of the 1st derivative peak-to-peak ESR amplitude against the microwave power that is read from a previously prepared file and initial trial parameter values for the Lorentzian and Gaussian line-widths and the microwave power (P_0) at saturation that are provided interactively. The corresponding theoretical amplitudes were obtained by numerical differentiation of the Voigt function (1). A non-linear least squares fit of the calculated saturation curve to the experimental data is performed. Output data consist of a graph of the experimental data and the

Fig. 9.15 Experimental (+) and theoretical (−) saturation curves for radicals trapped in an X-irradiated crystalline dosimeter material

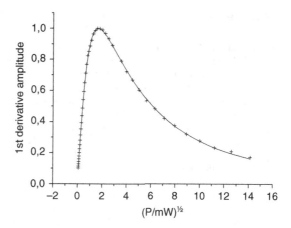

fitted saturation curve, exemplified in Fig. 9.15 below. The parameters with fitting error estimates are printed to a file of the type shown below.

Analysis with 45 points, 4 parameters

	Parameter	Error
Intensity:	.1614E+01	.5671E-02
Saturating power(mW):	.1289E+01	.2356E-01
Lorentzian width(G):	.3765E+00	.7793E-02
Gaussian width(G):	.2801E+01	
Exp. width(G):	.3000E+01	
Computed width(G):	.3000E+01	

$B1(G) = .4400E\text{-}01 * sqrt(P/mW)$

	Parameter	Error
T1(s):	.7424E-05	.2050E-06
T2(s):	.1743E-06	.3185E-08

| Modulation frequency = | .156E+04 Hz |
| Modulation amplitude = | .250E+00 G |

Here, the widths are the peak-to-peak values λ_L and λ_G of the first derivatives and can be expressed in terms of the widths of the absorption lines as:

$$\Delta B_L = \left(\sqrt{3}/2 \right) \lambda_L$$
$$\Delta B_G = \lambda_G / \sqrt{2}$$

$$(9.17)$$

The slow-passage condition (9.16) is approximately fulfilled:

$$0.245 \cdot 10^4 = \omega_m B_m (\text{G/s}) << \Delta B_L / T_1 = 0.621 \cdot 10^5$$

References

1. G. Foerster: Z. Naturforsch. **15a**, 1079 (1960).
2. M.L. Randolph: In '*Biological Applications of Electron Spin Resonance*' ed. by H.M. Swartz, J.R. Bolton, D.C. Borg, Wiley, New York, NY (1972), p. 119.
3. R. Aasa, T. Vänngård: J. Magn. Reson. **19**, 308 (1975).
4. G.A. Yordanov, V. Gancheva: In '*EPR of Free Radicals in Solids*' ed. by A. Lund, M. Shiotani, Kluwer Academic Publishers, Dordrecht (2003), Chapter 14.
5. S.Ya. Pshezhetskii, A.G. Kotov, V.K. Milinchuk, V.A. Roginski, V.I. Tupilov: '*EPR of Free Radicals in Radiation Chemistry*', Wiley, New York, NY (1974).
6. D.T. Burns, B.D. Flockhart: Phil. Trans. R. Soc. Lond. A **333**, 37 (1990).
7. D.F. Regulla, U. Deffner: Int. J. Appl. Radiat. Isot. **33**, 1101 (1982).
8. M. Ikeya: '*New Applications of Electron Paramagnetic Resonance, Dating, Dosimetry and Microscopy*', World Scientific, Singapore (1993).
9. Xenon: Spin Counting, http://www.bruker-biospin.com/spincount.html
10. J.A.Weil, J.R. Bolton: '*Electron Paramagnetic Resonance: Elementary Theory and Practical Applications*', 2nd Edition, Wiley, New York, NY (2007).
11. M. Iwasaki: J. Magn. Reson. **16**, 417 (1974).
12. V.Yu. Nagy: Anal. Chim. Acta **339**, 1 (1997).
13. S.S. Eaton, G.R. Eaton: Bull. Magn. Reson. **1**, 130 (1979.
14. V. Nagy: Appl. Radiat. Isot. **52**, 1039 (2000).
15. G.R. Eaton, S.S. Eaton, D. Barr, R. Weber: '*Quantitative EPR, A Practitioners Guide*', Wien, New York, Springer, Wien, New York (2010).
16. L.A. Bljumenfeld, W.W. Wojewodski, A.G. Semjonov: '*Die Anwendung der paramagetischen Resonanz in der Chemie*', Akademische Verlagsgesellschaft, Geest & Portig, Leipzig (1966).
17. P.B. Ayscough: '*Electron Spin Resonance in Chemistry*', Methuen & Co Ltd., London (1967).
18. C.P. Poole, H.C. Farach: '*Handbook of Electron Spin Resonance*', American Institute of Physics Press, New York, NY (1994).
19. R. Zamoramo-Ulloa, H. Flores-Llamas, H. Yee-Madeira: J. Phys. D: Appl. Phys. **25**, 1528 (1992).
20. A. Lund, S. Olsson, M. Bonora, E. Lund, H. Gustafsson: Spectrochim. Acta A **58**, 1301 (2002).
21. (a) T.A. Vestad, E. Malinen, A. Lund, E.O. Hole, E. Sagstuen: Appl. Radiat. Isot. **59**, 181 (2003). (b) T.A. Vestad, H. Gustafsson, A. Lund, E.O. Hole, E. Sagstuen: Phys. Chem. Chem. Phys. **6**, 3017 (2004). (c) T.A. Vestad: '*On the development of a solid-state, low dose EPR dosimeter for radiotherapy*', Dissertation, University of Oslo (2005).
22. (a) I. Miyagawa, W.J. Gordy: Chem. Phys. **32**, 255 (1960). (b) I. Miyagawa, K. Itoh: J. Chem. Phys. **36**, 2157 (1962).
23. E. Sagstuen, E.O. Hole, S.R. Haugedal, W.H. Nelson: J. Phys. Chem. **101**, 9763 (1997).
24. E. Sagstuen, A. Sanderud, E.O. Hole: Radiat. Res. **162**, 112 (2004).
25. P.-O. Samskog, G. Nilsson, A. Lund, T. Gillbro: J. Phys. Chem. **84**, 2819 (1980).
26. W.W. Bradshaw, D.G. Cadena, G.W. Crawford, H.A. Spetzler: Radiat. Res. **17**, 11 (1962).
27. D.F. Regulla: Appl. Radiat. Isot. **52**, 1023 (2000).
28. (a) K. Mehta, R. Girzikowsky: Radiat. Phys. Chem. **46**, 1247 (1995). (b) K. Mehta, R. Girzikowsky: Appl. Radiat. Isot. **52**, 1179 (2000).
29. (a) N.D. Yordanov, V. Gancheva: J. Radioanalyt. Nucl. Chem. **240**, 619 (1999); ibid. **245**, 323 (2000). (b) V. Gancheva, N.D. Yordanov, F. Callens, G. Vanhaelewyn, J. Raffi, E. Bortolin, S. Onori, E. Malinen, E. Sagstuen, S. Fabisiak, Z. Peimel-Stuglik: Radiat. Phys. Chem. **77**, 357 (2008).
30. A. Miller: Radiat. Phys. Chem. **42**, 731 (1993).
31. A. Miller, P.H.G. Sharpe: Radiat. Phys. Chem. **59**, 323 (2000).
32. P.H.G. Sharpe, A. Miller, J.P. Sephton, C.A. Gouldstone, M. Bailey, J. Helt-Hansen: Radiat. Phys. Chem. **78**, 473 (2009).

33. M. Lavalle, U. Corda, P.G. Fuochi, S. Caminati, M. Venturi, A. Kovács, M. Baranyai, A. Sáfrány, A. Miller: Radiat. Phys. Chem. **76**, 1502 (2007).
34. V. Nagy, S. Sholom, V. Chumak, M. Desrosiers: Appl. Radiat. Isot. **56**, 917 (2002).
35. R. Hayes, E. Haskell, A. Wieser, A. Romanyukha, B. Hardy, J. Barrus: Nucl. Instrum. Methods A **440**, 453 (2000).
36. M. Anton: Phys. Med. Biol. **51**, 5419 (2006).
37. (a) E.S. Bergstrand, E.O. Hole, E. Sagstuen: Appl. Radiat. Isot. **49**, 845 (1998). (b) E.S. Bergstrand, K.R. Shortt, C.K Ross, E.O. Hole: Phys. Med. Biol. **48**, 1753 (2003). (c) E.S. Bergstrand: '*Alanine dosimetry by EPR spectroscopy*', Dissertation, University of Oslo (2005).
38. M. Ikeya, G.M. Hassan, H. Sasaoka, Y. Kinoshita, S. Takaki, C. Yamanaka: Appl. Radiat. Isot. **52**, 1209 (2000).
39. T. Nakajima: Appl. Radiat. Isot. **46**, 819 (1995).
40. G.M. Hassan, M. Ikeya, S. Toyoda: Appl. Radiat. Isot. **49**, 823 (1998).
41. G.M. Hassan, M. Ikeya: Appl. Radiat. Isot. **52**, 1247 (2000).
42. S.K. Olsson, S. Bagherian, E. Lund, G. Alm Carlsson, A. Lund: Appl. Radiat. Isot. **50**, 955 (1999).
43. S. Olsson, E. Lund, A. Lund: Appl. Radiat. Isot. **52**, 1235 (2000).
44. A. Bartolotta, M.C. D'Oca, M. Brai, V. Caputo, V. De Caro, L.I. Giannola: Phys. Med. Biol. **46**, 461 (2001).
45. J.R. Morton, F.J. Ahlers, C.C.J. Schneider: Radiat. Prot. Dosimetry **47**, 263 (1993).
46. S.E. Bogushevich, I.I. Ugolev: Appl. Radiat. Isot. **52** 1217 (2000).
47. M.P. Baran, O.A. Bugay, S.P. Kolesnik, V.M. Maksimenko, V.V. Teslenko, T.L. Petrenko, M.F. Desrosiers: Radiat. Prot. Dosimetry **120**, 202 (2006).
48. M.P. Baran, M.O. Mazin, V.M. Maksimenko: Ukr. J. Phys. **52**, 676 (2007).
49. M.P. Baran, V.M. Maksimenko, V.V. Teslenko: J. Appl. Spectrosc. **75**, 15 (2008).
50. M. Danilczuk, H. Gustafsson, M.D. Sastry, E. Lund, A. Lund: Spectrochim. Acta A **69**, 18 (2008).
51. S. Murali, V. Natarajan, R. Venkataramani, Pusharja, M.D. Sastry: Appl. Radiat. Isot. **55**, 253 (2001).
52. H. Gustafsson, S. Olsson, A. Lund, E. Lund: Radiat. Res. **161**, 464 (2004).
53. H. Gustafsson, E. Lund, S. Olsson: Phys. Med. Biol. **53**, 4667 (2008).
54. L. Antonovic, H. Gustafsson, G. Alm Carlsson, Å. Carlsson Tedgren: Med. Phys. **36**, 2236 (2009).
55. G.A. Rinard, R.W. Quine, S.S. Eaton, G.R. Eaton: J. Magn. Reson. **156**, 113 (2002).
56. K. Komaguchi, Y. Matsubara, M. Shiotani, H. Gustafsson, E. Lund, A. Lund: Spectrochim. Acta A **66**, 754 (2007).
57. (a) H. Gustafsson, M. Danilczuk, M.D. Sastry, A. Lund, E. Lund: Spectrochim. Acta A **62**, 614 (2005). (b) M. Danilczuk, H. Gustafsson, M.D. Sastry, E. Lund, A. Lund: Spectrochim. Acta A **67**, 1370 (2007).
58. (a) M. Marrale, M. Brai, G. Gennaro, A. Triolo, A. Bartolotta: Radiat. Meas. **42**, 1217 (2007). (b) M. Marrale, G. Gennaro, M. Brai, S. Basile, A. Bartolotta, M.C. D'Oca: Radiat. Meas. **43**, 471 (2008).
59. E. Malinen, E. Waldeland, E.O. Hole, E. Sagstuen: Spectrochim. Acta A **63**, 861 (2006).
60. E. Lund, H. Gustafsson, M. Danilczuk, M.D. Sastry, A. Lund: Spectrochim. Acta A **60**, 1319 (2004).
61. M. Anton, H.-J. Selbach: Bruker Report **157/158**, 48 (2006).
62. K. Ohno: Magn. Reson. Rev. **11**, 275 (1987).
63. S. Schlick, K. Kruczala: In '*Advanced ESR Methods in Polymer Research*' ed. by S. Schlick, Wiley, Hoboken, NJ (2006).
64. H. Gustafsson, K. Kruczala, E. Lund, S. Schlick: J. Phys. Chem. B **112**, 8437 (2008).
65. M.V. Motyakin, S. Schlick: Macromolecules **35**, 3984 (2002).
66. A.R. Skinner: Appl. Radiat. Isot. **52**, 1311 (2000).

67. M. Ikeya: Nature **255**, 48 (1975).
68. V. Kirillov, S. Dubovsky: Radiat. Meas. **44**, 144 (2009).
69. E. Sagstuen, H. Theisen, T. Henriksen: Health Phys. **45**, 961 (1983).
70. http://www.iaea.org/programmes/nafa/d5/public/foodirradiation.pdf
71. N.D. Yordanov, V. Gancheva, R. Tarandjiiska, R. Velkova, L. Kulieva, B. Damyanova, S. Popov: Spectrochim. Acta A **54**, 2421 (1998).
72. J. Raffi, N.D. Yordanov, S. Chabane, L. Douifi, V. Gancheva, S. Ivanova: Spectrochim. Acta A **56**, 409 (2000).
73. M. Ukai, Y. Shimoyama: Appl. Magn. Reson. **29**, 1 (2005).
74. Y. Shimoyama, M. Ukai, H. Nakamura: Spectrochim. Acta A **63**, 888 (2006).
75. M. Ukai: JEOL News **39**(1), 27 (2004).
76. M. Ukai, H. Nakamura, Y. Shimoyama: Spectrochim. Acta A **63**, 879 (2006).
77. A.M. Portis: Phys. Rev. **91**, 1071 (1953).
78. T.G. Castner Jr.: Phys. Rev. **115**, 1506 (1959).
79. S. Schlick, L. Kevan: J. Magn. Reson. **22**, 171 (1976).
80. F. Bloch: Phys. Rev. **70**, 460 (1946).
81. J. Maruani: J. Magn. Reson. **7**, 207 (1972).
82. A. Lund, E. Sagstuen, A. Sanderud, J. Maruani: Radiat. Res. **172**, 753 (2009).

General Appendix G

Table G1 Fundamental constants[a]

Quantity	Symbol	Value	SI unit
Speed of light	c	2.99792458×10^{8}	m s^{-1}
Elementary charge	e	1.602176×10^{-19}	C
Faraday constant	$F = N_A e$	9.648534×10^{4}	C mol^{-1}
Boltzmann constant	K	1.38065×10^{-23}	J K^{-1}
Gas constant	$R = N_A k$	8.31447	J K^{-1} mol^{-1}
Planck constant	h	6.626069×10^{-34}	J s
	$\hbar = h/2\pi$	1.054572×10^{-34}	J s
Avogadro constant	N_A	6.02214×10^{23}	mol^{-1}
Atomic mass unit	u	1.66054×10^{-27}	kg
Mass			
electron	m_e	9.10938×10^{-31}	kg
proton	m_p	1.67262×10^{-27}	kg
neutron	m_n	1.67493×10^{-27}	kg
Vacuum permittivity	$\varepsilon_0 = 1/c^2\mu_0$	8.854188×10^{-12}	F m^{-1}

[a]CODATA recommended values of the fundamental physical constants 2006, National Institute of Standards and Technology, Gaithersburg, Maryland 20899-8420, USA; http://physics.nist.gov/cuu/Constants/

Table G2 Magnetic constants in SI units[a]

Quantity	Symbol	Numerical value	Unit
Magnetic constant	$\mu_0 = 4\pi \times 10^{-7}$	12.566371×10^{-7}	N A^{-2}
Bohr magneton	$\mu_B (\beta_e)$	$927.400915 \times 10^{-26}$	J T^{-1}
nuclear magneton	$\mu_N (\beta_N)$	$5.05078324 \times 10^{-27}$	J T^{-1}
electron g-factor	g_e	2.0023193043617	
electron gyromagnetic ratio	γ_e	$1.760859770 \times 10^{11}$	s^{-1} T^{-1}
μ_B/h	μ_B/h	13.9962464×10^{9}	Hz T^{-1}
μ_B/hc	μ_B/hc	46.6864515	m^{-1} T^{-1}
μ_N/h	μ_N/h	7.62259384	MHz T^{-1}

[a]CODATA recommended values of the fundamental physical constants 2006, National Institute of Standards and Technology, Gaithersburg, Maryland 20899-8420, USA; http://physics.nist.gov/cuu/Constants/

A. Lund et al., *Principles and Applications of ESR Spectroscopy*,
DOI 10.1007/978-1-4020-5344-3, © Springer Science+Business Media B.V. 2011

Table G3 Conversion factors for ESR coupling constants[a,b]

Unit	MHz	mT	cm^{-1}
MHz	1	0.07144771/g	$0.333564095 \times 10^{-4}$
mT	13.99625·g	1	4.668645×10^{-4} g
cm^{-1}	2.99792458×10^{4}	0.2141949×10^{4}/g	1

A coupling given in a unit of the 1st column is calculated in other units by multiplication with the factor in the corresponding row.

Calculations of g from measured values of microwave frequency ν_e (GHz) and resonance field B (T), of resonance field, and of nuclear frequency ν_N (MHz):

$$g = \frac{h}{\mu_B} \cdot \frac{\nu_e}{B} = 0.07144771 \frac{\nu_e(GHz)}{B(T)} \quad ,$$

$$B(T) = \frac{h}{\mu_B} \cdot \frac{\nu_e}{g} = 0.07144771 \frac{\nu_e(GHz)}{g}$$

$$\nu_N(MHz) = \frac{\mu_N}{h} \cdot g_N B = 7.62259384 \cdot g_N B(T)$$

[a]The factors were obtained from CODATA recommended values of the constants in Table G1; http://physics.nist.gov/cuu/Constants/
[b]P.J. Mohr, B.N. Taylor, D.B. Newell: Rev. Mod. Phys. **80**, 633 (2008); J. Phys. Chem. Ref. Data **37**, 1187 (2008).

Table G4 Other useful conversion factors

1 eV	1.60218×10^{-19} J, 96.485 kJ mol^{-1}, 8065.5 cm^{-1}
1 cal	4.184 J
1 atm	101.325 kPa
1 cm^{-1}	1.9864×10^{-23} J
1 D (Debye)	3.33564×10^{-30} C m
1 Å	10^{-10} m
1 T	10^{-4} G (or gauss)
1 L atm	= 101.325 J
$\theta/°C$	= T/K – 273.15

Table G5 Symbols, variables and units in ESR

Symbol	Name or description	Unit and/or value
A, a	Hyperfine coupling (splitting) constant	MHz, mT (milli-Tesla)
D	Zero-field splitting, Fine structure	cm^{-1}, MHz, mT
B (H)	External magnetic field	T (Tesla)
e	Electron charge	1.602176×10^{-19} A·s
g	g-factor	Dimensionless
G	Radiation yield	μmol·J^{-1}
h	Planck constant	6.626069×10^{-34} J·s
I	Nuclear spin angular momentum	J·s
I	Nuclear spin quantum number	Dimensionless
J	Heisenberg exchange coupling	cm^{-1}, MHz, mT

Table G5   (continued)

Symbol	Name or description	Unit and/or value
k	Boltzmann constant	1.380650×10^{-23} J·K^{-1}
L	Orbital angular momentum	J·s
l	Orbital quantum number	Dimensionless
m_e	Electron mass	0.910938×10^{-30} kg
m_I	Nuclear magnetic quantum number	Dimensionless
m_S	Electron magnetic quantum number	Dimensionless
P	Microwave power	J·s^{-1}
Q	Nuclear quadrupole coupling	cm^{-1}, MHz, mT
S	Electron spin angular momentum	J·s
S, s	Electron spin quantum number	Dimensionless
v	Speed	m·s^{-1}
λ	Spin-orbit coupling constant	cm^{-1}
μ	Magnetic moment	A·s
$\mu_B, (\beta_e)$	Bohr magneton	9.274009×10^{-24} J·T^{-1}
$\mu_N, (\beta_N)$	Nuclear magneton	5.050783×10^{-27} J·T^{-1}
ν	Frequency	Hz

Table G6   Abbreviations

Abbreviation	Name or Description
Electron Spin Resonance:	
1(2)D	One (two)-Dimensional
CW or cw	Continuous Wave
ELDOR	Electron-Electron Double Resonance
DEER	Dead-time free pulsed Electron-Electron double Resonance
EIE	ENDOR-Induced ESR (EPR)
ENDOR	Electron Nuclear Double Resonance
ESE	Electron Spin-Echo
EPR	Electron Paramagnetic Resonance
ESEEM	Electron Spin-Echo Envelope Modulation
ESR	Electron Spin Resonance
ESRI (EPRI)	ESR (EPR) Imaging
FID	Free-Induction Decay
FS, fs	Fine Structure
FT	Fourier Transformation
HFC, hfc	HyperFine Coupling
HFI, hfi	HyperFine Interaction
HYSCORE	Hyperfine Sublevel Correlation Spectroscopy
NQC, nqc	Nuclear Quadrupole Coupling
NQI, nqi	Nuclear Quadrupole Interaction
ODMR	Optically Detected Magnetic Resonance
NMR	Nuclear Magnetic Resonance
PELDOR	Pulse ELDOR
RF, rf	Radio Frequency
STESR	Saturation Transfer ESR
TRIPLE	Electron-Nuclear-Nuclear Triple Resonance
ZFI, zfi	Zero-Field Interaction
ZFS, zfs	Zero-Field Splitting

Table G6 (continued)

Abbreviation	Name or Description
Quantum Chemistry:	
ANADIP	Analytical Dipolar Calculation
B3LYP	Becke, Three-parameter, Lee-Yang-Parr
CB	Conduction Band
DFT	Density Functional Theory
ED	Electron Density
EHM	Extended Hückel Method
GTO	Gaussian Type Orbital
HF	Hartree-Fock
HOMO	Highest Occupied Molecular Orbital
HM	Hückel Method
INDO	Intermediate Neglect of Differential Overlap
J-T	Jahn-Teller
KS	Kohn and Sham
LCAO	Linear Combination of Atomic Orbitals
LUMO	Lowest Unoccupied Molecular Orbital
MNDO	Modified Neglect of Differential Overlap
MO	Molecular Orbital
NBMO	Non-Bonding Molecular Orbital
RHF	Restricted Hartree-Fock
SCF	Self-Consistent Field
SOMO	Singly Occupied Molecular Orbital
SD	Spin Density or Slater Determinant
TD-DFT	Time Dependent Density Functional Theory
UHF	Unrestricted Hartree-Fock
VB	Valence Band
Radiation Chemistry:	
Gy	Gray (radiation dose, J/kg)
rad	0.01 Gy (earlier used radiation dose)
LET	Linear Energy Transfer (J/m)
Chemistry:	
DMPC	O-(1,2-Dimyristoyl-sn-glycero-3-phosphoryl)choline
DPC	Dihydrosterculoyl Phosphatidylcholyne
DPPC	O-(1,2-Dipalmitoyl-X-glycero-3-phosphoryl)choline
DPPH	Diphenylpicrylhydrazyl
DTBN	Di-tert-butyl nitroxide
HDPE	High Density Polyethylene
HME	Hexamethylethane
i-PP	Isotactic Polypropylene
5-InVSL	2-[(-oxy-2,2,5,5-tetramethyl-3-pyrrolin-3yl)]Indian-1,3-dione
K26	26-residue peptide
LDPE	Low Density Polyethylene
LTA	Linde type A zeolite
MCP	Methylenecyclopropane
2-MTHF	2-Methyltetrahydrofuran
NODO	N-oxyl-4',4'dimethyl oxazolidine
PA	Polyacetylene
PASL	Phoshatidic Acid Spin Label

Table G6 (continued)

Abbreviation	Name or Description
PCL	Poly(ε-caprolactone)
PCSL	Phosphatidylcholine spin label
PE	Polyethylene
PGSL	Phosphatidylglycerol spin label
PMMA	Polymethylmethacrylate
PP	Polypropylene
PS	Polystyrene
PSSL	Phosphatidylserine Spin Label
PTFE	Polytetrafluoroethylene
SASL	Stearic Acid Spin Label
TEMPO	2,2,6,6-Tetramethylpiperidine-1-Oxyl
TEMPONE	2,2,6,6-tetramethylpiperidine-N-oxyl
TMM	Trimethylenemethane
TMS	Tetramethylsilane
UPEC	Urea-polyethylene complex
Others:	
MI	Matrix Isolation
MDSC	Modulation Differential Scanning Calorimeter
SAXS	Small Angle X-ray Scattering
WAXS	Wide Angle X-ray Scattering
WLF	Williams-Landel-Ferry

Table G7 Magnetic properties of stable isotopes[a,b,c]

Isotope	Atomic number (Z)	Natural abundance (%)	Nuclear spin (I)	Nuclear g-factor (g_N)
^1H	1	99.985	0.5	5.58569
^2H	1	0.0148	1	0.85744
^3He	2	0.00014	0.5	−4.25525
^6Li	3	7.5	1	0.82205
^7Li	3	92.5	1.5	2.17096
^9Be	4	100	1.5	−0.785
^{10}B	5	19.8	3	0.60022
^{11}B	5	80.2	1.5	1.79242
^{13}C	6	1.11	0.5	1.40482
^{14}N	7	99.63	1	0.40376
^{15}N	7	0.366	0.5	−0.56638
^{17}O	8	0.038	2.5	−0.75752
^{19}F	9	100	0.5	5.25773
^{21}Ne	10	0.27	1.5	−0.4412
^{23}Na	11	100	1.5	1.47839
^{25}Mg	12	10	2.5	−0.34218
^{27}Al	13	100	2.5	1.4566
^{29}Si	14	4.67	0.5	−1.1106
^{31}P	15	100	0.5	2.2632

Table G7 (continued)

Isotope	Atomic number (Z)	Natural abundance (%)	Nuclear spin (I)	Nuclear g-factor (g_N)
^{33}S	16	0.75	1.5	0.42911
^{35}Cl	17	75.77	1.5	0.54792
^{37}Cl	17	24.23	1.5	0.45608
^{39}K	19	93.26	1.5	0.26099
^{40}K	19	0.0117	4	−0.32453
^{41}K	19	6.73	1.5	0.14325
^{43}Ca	20	0.135	3.5	−0.37641
^{45}Sc	21	100	3.5	1.35906
^{47}Ti	22	7.4	2.5	−0.31539
^{49}Ti	22	5.4	3.5	−0.31548
^{50}V	23	0.25	6	0.55659
^{51}V	23	99.75	3.5	1.46836
^{53}Cr	24	9.5	1.5	−0.3147
^{55}Mn	25	100	2.5	1.3819
^{57}Fe	26	2.15	0.5	0.1806
^{59}Co	27	100	3.5	1.318
^{61}Ni	28	1.13	1.5	−0.50001
^{63}Cu	29	69.2	1.5	1.484
^{65}Cu	29	30.8	1.5	1.588
^{67}Zn	30	4.1	2.5	0.35031
^{69}Ga	31	60.1	1.5	1.34439
^{71}Ga	31	39.9	1.5	1.70818
^{73}Ge	32	7.8	4.5	−0.19544
^{75}As	33	100	1.5	0.95965
^{77}Se	34	7.6	0.5	1.0693
^{79}Br	35	50.69	1.5	1.40427
^{81}Br	35	49.31	1.5	1.51371
^{83}Kr	36	11.5	4.5	−0.2157
^{85}Rb	37	72.17	2.5	0.54125
^{87}Rb	37	27.83	1.5	1.83427
^{87}Sr	38	7	4.5	−0.24291
^{89}Y	39	100	0.5	−0.27484
^{91}Zr	40	11.2	2.5	−0.52145
^{93}Nb	41	100	4.5	1.3712
^{95}Mo	42	15.9	2.5	−0.3656
^{97}Mo	42	9.6	2.5	−0.3734
^{99}Ru	44	12.7	2.5	−0.249
^{101}Ru	44	17	2.5	−0.279
^{103}Rh	46	100	0.5	−0.1768
^{105}Pd	46	22.2	2.5	−0.256
^{107}Ag	47	51.83	0.5	−0.22725
^{109}Ag	47	48.17	0.5	−0.26174
^{111}Cd	48	12.8	0.5	−1.19043
^{113}Cd	48	12.2	0.5	−1.2454
^{113}In	49	4.3	4.5	1.22864
^{115}In	49	95.7	4.5	1.23129
^{115}Sn	50	0.38	0.5	−1.8377

Table G7 (continued)

Isotope	Atomic number (Z)	Natural abundance (%)	Nuclear spin (I)	Nuclear g-factor (g_N)
^{117}Sn	50	7.75	0.5	−2.00208
^{119}Sn	50	8.6	0.5	−2.09456
^{121}Sb	51	57.3	2.5	1.3455
^{123}Sb	51	42.7	3.5	0.72876
^{123}Te	52	0.89	0.5	−1.4736
^{125}Te	52	7	0.5	−1.7766
^{127}I	53	100	2.5	1.1253
^{129}Xe	54	26.4	0.5	−1.55595
^{131}Xe	54	21.2	1.5	0.46124
^{133}Cs	55	100	3.5	0.73785
^{135}Ba	56	6.59	1.5	0.55884
^{137}Ba	56	11.2	1.5	0.62515
^{138}La	57	0.089	5	0.74278
^{139}La	57	99.911	3.5	0.7952
^{141}Pr	59	100	2.5	1.6
^{143}Nd	60	12.2	3.5	−0.3076
^{145}Nd	60	8.3	3.5	−0.19
^{147}Sm	62	15.1	3.5	−0.2322
^{149}Sm	62	13.9	3.5	0.1915
^{151}Eu	63	47.9	2.5	1.389
^{153}Eu	63	52.1	2.5	0.6134
^{155}Gd	64	14.8	1.5	−0.1723
^{157}Gd	64	15.7	1.5	−0.2253
^{159}Tb	65	100	1.5	1.342
^{161}Dy	66	18.9	2.5	−0.189
^{163}Dy	66	24.9	2.5	0.266
^{165}Ho	67	100	3.5	1.192
^{167}Er	68	22.9	3.5	−0.1618
^{169}Tm	69	100	0.5	−0.466
^{171}Yb	70	14.4	0.5	0.9885
^{173}Yb	70	16.2	2.5	−0.27195
^{175}Lu	71	97.39	3.5	0.63943
^{176}Lu	71	2.61	7	0.452
^{177}Hf	72	18.6	3.5	0.2267
^{179}Hf	72	13.7	4.5	−0.1424
^{181}Ta	73	99.9877	3.5	0.67729
^{183}W	74	14.3	0.5	0.23557
^{185}Re	75	37.4	2.5	1.2748
^{187}Re	75	62.6	2.5	1.2878
^{187}Os	76	1.6	0.5	0.1311
^{189}Os	76	16.1	1.5	0.488
^{191}Ir	77	37.3	1.5	0.097
^{193}Ir	78	62.7	1.5	0.107
^{195}Pt	78	33.8	0.5	1.219
^{197}Au	79	100	1.5	0.09797
^{199}Hg	80	16.8	0.5	1.01177
^{201}Hg	80	13.2	1.5	−0.37348

Table G7 (continued)

Isotope	Atomic number (Z)	Natural abundance (%)	Nuclear spin (I)	Nuclear g-factor (g_N)
^{203}Tl	81	29.5	0.5	3.24451
^{205}Tl	81	70.5	0.5	3.2754
^{207}Pb	82	22.1	0.5	1.1748
^{209}Bi	83	100	4.5	0.938
^{235}U	92	0.72	3.5	−0.1

[a]P. Raghavan: At. Data Nucl. Data Tables 42, 189 (1989).
[b]http://ie.lbl.gov/toipdf/mometbl.pdf
[c]Values for the isotropic and anisotropic hyperfine couplings of the isotopes are reported in J.A. Weil, J.R. Bolton: 'Electron Paramagnetic Resonance: Elementary Theory and Practical Applications', 2nd Edition, J. Wiley (2007).

Index